AF815217

Green Energy and Technology

Climate change, environmental impact and the limited natural resources urge scientific research and novel technical solutions. The monograph series Green Energy and Technology serves as a publishing platform for scientific and technological approaches to "green"—i.e. environmentally friendly and sustainable—technologies. While a focus lies on energy and power supply, it also covers "green" solutions in industrial engineering and engineering design. Green Energy and Technology addresses researchers, advanced students, technical consultants as well as decision makers in industries and politics. Hence, the level of presentation spans from instructional to highly technical.

Indexed in Scopus.

Indexed in Ei Compendex.

Anuj K. Chandel · Suresh Sundaramurthy
Editors

Biorefineries in Circular Bioeconomy

Reactor Configurations Analysis

Springer

Editors
Anuj K. Chandel
Department of Biotechnology
Lorena School of Engineering (EEL)
University of São Paulo
Lorena, São Paulo, Brazil

Suresh Sundaramurthy
Department of Chemical Engineering
Maulana Azad National Institute
of Technology Bhopal
Bhopal, Madhya Pradesh, India

ISSN 1865-3529 ISSN 1865-3537 (electronic)
Green Energy and Technology
ISBN 978-981-95-3912-3 ISBN 978-981-95-3913-0 (eBook)
https://doi.org/10.1007/978-981-95-3913-0

© The Editor(s) (if applicable) and The Author(s), under exclusive license to Springer Nature Singapore Pte Ltd. 2025

This work is subject to copyright. All rights are solely and exclusively licensed by the Publisher, whether the whole or part of the material is concerned, specifically the rights of translation, reprinting, reuse of illustrations, recitation, broadcasting, reproduction on microfilms or in any other physical way, and transmission or information storage and retrieval, electronic adaptation, computer software, or by similar or dissimilar methodology now known or hereafter developed.
The use of general descriptive names, registered names, trademarks, service marks, etc. in this publication does not imply, even in the absence of a specific statement, that such names are exempt from the relevant protective laws and regulations and therefore free for general use.
The publisher, the authors and the editors are safe to assume that the advice and information in this book are believed to be true and accurate at the date of publication. Neither the publisher nor the authors or the editors give a warranty, expressed or implied, with respect to the material contained herein or for any errors or omissions that may have been made. The publisher remains neutral with regard to jurisdictional claims in published maps and institutional affiliations.

This Springer imprint is published by the registered company Springer Nature Singapore Pte Ltd.
The registered company address is: 152 Beach Road, #21-01/04 Gateway East, Singapore 189721, Singapore

If disposing of this product, please recycle the paper.

Foreword

Biorenewables production from different marginal feedstocks, exploring high-performance reactors, viz., percolation, plug-flow, and shrinking bed reactors, employing a mode of operations (batch, fed-batch, continuous, or semi-continuous), paves the way for the development of a circular bioeconomy. Further, reactor configurations prediction through AI/ML can revolutionize the smooth process operations and simulations. Finally, government regulations and policies to support circular economic initiatives are of utmost importance to strategize the transition of the petroleum economy to an advanced bioeconomy. The book entitled *Biorefineries in Circular Bioeconomy: Reactor Configurations Analysis* compiles 15 unique chapters on marginal feedstock utilization, high-performance reactors, customized bioreactors, process engineering challenges at various scales, ranging from the chemistry of the conversion processes to reactor and process design and optimization, and to enterprise-level economic and supply chain analysis. Economic production of biofuels and commodity chemicals seems challenging; however, it is not impossible to meet the growing demands of sustainable energy and chemicals. This book provides key aspects of biomass transformations into liquid fuel and value-added products of industrial significance through bioprocess engineering.

It is my pleasure to present this book to the scientific and academic community as a unique source of information on sustainable biotechnology. The editors of this book are among the main voices of biorefinery research. I thank all of the contributors for sharing their research and ideas with the scientific community using this unique platform of biomass conversion research.

Professor Dr. Durval Rodrigues Junior
Director
Lorena School of Engineering
Lorena, São Paulo, Brazil
durvalrj@usp.br

Preface

Biomass conversion is a promising "green" alternative to produce fuels, chemicals, and materials sustainably. Biorefineries, akin to petroleum refineries, will convert a complex biomass feedstock in a series of chemical and physical processing steps to obtain the desired products. Biomass is a primary shareholder in the biorefinery. Biomass processing at low cost, using effective conversion routes and a range of process configurations, plays a pivotal role in developing sustainable biorefineries. Although considerable emphasis has been given so far to the "upstream" conversion of biomass to intermediate platforms (sugars or syngas), progress in "downstream" conversion to chemicals and intermediates is still lagging. Development and optimization of biorefineries involves process engineering challenges at various scales, ranging from the chemistry of the conversion processes to reactor and process design and optimization, and to enterprise-level economic and supply chain analysis. Furthermore, embracing artificial intelligence (AI) and machine learning (ML) approaches could revolutionize the successful outcome of biorefineries.

This book comprises 15 well-written chapters in the field of bioenergy, biomass processing, bioreactor configurations, and energy systems for biorefinery processes. Particularly, we discuss recent biorefineries for first-generation (1G), 1.5-generation (1.5G), and second-generation (2G) ethanol and biochemicals production from lignocellulosic biomass and a wide variety of unexplored feedstocks. Different process parameters governing effective biomass pretreatment and production of renewable sugars through different bioreactor configurations under batch, fed-batch, and continuous modes of operation. In addition to production, life cycle assessment (LCA) and circular economy of the biorefinery process are also elaborated in some chapters. A chapter on reactor configurations prediction through AL/ML and optimizing through AI-driven simulations is also included in the book. Finally, Government regulations and policies to support circular economic initiatives, from strategies for the transition to advanced bioethanol production also discussed in the book.

Chapter "Feedstock Composition, Variation, and Availability in Various Geographical Regions in the World for Biofuels Production", written by **Brumano et al.**, provides a detailed analysis of the global landscape of feedstock availability for biofuels production, critically explores the challenges and opportunities region-wise, and offers valuable insights for optimizing biofuel production systems worldwide. In Chapter "Fruit and Vegetable Waste, Chitin, DDGs, Bread Waste, Broken Rice, Beetroot: Unexplored Feedstock for Biorefineries", **Singh et al.** made a comprehensive analysis of the underutilized biomass sources, including fruit and vegetable waste, chitin in crustacean shells, dried distillers' grains (DDGs), bread waste, broken rice, and beetroot. These feedstocks are available in plentiful amounts in various parts of the world, but have not been explored judiciously, such as sugarcane biomass, corn stover, and rice straw. They possess significant amounts of carbohydrates, protein, and bioactives, and their conversion into biofuels, bioplastics, enzymes, organic acids, and specialty chemicals offers unique strategic value, in turn adding a clear-cut value proposition in the renewable economy.

Chapter "Current Advancements in Algal Feedstock Cultivation and Processing Aiming 3G Biorefinery" written by **Rahul and Arumugam**, provides insights into 3G biorefineries utilizing algal biomass as sustainable energy producers. Algal biomass is a promising feedstock due to its fast growth rate, high photosynthetic efficiency, and lack of competition with food crops. The chapter explores the current progress on algal feedstock cultivation and downstream processing specifically designed for 3G biorefineries, in turn contributing to circular and sustainable industrial practices. Process parameters play an influential role in biomass processing for the fullest recovery of oligosaccharides and cellulosic sugars. **Bodhe et al.** in Chapter "Process Parameters Governing Effective Biomass Pretreatment and Enzyme-Mediated Oligosaccharide Recovery" discuss the effect of several pretreatment methods on the enzymatic hydrolysis of pretreated biomass to extract functional oligosaccharides (FOS). Additionally, consolidated bioprocessing (CBP) and simultaneous saccharification and fermentation (SSF) are discussed in this chapter to increase sugar yields and production efficiency, highlighting the importance of techno-economic analysis and life cycle assessment in a sustainable bioeconomy. In a similar vein, **Villar and co-authors** in Chapter "Key Analysis of Process Parameters for an Effective Enzymatic Hydrolysis for Cellulosic Sugar Production" have made a comparative account of process parameters on improved enzymatic hydrolysis for lignocellulosic sugars production from a variety of feedstocks. Several process parameters and other factors, for instance, high enzyme costs, product inhibition, and mass transfer limitations, particularly at high solids loadings, are summarized in this chapter.

In Chapter "Role of Counter-Current, Percolation, Plug-Flow, and Shrinking-Bed Reactor in Lignocellulosic Biomass Pretreatment and Enzymatic Hydrolysis", **García Bustos et al.** presented the application of these reactor technologies in the pretreatment and hydrolysis of lignocellulosic biomass. These processes are commonly implemented in various reactor configurations, including counter-current, percolation, plug-flow, and shrinking bed reactors operating under batch, semicontinuous, or continuous modes. **de Carvalho et al.**, in Chapter "Comparative Assessment of Batch, Fed-Batch, and Continuous Feeding of Lignocellulosic Biomass on Sugar Recovery", have made a comparative assessment of batch, fed-batch, and continuous feeding of lignocellulosic biomass on second-generation (2G) sugar recovery.

More specifically, this chapter provides an overview of reactor design, pretreatment strategies, and the interaction of feedstock characteristics that influence sugar recovery. Critical analysis of these process parameters and their advantages and disadvantages, and process limitations are summarized, aligning with the principles of circular bioeconomy. Continuing the theme of biomass processing, Chapter "Bioreactors Configuration in the Development of Process Consolidation for Biofuels and Biochemicals in Lignocellulosic Biorefineries", written by **Prado et al.**, competently evaluates the role of process parameters, viz., temperature control, pH, aeration, and agitation on the productivity of bioproducts. Nevertheless, process configuration strategies such as simultaneous saccharification and fermentation (SSF) and simultaneous saccharification and co-fermentation (SSCF) and consolidated bioprocessing may employ either single or separate reactors, sometimes integrated with membrane systems to enhance process efficiency. The special focus of this chapter is on the key process strategies, including SSF, SSCF, and CBP, aiming to increase the yield and productivity of biofuels and biochemicals.

Algal biomass is a sizeable feedstock for the production of biofuels, specialty biochemicals, and nutraceuticals, among others. Generally, the exploration of algal feedstock is considered in the broad spectrum of 3G biorefinery. **Cardozo and colleagues** in Chapter "Customized Bioreactors Development for Enhanced Fuels Production from Microalgal Feedstock" reviewed the customized bioreactors development for unleashing the potential of microalgal feedstock into enhanced fuels production. This chapter pertinently summarizes the key challenges for optimizing light distribution, gas exchange, and mixing of algal species in large-scale cultivation.

The second-generation sugar stream also has a considerable amount of inhibitors derived from carbohydrates and lignin, and other compounds present in the biomass cell wall. Detoxification of hydrolysates is a critical step to remove those inhibitors. Chapter "Detoxification Strategies to Remove Inhibitors from Lignocellulose Hydrolysates: Challenges and New Prospects" of **Murillo-Franco et al.** comprehensively reviews the physical, chemical, and enzymatic pretreatment methods to remove the inhibitors. This chapter categorically highlights the importance of selecting appropriate detoxification strategies based not only on technical performance but also on their economic and environmental feasibility. Chapter "Recent Developments in Ethanol Recovery After Fermentation of 1G, 1.5G, and 2G Sugars" of **José et al.** presents the advancements made in ethanol recovery after the fermentation of first-generation (1G), intermediate-generation (1.5G), and second-generation (2G) sugars. The chapter explores the conventional and emerging separation technologies, i.e., azeotropic and extractive distillation, adsorption, pervaporation, hybrid systems, and membrane technologies for ethanol recovery. Chapter "Life Cycle Assessment in Biorefineries: Foundations, Methodologies, and Future Directions", written by **Philippini and co-authors**, presents the fundamentals and critical analysis of life cycle assessment in biorefineries in light of the future directions. This chapter summarizes the historical development of LCA in the context of biorefineries, outlining its purpose, core methodology, and the evolution of key approaches. Economic analysis of any bioprocess is very important to gauge its viability at large-scale operations. Taking this into consideration, **Akhtar and Haroon** presented in Chapter

"Economic Analysis of Changes in Bioreactors Configuration Yielding Ethanol and Biochemicals", the outline on the economic analysis of changes in bioreactor configurations toward ethanol and biochemicals production. Most recent developments in AI and ML also seem to be a game-changer in developing economically sustainable chemicals and fuels at a commercial scale. Chapter "AI-Driven Simulations to Discover New Routes of Reactor Configurations in Biorefinery Process" of **Soni et al.** presents the ongoing convergence of AI technology, including machine learning, deep learning, reinforcement learning, and evolutionary algorithms, coupled with conventional process simulation scripts to transform the reactor design in biorefineries systems aligning smart manufacturing in the industry 4.0 paradigm.

Lastly, in Chapter "Government Regulations and Policies to Support Circular Economy Initiatives from Strategies for the Transition of 3G to Advanced Bioethanol Production", written by **Tripathi and co-authors**, they appraised the government regulations and policies to support the circular economy in the Indian context. As of mid-2025, India has already made considerable progress in blending 20% ethanol in gasoline. India can play a leading role in the development of a renewable economy, eventually having a noticeable impact on the advancement of the bioeconomy.

All 15 chapters are written by experts and enrich the scientific insights in the development of a sustainable economy.

The book critically appraises the important insights on smart bioreactors development, biorefinery, and sustainable bioeconomy. Editors are quite confident that the book will serve as a very useful source of information for researchers, academicians, and students of graduate and postgraduate levels in the field of biochemical engineering, environmental science/engineering, microbial biotechnology, AI, and other related disciplines.

Lorena, Brazil — Anuj K. Chandel
Bhopal, India — Suresh Sundaramurthy

Contents

Feedstock Composition, Variation, and Availability in Various Geographical Regions in the World for Biofuels Production

Larissa Pereira Brumano, Taciana Freire de Oliveira, Marcelo Chuei Matsudo, and Tales Alexandre Costa-Silva

Abstract The composition, variation, and availability of feedstocks for biofuel production vary significantly across different geographical regions, impacting both the feasibility and sustainability of biofuel technologies. This chapter explores diverse feedstock options utilized in biofuel production, including first-generation, second-generation, and third-generation sources, and examines how their availability is shaped by regional agricultural practices, environmental factors, and socioeconomic conditions. First-generation biofuels, derived from edible crops such as sugar, starch, and oils, have been widely used but face challenges relating to food security, land use competition, and environmental degradation. In contrast, second-generation and third-generation biofuels, sourced from nonfood feedstocks like algae, agricultural residues, and municipal waste, present a more sustainable alternative but encounter hurdles due to feedstock complexity and processing challenges. The regional variation in feedstock composition—driven by climate, land availability, and local agricultural systems—plays a crucial role in determining the success and scalability of biofuel production systems. Furthermore, developing advanced biotechnological processes is essential for enhancing the efficiency of feedstock conversion and improving the overall sustainability of biofuel production. Policy frameworks tailored to regional contexts are also vital in promoting the adoption of sustainable feedstocks and encouraging innovation in biofuel technologies. This chapter provides a comprehensive analysis of the global landscape of feedstock availability for biofuels, highlighting the challenges and opportunities inherent in regional differences and offering insights into the pathways for optimizing biofuel production systems worldwide.

L. P. Brumano (✉) · M. C. Matsudo · T. A. Costa-Silva
Center for Natural and Human Sciences, Federal University of ABC (UFABC), Santo André-SP, Brazil
e-mail: larissa.brumano@ufabc.edu.br

T. F. de Oliveira
Laboratory of Microbial Biomolecules, School of Pharmaceutical Sciences, University of São Paulo (USP), São Paulo-SP, Brazil

© The Author(s), under exclusive license to Springer Nature Singapore Pte Ltd. 2025

A. K. Chandel and S. Sundaramurthy (eds.), *Biorefineries in Circular Bioeconomy*, Green Energy and Technology, https://doi.org/10.1007/978-981-95-3913-0_1

Keywords Biomass · Lignocellulosic · Microalgae · Waste · Renewable energy · Sustainability · Circular economy · Bioeconomy

1 Introduction

Growing concerns about climate change, energy security, and the depletion of fossil fuel reserves are driving the global transition to sustainable energy solutions. In this context, biofuels have emerged as a crucial alternative, providing a renewable energy source with lower emissions, reducing reliance on petroleum-based fuels, and promoting economic and environmental resilience [1]. Moreover, biofuels—defined as energy derived from biomass and its refined products—align with the United Nations Sustainable Development Goals (SDGs), particularly SDG 7 (Affordable and Clean Energy), SDG 12 (Responsible Consumption and Production), and SDG 13 (Climate Action), thus reinforcing their strategic role in building a cleaner and more sustainable energy future [2].

The integration of biofuels into the global energy matrix aligns with the principles of the circular economy and bioeconomy, as these fuels utilize renewable biomass sources, agricultural residues, and even waste materials to produce energy while minimizing environmental impact [3]. According to the International Energy Agency (IEA), biofuel production is projected to grow at an annual rate of 4% under the Stated Policies Scenario (STEPS), reaching over 3 million barrels of oil equivalent per day (mboe/d) by 2030 and exceeding 4 mboe/d by 2050 [4].

Feedstock selection and conversion pathways significantly influence the total cost of biofuel production [5]. The use of edible feedstocks raises critical issues such as the food-versus-fuel debate and environmental concerns, including soil degradation, deforestation, and extensive use of arable land. These factors not only increase biofuel costs but also raise questions about the long-term sustainability of edible feedstocks. As a result, nonedible feedstocks are increasingly being explored as more sustainable and cost-effective alternatives, driving efforts to identify substitute raw materials that enhance biofuel sustainability [6].

The feasibility of large-scale biofuel adoption significantly depends on the availability, composition, and sustainability of feedstock, all of which are influenced by geographic factors, including the unique climatic, socioeconomic, and technological conditions of each region. Regional variations in feedstock characteristics drive global biofuel production strategies, shaped not only by inherent agronomic conditions but also by seasonal fluctuations and national incentive policies. Moreover, government policies play a crucial role in shaping market dynamics by directing investments, fostering technological innovations, and enhancing conversion efficiency; ultimately, these efforts lower production costs and boost the competitiveness of biofuels [7].

Biofuels are generally classified into four generations based on their feedstocks. First-generation biofuels are derived from edible crops, which can potentially impact the food supply. Second-generation biofuels utilize residual materials and nonedible energy crops; however, they face cost and technological challenges that hinder widespread commercialization. Third-generation biofuels, derived from microalgae, offer higher yields; however, further development is needed for large-scale commercialization. Lastly, fourth-generation biofuels utilize genetically modified organisms, providing the highest yields and the ability to replace fossil fuels [8]. Examples of biofuels produced from bio-sourced feedstocks include bioalcohols (methanol, ethanol, butanol), biochar, 2,5-dimethylfuran, biodiesel, biohydrogen, and biogas.

Currently, over 90% of liquid biofuels are derived from food-based feedstocks, mainly first-generation feedstocks, such as sugarcane, starch, and vegetable oils. Advanced biofuels made from nonfood feedstocks (second-generation and third-generation) are anticipated to increase in production, dependent on the development of sustainable feedstock sources, the execution of large-scale demonstration projects, and the establishment of stable regulatory frameworks. The long-term contribution of biofuels to the transportation fuel mix will depend on technological advancements and supportive public policies [4, 9].

In addition to conventional biomass sources, emerging research focuses on converting waste materials—such as animal waste, municipal organic waste, and food waste—into biofuels, which enhances sustainability while minimizing competition with food supplies. Waste-derived biofuels, including biomethane and bioethanol, not only improve waste management but also align with global efforts toward a circular bioeconomy [10].

Feedstock composition and geographical variations also impact conversion processes and economic viability, directly influencing process efficiency, techno-economic feasibility, and overall life cycle sustainability [5, 7]. Before large-scale implementation, it is crucial to assess how feedstock cultivation affects local socioeconomic conditions, food security, and biodiversity [11], with an emphasis on identifying sustainable practices suited to regional contexts.

In the following sections, feedstocks for biofuels are classified and discussed based on their composition and applications: sugar-rich and starch-rich feedstocks for bioethanol production, various oil-based sources for biodiesel production, lignocellulose-based feedstocks for second-generation biofuels, microalgae biomass for third-generation biodiesel production, and food waste for biomethane production. Given the critical role of feedstock characteristics and availability in global biofuel production, this chapter is structured into three main sections: (i) feedstock composition, classification, and applications; (ii) feedstock availability across geographical regions; and (iii) challenges and opportunities in global feedstock utilization. This integrated approach aims to provide a comprehensive perspective on biofuel production, emphasizing the importance of aligning sustainable practices with regional characteristics and contexts.

2 Feedstock Composition, Classification, and Applications

2.1 Sugar-Containing and Amylaceous Feedstock for First-Generation Bioethanol Production

The choice of biomass for biofuel production is a crucial factor as it influences the conversion process, costs, and product yield. The predominant bioethanol production is from first-generation feedstocks, which include sugar and starch crops.

Sugar-containing feedstocks (e.g., sugars from sugarcane, molasses, sugar beet, sweet sorghum, fruits, feed grains, food grains, and tubers) are sources of simple sugars that can be directly transformed into ethanol using basic technology and suitable fermentative microorganisms. Consequently, they are referred to as first-generation feedstocks. In Europe, North America, and France, sugar is primarily sourced from the beet sugar industry. In contrast, Brazil, Germany, and India rely mainly on sugarcane, sugar beet, and sweet sorghum as their primary sources [12]. Below are some of the most commonly used sugar-based raw materials for biofuel production:

- **Sugarcane (*Saccharum officinarum*)**: This four-carbon (C4) photosynthetic plant is primarily found in tropical and subtropical regions, such as Brazil, where it is used to produce bioethanol from sugarcane juice (sugar content: 11–15%) through the Melle-Boinot method. This sucrose-containing feedstock is less expensive than other raw materials used to produce bioethanol due to its straightforward processing and high productivity [13].
- **Sugar beet (*Beta vulgaris*)**: In addition to Europe, North America, and France, several other countries, including China, Kenya, Brazil, and Australia, also utilize this feedstock for biofuel production [12]. This sugar crop consists of 15–20% sugars, and beet molasses or syrup is commonly used for bioethanol production. The cost of ethanol production from this source is higher than from sugarcane due to the amount of chemicals and energy required.
- **Sweet sorghum (*Sorghum bicolor*)**: This highly drought-tolerant feedstock has a high sugar content in its stalks and can thrive in both temperate and tropical regions. This crop can be grown on marginal lands and requires less water and fertilizer than other sugar crops. Its Brix content is estimated to range from 14 to 23%, with sucrose making up around 75% and about 2.6% from fructose and glucose [14]. It is primarily used as feedstock for bioethanol production in China.

Alongside sugar-containing feedstocks, starch-rich options like corn, wheat, and potatoes remain predominant in industrial use, proving to be more economically viable than lignocellulosic materials [15]. While sugarcane is considered the most profitable feedstock for ethanol production, the USA stands as the world's largest ethanol producer, using corn as its primary feedstock, largely due to substantial government subsidies [16].

Starch consists of two glucan polymers called amylose and amylopectin, with proportions that vary significantly depending on the source plant. Amylose is a polymer of glucose molecules linked by -1,4 glycosidic bonds, while amylopectin features an additional branched structure with -1,6 glycosidic bonds, resulting in a higher molecular weight [17, 18]. This polysaccharide is the most abundant one in nature and is found in the seeds, fruits, and roots of plants [19], as well as in microorganisms such as microalgae [20, 21].

One of the world's best-known starch-rich plants is corn (of the genus *Zea*), which originated in Mexico but is found across all continents. Despite its nutritional properties that make it important for human and animal consumption, corn has significantly contributed to the world's ethanol production [16, 22], which justifies the search for alternative amylaceous sources. For instance, the starch-producing plant taioba (*Xanthosoma sagittifolium*) appears to be underutilized. Interestingly, despite the tuber's high starch content (around 65%), only the leaves are usually consumed [19]. Therefore, considering its relatively easy cultivation, this starch-rich plant holds potential as a raw material for ethanol production [23].

Cassava (*Manihot esculenta* and *Manihot glaziovii*) is a starch-rich plant containing up to 80% starch, showcasing excellent potential as a feedstock for bioethanol production [19, 24]. Native to South America, this plant is a vital source of carbohydrates for many developing countries across South America, Asia, and Africa. Its high carbohydrate yield per unit of land, drought tolerance, and minimal agrochemical needs make it easy to cultivate [24–26]. Additionally, bioethanol can be produced from the starch in cassava peels after hydrolysis catalyzed by sulfonated carbon [27], which would be particularly interesting for increasing the use of cassava as a food source. Importantly, to use amylaceous materials in alcoholic fermentation, this feedstock must first undergo (i) liquefaction of starch by α-amylase, followed by (ii) enzymatic saccharification of low molecular weight products to release glucose via glucoamylase. The α-amylase enzyme can be sourced from thermoresistant bacteria, such as *Bacillus licheniformis*, or recombinant bacteria like *Escherichia coli* or *Bacillus subtilis*, while glucoamylase can be obtained from fungal species, including *Aspergillus* and *Rhizopus* [28]. Furthermore, as will be discussed later regarding lignocellulose-based feedstocks, simultaneous saccharification and fermentation (SSF) could be effectively implemented, combining the enzymatic degradation of starch with the fermentation of glucose released from the hydrolysis of this polymeric carbohydrate [29].

2.2 *Different Oil Biosources for Biodiesel Production*

Biodiesel, an ester-based, oxygen-rich fuel that exhibits notable biodegradability and enhanced lubricity potential, is derived from various oil sources, including vegetable oils and animal fats. Based on their feedstock source, biodiesels can be categorized into three generations: first, second, and third. The primary feedstock for first-generation biodiesel production is derived from common energy crops such as soybeans, and over 95% of biodiesel is currently obtained from edible oils [30], which may contribute to global food insecurity. The second generation of biodiesel is sourced from nonedible vegetable oil, recycled oil, and waste animal fat (such as tallow, fish, chicken, beef, and pork), primarily from sources with 16–18 carbon chains, which are deemed ideal for biodiesel synthesis [31]. Finally, the third generation of biodiesel is produced from microorganisms with high lipid content, such as oleaginous algae, yeast, fungi, and bacteria [32].

- **Edible vegetable oil**: Oils obtained from sunflower, soybean, rapeseed, palm, coconut, canola, cottonseed, jojoba, rice, mustard, olive, castor, linseed, and animal fats. Nevertheless, one of the main criticisms of first-generation biofuels is the use of edible feedstocks and food crops, known as the "food vs. fuel" debate, which necessitates the use of cultivable land areas, fertilizers, and water resources.
- **Nonedible vegetal oil**: Examples of nonedible oil-based biodiesel: *Ricinus communis, Azadirachta indica, Carthamus tinctorius, Jatropha curcas, Sapindus mukorossi, Pongamia pinnata, Madhuca indica, Aleurites moluccanus, Hevea brasiliensis,* and *Calophyllum inophyllum.*
- **Recycled oil**: Many countries face significant environmental challenges due to the disposal of large quantities of waste oil. This oil is produced by both household and industrial sources and could serve as an alternative feedstock for biodiesel production. The primary challenges include the presence of various impurities, such as free fatty acids, degradation products, and polymers, along with compounds that cannot be converted into alkyl esters. Some authors reported that waste oil can contain 6,9% of nonconvertible compounds that would not produce esters during biodiesel production [33]. Additionally, an efficient collection system for waste oil is needed.
- **Animal fat**: The main sources of animal fat include fat generated from chicken and fish processing, lard, tallow, and fat obtained from the leather industry. A challenging aspect of the transesterification process is the high levels of free fatty acids and saturated fatty acids [34]. Ruminant and pig fats have a high content of saturated fatty acids (nearly 40% saturated fatty acids [SFA]), making them predominantly solid at ambient temperature. Consequently, preheating to 45 ºC is necessary for their use in a diesel engine [35].
- **Microbial sources**: High-lipid-containing microbial biomass can be utilized for biodiesel production, as oleaginous microorganisms (bacteria, fungi, and yeasts) can accumulate 20–25% of their biomass as lipids. The fungal strains *Trichoderma reesei*, *Cunninghamella echinulata*, and *Mucor circinelloides* were evaluated as sources of lipids for biodiesel production, as well as the yeasts *Yarrowia lipolytica*, *Lipomyces starkeyi*, and *Rhodosporidium toruloides* [36]. The lipid extraction mechanisms involve the selection of a cell-biomass disruption method and an extraction solvent. Lipid extraction methods include high-pressure homogenization (HPH), bead milling, ultrasound, microwave, pulsed electric field (PEF), osmotic shock, chemical hydrolysis, enzymatic hydrolysis, and subcritical water.

The feedstock's physicochemical properties are highly related to the oil content and fatty acid profiles. Edible oils, as well as nonedible oils, show the same most prominent compounds C18:1, C18:2, and C16:0. The high concentration of unsaturated fatty acid composition in biodiesel can decrease the emissions of smoke, carbon monoxide, and HC [37]. Table 1 shows different feedstocks used for biodiesel production. Independent of oil feedstock source, the most relevant fatty acids are saturated fatty acids (SFA): palmitic (16:0) and stearic (18:0) acids, monounsaturated fatty acids (MUFA): oleic acid (18:1), and polyunsaturated fatty acids (PUFA): linoleic (18:2).

The oil extraction technologies comprise mechanical presses (screw press or hydraulic press), Soxhlet extraction, microwave extraction, enzymatic oil extraction, and supercritical CO_2 extraction. However, the mechanical press is the most conventional oil extraction method, followed by Soxhlet extraction [30].

2.2.1 Oil Conversion Techniques

There are several methods for biodiesel production, including blending, thermal cracking/pyrolysis, microemulsification, and the most widely used technique, the transesterification process. Transesterification is the reaction of fatty acids or oils with alcohol, resulting in esters and the by-product glycerol. These processes can be performed using either homogeneous or heterogeneous catalysts:

- **Alkaline catalysts**: KOH, NaOH, alkaline metals, and alkaline metal carbonates. They are ideal for oils with low free fatty acids (FFAs), and their principal advantage is a high ester yield under moderate reaction conditions.
- **Acidic catalysts**: CH_3SO_3H, H_2SO_4, $C_2HF_3O_2$, $NaOCH_3$, $AlCl_3$, and H_3PO_4. Requires harsh conditions for the transesterification process, such as high temperatures (~100 °C) and pressures (~5 bar). It is often slower than alkaline catalysis, but is adequate for the efficient esterification of oil with a very high acid value (such as edible waste oil) and when the FFAs in the oil exceed 1% [44].
- **Enzymatic route**: Lipases obtained from different sources are also used for biodiesel production. The main advantage is the biodiesel conversion at moderate temperatures, pressures, and pH levels. However, the main disadvantages are the high cost of the biocatalyst and the lengthy reaction time. The enzyme immobilization process has been applied to enhance biocatalyst activity and stability, as well as to improve biodiesel production yields [45, 46].
- **NonCatalytic Production**: High temperatures and pressures are required for a noncatalytic reaction. An example of a noncatalytic reaction is a supercritical process (supercritical methanol, propanol, butanol, and ethanol).

2.3 Lignocellulose-Based Feedstocks for Second-Generation Biofuels

Lignocellulosic biomass is sourced from various origins, including energy crops, a wide range of grasses, soft and hard wood, agricultural waste (e.g., corn stover, sugarcane bagasse, and wheat straw), dead trees, forest residues, wood processing

Table 1 Percentage of oil content and chemical composition of different feedstocks used for biodiesel production

First-generation biodiesel	Edible oil source	Oil content (%)	*Main fatty acids
	Glycine max (Soybean)	15–20	C16:0/C18:1/C18:2
	Helianthus Annuus (Sunflower)	25–35	C16:0/C18:1/C18:2
	Brassica napus (Rapeseed)	38–46	C16:0/C18:1/C18:2
	Arecaceae (Palm)	30–60	C16:0/C18:1/C18:2
	Arachis Hypogaea (Groundnut)	40–44	C16:0/C18:1/C18:2
	Brassica napus (Canola)	40–45	C16:0/C18:1/C18:2
	Arachis Hypogaea (Peanut)	40–55	C16:0/C18:1/C18:2
	Olea europaea (Olive)	45–70	C16:0/C18:1/C18:2
	Caryocar brasiliense (Pequi)	30–40	C16:0/C18:1/C18:2
	Cocos nucifera (Coconut)	63–65	C16:0 / C18:1/C18:2
Second-generation biodiesel	Nonedible oil source	Oil content (%)	*Main fatty acids
	R. communis	49.2	C18:1/C18:2/C18:3
	A. indica	30–60	C18:0/C18:1/C18:2
	Crambe abyssinica	35–42	C16:0/C18:1/C18:2
	Moringa oleifera	30–40	C16:0/C18:0/C18:1
	H. brasiliensis	40–50	C18:1/C18:2/C18:3
	J. curcas	40	C16:0/C18:1/C18:2
	Attalea speciosa	65–68	C16:0/C18:1/C18:2
	Camelina sativa	30–40	C16:0/C18:1/C18:2
	C. inophyllum L	65–75	C18:0/C18:1/C18:2
	Madhuca longifolia	30–50	C16:0/C18:1/C18:2
	Simmondsia chinensis	40–50	C16:0/C16:1/C18:1
	Vernicia fordii	30–40	C18:0/C18:1/C18:2
	Nicotiana tabacum	33–43	C16:0/C18:1/C18:2
	Animal fat	*Main fatty acids	
	Poultry fat	C16:0/C18:1/C18:2	
	Pork lard	C16:0/C18:0/C18:1/C18:2	
	Beef tallow	C16:0/C18:0/C18:1	
	Mutton tallow	C16:0/C18:0/C18:1/C18:2	
	Duck tallow	C16:0/C18:1/C18:2	
	Fish oil	C16:0/C18:1/C18:2	
	Waste cooking oils	C16:0/C18:1/C18:2	

(continued)

Table 1 (continued)

Third-generation biodiesel	Oleaginous microbes	Oil content (%)	*Main fatty acids
	Rhodotorula mucilaginosa	16	C16:0/C18:0/C18:1
	Mucor indicus	30	C16:0/C18:0/C18:2
	Rhodotorula taiwanensis	60	C16:0/C18:0/C18:1
	Cryptococcus curvatus	43.1	C16:0/C18:0/C18:1
	Trichosporon cutaneum	45.4	C16:0/C18:0/C18:1
	Y. lipolytica	50	C16:0/C18:1/C18:2
	Wickerhamomyces anomalus	24	C16:0/C18:1/C18:2
	Aspergillus terreus	37.4	C16:0/C18:1/C18:2
	Cunninghamella elegans	33.6	C16:0/C18:1/C18:3
	Mortierella Isabelina	67	C16:0/C18:1/C18:2
	Mortierella vinacea	51.9	C16:0/C18:1/C18:2
	M. circinelloides	23.8	C16:0/C18:1/C18:2
	Thermomyces lanuginosus	21	C16:0/C18:1/C18:2
	Rhizopus oryzae	34.8	C16:0/C18:1/C18:2
	Mucor plumbeus	20.6	C16:0/C18:1/C18:2
	Scenedesmus obliquus	26 ~ 30	C16:0/C18:1/C18:2/C18:3
	Monoraphidium contortum	37 ~ 43	C16:0/C18:1/C18:2/C18:3
	Neochloris oleoabundans	≤ 50	C16:0/C18:1/C18:2/C18:3
	Botryococcus braunii	32 ~ 37	C16:0/C18:1/C18:2/C18:3

* C16:0 (Palmitic)–C18:0 (Stearic)–C16:1 (Palmitoleic)–C18:1 (Oleic)–C18:2 (Linoleic)–C18:3 (Linolenic) [31, 34, 35, 37–43]

residues, aquatic plants, algae, food industry residues, organic fractions of municipal waste, and others [31, 47, 48]. The harvesting and processing of agricultural products serve as excellent sources of lignocellulosic biomass for biofuel production. For example, the biofuel industries in the USA and Brazil use corn residue and bagasse as major feedstocks for bioethanol production, respectively [31].

Lignocellulose biomass is primarily composed of multiple layers of arranged biopolymers: cellulose (30–50%), hemicellulose (15–35%), and lignin (10–20%) [31, 48]. However, these values may vary depending on the different varieties of lignocellulosic sources, the plant's age, climatic conditions, and harvest season (Table 2).

- **Cellulose**: It is highly available in plants, grasses, woods, stalks, and is formed by linearly organized D-anhydroglucopyranose units covalently linked by β-(1, 4)

Table 2 Composition of different lignocellulosic biomass materials

Biomass	Cellulose (%)	Hemicellulose (%)	Lignin (%)	Reference
Corn stover	38–40	24–26	7–19	[49]
Corn cob	45	35	15	[50]
Sugarcane (bagasse)	50	25	25	[47]
Rice husk	25–35	18–21	26–31	[48]
Rice straw	2 8–36	23–28	12–14	[49]
Wheat straw	33–38	26–32	17–19	[49]
Grasses	25–40	25–50	10–30	[51]
Softwoods	42	27	28	[49]
Hardwoods	40–55	20–40	18–25	[51]
Sorghum straw	32	24	13	[52]
Sweet sorghum	34–45	18–27	14–21	[49]
Barley straw	33.25	20.36	17.3	[47]
Wheat straw	33–38	26–32	17–19	[49]
Corn cob	42–45	35–39	14–15	[48]
Bamboo	26–43	15–26	21–31	[53]
Coconut husk	21.26	17.33	46.36	[47]
Banana waste	13.2	14.8	14.0	[53]
Oat straw	31–37	27–38	16–19	[53]
Corn fiber	14.28	16.8	8.4	[48]
Soybean straw	35.3	16.9	21.8	[47]
Cassava peel	9.71	32.36	16.89	[47]
Groundnut shell	38.31	27.62	21.10	[47]
Jute fiber	45–53	18–21	21–26	[51]
Cotton stalk	31	11	30	[51]
Coffee pulp	18.8	46.3	35	[53]
Prairie cord grass	36.7	16.81	20.96	[52]

glycosidic bonds, responsible for the strong crystalline structure and amorphous regions within the lignocellulosic biomass united principally by hydrogen bonds (crystalline regions).

- **Hemicellulose**: Amorphous and highly branched structure, composed of a heterogeneous group of polysaccharides accounting for almost 30% lignocellulosic biomass content → D-xylose, D-mannose, D-glucose, D-galactose, and other glycosyls.
- **Lignin**: A Complex heteropolymer of phenylpropanoid monomers with a rigid aromatic nature, responsible for binding the hemicellulose and cellulose fibers to the cell wall. It is also responsible for the recalcitrant character of lignocellulosic

biomass (mechanical strength, hydrophobicity, and indigestibility of plant cell walls) and is a major physical barrier to the enzymatic hydrolysis process.

2.3.1 Pretreatment Methods for Lignocellulosic Biomass Processing

The complex, cross-linked structure and recalcitrant nature of lignocellulose arise from various intrapolymer and interpolymer linkages, including hydrogen, ether, and ester bonds [52]. Consequently, due to the recalcitrant characteristics of lignocellulosic biomass (such as the crystallinity of cellulose, low digestibility of hemicellulose, and the highly crosslinked structure of lignin), an appropriate pretreatment process—whether chemical, biological, and/or mechanical—is necessary to modify the macromolecular structure, enhance digestibility, and facilitate the fractionation of its chemical bonds [47, 52].

The following steps are essential for deconstructing the compact structure of lignocellulose and efficient bioconversion of biomass recalcitrance into biofuel: pretreatment process → hydrolysis → fermentation → product separation → distillation → biofuel. In this scenario, pretreatment accounts for approximately 20–40% of the total processing cost [47, 48]. Each pretreatment process has its advantages and disadvantages, and several parameters influence the choice of pretreatment technique and the cellulosic saccharification yield, including material source, crystallinity, degree of polymerization (DP), glucose content, accessible volume, pore volume, accessible surface area, and particle size [48, 49]. On the other hand, the best pretreatment technique relies on both economic implications and its environmental aspects.

In the lignocellulose biorefinery process, the pretreatment processes can be divided into two generations. First-generation pretreatment processes are straightforward, consisting of washing, grinding, pressing, and juice filtration. The second generation of pretreatment strategies includes mechanical, physical, chemical, physicochemical, and biological/enzymatic methods. For example, the enzymatic/chemical hydrolysis of starchy raw material is used for saccharification. Below, different pretreatment methods for converting lignocellulosic biomass into biofuel are presented, highlighting their advantages and disadvantages.

- **Physical pretreatment**: a combination of drying, chipping, grinding, fragmenting, or milling. Recently emerged techniques are repeated freeze–thaw cycles (freezing), microwave, and ultrasound assistance, which lead to the rupture of cell walls, delignification, and surface erosion [31]. Equipment: ball mills, irradiation, roll mills, pin mills, centrifugal mills, hammer mills, colloidal mills, vibrio energy milling, and extruders. Function: size reduction and increase the surface area to reduce the crystallinity of the lignocellulose biomass. Cons: high energy consumption (mechanical) and high cost, temperature, and pressure (microwave and ultrasound).
- **Chemical pretreatment**: acidic and alkaline hydrolysis, green solvents like supercritical fluid, ionic liquid, and deep eutectic solvent, ammonia pretreatment, ozonolysis, wet oxidation, oxidative delignification, steam treatment, liquid hot

water, ammonia fiber explosion, ammonia recycle percolation, and more. Critical parameters to consider include solid content, acid/base concentration, temperature, and residence time. These methods aim to remove lignin and hemicellulose, decrease the crystallinity of the biomass, reduce the degree of polymerization, and increase porosity (through an alkaline solution), thereby disrupting the lignin layer and enhancing enzyme accessibility to cellulose during hydrolysis. Downsides include equipment corrosion, the release of toxic chemicals, environmental pollution, and high costs associated with the use of organic solvents.
- **Physiochemical pretreatment**: a combination of physical and chemical pretreatment, such as steam explosion (160–240 °C/0.69–4.83 MPa/10 s–1 min) to obtain monomeric and oligomeric sugars. Other important techniques include hydrothermal pretreatment (which allows solvent penetration in recalcitrant biomass), torrefaction (which occurs in an inert atmosphere at 200–300 °C), ammonium fiber explosion (AFEX), wet oxidation, and ionic liquid pretreatment. Function: increase of biomass porosity, lignin removal, hemicellulose solubilization, removal of hydroxyl groups, and improvement of the cellulose digestibility [54]. Cons: high process costs and harsh conditions (high temperature, pressure, and water consumption).
- **Biological pretreatment**: delignification of biomass by enzymatic hydrolysis (utilization of isolated enzymes) or specific strains of microorganisms (source of the enzymes), such as bacteria or filamentous fungi. Examples of bacterial genera: *Erwinia, Sphingomonas, Microbispora, Clostridium, Agrobacterium, Cellulomonas, Thermomonospora, Acetobacter, Bacillus, Ruminococcus, Bacteroides, Acetovibrio*, and *Streptomyces*. Some examples of fungi that are well known for producing lignocellulosic biomass degradation enzymes: brown-rot fungi (*Serpula lacrymans, Coniophora puteana*), white-rot fungi (*Phanerochaete chrysosporium, Ceriporia lacerata*), soft-rot fungi (*Paecilomyces* sp., *Cadophora* spp.), *Sclerotium rolfsii, Phanerochaete chrysosporium, Aspergillus* spp., *Trichoderma* spp., *Penicillium spp., Talaromyces* spp. [31, 54]. Function: decrease the crystallinity/polymerization and improve the surface area; lignin removal results in improved digestibility of cellulose. Cons: long process (microorganisms), high cost (enzyme production and purification), and cellulose degradation.

In biological pretreatment, pure enzymes can be used to enhance the degradation of lignocellulose. Cellulases facilitate the conversion of cellulose (polymeric units: β-Glucan) into monomeric units (D-glucopyranose) by disrupting the β-1,4-glycosidic bonds.

- **Endoglucanase/cellobiohydrolase (EC 3.2.1.4)**: Targets regions of low crystallinity, creating free chain ends
- **Exoglucanase (EC 3.2.1.91)**: Removes cellobiose units from the ends of free chains
- **β-glucosidase (EC3.2.1.21)**: Hydrolyzes cellulooligosaccharides and cellobiose to produce β-D-glucopyranose
- **Hemicellulases**: Hemicellulose (polyxylose, galactoglucomannan [Gal-Glu-Man], glucomannan [Glu-Man]) degradation occurs through the disruption of

β-1,4-glycosidic bonds and β-1,6-glycosidic bonds, producing the monomeric units: L-arabinose, galactose, glucuronic acid, D-xylose, and mannose. The hemicellulase system typically includes xylanase, esterase, and xylosidase
- **Lignin degradation system**: Major lignolytic enzymes for lignin (aromatic polymer) degradation: laccases, manganese peroxidases, and lignin peroxidases. Monomeric units: guaiacylpropane, syringylpropane, p-hydroxyphenylpropane

The production of lignocellulosic biofuels continues to face significant challenges, mainly due to the recalcitrant nature of biomass, which hampers the efficient release of fermentable sugars, and the high costs associated with the equipment required to achieve satisfactory yields. To address these limitations, a variety of integrated bioprocessing strategies have been developed, including separate hydrolysis and fermentation (SHF), simultaneous saccharification and fermentation (SSF), simultaneous saccharification and co-fermentation (SSCF), simultaneous saccharification, filtration, and fermentation (SSFF), and consolidated bioprocessing (CBP). These strategies aim to improve process efficiency, reduce operational costs, and enhance overall productivity. Technologies focused on process integration and consolidation have garnered particular attention and will be explored in greater detail in the following chapters.

2.4 Microalgae Biomass for the Production of Third-Generation Biodiesel

Microalgae represent a significant sustainable source of feedstock to meet the increasing energy demand. Microalgae can be described as photosynthetic microorganisms that produce oxygen, including both eukaryotic microalgae and cyanobacteria [55]. Microalgae biomass is known to have a high content of lipids, depending on the species, producing different kinds of these biomolecules.

Several studies have been conducted to identify the optimal attributes for species that combine high growth rates and lipid content, a suitable fatty acid profile, as well as ease of biomass harvest and lipid extraction [56]. Additionally, efforts have also been dedicated to bioprocesses that stimulate oil accumulation within the microalgae cells. Avila-Leon et al. (2020), for instance, observed an increase in lipid content from 28.9% (standard condition) to 52.0% with the addition of sodium thiosulfate; however, this increase in oil content was accompanied by a decrease in biomass production [41]. In fact, stress conditions, such as nutrient limitation, appear to be the most significant factor influencing lipid accumulation [57]. On the other hand, Choi et al. (2023) found a microalga species, *Monoraphidium contortum*, that does not exhibit this same behavior pattern, presenting the same lipid content regardless of nitrogen limitation or abundance [40].

When selecting a specific biomass as feedstock for biodiesel production, it is important to assess not only the lipid content but also the fatty acid profile, as it greatly influences the performance of this biofuel. Table 1 displays the average lipid content and the main fatty acids found in various oil sources (microalgae, bacteria, yeasts, and plants).

Two key properties of fatty acids influence the characteristics of the resulting biodiesel: carbon chain length and the degree of unsaturation [58, 59]. For instance, longer carbon chains increase the cetane number, heat of combustion, viscosity, and melting point, whereas higher levels of unsaturation reduce these parameters. Additionally, saturated fatty acids enhance oxidative stability, while unsaturated fatty acids may improve the melting point [59, 60]. In this context, it is essential to note that when choosing a microalgae species, those containing high amounts of polyunsaturated fatty acids should be avoided [58]. Indeed, according to the biodiesel specifications in EN 14,214 (European Standard), the content of polyunsaturated C18:3 (linolenic acid) is limited to 12%, and fatty acid methyl esters (FAME) with four or more unsaturations are restricted to a maximum of 1% [61], primarily concerning oxidative stability.

2.5 *Food Waste for the Production of Biomethane*

Industrialization and the ongoing rise in urbanization have also contributed to increased waste production. In 2016, global municipal solid waste production was around 2 billion tons annually, with projections suggesting it could reach 3.4 billion tons by 2050 [61]. Currently, food waste accounts for approximately 1.6 billion tons, often ending up in landfills or being incinerated without adequate control [62]. This situation creates a pressing need for improvements in waste collection and separation, opening the door for utilizing organic waste in bioenergy production.

The composition of food waste is quite heterogeneous, depending on its origin, including supermarkets, street markets, hotels, restaurants, and households [63]. It is usually composed of carbohydrates (mainly due to rice and vegetables), proteins (due to animal-derived products), lipids, and traces of inorganic nutrients [64]. In addition, it may contain a high water content (approximately 80%), which makes it prone to biological biodegradation and difficult to store [62].

With the possibility of extracting the lipid fraction from food waste, this oil could be used to produce biodiesel. For example, after drying food waste produced by a restaurant, Carmona-Cabello et al. [63] extracted the lipid fraction using the Soxhlet method and produced biodiesel through alkali-catalyzed transesterification. In this study, specifically, they found that the oil derived from solid food waste could be used for biodiesel production, developing a more sustainable technology (compared to the combustion of fossil fuels); however, it was recommended to blend it with diesel fuel due to characteristics that do not fulfill the EN14214 standard, such as oxidation stability and glyceride content.

Paritosh et al. (2017) suggest, in turn, that anaerobic digestion of food waste is one of the most promising and eco-friendly solutions for food waste management, as it produces methane, contributing to the increasing energy requirement [64]. By anaerobic digestion, organic matter is converted to methane and carbon dioxide, in the absence of oxygen, through the combination of the following phases [62, 64]:

- **Enzymatic hydrolysis**: Polymeric compounds are broken down by microbial hydrolytic enzymes for releasing oligomer or monomeric units (sugar, amino acids, and fatty acids); hydrolytic bacteria genera that could be mentioned are *Bifidobacterium, Lactobacillus, Megasphaera, Propionibacterium,* and *Sphingomonas.*
- **Acidogenesis**: The monomeric hydrolytic products are converted to volatile fatty acids, carbon dioxide, hydrogen, and ammonia by fermentation; the main bacterial genera responsible for this phase are: *Ruminococcus*, *Paenibacillus*, and *Clostridium.*
- **Acetogenesis**: Acid phase products are converted into acetate and hydrogen; the main acetogenic bacteria are from the genera *Desulfovibrio, Aminobacterium*, and *Acidaminococcus.*
- **Methanogenesis**: By the action of methanogens, *Archaea* microorganisms (*Methanolobus, Methanohalophilus, Methanococcoides, Halomethanococcus, Methanoculleus*, among other genera), which consume acetic acid, hydrogen, and carbon dioxide produced in the previous phases to generate methane. This biogas can be produced by the fermentation of acetic acid or by reducing carbon dioxide.

Heterogeneity in food waste composition poses a challenge, as substrates with high carbohydrate and protein content have quicker conversion rates but lower biogas yields. Conversely, lipid-rich substrates require longer retention times due to their lower biodegradability, but provide higher biogas yields [65]. Additionally, the C/N ratio has a strong influence on biodigestion performance and, consequently, methane production. Therefore, several studies have been conducted that combine food waste with other substrates, such as fruit and vegetable waste [67], cow manure [68], and cellulose-based bioplastics [66], and paper waste [67].

3 Geographic Distribution of Biofuel Feedstock

The availability and characteristics of biofuel feedstocks play a crucial role in shaping the environmental and economic sustainability of biofuel production. Although different raw materials can be used for biofuel production (Fig. 1), feedstock availability is not uniform globally, being significantly influenced by a combination of environmental, economic, and policy-driven factors [8]. Climate conditions, soil characteristics, and land availability determine the suitability of specific crops for biofuel production in different regions. Meanwhile, government policies, infrastructure, and market demand influence their large-scale cultivation and commercialization [7]. Additionally, comparing these impacts across regions remains challenging, as many estimates fail to account for key factors such as geographic location, climate, soil type, and specific agricultural management practices for each crop [68].

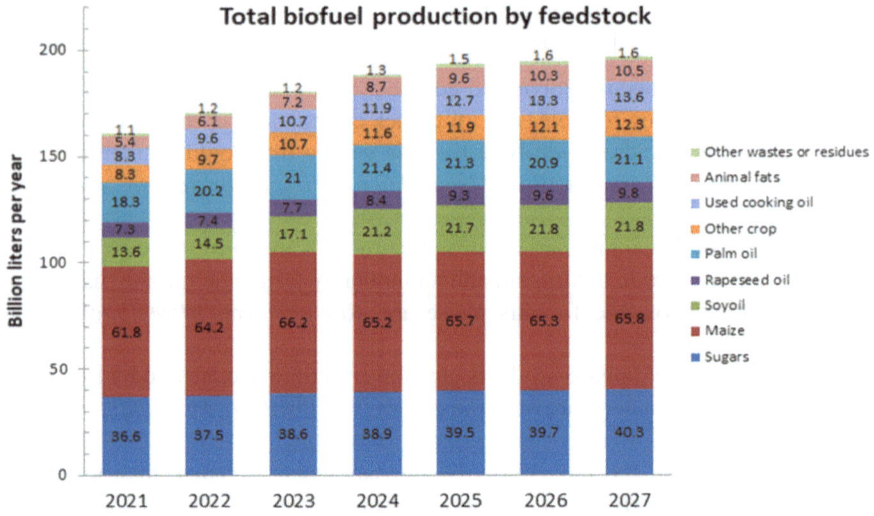

Fig. 1 Total biofuel production by feedstock. Data from IEA, 2022 [69]

Regional variations in feedstock availability and properties significantly impact biofuel conversion processes, creating both challenges and opportunities for optimization [70]. Understanding these variations is crucial for selecting suitable feedstocks and ensuring a sustainable and efficient biofuel supply chain. Moreover, effective land use planning and enforcement are necessary to mitigate potential adverse environmental impacts from expanded cultivation [68].

Despite the growing demand for biofuels in the transport sector, it does not guarantee investments in national production. Since biofuels are typically more expensive than fossil fuels, their production and consumption rely on policy support, including tax exemptions and blending targets [70]. In this context, biofuel policies are crucial for the development and expansion of both regional and national biofuel markets. In particular, some studies have demonstrated that countries that implement a combination of market-pull and technology-push policy instruments have been the most successful in enhancing biofuel production and consumption [71].

Within this framework, integrating renewable energy technologies with development policies has proven to be a reliable and cost-effective approach to expanding access to modern energy services. This integration not only strengthens energy security but also promotes broader social and economic development. Against this backdrop, this section examines the key factors influencing the geographical availability of primary biofuel feedstocks worldwide, including prevailing policy frameworks and market conditions.

3.1 Major Production Regions and Their Biofuel Feedstocks

3.1.1 Latin America and the Caribbean

Biofuels hold significant potential in the emerging economies of Latin America and the Caribbean (LAC), Africa, and Asia, where increasing demand for sustainable energy, abundant natural resources, and available land create favorable conditions for biofuel production [72]. Despite the continued reliance on fossil fuels, the energy sector in Latin America and the Caribbean maintains one of the lowest CO_2 emission intensities worldwide, mainly due to the widespread use of biofuels in transportation, with the region accounting for approximately 25% of global liquid biofuel production [73]. Currently, biofuel production in LAC is primarily driven by conventional ethanol and biodiesel, primarily for road transportation. However, with the increasing global demand for sustainable aviation fuel, the region is well-positioned to emerge as a key supplier of biojet kerosene [74].

Brazil is the largest producer and consumer of liquid biofuels in Latin America and the second-largest producer globally [73]. This success can be attributed to the synergy among the global sugar market, government support, electricity generation, and favorable geographic conditions [75]. Brazil's institutional and technological capabilities for large-scale renewable energy use began developing in the 1970s, driven by significant government interventions aimed at increasing ethanol demand [70]. The country primarily relies on sugarcane as a feedstock for bioethanol, as the crop thrives in tropical climates with abundant rainfall and sunlight [76]. Additionally, Brazil's sugarcane industry benefits from a well-established agricultural production system and advanced processing infrastructure, making it one of the most efficient feedstocks for bioenergy [77].

With the emergence of second-generation bioenergy platforms to convert lignocellulose materials to liquid fuels, the energy output from sugarcane could increase significantly. Brazil is the world's largest producer of sugarcane and holds the most significant global potential for cellulosic feedstock utilization, particularly due to its abundant and low-cost sugarcane residues, such as bagasse [78, 79]. Bagasse is combusted to generate electricity, which powers sugarcane mills and bioethanol production facilities, with surplus energy supplied to the national grid. As a result, Brazil produces both first- and second-generation bioethanol and, together with the United States, contributes to 85% of global bioethanol production [80].

Alongside Brazil, Argentina and Colombia are emerging as key producers of biofuels. In Argentina, the primary feedstocks for ethanol production are sugarcane and corn. However, ethanol yields from sugar-based feedstocks tend to be higher than those from starch-based alternatives [81]. The Caribbean, Honduras, Nicaragua, and Panama also possess considerable growth potential, bolstered by favorable climatic conditions, a developing sugarcane sector, and government incentives aimed at reducing dependence on imported oil [70].

Biodiesel production in the region is also significant, with Brazil, Argentina, and Colombia having the largest production capacity [80]. The key feedstocks include plant oils, such as soybean oil and palm oil. Among these, palm oil yields higher biodiesel efficiency compared to soybean oil. However, its expansion is limited by specific climatic conditions, such as high temperatures and humidity, as well as environmental concerns related to deforestation [73]. In Brazil, soybean is the primary feedstock for biodiesel production, with cultivation expanding across multiple regions [82]. Meanwhile, palm oil production is concentrated in the northern part of the country, where the climate is similar to that of Colombia, the largest producer of palm oil in America, creating favorable conditions for its cultivation [81, 83].

Government policies have strongly supported the development of the biofuels sector in Latin America and the Caribbean. Several countries in the region have implemented biofuel mandates, tax incentives, and subsidies to stimulate production and consumption. For instance, Brazil established the RenovaBio program, which stimulates biofuel production through carbon intensity reduction targets and the issuance of decarbonization credits [71]. Argentina enforces mandatory biodiesel and ethanol blending requirements, with ethanol blends reaching up to 12% in gasoline. Similarly, Colombia mandates a 10% ethanol blend in gasoline and has introduced policies to support palm oil-based biodiesel production [80]. These policies play a pivotal role in nurturing a stable biofuels market, attracting investment, and positioning the region as a key player in the global biofuel supply.

3.1.2 North America

North America is home to two of the world's leading oil producers, the United States and Canada. At the same time, increasing support for clean energy initiatives has accelerated the region's transition toward sustainable energy, contributing significantly to global decarbonization efforts [84]. The United States is at the forefront of the transition to renewable energy, implementing a range of policies and mandates under the Renewable Fuels Standard (RFS) Program, which requires the annual incremental blending of renewable fuels into transportation fuels. Similarly, Canada has reinforced its commitment to biofuels through the Renewable Fuels Regulations, designed to reduce greenhouse gas emissions and promote the growth of the biofuel industry [85].

Within this framework, North America plays a pivotal role in biofuel production, leveraging abundant feedstock resources, supportive policies, and technological advancements. The primary biofuels produced in the region include ethanol, predominantly derived from corn, as well as biodiesel and renewable diesel, which are made from vegetable oils, animal fats, waste oils, and greases [86].

The United States stands as the world's leading producer of fuel ethanol, which is primarily sourced from maize, utilizing plant starches and sugars, particularly corn starch [87]. Moreover, the development of bioethanol derived from lignocellulosic materials is progressing, primarily in the Corn Belt and Southeastern U.S. regions, where crop residues such as corn stover are abundant. These residues represent some of the most readily available bioenergy feedstocks due to their high availability and the familiarity of producers with their use [88]. Additionally, in the Corn Belt, maize is often rotated with soybean, which requires minimal nitrogen (N) addition. Thus, the soybean crop benefits from residual N left by the preceding maize crop, in a symbiotic relationship with nitrogen-fixing rhizobia [89].

Canada, the second-largest country in the world, possesses vast biomass resources derived from its extensive forests and well-developed agricultural lands, presenting significant potential for biofuel production. In Canada, first-generation ethanol is primarily produced from starch-based feedstocks, with wheat and corn serving as the dominant sources. Approximately 25% of the country's grain corn production is allocated to biofuel feedstock, while wheat accounts for around 3% of its total output for biofuel [90]. Other notable Canadian crops for commercial biofuel production include oilseeds, such as canola and soybeans, for biodiesel production [91].

An emerging biofuel technology utilizes microalgae as a feedstock. Projections for renewable energy development in North America suggest that biomass-derived fuels, including those sourced from brown algae, will play a crucial role, particularly in the transportation sector. These biofuels, such as bioethanol and biodiesel, are integrated into optimized biorefinery systems designed to maximize the utilization of all components of brown algae [92]. However, large-scale commercial cultivation of brown algae remains challenging due to high production costs and fluctuating market demand.

The Americas encompass a vast range of environmental contrasts, spanning from the Arctic to tropical regions and the sub-Antarctic, with diverse soils, landforms, and climatic conditions [93]. Within this context, North America offers a diverse array of biofuel feedstocks, ranging from conventional crops such as corn and soybeans to emerging sources like algae and lignocellulosic residues. Ongoing technological advancements and supportive policy frameworks will continue to shape the sector, enhancing both the efficiency and sustainability of biofuel production.

3.1.3 Europe

Europe is at the forefront of clean energy innovation and continues to advance its transition toward sustainable biofuels. The European Union (EU) remains a major player in the global biofuels market, with biofuels accounting for approximately 6.3% of the region's total transport fuel consumption in 2020 [94]. To further promote biofuel production, the EU has implemented various policy instruments, most notably the Renewable Energy Directive (RED II). This directive mandates that, by 2030, fuel suppliers must ensure that at least 14% of the energy used in road and rail transport is derived from renewable sources [95].

Biofuel feedstock selection within Europe is influenced by both regional availability and regulatory constraints. In this regard, the Renewable Energy Directive imposes stringent sustainability criteria, prohibiting biofuel production on land classified as possessing high biodiversity value, significant carbon stocks, or peatland [96]. These regulations aim to ensure that biofuel development does not contribute to environmental degradation while supporting the EU's broader decarbonization objectives [94].

Most biofuels produced in the EU consist of biodiesel derived from vegetable oils (approximately 75%) and bioethanol obtained from sugar-based and starch-based crops (around 25%). Regarding biodiesel production, rapeseed oil is the predominant raw material, accounting for 38% of total feedstock. Its widespread adoption is attributed to both its domestic availability and the superior winter stability of rapeseed methyl ester (RME) compared to other biodiesel feedstocks [97]. Soybean and palm oil are also used; however, due to sustainability concerns, the share of biodiesel produced from palm oil is projected to decline from 21 to 8% by 2032 [98]. Additionally, the increasing use of waste oils and animal fats has contributed to reducing reliance on food crops, thereby enhancing the sector's sustainability [99].

In terms of bioethanol production, the primary feedstocks include cereals, such as wheat and corn, which account for approximately 50% of total ethanol feedstock use, and sugar beet, which contributes 39%. Corn, primarily imported from Ukraine, is mainly utilized in Central Europe, the Netherlands, and Spain. Wheat-based ethanol is predominantly produced in Germany, France, and the United Kingdom, while sugar beet serves as a key feedstock in France, Germany, the United Kingdom, Czechia, and Belgium [100].

Advanced biofuels derived from lignocellulosic feedstocks currently represent a small share of total biofuel production, but they have received increasing attention. This shift is primarily driven by the principles of the circular economy, which aim to minimize environmental and biodiversity impacts and prevent additional land-use demand, as mandated by the Renewable Energy Directive [100, 101].

3.1.4 Sub-Saharan Africa

Sub-Saharan Africa (SSA) is a vast and diverse region, encompassing 19% of the world's agricultural land and hosting 14% of the global population, while also experiencing the highest global population growth rate [98]. Currently, oil accounts for over 99% of fuel consumption in road transport across Africa; however, this share is projected to decline by 2030, primarily due to policies promoting the use of biofuels. Biofuel policies vary across African countries, reflecting differences in regulatory frameworks and national priorities. However, biofuel development remains a contentious issue in SSA, mainly due to concerns regarding food security and environmental sustainability [11].

Despite the region's significant potential for expanding biofuel production, given the size of its agricultural sector, the biofuels contribute only about 0.1% of road transport energy consumption [102]. Bioethanol and biodiesel are the most widely recognized biofuels in the SSA region, and various feedstocks can be used for their production. For ethanol, sugarcane and molasses have attracted the most interest, while cassava and sweet sorghum are also utilized in some countries. In South Africa, Nigeria, Senegal, Ghana, Malawi, Cameroon, and the Democratic Republic of the Congo, sugarcane is the primary feedstock. Producing bioethanol from sugarcane is particularly appealing, as sugarcane is cultivated on a large scale in Africa, and the expansion of sugar production remains vital for the continent's economy [11].

The production of biodiesel in Sub-Saharan Africa is still in its early stages, despite many countries in the region possessing sufficient agricultural resources and favorable temperatures for cultivating feedstocks such as Jatropha and other crops [103]. Jatropha oil is known for its high biodiesel productivity. This plant thrives in high-temperature climates and can be irrigated with wastewater, making it particularly well-suited for biodiesel production in Egypt. As a result, it has already become a key feedstock in both Asia and Africa [104]. In contrast, South Africa, which has a more temperate climate, has incorporated soybean, rapeseed, and sunflower seeds into its biofuel strategy for biodiesel production [105].

Although maize and other grains are commonly used for ethanol production in the United States and Europe, they are unlikely to be viable feedstocks in most African countries due to high production costs and limited demand for the by-product feed. Similarly, soybean, rapeseed, and sunflower seeds are generally not well-suited to tropical climates and have not gained widespread attention as biofuel feedstocks in most countries in Africa [105].

3.1.5 Asia

In recent years, several Asian countries have implemented strategies and policies to promote biofuel development, including blending mandates, subsidies, fiscal incentives, and tax exemptions [106]. Ethanol and biodiesel production in Asia account for 9.7% and 32.6% of the global output, respectively. The production supply not only serves the domestic and regional markets, but it is also exported to other Asian and European countries [107]. Among the largest biofuel-producing nations in the region, Thailand, China, and India primarily focus on ethanol production, whereas Indonesia and Malaysia lead in biodiesel production [108].

Southeast Asia is a major producer of first-generation biofuels. However, the large-scale expansion of biofuel production raises concerns regarding food security, land use, deforestation, and environmental impacts, underscoring the need for alternative feedstocks [109]. In Indonesia, Malaysia, and Thailand, biodiesel production primarily relies on palm oil, with Indonesia and Malaysia ranking as the world's largest producers [110]. While domestic demand for palm oil has increased due to national biodiesel programs, exports have declined in response to trade barriers and regulatory measures imposed by the United States and the European Union aimed at discouraging its use [107].

Ethanol production remains limited in Indonesia and Malaysia; however, Thailand has a well-established ethanol industry, benefiting from the country's extensive cultivation of sugarcane and cassava. Additionally, the Philippines promotes both ethanol and biodiesel production [110]. As the world's largest producer of coconut oil, the country is uniquely positioned to produce coconut methyl ester, a distinct form of biodiesel. Moreover, ethanol production in the Philippines is primarily derived from sugarcane and molasses [108, 110].

In China and India, large-scale biofuel production has progressed slowly, primarily due to regulations that limit competition with food production [111]. China's biofuel policy focuses mainly on ethanol. In 2021, the country's fuel ethanol production accounted for only about 3% of global output. Notably, more than 80% of China's fuel ethanol was derived from grains such as corn, wheat, and rice, while approximately only 10% originated from cassava or sugarcane. Moreover, cellulosic ethanol production is currently limited to a small scale in China [112]. The supply of Chinese corn, however, is subject to various regulations and current events [111].

Chinese biodiesel production is minimal compared to its ethanol production and faces challenges due to the limited availability of suitable organic feedstocks, such as soy and palm oil. Nearly all biodiesel production relies on animal fats or waste vegetable oils, often resulting in lower-quality products. The sector is further constrained by the reliance on imported feedstocks, underdeveloped policies for biodiesel consumption, and insufficient financial support for farmers [110, 112].

In India, biofuel production is restricted by feedstock regulations, including a prohibition on using sugarcane juice for ethanol production. This country is the world's second-largest producer of sugarcane after Brazil, but Indian ethanol is produced almost exclusively from molasses, a byproduct of domestically produced sugarcane [111]. Regarding biodiesel, India does not use edible oils for production but instead explores alternatives such as industrial vegetable oils, including Jatropha [111]. Currently, biodiesel blending levels remain low at just 0.1%. Still, this scenario could escalate rapidly due to potential technology cooperation agreements within the Global Biofuels Alliance, which has been signed with Brazil and the United States. This agreement promotes technology investment and collaboration aimed at reducing fossil fuel dependence, accelerating the adoption of biofuels, and contributing to significant environmental benefits [107].

3.1.6 Oceania

In the Oceania region, research on biofuel adoption remains limited [113]. Australia's energy scenario remains dominated by oil, coal, and natural gas, with the integration of biofuels in the transport sector remaining minimal [114]. The country's diverse geography, which encompasses arid and semi-arid regions throughout the inner continent, tropical rainforests in the northeast, and mountain ranges in the southeast, supports a wide range of biodiversity [115]. Australia possesses substantial feedstock resources that can be considered for biofuel production, including grains and molasses for bioethanol, as well as canola, tallow, and used cooking oil for biodiesel. However, only a small fraction of these resources is utilized domestically for biofuel production. Notably, canola is not used for biodiesel production within Australia; instead, it is exported in large quantities to the European Union for this purpose [116]. This limited domestic utilization is partly attributed to the absence of a national biofuels program with uniform goals and mandates, as well as the lack of significant policy support in the few states with the technology to produce and implement biofuels [117].

Within this framework, several research projects in Australia have explored second-generation and advanced biofuels using nontraditional feedstocks, including lignocellulosic materials. Considering the country's vast biomass resources, including agricultural and forestry residues, bagasse, and other underutilized feedstocks for biofuel production, biofuels have the potential to optimize the use of arid and marginal lands while mitigating fossil fuel market volatility and reducing environmental pollution [118]. However, commercial-scale biofuel production from these resources requires advanced technological maturity and industry expertise [117].

Similarly, in New Zealand, the most promising opportunity for expanding biofuel production lies in utilizing nonfood feedstocks, particularly planted forestry on nonarable land. Moreover, the country has a predominantly temperate maritime climate, with 40% of its land area dedicated to agriculture, primarily consisting of permanent grasslands, and 37% covered by forests. Consequently, New Zealand possesses sufficient residual biomass to initiate biofuel production and has the potential to further develop its biofuel feedstock systems [119]. Currently, biofuel use in New Zealand's transport sector is relatively insignificant. The development of liquid biofuels in the country will require the deployment of advanced technologies that utilize lignocellulosic feedstocks, primarily wood [120]. Additionally, other sources, such as vegetable processing and fruit waste—including maize, potatoes, and kiwifruit—have been identified as potential biofuel feedstocks [121].

4 Challenges and Opportunities in Global Feedstock Utilization

With the global population's growth, energy demand continues to rise, driven by the urgent need to reduce fossil fuel dependence and mitigate climate change. The development of new technologies for biofuels and bioproducts production is central to reducing environmental impact and creating a sustainable economy. However, despite the need for alternatives to fossil fuels, several studies have explored the social and ecological issues associated with the global expansion of biofuel production, including its impacts on food security, greenhouse gas emissions, large-scale monoculture systems, and poverty, highlighting both risks and opportunities [122, 123].

First-generation biofuels are derived from edible energy crops such as sugar, starch, and oil-based crops. These biofuels initially showed potential in reducing dependence on conventional fossil fuels and lowering greenhouse gas (GHG) emissions associated with their combustion [124]. However, large-scale production has raised significant environmental concerns, particularly regarding soil degradation, water resource depletion, and biodiversity loss [125]. Another major risk associated with biofuel production is the widespread cultivation of monocultures, which can have detrimental effects on biodiversity and water conservation, especially when natural terrestrial ecosystems are converted for biofuel feedstock cultivation.

A notable example is the expansion of palm oil plantations for biodiesel production, which can contribute to deforestation, biodiversity loss, and substantial carbon emissions from the destruction of carbon-rich ecosystems [126].

Moreover, first-generation biofuels are often associated with concerns regarding food security. The large-scale shift of food crop producers toward cultivating raw materials for biofuel production could result in food shortages, even in countries with abundant agricultural land. This trend is further exacerbated by economic incentives that encourage farmers to prioritize biofuel crops over food crops, resulting in food security issues [127]. Additionally, biofuel production relies heavily on arable land, creating a direct conflict with the need to increase food production for a growing global population.

Addressing this challenge requires prioritizing the selection of biofuel crops with high productivity on marginal lands, along with strong resilience and adaptability traits [128]. Another promising approach is the development of advanced biofuels derived from nonfood feedstocks, such as algae, agricultural residues, and municipal solid waste. This alternative is considered more ecologically sustainable than first-generation biofuels and offers significant environmental benefits [122].

The most pressing challenge is the complexity and variability of these feedstocks, which, once brought into bioproduction systems, can disrupt biological processes, deactivate catalysts, and consequently compromise product quality. The difficulty of processing and extracting energy from these feedstocks is significant due to their recalcitrant nature. Furthermore, much of the carbon in future feedstocks (e.g., cellulose, hemicellulose, and lignin) is less biologically accessible than the carbon in starch, sugars, and oils that earlier bioeconomy successes relied upon. The difficulty of collecting and transporting feedstocks only adds to these challenges because many feedstocks cannot be pipelined like petroleum and other liquids. The successful utilization of future feedstocks will require a concerted effort to overcome their inherent complexity, recalcitrance, and cumbersome collection [3]. Overcoming the complex nature of these feedstocks will require substantial research to identify and develop more efficient pretreatment and conversion processes.

Additionally, the role of policy in driving biofuel expansion is crucial to the success of these energy alternatives, with diverse approaches being implemented globally to address the complex challenges associated with biofuel production and consumption. Policies supporting biofuel expansion generally fall into two categories: technology-push and market-pull policies [71]. The choice and effectiveness of these policies vary across countries, as they are influenced by differences in economic and energy development stages, as well as natural resource availability, which shape national priorities and approaches to clean energy transitions [73]. However, those nations that have implemented a combination of both market-pull and technology-push policies have been the most successful in increasing biofuel production and consumption [71].

Technology-push policies primarily focus on reducing costs and fostering the development of early-stage technologies, such as research and development (R&D), until they reach commercialization. This includes advancements in synthetic biology, precision fermentation, and microbial engineering, which are creating novel pathways for utilizing diverse feedstocks. This approach is particularly important for emerging technologies, including microalgae cultivation, second- and third-generation biofuels, and the use of new waste streams as feedstock. In contrast, market-pull policies create demand and incentivize the adoption of biofuels through mechanisms such as blending mandates, tax incentives, and carbon pricing. By aligning technological innovation with market needs, these combined approaches can accelerate the transition towards sustainable energy systems, ensuring that advancements in biofuel production translate into tangible economic and environmental benefits. This policy not only supports the scaling of novel technologies but also fosters a more resilient and diversified energy matrix, reducing dependence on fossil fuels and promoting long-term sustainability.

Addressing these challenges is critical to advancing several United Nations Sustainable Development Goals (SDGs), including Zero Hunger (SDG 2), Clean Water and Sanitation (SDG 6), Affordable and Clean Energy (SDG 7), Climate Action (SDG 13), and Life on Land (SDG 15). In this context, a comprehensive and adaptive approach to feedstock management is therefore essential to ensure the sustainability and long-term viability of biofuel production.

In summary, the production of biofuels offers both substantial challenges and significant opportunities in the transition toward a more sustainable energy paradigm. While biofuels hold the potential to reduce dependency on fossil fuels, mitigate climate change, and enhance energy security, they also present complex issues related to food security, environmental degradation, and social equity. Addressing these challenges requires a comprehensive and integrated approach that prioritizes the selection of sustainable feedstocks, the development of advanced biotechnological processes, and the implementation of robust policy frameworks. In parallel, policymakers must design and implement strategies that foster both technological advancements and market adoption, ensuring that biofuel production can scale effectively while minimizing competition with food production and land use. Innovative practices, such as biorefineries that integrate biofuel production with other industrial outputs like bioplastics and biochemicals, hold great promise for enhancing the overall sustainability of the bioeconomy. Collaborative efforts among governments, industry stakeholders, and academic researchers will be essential to align technological innovation with societal needs and environmental goals.

References

1. El-Araby R (2024) Biofuel production: exploring renewable energy solutions for a greener future. Biotechnol Biofuels Bioprod 17:129. https://doi.org/10.1186/s13068-024-02571-9
2. Hassan Q, Viktor P, J. Al-Musawi T et al (2024) The renewable energy role in the global energy Transformations. Renew Energy Focus 48:100545. https://doi.org/10.1016/j.ref.2024.100545
3. RedCorn R, Croft G, McGill R, Reich J (2023) Feedstocks of the future for a circular U.S. bioeconomy: a summary from a stakeholder convening

4. International energy agency (2023) World energy outlook 2023
5. Karimi M, Simsek H, Kheiralipour K (2025) Advanced biofuel production: a comprehensive techno-economic review of pathways and costs. Energy Convers Manag X 25:100863. https://doi.org/10.1016/j.ecmx.2024.100863
6. Dhiman S, Mukherjee G (2021) Present scenario and future scope of food waste to biofuel production. J Food Process Eng 44. https://doi.org/10.1111/jfpe.13594
7. Williams CL, Westover TL, Emerson RM et al (2016) Sources of biomass feedstock variability and the potential impact on biofuels production. BioEnergy Res 9:1–14. https://doi.org/10.1007/s12155-015-9694-y
8. Sarwer A, Hussain M, Al-Muhtaseb AH et al (2022) Suitability of biofuels production on commercial scale from various feedstocks: a critical review. ChemBioEng Rev 9:423–441. https://doi.org/10.1002/cben.202100049
9. Jeswani HK, Chilvers A, Azapagic A (2020) Environmental sustainability of biofuels: a review. Proc R Soc A Math Phys Eng Sci 476. https://doi.org/10.1098/rspa.2020.0351
10. Bakari R, Asha R, Hossein M et al (2024) Converting food waste to biofuel: a sustainable energy solution for Sub-Saharan Africa. Sustain Chem Environ 7:100126. https://doi.org/10.1016/j.scenv.2024.100126
11. Jha P, Schmidt S (2021) State of biofuel development in sub-Saharan Africa: how far sustainable? Renew Sustain Energy Rev 150:111432. https://doi.org/10.1016/j.rser.2021.111432
12. Devi A, Bajar S, Sihag P et al (2023) A panoramic view of technological landscape for bioethanol production from various generations of feedstocks. Bioengineered 14:81–112. https://doi.org/10.1080/21655979.2022.2095702
13. Abdul Kareem Joyia M, Ahmad M, Chen Y-F et al (2024) Trends and advances in sustainable bioethanol production technologies from first to fourth generation: a critical review. Energy Convers Manag 321:119037. https://doi.org/10.1016/j.enconman.2024.119037
14. Mathur S, Umakanth AV, Tonapi VA et al (2017) Sweet sorghum as biofuel feedstock: recent advances and available resources. Biotechnol Biofuels 10:146. https://doi.org/10.1186/s13068-017-0834-9
15. Mojovic L, Pejin D, GrujiĆ O et al (2009) Progress in the production of bioethanol on starch-based feedstocks. Chem Ind Chem Eng Q 15:211–226. https://doi.org/10.2298/CICEQ0904211M
16. Manochio C, Andrade BR, Rodriguez RP, Moraes BS (2017) Ethanol from biomass: a comparative overview. Renew Sustain Energy Rev 80:743–755. https://doi.org/10.1016/j.rser.2017.05.063
17. Apriyanto A, Compart J, Fettke J (2022) A review of starch, a unique biopolymer—Structure, metabolism and in planta modifications. Plant Sci 318:111223. https://doi.org/10.1016/j.plantsci.2022.111223
18. Duarte S de P de A, Teixeira-costa BE, Rosário RC et al (2024) Valorization of Taioba products and by-products : focusing on starch. Foods 13:2415. https://doi.org/10.3390/foods13152415
19. Farias F de AC, Moretti MMS, Costa MS et al (2020) Structural and physicochemical characteristics of taioba starch in comparison with cassava starch and its potential for ethanol production. Ind Crop Prod 157:112825. https://doi.org/10.1016/j.indcrop.2020.112825
20. Di Caprio F, Amenta S, Francolini I et al (2023) Microalgae biorefinery: optimization of starch recovery for bioplastic production. ACS Sustain Chem Eng 11:16509–16520. https://doi.org/10.1021/acssuschemeng.3c04133
21. Carnovale G, Leivers S, Rosa F et al (2022) Starch-rich microalgae as an active ingredient in beer brewing. Foods 11:1449. https://doi.org/10.3390/foods11101449
22. Buckler ES, Stevens NM (2005) Maize origins, domestication, and selection. In: Motley TJ, Zerega N, Cross H (eds) Darwin's harvest–origins, evolution, and conservation of crops. Columbia University Press, New York, pp 67–90
23. Winara A, Fauziyah E, Widiyanto A et al (2022) Assessing the productivity and socioeconomic feasibility of cocoyam and teak agroforestry for food security. Sustain 14:11981. https://doi.org/10.3390/su141911981

24. Pradyawong S, Juneja A, Sadiq MB et al (2018) Comparison of cassava starch with corn as a feedstock for bioethanol production. Energies 11:3467. https://doi.org/10.3390/en11123476
25. Muiruri SK, Ntui VO, Tripathi L, Tripathi JN (2021) Mechanisms and approaches towards enhanced drought tolerance in cassava (Manihot esculenta). Curr Plant Biol 28:100227. https://doi.org/10.1016/j.cpb.2021.100227
26. Colombo C, Second G, Charrier A (2000) Diversity within American cassava germ plasm based on RAPD markers. Genet Mol Biol 23:189–199. https://doi.org/10.1590/S1415-47572000000100034
27. Mardina P, Irawan C, Putra MD et al (2021) Bioethanol production from cassava peel treated with sulfonated carbon catalyzed hydrolysis. J Sci Appl Chem 24:1–8. https://doi.org/10.14710/jksa.24.1.1-8
28. Sánchez ÓJ, Cardona CA (2008) Trends in biotechnological production of fuel ethanol from different feedstocks. Bioresour Technol 99:5270–5295. https://doi.org/10.1016/j.biortech.2007.11.013
29. Cardona CA, Sánchez ÓJ (2007) Fuel ethanol production: process design trends and integration opportunities. Bioresour Technol 98:2415–2457. https://doi.org/10.1016/j.biortech.2007.01.002
30. Abdul Hakim Shaah M, Hossain MS, Salem Allafi FA et al (2021) A review on non-edible oil as a potential feedstock for biodiesel: physicochemical properties and production technologies. RSC Adv 11:25018–25037. https://doi.org/10.1039/D1RA04311K
31. Guldhe A, Singh B (2022) Novel feedstocks for biofuels production, 1st ed. Springer Nature Singapore, Singapore
32. Ramos M, Dias APS, Puna JF et al (2019) Biodiesel production processes and sustainable raw materials. Energies 12:4408. https://doi.org/10.3390/en12234408
33. Fonseca JM, Teleken JG, de Cinque AV, da Silva C (2019) Biodiesel from waste frying oils: methods of production and purification. Energy Convers Manag 184:205–218. https://doi.org/10.1016/j.enconman.2019.01.061
34. Chintagunta AD, Zuccaro G, Kumar M et al (2021) Biodiesel production from lignocellulosic biomass using oleaginous microbes: prospects for integrated biofuel production. Front Microbiol 12. https://doi.org/10.3389/fmicb.2021.658284
35. Toldrá-Reig F, Mora L, Toldrá F (2020) Trends in biodiesel production from animal fat waste. Appl Sci 10:3644. https://doi.org/10.3390/app10103644
36. Sindhu R, Binod P, Pandey A et al (2019) Biofuel production from biomass. In: Current developments in biotechnology and bioengineering. Elsevier, pp 79–92
37. Singh D, Sharma D, Soni SL et al (2020) A review on feedstocks, production processes, and yield for different generations of biodiesel. Fuel 262:116553. https://doi.org/10.1016/j.fuel.2019.116553
38. Zheng Y, Yu X, Zeng J, Chen S (2012) Feasibility of filamentous fungi for biofuel production using hydrolysate from dilute sulfuric acid pretreatment of wheat straw. Biotechnol Biofuels 5:50. https://doi.org/10.1186/1754-6834-5-50
39. Matsudo MC, Sant´Anna CL, Pérez-Mora LS et al (2021) Evaluation of green microalgae isolated from central and north coast of Sao Paulo as source of oil. J Biotechnol Biodivers 9:012–019. https://doi.org/10.20873/jbb.uft.cemaf.v9n1.matsudo
40. Choi N, Nunes IVO, Ohira GOM et al (2023) Evaluation of Monoraphidium contortum for the tertiary treatment of dairy industry wastewater and biomass production with nitrogen supplementation. Bioprocess Biosyst Eng 46:265–271. https://doi.org/10.1007/s00449-022-02831-8
41. Avila-león IA, Matsudo MC, Ferreira-camargo LS et al (2020) Evaluation of Neochloris oleoabundans as sustainable source of oil-rich biomass. Brazilian J Chem Eng. https://doi.org/10.1007/s43153-020-00011-3
42. Pérez-Mora LS, Matsudo MC, Cezare-Gomes EA, Carvalho JCM (2016) An investigation into producing Botryococcus braunii in a tubular photobioreactor. J Chem Technol Biotechnol 91:3053–3060. https://doi.org/10.1002/jctb.4934

43. Matsudo MC, Fukui de Sousa T, Carvalho JCM (2020) Bicarbonate supplementation improves biomass productivity in Scenedesmus obliquus cultivation. Res J Biotechnol 15:27–42
44. Maheshwari P, Haider MB, Yusuf M et al (2022) A review on latest trends in cleaner biodiesel production: role of feedstock, production methods, and catalysts. J Clean Prod 355:131588. https://doi.org/10.1016/j.jclepro.2022.131588
45. Bento HBS, Reis CER, Pinto PA et al (2022) Continuous synthesis of biodiesel from outstanding kernel oil in a packed bed reactor using burkholderia cepacia lipase immobilized on magnetic nanosupport. Catal Letters 152:2434–2444. https://doi.org/10.1007/s10562-021-03826-y
46. Costa-Silva TA, Carvalho AKF, Souza CRF et al (2017) Enzymatic transesterification of coconut oil using chitosan-immobilized lipase produced by fluidized-bed system. Energy Fuels 31:12209–12216. https://doi.org/10.1021/acs.energyfuels.7b02033
47. Saravanan A, Kumar PS, Jeevanantham S et al (2022) Recent advances and sustainable development of biofuels production from lignocellulosic biomass. Bioresour Technol 344:126203. https://doi.org/10.1016/j.biortech.2021.126203
48. Jatoi AS, Abbasi SA, Hashmi Z et al (2023) Recent trends and future perspectives of lignocellulose biomass for biofuel production: a comprehensive review. Biomass Convers Biorefinery 13:6457–6469. https://doi.org/10.1007/s13399-021-01853-8
49. Barakat A, Mayer-Laigle C, Solhy A et al (2014) Mechanical pretreatments of lignocellulosic biomass: towards facile and environmentally sound technologies for biofuels production. RSC Adv 4:48109–48127. https://doi.org/10.1039/C4RA07568D
50. Prasad S, Ingle AP (2019) Impacts of sustainable biofuels production from biomass. In: Rai M, Ingle AP (eds) Sustainable Bioenergy. Elsevier, pp 327–346
51. Adewuyi A (2022) Underutilized lignocellulosic waste as sources of feedstock for biofuel production in developing countries. Front Energy Res 10:1–21. https://doi.org/10.3389/fenrg.2022.741570
52. Roy R, Rahman S, Raynie DE (2020) Recent advances of greener pretreatment technologies of lignocellulose. Curr Res Green Sustain Chem 3:100035. https://doi.org/10.1016/j.crgsc.2020.100035
53. Sánchez C (2009) Lignocellulosic residues: biodegradation and bioconversion by fungi. Biotechnol Adv 27:185–194. https://doi.org/10.1016/j.biotechadv.2008.11.001
54. Rodionova MV, Bozieva AM, Zharmukhamedov SK et al (2021) A comprehensive review on lignocellulosic biomass biorefinery for sustainable biofuel production. Int J Hydrogen Energy 47:1481–1498. https://doi.org/10.1016/j.ijhydene.2021.10.122
55. Raven JA, Giordano M (2014) Algae. Curr Biol 24:590–595. https://doi.org/10.1016/j.cub.2014.05.039
56. Greenwell HC, Laurens LML, Shields RJ et al (2010) Placing microalgae on the biofuels priority list: a review of the technological challenges. J R Soc Interface 7:703–726. https://doi.org/10.1098/rsif.2009.0322
57. Rodrigues-Sousa AE, Nunes IVO, Muniz-Junior AB et al (2021) Nitrogen supplementation for the production of Chlorella vulgaris biomass in secondary effluent from dairy industry. Biochem Eng J 165. https://doi.org/10.1016/j.bej.2020.107818
58. Stansell GR, Gray VM, Sym SD (2012) Microalgal fatty acid composition: implications for biodiesel quality. J Appl Phycol 24:791–801. https://doi.org/10.1007/s10811-011-9696-x
59. Knothe G (2005) Dependence of biodiesel fuel properties on the structure of fatty acid alkyl esters. Fuel Process Technol 86:1059–1070. https://doi.org/10.1016/j.fuproc.2004.11.002
60. Knothe G (2008) Designer biodiesel: optimizing fatty ester composition to improve fuel properties, pp 1358–1364
61. Sharma KD, Jain S (2020) Municipal solid waste generation, composition, and management: the global scenario. Soc Responsib J 16:917–948. https://doi.org/10.1108/SRJ-06-2019-0210
62. Pour FH, Makkawi YT (2021) A review of post-consumption food waste management and its potentials for biofuel production. Energy Rep 7:7759–7784. https://doi.org/10.1016/j.egyr.2021.10.119

63. Carmona-cabello M, Leiva-candia D, Castro-cantarero JL et al (2018) Valorization of food waste from restaurants by transesteri fi cation of the lipid fraction. Fuel 215:492–498. https://doi.org/10.1016/j.fuel.2017.11.096
64. Paritosh K, Kushwaha SK, Yadav M et al (2017) Food waste to energy: an overview of sustainable approaches for food waste management and nutrient recycling. https://doi.org/10.1155/2017/2370927
65. Esposito G, Frunzo L, Giordano A et al (2012) Anaerobic co-digestion of organic wastes. Rev Environ Sci Bio/Techn 11:325–341. https://doi.org/10.1007/s11157-012-9277-8
66. Kosheleva A, Gadaleta G, De GS et al (2023) Co-digestion of food waste and cellulose-based bioplastic: from batch to semi-continuous scale investigation. Waste Manag 156:272–281. https://doi.org/10.1016/j.wasman.2022.11.031
67. Li L, Bu Y, Feng W et al (2024) Biomethane recovery and prokaryotic shifts in anaerobic co-digestion of food waste and paper waste in organic fraction of municipal solid waste: effect of paper content. Bioresour Technol 406. https://doi.org/10.1016/j.biortech.2024.130964
68. Johnston M, Foley JA, Holloway T et al (2009) Resetting global expectations from agricultural biofuels. Environ Res Lett 4:014004. https://doi.org/10.1088/1748-9326/4/1/014004
69. IEA (2022) Total biofuel production by feedstock, main case, 2021–2027. IEA, Paris
70. Kline KL, Oladosu GA, Wolfe AK et al (2008) Biofuel feedstock assessment for selected countries. Oak Ridge, TN (United States)
71. Ebadian M, van Dyk S, McMillan JD, Saddler J (2020) Biofuels policies that have encouraged their production and use: an international perspective. Energy Policy 147:111906. https://doi.org/10.1016/j.enpol.2020.111906
72. Souza G mendes, Maciel Filho R, Nogueira LAHo et al (2023) Biofuels in emerging markets: potential for sustainable production and consumption
73. IEA (2024) World energy outlook 2024. IEA, Paris
74. IEA (2023) Latin America energy outlook
75. Nass LL, Pereira PAA, Ellis D (2007) Biofuels in Brazil: an overview. Crop Sci 47:2228–2237. https://doi.org/10.2135/cropsci2007.03.0166
76. Woiciechowski AL, Medeiros ABP, Rodrigues C et al (2016) Feedstocks for Biofuels. In: Socool C, Brar S, Faulds C, L. R (eds.) Green fuels technology. Springer, Cham, pp 15–39
77. Poppe MK, Nogueira LAHo, Macedo I de C (2017) Second-generation sugarcane bioenergy and biochemicals: advance low-carbon fuels for transport and industry. Center for Strategic Studies and Management, Brasilia
78. Vandenberghe LPS, Valladares-Diestra KK, Bittencourt GA et al (2022) Beyond sugar and ethanol: the future of sugarcane biorefineries in Brazil. Renew Sustain Energy Rev 167:112721. https://doi.org/10.1016/j.rser.2022.112721
79. Ramos NP, Mendes NC, Folegatti MIS et al (2024) Regionalização dos perfis "típicos" da produção de cana-de-açúcar para uso no RenovaBio. Jaguariúna
80. Vega LP, Bautista KT, Campos H et al (2024) Biofuel production in Latin America: a review for Argentina, Brazil, Mexico, Chile, Costa Rica and Colombia. Energy Rep 11:28–38. https://doi.org/10.1016/j.egyr.2023.10.060
81. Canabarro NI, Silva-Ortiz P, Nogueira LAH et al (2023) Sustainability assessment of ethanol and biodiesel production in Argentina, Brazil, Colombia, and Guatemala. Renew Sustain Energy Rev 171:113019. https://doi.org/10.1016/j.rser.2022.113019
82. CEPEA PIB da cadeia de soja e biodiesel
83. Rose N (2015) From feedstock to fuel–the Latin American experience
84. IEA energy systems of North America
85. Martin M, Lazarevic D (2015) A review of North American biofuel production, policies and research
86. Ramsey S (2025) U.S. bioenergy statistics. In: USDA Econ Res Serv
87. U.S. department of energy biofuel basics
88. Mitchell RB, Schmer MR, Anderson WF et al (2016) Dedicated energy crops and crop residues for bioenergy feedstocks in the central and Eastern USA. BioEnergy Res 9:384–398. https://doi.org/10.1007/s12155-016-9734-2

89. Connelly R (2014) Second-generation biofuel from high-efficiency algal-derived biocrude. In: Gupta VK, Tuohy MG, Kubicek CP et al (eds.) Bioenergy Research: advances and applications. Elsevier, pp 153–170
90. Littlejohns J, Rehmann L, Murdy R et al (2018) Current state and future prospects for liquid biofuels in Canada. Biofuel Res J 5:759–779. https://doi.org/10.18331/BRJ2018.5.1.4
91. Saini R, Osorio-Gonzalez CS, Brar SK, Kwong R (2021) A critical insight into the development, regulation and future prospects of biofuels in Canada. Bioengineered 12:9847–9859. https://doi.org/10.1080/21655979.2021.1996017
92. Kouhgardi E, Zendehboudi S, Mohammadzadeh O et al (2023) Current status and future prospects of biofuel production from brown algae in North America: progress and challenges. Renew Sustain Energy Rev 172:113012. https://doi.org/10.1016/j.rser.2022.113012
93. Gollany HT, Titus BD, Scott DA et al (2015) Biogeochemical research priorities for sustainable biofuel and bioenergy feedstock production in the Americas. Environ Manage 56:1330–1355. https://doi.org/10.1007/s00267-015-0536-7
94. European comission–directorate-General for energy renewable energy directive
95. European comission–joint research centre renewable energy–Recast to 2030 (RED II)
96. Stattman SL, Gupta A, Partzsch L, Oosterveer P (2018) Toward sustainable biofuels in the European Union? Lessons from a decade of hybrid biofuel governance. Sustainability 10:4111. https://doi.org/10.3390/su10114111
97. IEA bioenergy TCP task 39 (2023) IEA bioenergy T39–Biofuel news
98. OECD/FAO (2023) OECD-FAO agricultural outlook 2023–2032. OECD
99. Chrysikou LP, Dagonikou V, Dimitriadis A, Bezergianni S (2019) Waste cooking oils exploitation targeting EU 2020 diesel fuel production: environmental and economic benefits. J Clean Prod 219:566–575. https://doi.org/10.1016/j.jclepro.2019.01.211
100. Puricelli S, Cardellini G, Casadei S et al (2021) A review on biofuels for light-duty vehicles in Europe. Renew Sustain Energy Rev 137:110398. https://doi.org/10.1016/j.rser.2020.110398
101. Prussi M, Panoutsou C, Chiaramonti D (2022) Assessment of the feedstock availability for covering EU alternative fuels demand. Appl Sci 12:740. https://doi.org/10.3390/app12020740
102. IEA (2022) Africa energy outlook 2022
103. Muhammad KM, Adeyemi MM, Jacob J et al (2025) Biodiesel production in Africa from non-edible sources: sources, production, properties and policies. Sustain Chem Environ 9:100201. https://doi.org/10.1016/j.scenv.2024.100201
104. Elgharbawy AS, Sadik WA, Sadek OM, Kasaby MA (2021) A review on biodiesel feedstocks and production technologies. J Chil Chem Soc 66:5098–5109. https://doi.org/10.4067/S0717-97072021000105098
105. Donald M (2011) Biofuels in Africa: opportunities, prospects, and challenges
106. Takeuchi K, Shiroyama H, Saito O, Matsuura M (2018) Biofuels and sustainability. Springer Japan, Tokyo
107. IEA (2024) Biofuels in emerging markets of Africa and Asia—An overview of costs and greenhouse gas savings
108. Yan J, Lin T (2009) Biofuels in Asia. Appl Energy 86:S1–S10. https://doi.org/10.1016/j.apenergy.2009.07.004
109. Chanthawong A, Dhakal S (2016) Liquid biofuels development in Southeast Asian countries: an analysis of market, policies and challenges. Waste Biomass Valorizat 7:157–173. https://doi.org/10.1007/s12649-015-9433-9
110. Zhou A, Thomson E (2009) The development of biofuels in Asia. Appl Energy 86:S11–S20. https://doi.org/10.1016/j.apenergy.2009.04.028
111. Beckman J, Gooch E, Gopinath M, Landes M (2018) Market impacts of China and India meeting biofuel targets using traditional feedstocks. Biomass Bioenerg 108:258–264. https://doi.org/10.1016/j.biombioe.2017.11.018
112. IEA (2023) Biofuels production and development in China
113. Vinod AD, Prasad R, Singh P et al (2025) Biofuels for decarbonizing land transport sector: a case study on people's perspective and policy in Fiji. Renew Energy 240:122156. https://doi.org/10.1016/j.renene.2024.122156

114. IEA (2024) Implementation of bioenergy in Australia–2024 update
115. Glover J, Johnson H, Lizzio J, et al (2008) Australia's crops and pastures in a changing climate–can biotechnology help?
116. Biki Z (2022) Australia: biofuels annual
117. Puri M, Abraham RE, Barrow CJ (2012) Biofuel production: prospects, challenges and feedstock in Australia. Renew Sustain Energy Rev 16:6022–6031. https://doi.org/10.1016/j.rser.2012.06.025
118. Hazrat MA, Rasul MG, Khan MMK (2015) Biofuel: an Australian perspective in abating the fossil fuel vulnerability. Procedia Eng 105:628–637. https://doi.org/10.1016/j.proeng.2015.05.041
119. IEA (2021) Implementation of bioenergy in New Zealand–2021 update
120. IEA (2024) Biofuels production and development in New Zealand
121. Acharya V, Young BR (2008) A review of the potential of bio-ethanol in New Zealand. Bull Sci Technol Soc 28:143–148. https://doi.org/10.1177/0270467607313952
122. Padder SA, Khan R, Rather RA (2024) Biofuel generations: New insights into challenges and opportunities in their microbe-derived industrial production. Biomass Bioenerg 185:107220. https://doi.org/10.1016/j.biombioe.2024.107220
123. Mohr A, Raman S (2013) Lessons from first generation biofuels and implications for the sustainability appraisal of second generation biofuels. Energy Policy 63:114–122. https://doi.org/10.1016/j.enpol.2013.08.033
124. Rodionova MV, Poudyal RS, Tiwari I et al (2017) Biofuel production: challenges and opportunities. Int J Hydrogen Energy 42:8450–8461. https://doi.org/10.1016/j.ijhydene.2016.11.125
125. Moodley P (2021) Sustainable biofuels: opportunities and challenges. In: Sustainable Biofuels. Elsevier, pp 1–20
126. Malik MAI, Zeeshan S, Khubaib M et al (2024) A review of major trends, opportunities, and technical challenges in biodiesel production from waste sources. Energy Convers Manag X 23:100675. https://doi.org/10.1016/j.ecmx.2024.100675
127. Pulyaeva VN, Kharitonova NA, Kharitonova EN (2020) Advantages and disadvantages of the production and using of liquid biofuels. IOP Conf Ser Mater Sci Eng 976:012031. https://doi.org/10.1088/1757-899X/976/1/012031
128. Russo C, Cirillo V, Pollaro N et al (2025) The global energy challenge: second-generation feedstocks on marginal lands for a sustainable biofuel production. Chem Biol Technol Agric 12:10. https://doi.org/10.1186/s40538-025-00729-7

Fruit and Vegetable Waste, Chitin, DDGs, Bread Waste, Broken Rice, Beetroot: Unexplored Feedstock for Biorefineries

Amit Singh, Pradeep Khiriya, Gagan Kant Tripathi, Priyavand Bundela, and Purnima Swarup Khare

Abstract The increased demand for circular economy and bio-based solutions to address waste issues has triggered increasing interest in alternative feedstocks for biorefineries. Among the alternative feedstocks, underutilized biomass sources, including fruit and vegetable waste, chitin in crustacean shells, dried distillers' grains (DDGs), bread wastes, broken rice, and beetroot, are promising opportunities for the development of bio-based products. They possess significant amounts of carbohydrates, protein, and bioactives, which are amenable to the prospect of utilizing them to develop high-potential feedstocks convertible to biofuels, bioplastics, enzymes, organic acids, and high-end chemicals. Valorization of the same feedstocks not only reduces the environmental footprint of food and agro-food waste but also enhances resource utilization efficiency. Technical challenges such as composition heterogeneity, efficient pretreatment, and economically viable conversion procedures, however, need to be addressed. New generation biorefinery configurations, novel bioconversion technologies, and regulatory facilitation need to be developed to promote these alternative feedstocks' sustainable use in the future.

Keywords Bioconversion · Biorefinery · Fruit · Feedstocks · Beetroot

1 Introduction

Along with the increasing population of the world, their needs are also increasing, due to which the limited amount of minerals present in nature, such as water, coal, metal, etc., will decrease over time, which will make it difficult for humans to live. In view of all these circumstances, new sources are being discovered and are also being used. With the increasing population, the things needed by them are not being

A. Singh · P. Khiriya · G. K. Tripathi (✉) · P. Bundela · P. S. Khare
School of Nanotechnology, Rajiv Gandhi Proudyogiki Vishwavidyalaya (RGPV), Bhopal, Madhya Pradesh, India
e-mail: gagankanttripathi@gmail.com

© The Author(s), under exclusive license to Springer Nature Singapore Pte Ltd. 2025
A. K. Chandel and S. Sundaramurthy (eds.), *Biorefineries in Circular Bioeconomy*, Green Energy and Technology, https://doi.org/10.1007/978-981-95-3913-0_2

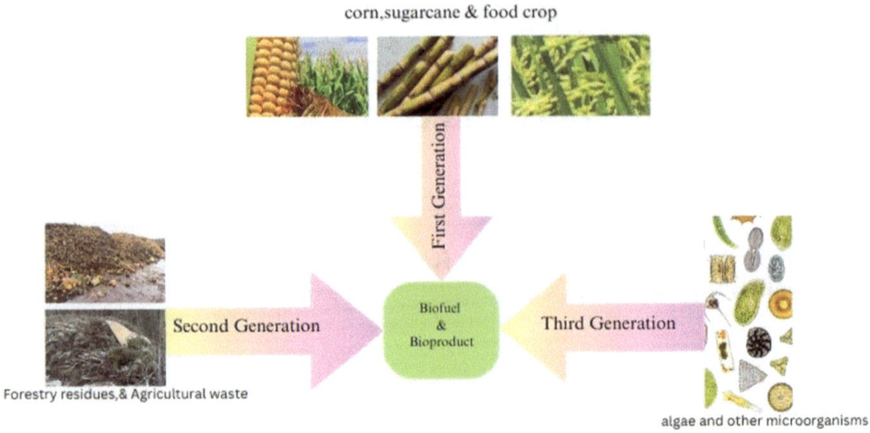

Fig. 1 Schematic diagram shows the different generations of feedstock

consumed as much as they are being misused. In this chapter, we are going to study those things in detail.

Our atmosphere is teeming with waste. Have you ever thought that we can use that waste and convert it into some new energy? Yes, in today's era, everything is possible. Today, can we convert food waste, vegetable waste, and waste generated during farming, like straw, rice husks, wheat crop waste, etc., into energy? The answer is yes; we can do it. For this, various types of methods are available in the market, and they are being used rapidly in today's time, through which efforts are being made to almost meet our energy shortage. Figure 1 shows a schematic diagram of the different generations of feedstock.

1.1 Importance

There are mainly 3 types of importance:

- Renewable energy source: Through biorefineries, we generate various renewable energy sources that reduce human dependence on nature and thus reduce nature's consumption of fossil fuels.
- Sustainable resource utilization: Biorefineries utilize biomass, a renewable resource, to produce various products, reducing waste and promoting sustainable resource utilization.
- Economic benefits: If seen from the point of view of economic benefits, biorefinery provides employment opportunities and makes the economy very strong. It also provides business opportunities.

1.2 Role in Sustainable Resource Utilization

- Biomass conversion: Biorefineries convert different types of biomass into a variety of bioproducts such as biofuels, biochemicals, and bioproducts.
- Value chain creation: Biorefineries create a value chain by producing multiple products from biomass, increasing the overall efficiency and profitability.
- Waste reduction: Biorefineries utilize waste biomass, reducing waste disposal costs and environmental impacts.

1.3 Types of Biorefineries

- First-generation biorefineries: There are some examples of first-generation biorefineries that utilize food crops, such as corn and sugarcane, to produce biofuels.
- Second-generation biorefineries: There are some examples of second-generation biorefineries that utilize nonfood biomass, such as agricultural waste and forestry residues, to produce biofuels and biochemicals.
- Third-generation biorefineries: There are some examples of third-generation biorefineries that utilize algae and other microorganisms to produce biofuels and biochemicals.

1.4 Challenges with Conventional Feedstocks

The use of conventional feedstocks, such as corn, sugarcane, and vegetable oils, in biorefineries presents several economic, environmental, and social challenges. One of the primary concerns is the competition between biofuel production and food security. Large-scale utilization of food crops for biofuel generation can lead to an increase in food prices and exacerbate food shortages, particularly in developing countries [1]. Additionally, the cultivation of these crops requires substantial amounts of land, water, and fertilizers, which contribute to deforestation, biodiversity loss, and soil degradation [2]. The excessive use of fertilizers and pesticides in conventional feedstock farming also leads to water pollution and eutrophication, affecting aquatic ecosystems and water quality [3].

Another significant challenge is the economic feasibility of using traditional feedstocks in biorefineries. The fluctuating prices of agricultural commodities and the dependency on seasonal crop yields make biofuel production from conventional sources economically unstable [4]. The high production costs associated with refining food-based feedstocks into biofuels further limit their commercial viability when compared to fossil fuels. Moreover, the energy-intensive processes required to convert these feedstocks into biofuels often result in a high carbon footprint, diminishing the environmental benefits they are intended to provide [5].

Policymakers and researchers have increasingly emphasized the need for alternative, nonfood-based feedstocks to address these challenges. Waste-derived materials such as fruit and vegetable waste, chitin, DDGs, bread waste, broken rice, and beetroot offer a more sustainable and economically viable option for biorefineries. By repurposing agricultural and food waste, it is possible to reduce environmental impacts, lower production costs, and minimize the conflict between biofuel production and food security. Continued research and advancements in biorefinery technology are essential to overcoming the limitations of conventional feedstocks and promoting a circular bioeconomy [6].

1.5 Need for Alternative Feedstocks: Justification for Exploring Unconventional Waste Materials as Feedstock

The increasing global demand for sustainable energy and bioproducts has necessitated the exploration of alternative feedstocks for biorefineries. Conventional feedstocks, such as sugarcane, corn, and vegetable oils, have raised concerns related to food security, land-use change, and environmental degradation. Additionally, these crops require significant water and fertilizer inputs, contributing to soil depletion, greenhouse gas emissions, and biodiversity loss. In contrast, unconventional feedstocks such as fruit and vegetable waste, chitin, dried distillers' grains with soluble (DDGs), bread waste, broken rice, and beetroot offer a sustainable and cost-effective alternative. These waste materials are abundant, often underutilized, and rich in valuable biochemical compounds that can be efficiently converted into biofuels, bioplastics, and high-value biochemicals [6].

Using waste-derived feedstocks in biorefineries aligns with the principles of the circular economy, where waste is repurposed to create value-added products, thereby reducing environmental pollution and resource wastage. For instance, fruit and vegetable waste, which accounts for nearly 50% of global food waste, is an excellent source of fermentable sugars, organic acids, and bioactive compounds that can be transformed into bioethanol, biogas, and biodegradable polymers [7]. Similarly, chitin from seafood processing waste can be converted into chitosan, a biopolymer with applications in medicine, wastewater treatment, and bioplastic production [8]. These alternative feedstocks not only reduce reliance on food-based resources but also provide economic benefits by turning waste management challenges into revenue-generating opportunities.

1.6 Utilization of Bread Waste for Bioethanol Production

A research study by Lin et al. [9] revealed the potential of utilizing bread waste as a source for bioethanol production. Bread waste is rich in fermentable sugars and was hydrolyzed with the enzymes amylase and glucoamylase to convert the starch into glucose. The resulting hydrolysate underwent fermentation with Saccharomyces cerevisiae, yielding ethanol levels that were comparable to those derived from conventional sugar-based feedstocks. The research indicated that fermenting bread waste not only offered a cost-effective alternative to traditional raw materials but also addressed challenges related to food waste disposal. Furthermore, the study emphasized that incorporating bread waste bioconversion into current bioethanol production facilities could improve both sustainability and economic viability.

2 Different Feedstocks for Biorefineries

Feedstock

The biomass utilized in a biorefinery is referred to as the feedstock. Historically, biorefineries have relied on dedicated feedstocks like corn and sugarcane. Nevertheless, there is an increasing focus on using waste and residual biomass as feedstock. This shift is favored as it presents a more sustainable alternative that does not interfere with food production.

2.1 Fruit and Vegetable Waste: A Promising Feedstock for Biorefineries

Waste from fruits and vegetables (FVW) constitutes one of the largest segments of worldwide food waste, significantly contributing to environmental pollution and the loss of resources. [10] The Food and Agriculture Organization reports that approximately 1.3 billion tons of food are wasted each year, with losses of fruits and vegetables representing about 40–50% of this amount. These types of waste contain a high level of biodegradable organic matter, which includes carbohydrates, fibers, organic acids, proteins, and vital micronutrients, positioning them as a promising raw material for biorefineries. In contrast to traditional feedstocks like corn or sugarcane, fruit and vegetable waste does not compete with food supply, is low-cost, and is readily available from food processing sectors, markets, and households [11]. Figure 2 shows a conceptualized food waste biorefinery to produce value-added products, animal feed, and energy.

The significant carbohydrate levels in fruit and vegetable waste (FVW) enable its transformation into bioethanol through enzymatic hydrolysis and fermentation, serving as a substitute for fossil fuels [12]. Moreover, the anaerobic digestion of

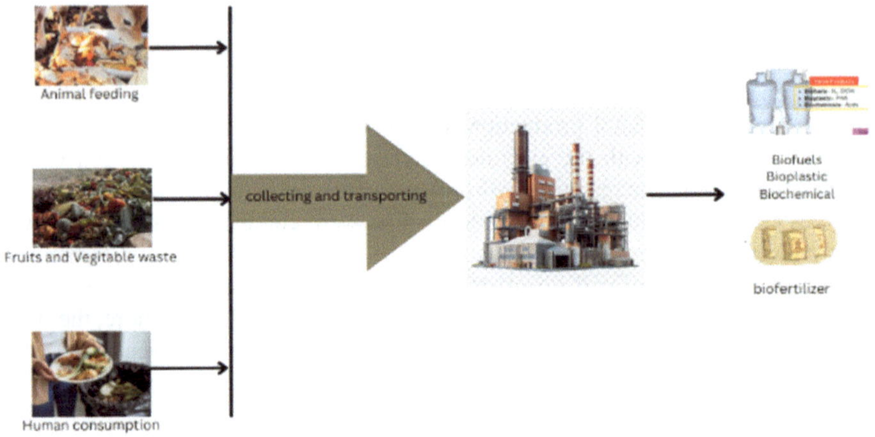

Fig. 2 Conceptualized food waste biorefinery to produce value-added products, animal feed, and energy

FVW generates biogas, which can act as a renewable energy source. Additionally, other valuable bioproducts like organic acids, enzymes, and bioactive compounds can be extracted from these residues for applications in the food, pharmaceutical, and cosmetic sectors [13]. Furthermore, the peels and seeds of certain fruits, including citrus and pomegranate, are rich in antioxidants, polyphenols, and dietary fibers, making them suitable for incorporation into functional food products [14]. Although it holds promise, effectively using fruit and vegetable waste encounters obstacles like seasonal differences in composition, logistical issues regarding collection and storage, and the necessity for economic conversion technologies. Nonetheless, progress in biotechnology and waste management approaches is transforming FVW into a practical and sustainable resource for the bioeconomy.

2.2 *Chitin: A Sustainable Feedstock for Biorefineries*

Chitin is the second most abundant biopolymer on Earth, primarily found in the exoskeletons of crustaceans (shrimp, crab, and lobster), insect cuticles, and fungal cell walls. It is a nitrogen-rich polysaccharide composed of N-acetylglucosamine (GlcNAc) units, making it a valuable resource for biotechnological applications. Every year, seafood industries generate millions of tons of chitin-rich waste, which is largely underutilized or disposed of in landfills, leading to environmental pollution [15]. Instead of treating chitin-containing waste as a disposal problem, its valorization into high-value bioproducts presents a sustainable solution aligned with circular economy principles. Figure 3 shows a schematic diagram of chitin-based materials.

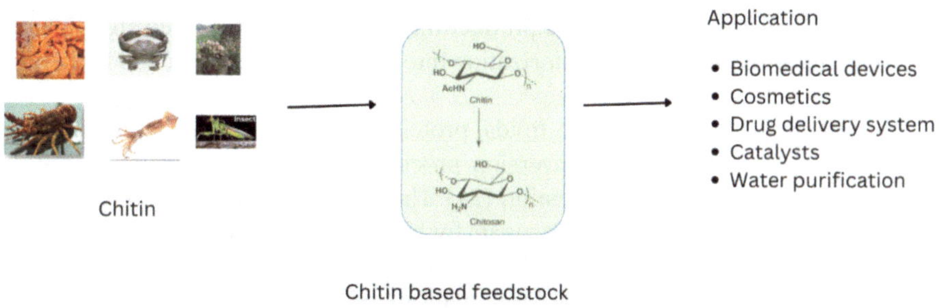

Fig. 3 Schematic diagram showing chitin-based materials

Chitin can be converted into chitosan, a deacetylated derivative widely used in medicine, agriculture, food packaging, and wastewater treatment due to its biocompatibility, antimicrobial properties, and biodegradability [16]. Additionally, enzymatic or microbial hydrolysis of chitin produces N-acetylglucosamine (GlcNAc), which serves as a precursor for pharmaceuticals, nutraceuticals, and cosmetics. Chitin-based nanomaterials have also shown great potential in developing biodegradable plastics, wound dressings, and drug delivery systems (G) [17]. Moreover, its applicaltion in bioethanol and biogas production is gaining attention, as chitin residues can be used as a carbon and nitrogen source for microbial fermentation [18]. However, challenges such as the complexity of chitin extraction, high processing costs, and the need for environmentally friendly deacetylation methods must be addressed to optimize its industrial applications.

2.2.1 Chitin Valorization for Bioplastics Production

A study conducted by Ifuku et al. [19] explored the use of chitin nanofibers derived from crab shells to develop biodegradable polymer composites. The researchers extracted chitin nanofibers using a mechanical disintegration method and incorporated them into polylactic acid (PLA)-based bioplastics. The resulting biocomposite materials exhibited improved mechanical strength, thermal stability, and biodegradability compared to conventional petroleum-based plastics. This study highlighted the potential of chitin as a renewable and eco-friendly alternative to synthetic polymer reinforcements in the packaging and biomedical industries.

2.3 Distillers Dried Grains with Soluble (DDGs)

Distillers dried grains with soluble (DDGs) are the co-products of ethanol production from cereal grains such as corn, wheat, and sorghum. Traditionally used as animal feed due to their high protein and fiber content, DDGs have gained attention as an underutilized feedstock for biorefineries [20]. With the global bioethanol industry expanding, the production of DDGs has increased significantly, creating opportunities for their valorization into biofuels, bioplastics, and high-value chemicals.

DDGs contain residual starch, lipids, proteins, and lignocellulosic components, making them suitable for bioconversion processes such as enzymatic hydrolysis, fermentation, and anaerobic digestion [21] The residual starch and fiber in DDGs can be hydrolyzed into fermentable sugars for bioethanol production, while lipids can be extracted for biodiesel synthesis [22] Furthermore, the proteins in DDGs can serve as a nitrogen source for microbial fermentation, aiding the production of bio-based chemicals such as lactic acid, butanol, and succinic acid. Advanced technologies have also explored the use of DDGs in biodegradable plastics and composite materials.

Despite these opportunities, challenges remain in utilizing DDGs efficiently. Variability in composition, seasonal fluctuations, and the presence of anti-nutritional factors can affect their suitability for certain applications [23]. However, ongoing research in pretreatment methods and bioprocess optimization continues to improve the feasibility of DDGs as a sustainable feedstock for biorefineries.

2.4 *Conversion of DDGs into Bioplastics*

A study by Liu et al. (2019) investigated the potential of DDGs as a raw material for producing biodegradable plastics. The researchers extracted protein and fiber from DDGs and blended them with polyhydroxyalkanoates (PHA) to create bio-based plastic films. The resulting bioplastic exhibited good mechanical strength and biodegradability, demonstrating its potential as an eco-friendly alternative to petroleum-based plastics. This study highlighted the value of DDGs in reducing plastic pollution and promoting a circular bioeconomy.

2.5 *Bread Waste: A Valuable Feedstock for Biorefineries*

Bread waste is a significant component of global food waste, primarily generated by bakeries, supermarkets, and households. It is estimated that nearly *900,000 tons of bread are wasted annually in the European Union alone* (Mirabella et al. 2014) [24]. Given its high starch content and relatively low lignin and cellulose levels, bread waste serves as an excellent feedstock for bioconversion into biofuels, organic acids, and other value-added products. Instead of being discarded or used as low-value animal feed, bread waste can be upcycled into sustainable bioproducts, reducing food waste and promoting a circular bioeconomy.

The high carbohydrate content in bread waste makes it particularly suitable for *bioethanol production* through enzymatic hydrolysis and fermentation. Studies have demonstrated that amylase and glucoamylase enzymes effectively break down bread starch into fermentable sugars, which are then converted into ethanol by *Saccharomyces cerevisiae* (Jagannath et al. 2013) [25]. Additionally, bread waste can be used to produce *biogas* through anaerobic digestion, offering a renewable energy source for industries and households (Kiran et al. 2014) [26]. Beyond energy applications, bread waste is also a potential raw material for the production of lactic acid, a precursor for biodegradable plastics such as polylactic acid (PLA), which is widely used in packaging and medical applications.

Despite these promising applications, certain challenges need to be addressed. The presence of preservatives and additives in commercial bread can affect microbial fermentation efficiency. Furthermore, the logistics of collecting and storing bread waste require efficient waste management systems to ensure a consistent and high-quality feedstock supply. Nonetheless, advancements in pretreatment and fermentation technologies are making bread waste a viable and sustainable resource for biorefineries.

2.5.1 Bioethanol Production from Bread Waste

A study by Kodogiannis et al. (2017) [27] explored the conversion of surplus bread into bioethanol using enzymatic hydrolysis and yeast fermentation. The researchers found that pretreated bread waste yielded up to 0.48 g of ethanol per gram of glucose, demonstrating its high fermentation efficiency. The study concluded that bakeries and food industries could integrate bread waste valorization into their supply chains to create sustainable biofuels, reducing both food waste and fossil fuel dependency.

2.6 Broken Rice: An Underutilized Feedstock for Biorefineries

Broken rice is a by-product generated during rice milling, where grains break due to mechanical stress or improper handling. While it is often used in animal feed, brewing, and as an ingredient in lower-grade food products, its potential as a sustainable feedstock for biorefineries remains largely unexplored. Given its high starch content (70–80%), broken rice can be efficiently converted into bioethanol, bioplastics, organic acids, and value-added chemicals (Lu et al. 2016) [28]. Instead of being considered a low-value residue, broken rice can be valorized into bio-based products, contributing to waste minimization and circular bioeconomy principles. Figure 4 shows a diagram of broken rice for biorefineries.

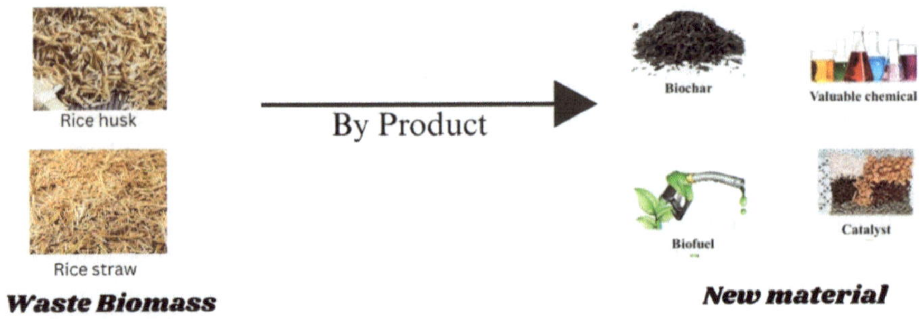

Fig. 4 Diagram of broken rice for biorefineries

2.6.1 Potential Applications of Broken Rice in Biorefineries

(a) **Bioethanol production**: Broken rice is an excellent raw material for bioethanol production due to its high carbohydrate content. The starch in broken rice can be enzymatically hydrolyzed into fermentable sugars using α-amylase and glucoamylase, followed by fermentation with Saccharomyces cerevisiae to produce ethanol. Research has shown that bioethanol yields from broken rice are comparable to those obtained from sugarcane and corn, making it a competitive alternative for renewable energy (Pham et al. 2018) [29].

(b) **Bioplastics and Biodegradable Films**: Broken rice starch can be modified to create biodegradable plastics and edible films. Studies have explored the use of rice starch in producing polylactic acid (PLA) composites, which have applications in food packaging and medical materials (Tang et al. 2019) [30]. Additionally, rice starch-based films exhibit excellent biodegradability, transparency, and flexibility, making them a sustainable alternative to petroleum-based plastics.

(c) **Lactic acid and other biochemicals**: Lactic acid, a key precursor for biodegradable plastics, can be produced from broken rice through microbial fermentation. Bacteria such as *Lactobacillus casei* efficiently convert rice hydrolysates into lactic acid, which is then used in bioplastics, pharmaceuticals, and food industries (Orozco et al. 2020) [31]. Furthermore, broken rice can be utilized to produce succinic acid, citric acid, and enzymes through microbial bioconversion.

(d) **Biohydrogen and biogas production**: Broken rice can also serve as a feedstock for anaerobic digestion, producing biohydrogen and biogas. This process converts starch-rich waste into clean energy sources, reducing reliance on fossil fuels and contributing to sustainable energy development (Li et al. 2017) [32].

(e) **Bioethanol production from broken rice**: A study by Pham et al. (2018) [29] investigated the feasibility of producing bioethanol from broken rice using enzymatic hydrolysis and yeast fermentation. The researchers found that broken rice yielded 0.45 g of ethanol per gram of fermentable sugar, which is comparable to ethanol production from sugarcane molasses. The study concluded that broken rice could serve as an economically viable and sustainable feedstock for bioethanol production, particularly in rice-producing countries where surplus broken rice is available.

2.7 *Beet Root: A Potential Feedstock for Biorefineries*

Beetroot, also known as *Beta vulgaris*, is a highly cultivated crop with significant health and industrial benefits, particularly for its extensive use in the production of sugar. All parts of the beet, including the leaves, roots, and peels, are used. Rich in sugars, fiber, bioactive compounds, and pigments, these residues can serve as valuable feedstocks for biofuels, bioplastics, organic acids, and high-value biochemicals (Slukova et al. 2020) [33]. Instead of discarding beet waste or using it as low-value animal feed, it can be converted into sustainable bioproducts that greatly contribute to the economy. This awareness can help people use these wastes instead of throwing them away, without harming the environment, and convert them into fuel.

2.7.1 Potential Applications of Beetroot in Biorefineries

(a) **Bioethanol production**: Beetroot contains high levels of fermentable sugars (glucose, fructose, and sucrose), making it a promising feedstock for bioethanol production. Fermentation of beetroot juice with Saccharomyces cerevisiae yields ethanol efficiently, with conversion rates comparable to sugarcane and molasses (Wojtusik et al. 2019) [34]. Additionally, beet pulp, a byproduct of sugar extraction, can be hydrolyzed and fermented to enhance bioethanol yields.

(b) **Biogas and biohydrogen production**: Beetroot residues are rich in organic matter, making them ideal for anaerobic digestion to produce biogas and biohydrogen. Methane-rich biogas generated from beet waste can be used as a renewable energy source, reducing reliance on fossil fuels (Mizgier et al. 2018) [35].

(c) **Bioplastics and biodegradable films**: The high pectin and fiber content in beetroot makes it an excellent raw material for biodegradable plastics and edible films. Studies have explored the use of beet pulp-derived cellulose and hemicellulose in biopolymer synthesis, which can replace petroleum-based plastics in packaging applications (Gómez et al. 2021) [36].

(d) **Lactic acid and organic Acid production**: Fermentation of beetroot hydrolysates can yield lactic acid, succinic acid, and citric acid, which are key precursors for biodegradable plastics, pharmaceuticals, and food additives (Jörgensen et al. 2022) [37]. The high sugar content in beetroot makes it an excellent substrate for microbial fermentation.

(e) **Natural pigments and antioxidants**: Beetroot is rich in betalains, natural pigments used in food coloring, cosmetics, and pharmaceuticals. Extracting these compounds from beet waste can provide an eco-friendly alternative

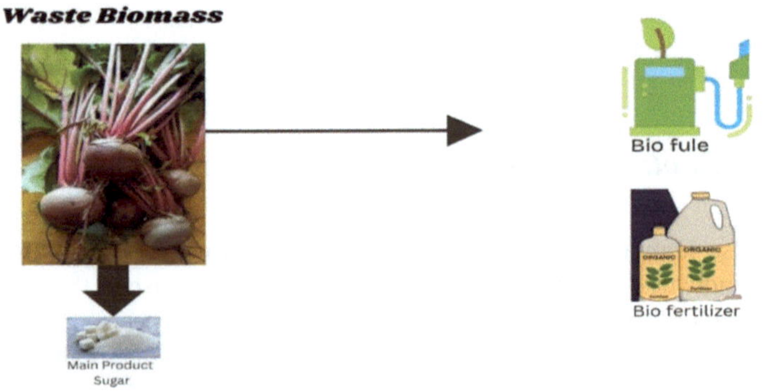

Fig. 5 Schematic diagram of beetroot

to synthetic dyes while adding economic value to the biorefinery process (Sadowska-Bartosz & Bartosz, 2021) [38].

2.7.2 Bioethanol Production from Beetroot Waste

A study by Wojtusik et al. (2019) [34] investigated the potential of beetroot juice and beet pulp for bioethanol production. The research found that beetroot juice fermentation yielded up to 0.47 g of ethanol per gram of sugar, comparable to ethanol production from traditional sugar sources like sugarcane and molasses. The study concluded that beetroot waste could serve as a cost-effective and sustainable feedstock for bioethanol production, especially in sugar beet-growing regions. Figure 5 represents the schematic diagram of beetroot.

3 Benefits of Using Unexplored Feedstocks for Biorefineries

In bio refinery, fruits and vegetables that we throw away as garbage, chitin distillers' dry grains with solubles (DDGs), bread waste, broken rice, and beet wastes are used in unconventional ways and provide economic and environmental benefits. By using these as raw materials, we can convert them into biofuels, biochemicals, etc. This can reduce dependence on natural fossil fuels, thus increasing the biocircular economy.

3.1 Environmental Benefits

(a) **Reduction in food and agricultural waste**: A significant portion of global food production is wasted, contributing to environmental pollution and greenhouse gas emissions (FAO, 2019) [39]. By repurposing food and agricultural residues into biofuels, bioplastics, and biochemicals, these feedstocks help mitigate landfill accumulation and methane emissions from decomposing organic matter.

(b) **Lower carbon footprint**: Unlike fossil fuel-based feedstocks, these waste materials are biodegradable and renewable. Their utilization in biorefineries reduces carbon dioxide emissions associated with petroleum-based plastics, synthetic chemicals, and nonrenewable energy sources (Sadh et al. 2018) [40].

(c) **Sustainable resource utilization**: Using waste-based feedstocks ensures efficient utilization of agricultural byproducts, decreasing land and water usage required for growing dedicated biomass crops such as corn and sugarcane for biofuel production.

3.2 Economic Benefits

(a) **Cost-Effective Raw Materials**

Since these feedstocks are often low-cost or free by-products from food processing, breweries, or agriculture, they offer a financially viable alternative to expensive traditional resources (Mirabella et al. 2014) [24]. Utilizing such materials reduces raw material costs for industries producing biofuels, biochemicals, and biomaterials.

(b) **Revenue generation for industries**

Food manufacturers, breweries, and agricultural businesses can generate additional revenue streams by selling waste as feedstock to biorefineries instead of disposing of it. For example, distillers can sell DDGs for bioethanol production rather than solely for livestock feed.

(c) **Job creation and economic growth**

Developing biorefineries focused on waste valorization fosters job opportunities in waste collection, processing, and biomanufacturing, stimulating economic development in rural and urban areas.

3.3 Technological and Industrial Benefits

(a) **High bioconversion efficiency**:

These feedstocks contain high levels of fermentable sugars, proteins, fibers, and bioactive compounds, making them excellent substrates for fermentation, anaerobic digestion, and enzymatic hydrolysis (Kumar et al. 2017) [41]. Their chemical composition allows efficient conversion into biofuels, organic acids, and polymers.

(b) **Diverse product portfolio**

Using alternative feedstocks enables the production of a wide range of bio-based products, including:

- Biofuels (bioethanol, biodiesel, biogas)
- Biodegradable plastics (PLA, PHA)
- Organic acids (lactic acid, citric acid)
- Functional proteins and enzymes
- Natural pigments and antioxidants (betalains from beetroot)

(c) **Reduced competition with food crops**

Unlike first-generation biofuels derived from edible crops (e.g., corn and sugarcane), these waste-based feedstocks do not compete with food supply chains, addressing concerns about food security and land use.

3.4 *Case Studies Supporting Benefits*

Bioethanol from *Bread Waste, a study* by Kodogiannis et al. (2017) [27] found that surplus bread could be converted into bioethanol with high fermentation efficiency, offering bakeries an eco-friendly waste management solution while producing renewable energy.

Bioplastics from DDGs Research by Liu et al. (2019) [42] demonstrated that distillers dried grains could be used to develop biodegradable plastic films, reducing reliance on petroleum-based plastics.

Biogas from *Fruit & Vegetable Waste,* a study by Kiran et al. (2014) [26] showed that anaerobic digestion of fruit and vegetable waste produced high methane yields, making it a sustainable biogas feedstock for energy generation.

Incorporating unexplored feedstocks into biorefineries provides a win–win solution by reducing environmental impact, enhancing economic viability, and promoting technological advancements. These feedstocks offer an efficient, cost-effective, and sustainable pathway to transition from a fossil-based economy to a bio-based circular economy. Table 1 shows the chemical composition comparison of unexplored biorefinery feedstocks.

4 Applications of Unexplored Feedstocks in Biorefineries

Unconventional feedstocks such as fruit and vegetable waste, chitin, distillers dried grains with soluble (DDGs), bread waste, broken rice, and beetroot have diverse applications in biorefineries. These feedstocks can be processed into biofuels, bioplastics, biochemicals, functional compounds, and animal feed, making them valuable resources for sustainable industries.

Table 1 Chemical composition comparison of unexplored biorefinery feedstocks

Feedstock	Carbohydrates (%)	Proteins (%)	Lipids (%)	Fibers (%)	Ash (%)	Key references
Fruit and vegetable waste	50–70 (sugars, cellulose, hemicellulose)	1–5	< 1	5–15	2–5	Mirabella et al. 2014; Kaur et al. 2021
Chitin (from crustaceans/ fungi)	10–20 (residual sugars, acetyl groups)	5–10	1–3	60–80 (chitin/chitosan)	10–15	Kumari et al. 2022; Muzzarelli, 2010
DDGs (distillers dried grains)	15–25 (residual starch, fiber)	25–30	10–12	30–40	4–6	Rosentrater&Muthukumarappan, 2006; Liu, 2011
Bread waste	55–70 (starch, sugars)	8–12	2–4	2–5	2–4	Sirohi et al. 2020; Kibler et al. 2018
Broken rice	75–85 (mainly starch)	6–8	0.5–1.5	1–3	1–2	Ho& Abdul Rahman, 2019; Pathak et al. 2018
Beetroot	65–75 (sucrose, fiber)	1–2	< 1	5–10	1–2	Kujala et al. 2002; Kapadia & Bahadur, 2022

4.1 *Biofuels and Bioenergy Production*

(a) **Bioethanol**

- Feedstocks used: Fruit and vegetable waste, bread waste, broken rice, beetroot, DDGs.
- Process: These feedstocks contain fermentable sugars and starch that can be enzymatically hydrolyzed and fermented by *Saccharomyces cerevisiae* to produce bioethanol.
- Application: Bioethanol is blended with gasoline to reduce carbon emissions in transportation.
- A study by Pham et al. (2018) [29] showed that broken rice produced 0.45 g of ethanol per gram of fermentable sugar, making it a viable alternative to corn-based ethanol.

(b) **Biogas and Biohydrogen**

- Feedstocks used: Fruit and vegetable waste, bread waste, beetroot pulp, DDGs.
- Process: Anaerobic digestion of organic matter by microbes produces biogas (methane) and biohydrogen.
- Application: Biogas is used for electricity, heating, and transportation, reducing dependence on fossil fuels.
- Kiran et al. (2014) [26] demonstrated that fruit and vegetable waste digestion yielded high methane output, making it a suitable feedstock for renewable energy production.

4.2 *Bioplastics and Sustainable Packaging*

(a) **Starch-Based Bioplastics**

- Feedstocks used: Broken rice, bread waste, DDGs.

- Process: Starch is extracted, modified, and polymerized to produce biodegradable films and packaging materials.
- Application: Used in food packaging, shopping bags, and agricultural mulch films as a sustainable alternative to petroleum-based plastics.
- Research by Tang et al. (2019) [30] found that broken rice starch produced biodegradable plastic films with excellent mechanical and barrier properties.

(b) **Chitin-based bioplastics**

- Feedstocks used: Chitin (from seafood waste).
- Process: Chitin is converted into chitosan, which is blended with polymers to create biodegradable plastics.
- Application: Used in medical sutures, wound dressings, water purification, and food packaging.

- Liu et al. (2021) [42] developed a chitosan-based bioplastic with antimicrobial properties, making it ideal for food preservation.

4.3 Biochemicals and High-Value Products

(a) **Organic acid production**

- Feedstocks used: Beetroot, fruit and vegetable waste, DDGs, broken rice.
- Process: Microbial fermentation converts sugars into lactic acid, succinic acid, and citric acid.
- Application: Used in food preservation, cosmetics, and biodegradable plastic production.
- Orozco et al. (2020) [31] demonstrated that beetroot hydrolysates could efficiently produce lactic acid, a key component in bioplastics.

(b) **Natural pigments and antioxidants**

- Feedstocks used: Beetroot, fruit, and vegetable waste.
- Process: Extracted bioactive compounds such as betalains (from beetroot) and flavonoids (from fruit peels).
- Application: Used in natural food colorants, cosmetics, and pharmaceuticals.
- Sadowska-Bartosz & Bartosz (2021) [40] showed that beetroot pigments have strong antioxidant properties, making them suitable for functional foods and nutraceuticals.

4.4 Animal Feed and Agricultural Applications

(a) **High-protein animal feed**

- Feedstocks used: DDGs, bread waste, broken rice, fruit and vegetable waste.
- Process: Waste materials rich in protein and fiber are processed into livestock feed supplements.
- Application: Used in poultry, cattle, and aquaculture feed.
- Liu et al. (2019)[43] found that DDGs contain 30–40% protein, making them an excellent alternative to soybean meal in animal feed.

(b) **Soil enhancers and composting**

- Feedstocks used: Fruit and vegetable waste, chitin, beetroot waste.
- Process: Organic residues are composted to enrich soil with nutrients and microbial activity.
- Application: Used in organic farming and sustainable agriculture.
- Sadh et al. (2018) [13] found that applying fruit and vegetable waste compost improved soil fertility and crop yield.

4.5 *Pharmaceutical and Biomedical Applications*

a. **Chitin and chitosan for wound healing**

- Feedstocks used: Chitin from seafood waste.
- Process: Chitin is processed into chitosan, which has antimicrobial and biocompatible properties.
- Application: Used in wound dressings, drug delivery systems, and tissue engineering.
- Goy et al. (2016) [44] found that chitosan-based wound dressings promoted faster healing and reduced infections compared to conventional bandages.

5 Challenges in Utilizing Unexplored Feedstocks for Biorefineries

While fruit and vegetable waste, chitin, distillers dried grains with soluble (DDGS), bread waste, broken rice, and beetroot offer significant potential as feedstocks for biorefineries, several challenges hinder their large-scale adoption. These challenges can be categorized into technical, economic, environmental, and regulatory barriers.

5.1 *Technical Challenges*

(a) **Heterogeneous composition and variability**

- These waste materials exhibit high variability in composition due to differences in source, season, and processing methods.
- Inconsistent sugar, starch, protein, and fiber content make it difficult to optimize conversion processes.
- Example: Fruit and vegetable waste varies significantly in moisture content, affecting bioethanol and biogas yields (Mirabella et al. 2014) [24].

(b) **High moisture content and storage issues**

- Many of these feedstocks (especially fruit and vegetable waste, beetroot, and bread waste) have a high-water content (60–95%), leading to rapid microbial degradation and spoilage.
- Solution needed: Efficient drying, pretreatment, or immediate processing is required to prevent loss of valuable organic compounds.

(c) **Complex pretreatment and processing requirements**

- Unlike conventional feedstocks (e.g., sugarcane and corn), these waste-based feedstocks require intensive pretreatment to break down fibers, starches, and proteins.
- Pretreatment methods (e.g., enzymatic hydrolysis, and acid treatment) add to the cost and complexity.
- Example: Chitin from seafood waste requires deproteinization and demineralization before it can be processed into chitosan-based bioproducts (Goy et al. 2016) [44].

5.2 *Economic Challenges*

(a) **High collection, transportation, and storage costs**

- Unlike large-scale agricultural crops (e.g., corn and sugarcane), waste feedstocks are scattered across multiple sources (restaurants, food processing plants, breweries, and households).
- Collection and transportation logistics increase costs, especially for perishable waste.
- Example: Bread waste and broken rice require efficient supply chain networks to ensure they are collected before spoilage.

(b) **Low market value and economic viability**

- Some waste feedstocks (e.g., DDGs and bread waste) have existing uses in animal feed, making them less attractive for bio-based applications.
- Developing new biorefineries and processing facilities requires significant capital investment with uncertain returns.
- Solution needed: Government incentives or subsidies to support investment in waste valorization technologies.

5.3 *Environmental Challenges*

(a) **Waste contamination and safety issues**

- Food waste can contain pathogens, mycotoxins, pesticides, and heavy metals, posing risks for human and animal health.
- Example: Bread waste may contain preservatives that could interfere with microbial fermentation processes.
- Solution needed: Strict quality control and sorting mechanisms before processing into bio-based products [42, 44–48].

(b) **Competition with other waste utilization pathways**

- Some waste feedstocks are already used in composting, livestock feed, and biofertilizers.
- Example: DDGs are widely used as high-protein animal feed, making biofuel conversion less competitive unless it offers higher economic benefits (Liu et al. 2019) [42, 45–48].

(c) **Land and water use for processing facilities**

- Setting up new biorefineries requires land, water, and energy resources, potentially causing environmental trade-offs.
- Example: Large-scale chitin extraction from seafood waste can result in acidic wastewater pollution, requiring sustainable disposal strategies.

5.4 Regulatory and Policy Challenges

(a) **Lack of standardized regulations and policies**

- Unlike first-generation biofuels, waste-derived biofuels and bioproducts lack clear regulatory frameworks in many countries.
- Solution needed: Policy support, such as subsidies, tax incentives, and mandates for waste-based biofuels and biochemicals.

(b) **Food versus fuel debate**

- Some stakeholders argue that diverting food waste for biofuels may reduce food availability or increase feed costs for livestock industries.
- Example: If broken rice is used for bioethanol instead of animal feed, it might impact livestock farming economics.

(c) **Public perception and industry adoption**

- Consumers and industries are hesitant to adopt products made from waste due to safety concerns and a lack of awareness.
- Solution needed: Stronger public education, certification, and industry collaborations to promote waste-based biorefineries [49].

6 Conclusion

The utilization of unconventional feedstocks such as fruit and vegetable waste, chitin, distillers dried grains with soluble (DDGs), bread waste, broken rice, and beetroot presents a promising opportunity for sustainable biorefineries. These materials, often regarded as waste, can be transformed into valuable products, including biofuels, bioplastics, biochemicals, animal feed, and pharmaceuticals, contributing to a circular bioeconomy. However, challenges such as feedstock variability, high moisture content, logistical constraints, regulatory gaps, and competition with existing waste utilization pathways must be addressed to enhance feasibility. Innovations in pretreatment technologies, efficient collection systems, supportive government policies, and public awareness initiatives are crucial to overcoming these barriers. By integrating these feedstocks into biorefinery systems, industries can reduce waste pollution, mitigate reliance on fossil-based resources, and promote environmental sustainability, ultimately advancing the transition toward a more resource-efficient and eco-friendly bioeconomy.

Acknowledgements We sincerely thank the Vice Chancellor of RGPV and all relevant authorities for providing the necessary facilities and support that enabled the successful completion of this book chapter. Their commitment to fostering an environment of research and academic excellence has been instrumental. We also acknowledge the collaborative efforts of our team members, whose dedication and contributions were vital throughout this project.

References

1. FAO (2020) The state of food security and nutrition in the world. Food and Agriculture Organization of the United Nations
2. Tilman D et al (2019) Benefits and costs of biofuels and the biorefinery concept. Nat Sustain 2(5):310–320
3. Pimentel D et al (2018) Environmental and economic costs of fertilizer use in conventional agriculture. Ecol Econ 150:54–60
4. Mohr A, Raman S (2018) Socio-economic and political considerations of biofuel development. Energy Policy 122:120–130
5. Searchinger T et al (2019) The carbon opportunity cost of using productive land for biofuels. Sci Adv 5(9):eaax1392
6. Chen H et al (2021) Waste-derived feedstocks for sustainable biorefineries. Renew Sustain Energy Rev 137:110464
7. Girotto F et al (2015) Food waste generation, management, and solutions: a review. Waste Manage 45:32–41
8. Sharma S et al (2020) Chitin and chitosan: production, properties, and applications in biorefineries. Carbohyd Polym 250:116885
9. Lin CSK et al (2020) Bioethanol production from bakery waste: a sustainable approach for waste reduction. Biores Technol 303:122931
10. FAO (2021) The state of food and agriculture: food loss and waste. Food and Agriculture Organization of the United Nations
11. Girotto F, Alibardi L, Cossu R (2015) Food waste generation, management, and solutions: a review. Waste Manage 45:32–41
12. Lin CSK, Pfaltzgraff LA, Herrero-Davila L, Mubofu EB, Abderrahim S, Clark JH, Luque R (2013) Food waste as a valuable resource for the production of chemicals, materials, and fuels. Energy Environ Sci 6(2):426–464
13. Sadh PK, Duhan S, Duhan JS (2018) Agro-industrial wastes and their utilization using solid state fermentation: a review. Biores Bioproc 5(1):1–15
14. Pathak PD, Mandavgane SA, Kulkarni BD (2017) Fruit peel waste as a resource for value-added products: a review. J Clean Prod 143:445–467
15. Sharma S, Nehru B, Kaur R (2020) Chitin-based biopolymers for sustainable applications: challenges and future prospects. Environ Sci Pollut Res 27(17):20950–20966
16. Kumar ABV, Varadarajan N, Rath PK, Viswanath P (2020) Chitin and chitosan: structure, properties, and applications in biomedical engineering. Int J Biol Macromol 164:2406–2421

17. Gopi S, Kontopoulou M, Sabliov CM (2018) Chitin and chitosan-based nanomaterials: chemistry, properties, and applications. Chem Rev 118(3):1205–1256
18. ranaz I, Acosta N, Civera C, Elorza B, Mingo J, Castro C, Galed G (2021) Chitosan: an overview of its properties and applications. Polymers 13(19)
19. Ifuku S, Nogi M, Yoshioka M, Morimoto M, Yano H, Saimoto H (2011) Fibrillation of dried chitin into 10–20 nm nanofibers by a simple grinding method under acidic conditions. Carbohyd Polym 84(2):762–764
20. Liu K (2011) Chemical composition of distillers grains, a review. J Agric Food Chem 59(5):1508–1526
21. Singh V, Johnston DB, Naidu K, Rausch KD (2015) Advancements in the utilization of distillers grains for biofuels and bioproducts. Ind Biotechnol 11(3):140–148
22. Bothast RJ, Schlicher MA (2005) Biotechnological processes for conversion of corn into ethanol. Appl Microbiol Biotechnol 67:19–25
23. Mosier NS, Ladisch MR (2009) Fractionation of DDGS to maximize value-added processing. Biores Technol 100(12):3245–3250
24. Mirabella N, Castellani V, Sala S (2014) Current options for the valorization of food manufacturing waste: a review. J Clean Prod 65:28–41
25. Jagannath A, Kalaiselvan A, Manimaran A, Ramasamy D (2013) Bread waste utilization for bioethanol production with Saccharomyces cerevisiae. Int J Environ Sci Technol 10(6):1283–1290
26. Kiran EU, Trzcinski AP, Ng WJ, Liu Y (2014) Bioconversion of food waste to energy: a review. Fuel 134:389–399
27. Kodogiannis VS, Lygouras JN, Tarus B (2017) Ethanol production from food waste: Experimental investigation and intelligent modeling. Energy 134:915–929
28. Lu Z, Yang J, Yuan X, Rao C (2016) Rice starch-based bioplastics: development, properties, and applications. Carbohyd Polym 150:411–421
29. Pham TPT, Kaushik R, Parshetti GK, Mahmood R, Balasubramanian R (2018) Rice-based ethanol production: a sustainable approach for biofuel development in Asia. Renewable Energy 121:178–185
30. Tang X, Chen H, Dou J (2019) Biodegradable rice starch-based films: characterization and application in food packaging. Food Chem 292:115–123
31. Orozco RL, Audic JL, Chauvet M (2020) Lactic acid production from broken rice hydrolysates using Lactobacillus casei. Biores Technol 318:124119
32. Li Y, Jin W, Mu C, Wang X (2017) Biohydrogen production from broken rice using a two-stage fermentation process. Int J Hydrogen Energy 42(5):2891–2899
33. Sluková M, Poláková J, Plíšková J (2020) Utilization of sugar beet residues in sustainable biorefineries. Food Sci Biotechn 29(2):173–182
34. Wojtusik M, Kamiński J, Zięba M (2019) Ethanol production from sugar beet juice and beet pulp: comparative study of fermentation efficiency. Ind Crops Prod 132:12–19
35. Mizgier P, Wachowska H, Słomczyńska D (2018) Biogas production from sugar beet pulp: optimization and scale-up. Renewable Energy 129:271–279
36. Gómez B, Navarro R, Gómez C (2021) Beet pulp-derived bioplastics: development and applications in food packaging. J Appl Polym Sci 139(3):51097
37. Jörgensen H, Eriksen PE, Lübeck M (2022) Microbial production of organic acids from sugar beet residues. Biores Technol 345:126510
38. Sadowska-Bartosz M, Bartosz G (2021) Betalains: natural plant pigments with potential health benefits. Antioxidants 10(9):1467
39. FAO (2019) The state of food and agriculture: food loss and waste reduction. Food and Agriculture Organization of the United Nations
40. Sadh PK, Duhan S, Duhan JS (2018) Agri-food waste: nutritional, biotechnological, and therapeutic potential. Biores Technol 250:159–173
41. Kumar A, Samadder SR, Handa H (2017) Biorefineries and food waste: sustainability and economic aspects. Renew Sustain Energy Rev 80:609–629

42. Liu S, Yang Y, Xie Y, Cui W (2019) Utilization of distillers dried grains with solubles (DDGs) for bio-based plastics: a sustainable approach. J Clean Prod 234:761–769
43. Liu Y, Chen X, Zhang Z (2021) Development of chitosan-based antimicrobial films for food packaging applications. Int J Biol Macromol 183:1398–1405
44. Goy RC, Britto D, Assis OB (2016) Chitosan-based wound dressings: antimicrobial properties and applications. Carbohyd Polym 144:414–425
45. Tripathi GK (2019) Engineered nanomaterials and their properties: a review. Biosci Biotechnol Res Commun 12:764–771
46. Tripathi GK, Rathore H, Chavali M, Rathore D (2021) Nanotechnology for mitigating impact of COVID-19. J App Sci Eng Techn Educat 3(2):171–180
47. Tripathi GK, Khare V, Bundela P, Khiriya PK, Khare PS, Dixit P, Dixit M (2024) Nanomaterials in ultra-high definition displays. In: Handbook of nanomaterials, volume 1, pp 81–99. Elsevier
48. Tripathi GK, Soni A, Singh P, Bundela P, Khiriya P, Khare PS, Dixit P, Sundaramurthy S (2024) Advanced conversion technologies for PPEs and their recent research trends. Adv Mat Convers Techn Personal Prot Equ Used COVID-19 Pand, pp 53–71
49. Cakmak EK et al (n.d.) Circular city concept for future biorefineries. https://doi.org/10.1016/B978-0-323-90178-9.00009

Current Advancements in Algal Feedstock Cultivation and Processing Aiming 3G Biorefinery

S. Rahul and A. Arumugam

Abstract The inexorable shift toward renewable energy sources has catapulted third-generation (3G) biorefineries, which utilize algal biomass, into the spotlight as the greenest and most sustainable energy producers. Algal biomass is a top-performing feedstock because of its fast growth rate, high photosynthetic efficiency, and nonfood crop competition. This chapter comprehensively explores the current progress in algal feedstock cultivation and downstream processing specifically designed for 3G biorefineries. It first briefly introduces algal taxonomy, pointing out the potential of microalgae as well as macroalgae. The efforts made to find new approaches in cultivation systems, such as new photobioreactor designs, wastewater-based systems, and the employment of industrial flue gases in order to sequester CO_2, have been examined. Besides, the chapter delves into possible methods of improving the nourishment conditions to the utmost by optimizing light intensity, temperature, materials, and biostimulants. Some of the most recent achievements in energy-saving harvesting and dewatering technologies—such as filtration, flotation, and membrane-based separations—are included. Several different routes for the processing of algal biomass into renewable transportation fuels, including biodiesel and bioethanol, as well as biogas, have been developed; moreover, the high-value coproduct extracts such as pigments and bioplastics point out the integrated and holistic nature of 3G biorefineries. Such an in-depth review and coverage of these topics vividly illustrate the promising future of algae-based systems in cultivating the bioeconomy and substantially contributing to circular and sustainable industrial practices.

Keywords Algae · Biorefinery · Biofuel · Bioreactors · Photobioreactor · Algal cultivation

S. Rahul · A. Arumugam (✉)
Advanced Biorefinery & Catalysis [ABC] Lab Centre for Bioenergy, School of Chemical and Biotechnology, SASTRA Deemed University, Thirumalaisamudram, Thanjavur, India
e-mail: aruchemxl13@gmail.com

© The Author(s), under exclusive license to Springer Nature Singapore Pte Ltd. 2025

A. K. Chandel and S. Sundaramurthy (eds.), *Biorefineries in Circular Bioeconomy*, Green Energy and Technology, https://doi.org/10.1007/978-981-95-3913-0_3

1 Introduction

Biorefinery is a practical method that enables the conversion of biomass into a wide range of highly value-added products such as biofuel and biochemical. The biological resources such as agricultural residues, organic wastes, and algal residues have been efficiently used for value generation [15]. It maximizes resource utilization and waste reduction through the valorization of multiple products, and the biorefinery approach diminishes the reliance on fossil fuels and the eventual impacts on the environment. Biorefinery systems can be grouped depending on their feedstock and classified into first-generation, second-generation, and third-generation (3G) biorefineries. A first-generation biorefinery uses crops (sugarcane, vegetable oil, corn) for energy. A second-generation biorefinery utilizes nonedible crops or nonedible portions of edible crops and draws attention to the impact of land space that can be utilized and the lignocellulosic structure. The third-generation system is algae-based biomass that does not need large space for cultivation and can be grown in several types of industrial wastewater [40].

Algae are truly unique in their chemical makeup, containing lipids, proteins, carbohydrates, pigments, and many bioactive compounds that can yield multiple product streams from one feedstock. Recent advances have led to the ability to grow algae from wastewater, which is nutrient-dense, thereby decreasing the costs of production and improving sustainability [69]. An algal biorefinery can provide energy and non-energy bioproducts such as biodiesel, bioethanol, bioplastics, pigments, and animal feed; this diversification can improve both financial and economic viability while providing circular bioeconomy strategies for integrated low-waste processing options [86]. Algae are so versatile, with hundreds of thousands of strains and the ability to produce different categories of compounds. Since there are thousands of species (thus strains), one can select the species of algae to grow to obtain the desired end products (such as pigments, antioxidants, enzymes, and specialty oils). Algal metabolic pathways can also be altered to promote the buildup of certain metabolites, or the metabolic activity of an algal species can be altered by changing nutrient regimes or applying stress [78]. An extra advantage of algae over alternative biomass feedstocks is feedstock decentralization. Algal cultivation systems can be housed anywhere with access to water, sunlight, and nutrients: urban areas, coastal areas, desert areas, and so on. Industrial synergies can be planned with algae. These photosynthetic algae can use CO_2 or waste heat generated from an industrial process to enhance their growth [3]. There might also be potential advantages that these algae can offer over the traditional crops such as a very short generation/growing period, and they can be cultivated in a closed loop system, providing a higher level of control over the quality and consistency of the feedstock compared with traditional crops. The application of algae as a feedstock has the potential to boost industrial scalability, robustness, and productivity. All things considered, algae provide next-generation biorefineries with a creative and strategic feedstock choice [34].

In the framework of algal biorefineries, the need for integrated and sustainable approaches becomes very crucial. There is a world of opportunity for processing and cultivating algae, but corresponding to scalability, it is important to resolve the high-energy input, nutrient requirements, and economic issues. These types of extraction processes for single products, such as producing solely biodiesel from algae, are too costly to justify from an environmental point of view [34]. Thus, an integrated biorefining approach is warranted for maximizing algal biomass toward the coproduction of biofuels, bioplastics, pigments, proteins, and bioactive compounds. Wastewater can be used for cultivating algae along with industrial CO_2 absorption; these are examples of sustainable technologies that reduce input costs and increase benefits to the environment through nutrient recycling and carbon sequestration. Efficiency gains and waste reduction are achieved when extraction and conversion technologies are integrated along a single loop. Algal biorefineries further grant an increase in circularity when connected with renewable energy (for instance, photovoltaic-powered photobioreactors) and through the utilization of leftover biomass in anaerobic digestion or composting [86].

2 Types of Algal Feedstock

Based on their size and shape, algae can be divided into two groups: microalgae and macroalgae. Microalgae are simple filamentous organisms that are unicellular. They can be found in freshwater and saltwater habitats. Microalgae are further divided into three groups, which include diatoms, green algae, and blue-green algae (cyanobacteria). Macroalgae are multicellular and naked, and they are also visible to the human eye. Most macroalgae are found in marine environments. Macroalgae are divided into three groups: green, red, and brown algae [49]. Some microalgal strains have been proven to possess outstanding potential as candidates for biorefineries based on high productivity, biochemical diversity, and applicability across a wide range of growth conditions. Some of the best microalgal species include *Chlorella vulgaris, Scenedesmus obliquus, Nannochloropsis gaditana, Chaetoceros muelleri,* and *Arthrospira platensis* (Spirulina). These species have shown good growth and are capable of yielding valuable macromolecules, lipids, carbohydrates, and proteins. For instance, *C. vulgaris* yields 25% lipids and more than 50% protein dry-weight for both nutraceutical and biofuel markets. *S. obliquus* has even higher capacities of up to 55% for total lipid production, in addition to being termed to have high protein content, making it suitable for application in the production of biodiesel feed as well as animal feed purposes [11]. *N. gaditana* has good lipid yields and is most commonly chosen for the purpose of biodiesel and omega-3 fatty acid production. The lipid content in *C. muelleri* is optimized, along with its protein content, which makes it ideal for concurrent processing in a biorefinery for the production of coproducts from a single biomass source. *A. platensis* contains a very high protein content of up to 70% and can be used in food, animal feed, and specialty chemicals [67].

3 Recent Advancements in Algal Cultivation

3.1 Novel Photobioreactor Designs and Configurations

In the past few years, there have been some interesting advances regarding the design and configuration of photobioreactors (PBRs), which has caused significant changes in algal cultivation applicable for biofuels, nutraceuticals, wastewater treatment, and carbon sequestration [2]. While open-pond systems are inexpensive, they fail to provide adequate protection from contamination, do not maintain the optimal environment for culture growth, do not allow for extensive biomass productivity, have low yields compared to other methods, and can become breeding sites for insects [28]. Closed photobioreactors, on the other hand, have emerged as the preferred quality standard for high-value applications in algal cultivation since they increase control over environmental aspects of the culture, including light, temperature, gaseous exchange, and contamination. Perhaps the most creative developments are the V-shaped photobioreactors, which specifically target the issue of excess light exposure or light oversaturation, a problem that can plague equatorial cultures receiving high solar radiation (Fig. 1). The V-shaped design fails to resemble any other culture growth assembly in the sense that it dilutes the incoming sunlight while ensuring proper distribution across the reactor, which in turn helps prevent photoinhibition and can help maintain some level of consistent light exposure across the algal culture [21]. Other modeling studies and a few pilot projects have illustrated that these reactors can produce theoretical biomass productivities which may be significantly higher than those illustrated in the literature relative to traditional flat-panel or horizontal designs, and they can achieve a photosynthetic efficiency of 3.3%. The V-shaped reactors have great merit regarding the performance demonstration of larger batch cultures of robust fast-growing strains such as *Chlorella sorokiniana*, specifically in a tropical region [22].

Another typical kind is the flat-panel photobioreactor, with its large surface-area-to-volume ratio, an imperative condition for maximizing light penetration and ensuring temperature control (Fig. 2) [38]. For the best setup to capture light, these flat-panel photobioreactors are kept either vertical or at a slight tilt. They are mostly made out of transparent materials such as acrylic and glass, thus promoting access to maximum light. Flat-panel systems are advantageous because they are scalable and require less maintenance, making them an industrial as well as a research application [8]. However, flat-panel photobioreactors have problems with biofilm accumulation on internal surfaces, thus further reducing light and productivity. Thus, newer designs are accompanied by different design materials that mitigate biofilm formation, enable automatic maintenance, and provide antifouling coatings [2].

The reactors have gained quite a reputation because they mix energy efficiently; furthermore, they can also exchange gas effectively. Air (or carbon dioxide) is injected into the reactors instead of relying on mechanical stirrers. This creates a gentle circulation, and also the circulation minimizes shear stress on the delicate algal cells. Sometimes, they integrate internal baffles or spiral inserts. These additions work to

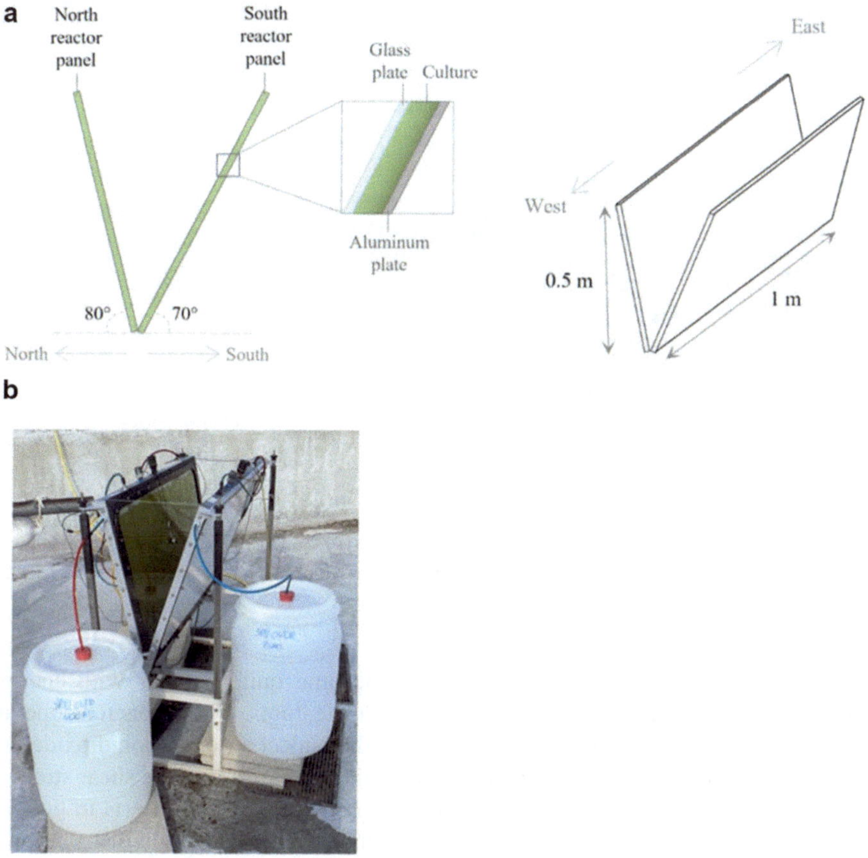

Fig. 1 The design of the V-shaped photobioreactor (**a**) and a photograph of the prototype at AlgaePARC Bonaire (**b**) [21]. © 2024 Chin-On, de Boer, van de Voort, Camstra, Barbosa, Wijffels, and Janssen. Under a Creative Commons license

improve turbulence as well as improve light distribution. These designs are especially helpful for microalgal species that are sensitive to shear. Because the design is simple, they can be scaled up and maintained more easily.

Tubular photobioreactors are yet another significant breakthrough, and they consist of transparent tubes arranged horizontally, vertically, or in a helical pattern. The systems are very flexible in terms of size and can be run in continuous mode, which makes them very appealing for heavy industrial use (Fig. 3). Tubular PBRs enable very efficient mixing and gas exchange via airlift pumps or mechanical agitation, but at the same time, they are vulnerable to problems like oxygen buildup and biofilm formation. However, innovations in tube materials, flow dynamics, and cleaning procedures have solved most of these problems and thus, it is possible to have more stable and productive operation in the long run [36].

Fig. 2 The flat-panel photobioreactor system [63]. © 2019 The Authors. Published by Elsevier B.V. Under a Creative Commons license. With permission from Elsevier. Copyright (2019), License Number—6,053,180,011,794

Airlift and bubble column reactors have become quite popular with energy-efficient mixing and gas exchange as their lead features. These reactors operate without the use of mechanical stirrers; instead, they inject air or carbon dioxide that runs through the culture, thus allowing gentle mixing that lessens the shear stress of algal cells. It is a good practice to put internal baffles or helical inserts to go with these reactors, and as a result, turbulence gets enhanced and the light distribution is better. These designs are a perfect match for cultivating sensitive or shear-intolerant microalgal species, and their simplicity also makes them very easy to scale and maintain [74].

Current research with commercial deployment puts hybrid with modular photobioreactor systems at the forefront. Combining the strengths of different configurations like flat-panel and airlift technologies lets these systems achieve greater operational flexibility with higher photosynthetic efficiencies [45]. Modular designs allow for easier scaling, and also redundancy and maintenance in that multiple smaller units are operated in parallel. This approach allows adapted growth conditions for algal strains that differ or end products that are desired, like proteins or lipids.

Modern PBR design trends greatly aim toward integrating algal cultivation with processes that treat wastewater and capture carbon. As microalgae use nutrients present in wastewater, valuable biomass production and water purification can occur at the same time [2]. PBRs can be coupled directly with flue gas streams derived from power plants or industrial facilities. This capture as well as utilization of carbon dioxide is able to reduce greenhouse gas emissions. For operating modern photobioreactors, it is necessary to have advanced controlling and operating techniques. Factors like nutrient concentration, light intensity, and mixing rates can be monitored

Fig. 3 Tubular photobioreactor system for algal cultivation [33]. © 2014 by the authors; licensee MDPI, Basel, Switzerland. Under the terms and conditions of the Creative Commons Attribution license

and controlled using artificial intelligence and machine learning tools [76]. These necessary controls allow for greater production with fewer resources and less operational costs by reacting quickly to changes in the environment. Continuous automation becomes a necessity to maintain consistent operating performance, especially when working on a large-scale or outdoor environments where manual operation is unsuitable [82].

Though these reactors have multiple advantages, there are several challenges associated with them. The major issue associated with this reactor is biofouling. The accumulation of algal cells over the surface of the reactor leads to biofouling. This affects light penetration and lowers the overall efficiency of the system [100]. Several solutions have been proposed by researchers to overcome these issues, including applying antifouling coatings on the reactor surface, having mechanical wipers, and regularly washing the reactor to keep the surface clean. Another major issue is maintaining the reactor temperature, mainly in outdoor and large-scale systems. Installation of cooling units, insulation systems, and heat exchange systems is a technical approach to overcome this issue [2, 100].

Cost and energy consumption are the primary determinants of closed photobioreactors' suitability for industrial usage. However, these are more expensive to build and run than open ponds, which provide minimal control. Large-scale in situ algae culture requires ongoing innovation in design materials, process integration, and energy-efficient operation to be economically competitive. Scaling up from pilot or lab-scale units to industrial levels poses further challenges. To prevent productivity loss while scaling up from a small-scale reactor to a large scale, several factors like mass transfer, light distribution, and hydrodynamics need to be controlled. Modular and hybrid strategies, together with sophisticated modeling and simulation methods, are addressing these scale-up issues [29].

3.2 Wastewater-Based Cultivation Systems

Nowadays, the utilization of wastewater for cultivating the algae becomes a dual-benefit strategy to reduce the biomass cultivation cost and to remove the pollutants from the wastewater. Conventionally, algal cultivation depends on synthetic mediums rich in nutrient contents [9]. The cost of these mediums is expensive and found to be unsustainable for large-scale operation. However, the wastewater from domestic, agricultural, and industrial regions acts as a low-cost and nutrient-rich alternative. The organic carbon, nitrogen, and phosphorus content in this wastewater is found to be high, and it is suitable for algal cultivation.

The removal of pollutants from wastewater by algae occurs mainly via two biological mechanisms: biosorption and bioaccumulation [7, 10]. In biosorption, the pollutants are bound to the surface or inside the algal cell. This forms a complex to reduce the availability of these pollutants in the environment. Bioaccumulation involves the absorption of contaminants into algal cells and their utilization for cellular metabolisms. Diatoms are a microalgae type that has the ability to remove heavy metals from wastewater via both active and passive mechanisms, as they have siliceous cell walls and high productivity. Various marine and freshwater microalgae prove their effectiveness in removing hazardous contaminants, even endocrine-disrupting compounds, from the wastewater. The strains like *Nannochloropsis oculata, Scenedesmus quadricauda, Chlorella* spp., *Dunaliella salina,* and *Phaeocystis globosa* have been studied for wastewater remediation [72]. Several strains, like *Chlamydomonas, Oscillatoria,* and *Spirulina,* have been found to be effective in removing pollutants while producing high biomass. The removal efficiency of the algae and their tolerance toward wastewater vary depending on the species. Species like *Euglena* and *Stigeoclonium* are resilient to organic toxins and can survive under harsh conditions [1]. *Scenedesmus* sp. has been found to remove heavy metals like molybdenum. Studies on *Chlamydomonas* for lead removal, and other species, like *Oscillatoria* and *Chlorell*a, have been effective in removing cadmium, zinc, and copper from wastewater [72]. These processes are facilitated by a suite of enzymes that catalyze pollutant transformation into less harmful intermediates.

Algal biomass produced here has the potential to be converted into biofuels, biodiesel, bioethanol, animal feeds, fertilizers, and high-value biochemicals. The algal cultivation integrated with wastewater treatment systems is an example of a circular bioeconomy because it uses wastes for producing clean energy and value-added products. While the present technologies seem promising, it is worthwhile to conduct research continuously in enhancing pollutant removal, selecting and optimizing algal strains, and furthering improvements to reactor design. Tackling such challenges can lead to algal production and uses becoming major considerations for industrial and environmental applications. There should be an expectation that these wastewater-fed algal systems will play important roles in the long-term interest of sustainable energy production and environmental conservation.

3.3 Use of Industrial Flue Gas

The direct conversion of industrial flue gases into algae is a new technology that could become an effective means for CO_2 mitigation on a large scale with a clean environment. Microalgae can use CO_2 as a carbon source during growth, and carbon constitutes nearly 50% of their total dried biomass. This ability gives photosynthetic microorganisms the advantage they need to weigh with industrial gas producers, such as chemical refineries, cement factories, and power plants [62]. These businesses can produce useful biomass by growing algae and lowering their environmental impact through flue gas emission reductions. Some strains of algae can thrive in open raceway ponds or closed photobioreactors by receiving carbon dioxide-rich flue gases directly. Thus, these systems strive to reduce CO_2 emissions via phytoplankton biomass for biofuels and bioproducts [51]. Their rate of carbon fixation is much higher than most land plants, many times on an area basis. Another benefit is that microalgae do not race with crops for land or freshwater. Experiments have shown that green algae like chlorella can handle high concentrations of carbon dioxide, up to 15%, as well as other dangerous chemicals like sulfur dioxide and nitrogen that come from manufactories. Besides getting relief from CO_2, this approach also uses it to induce other useful goods [56]. India's warm climate and abundant sunlight make it a good place for these systems since they help algae grow rapidly. This makes it possible to put up decentralized CO_2 prisoner systems in artificial areas all throughout the country. By combining these systems with wastewater treatment units, we can make them indeed more. This will help break two environmental problems formerly carbon emigrations and water pollution [24]. Still, it is pivotal to know that algae's capability to fix CO_2 does not always mean that carbon will stay in the ground for a long time. The biomass needs to be converted into stable or non-releasable composites for that to be. The main benefit for the terrain comes from replacing fossil energies with long-continuing accouterments that trap carbon. Therefore, the performance of algal systems depends not only on how well they gather CO_2 but also on how they turn it into other effects and how they regulate their lifetime [24, 73].

4 Factors Influencing Algal Growth and Secondary Metabolites

The algae's growth rate and secondary metabolite production are affected by nutrient availability and environmental conditions. Nitrogen, phosphorus, iron, sulfur, and magnesium are particularly important for the growth of algae and the production of metabolites [26]. Algal cells need a source of nitrogen to make proteins and pigments. Nitrogen-limiting conditions affect the buildup of secondary metabolites, such as lipids, antioxidants, and polysaccharides. Proper nutrient management, which includes accurate dosing and monitoring, not only increases biomass yield but also changes the production of useful secondary metabolites [61, 91].

Another important factor is the temperature of the medium and the light source. To help photosynthesis and growth, it is important to maintain an optimal quantum of light. Changing the light intensity and the light–dark cycle can help induce the production of secondary metabolites as high as possible [66, 70]. Temperature can have an effect on the metabolic flux. At low temperatures, certain adipose acids can accumulate inside the cells. Raising the temperature can help secondary metabolite production reach its maximum. By adjusting these factors, both the growth and secondary metabolite production of the algae can be managed [70].

Microalgae-based biostimulants, like amino acids, phytohormones, and antioxidants, have been shown to boost algal growth, nutrient input, and stress resistance. These chemicals can boost defense mechanisms, make photosynthesis more effective, and stimulate secondary metabolite production, especially under limited water supply or other stress conditions [25]. Using biostimulants as seed treatments, foliar sprays, or in the culture medium is a promising way to increase the productivity of algae and the quantum of high-value metabolites they produce [96]. In short, using nutrient control, environmental revision, and biostimulant operation together is a complete way to get the stylish algal growth and secondary metabolite product for a variety of artificial and agrarian uses [81].

5 Harvesting and Dewatering Technologies of Algae

Algal harvesting and dewatering technologies are very energy-intensive because the algal culture is highly diluted and must be filtered extensively before it can be used to produce concentrated microalgal biomass paste of high value. These are the steps that convert the culture of algae that is spread very thin (less than 1 g of algae per liter) into a dense pulp and carry out the removal of the liquid in which the algae were grown. Microalgae are usually smaller than 30 microns, possess a natural negative charge, and reproduce very fast. This makes the procedure more complex and hence energy usage goes up [101].

Scientists have been trying to make the process easier through various physical, chemical, and even biological approaches (Fig. 4) [48]. Filtration is a primary strategy that resorts to the use of obstacles such as micro- or ultrafiltration membranes. Such membranes, that retain the cells, however, do not get rid of the chemicals that are present in them. From time to time during the operation of these membranes, deposits tend to accumulate, leading to partial or total blockage of the pores [58]. Updating with novel materials such as hydrophilic coatings or nanofiber sheets seems to be able to take good advantage of the improvement of flow rates and remain for a long time. In practical use, crossflow systems with polyvinylidene fluoride (PVDF) or polyethersulfone (PES) membranes can often recover more than 100% of the biomass by filtering water multiple times [47].

Centrifugation, a physical technique, is the process of separation of microalgae based on cell density. The process is suitable for cell recovery at high rates, but it is energy-consuming and can damage fragile algal cells due to high shear forces [30, 54]. Chemical flocculation is the process of using chemical substances like ferric chloride, aluminum sulfate, or biochemicals such as chitosan to coalesce the microalgae into

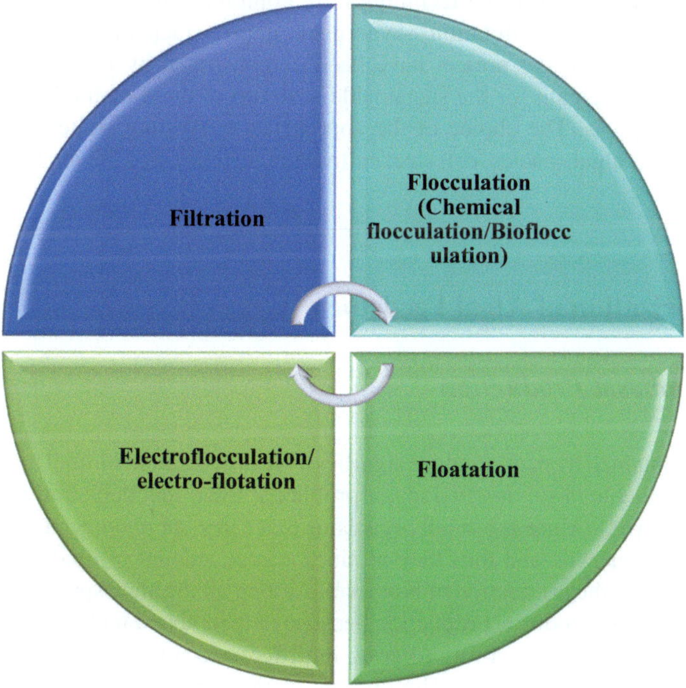

Fig. 4 Different types of algal harvesting methods

larger masses that can either be deposited or filtered more easily. The drawback of this method is that the chemical residues can be a source of contamination for the biomass, and additional purification is necessary [95]. By contrast, bioflocculation, using microbial or plant-based polymers, is a more environmentally friendly option. The use of bioflocculants such as *Moringa oleifera* seed extract and the assistance of fungal flocculation are some of the examples that provide results of more than 90%, and at the same time, they are nontoxic and do not cause environmental problems [60].

The flotation process involves the generation of fine air or gas bubbles. The bubbles without the algal cells get the algal cells to the surface for collection. Variants of this method include dissolved air flotation, dispersed air flotation, and buoy-bead flotation [57]. One of the advantages of these methods lies in their easiness and minimal chemical usage, while the recovery efficiencies usually surpass 90% [89]. Some of the recent energy-intensive but environmentally friendly methods for algal biomass harvesting include electroflocculation and electrocoagulation, which employ electric currents to break up algal suspensions and facilitate the formation of floc. These methods are free from chemical additives; however, they may require high energy and expensive electrodes to work. Magnetic harvesting, which is based on the use of nanoparticles like Fe_3O_4, has also become one of the most promising methods, as it allows for a high recovery rate with a short contact time and also for magnetic recycling.

Besides that, two or more techniques are combined; for example, flocculation followed by filtration or flotation—synergistic improvements in biomass yield and cost-effectiveness of the project have been demonstrated by these methods [71]. Challenges still remain for the sustainable scale-up of these methods despite technological progress. The choice of the appropriate harvesting strategy depends on factors such as algal species, cell size, cultivation medium, and the intended biomass application.

6 Valorization of Algal Feedstock

6.1 Biodiesel Production

Biodiesel obtained from microalgae signifies a third-generation biofuel that is considered quite reliable in solving two major problems: fossil fuel depletion and environmental harm [68]. Algae are small organisms that carry out photosynthesis and have a high rate of growth, and thus they produce a large amount of lipid, which makes them perfect for the production of biodiesel. Among all the algal species, microalgae are best suited for biodiesel production because of their abundant lipid profiles, rapid biomass production, and ability to grow in different conditions including contaminated water and saline water. The process of producing biodiesel-utilizing algae involves several steps that are interconnected: cultivation, harvesting of biomass, cell disruption, lipid extraction, and transesterification. The yield and quality of biodiesel are strongly dependent on the microalgal strain chosen and the conditions of cultivation. Parameters that are of great importance for biomass productivity and lipid storage are light intensity, time of exposure to light, temperature, pH, nutrients, and the degree of aeration. As an example, the lack of nitrogen or phosphorus can act as a trigger for lipid synthesis, especially the production of neutral lipids such as triacylglycerols (TAGs), which are the main materials for biodiesel. After the stage of cultivation, lipid-rich biomass undergoes a series of pretreatment operations to open up the rigid cell walls. These operations employ different energy sources such as sonication, microwave-assisted treatment, autoclaving, and bead milling for complete cell disruption. The most important point is that this process facilitates the release of intracellular lipids and hence the next step of extraction and conversion processes is done more efficiently.

Lipid is generally removed from the material using solvent-based methods such as the Bligh and Dyer, Folch, or Soxhlet techniques [32, 64]. The withdrawn lipids are later on carried through the process of transesterification which is a vital reaction where the triglycerides, in the presence of a catalyst, react with alcohol (usually methanol) to form fatty acid methyl esters (FAMEs), which are the components of biodiesel. This reaction may be accomplished through either of the two paths. The direct transesterification method is the utilization of wet biomass with no lipid extraction, thus, it is more time- and cost-efficient [75]. On the other hand, the indirect transesterification process involves initial lipid extraction followed by conversion and is usually capable of producing higher-purity biodiesel [59].

The role of catalysts in the transesterification process is very important. Homogeneous catalysts such as sulfuric acid and sodium hydroxide are agents with high activity that are commonly used, but they can also cause soap formation, which necessitates a downstream purification process. The heterogeneous catalysts, which are composed of metal oxides like CaO or ZnO, emphasize advantages such as the possibility of reuse and a minimum amount of soap production; however, they may require higher reaction temperatures [92]. Biocatalysts, such as enzymes (e.g., lipases) or catalysts of biological origin obtained from sources like eggshells or mollusk shells, are presented as green alternatives. They work at lower temperatures and thus have a lower environmental impact.

The use of nanoscale technologies has also contributed to the improvement of the transesterification procedure. Bionanocatalysts, which are formed through algae or plant extracts besides metal nanoparticles, have proven to be highly effective and reusable [85]. These achievements have dramatically propelled the biodiesel yield and green process. Despite advancements, biodiesel production from algae faces challenges including high cultivation and processing costs, scalability issues, and the need for integrated biorefinery approaches. Nonetheless, ongoing research continues to optimize each stage of the process, aiming to make algal biodiesel a commercially viable and environmentally sustainable alternative to fossil fuels [16].

Diverse studies have thus far explored several algal species that are ideal for the production of biodiesel with properties such as species growth, lipid content, and fatty acid type being used as a basis for identification. Microalgae, in particular,

Table 1 Biodiesel production from various algal species

Algae	Catalyst	Catalyst concentration	Temperature (°C)	Time (min)	Biodiesel yield (% dry wt)	Reference
C. vulgaris	NaOH	0.5	45	15	101 +_ 2.7	[18]
C. pyrenoidosa	H_2SO_4	1–3	90	30	86.74	[20]
Chlamydomonas sp.	NaOH	0.5	45	15	> 80	[18]
C. pyrenoidosa	H_2SO_4	5	90	60	46.54	[94]
Euglena sanguinea	$Ca(OCH_3)_2$	5	70	75	94.83	[5]
S. obliquus	WO_3/ZrO_2	5	50	180	94.58	[42]
Spirulina platensis	MgO/ZSM-5	3	75	60	92.1	[83]
Chlorella sp.	LiOH–pumice	20	80	180	47	[31]

have been thoroughly researched as third-generation biofuel feedstocks, due to their incredibly rapid growth, very high photosynthetic efficiency, and capacity to produce large amounts of neutral lipids, especially triacylglycerols (TAGs), under stressful conditions. Among others, *C, vulgaris* and *Chlorella pyrenoidosa* are the most widely examined species. According to the literature, it was found that the species *C. vulgaris* cultivated in photoheterotrophic mode by using wastewater gave biodiesel yields that were better than 84%, with the content of monounsaturated fatty acid being high, which is an important quality for biodiesel. In the absence of nitrogen, *C. pyrenoidosa* was able to produce as much as 64.9% lipid, and the yield of biodiesel was at the level of 46.5% [35]. Adaptability and high lipid productivity are characteristics of *S. obliquus* and *Scenedesmus abundans*, respectively. It has been demonstrated in various studies that lipid yields of over 60% can be achieved under heterotrophic conditions, and biodiesel yields in the range of 60–86% are possible, depending on the cultivation and processing strategies adopted [41, 90]. In an example of a closed raceway pond, *S. obliquus* gave 13.12 g/m^2/day biomass and 1.39 g/m^2/day lipid productivity [6]. This marine microalga is highly valued for its ability to accumulate lipids and achieve high fatty acid methyl ester (FAME) yields. *N. gaditana* was investigated under both direct and indirect causes of transesterification, and the maximum biodiesel yield obtained was 92.2% when using ethanol as a solvent and acid catalysts as a catalyst [53]. Table 1 shows the various other algal species utilized for biodiesel production.

6.2 Biogas Production

The major concept of generating biogas from algal biomass is anaerobic digestion (AD), which is one of the processes for producing energy from biomass. In this process, microorganisms break down the organic material without oxygen resulting in a biogas consisting mainly of methane (CH_4) and carbon dioxide (CO_2). Some microalgae including *Chlorella, Scenedesmus*, and *Nannochloropsis* are high in protein and lipids, but macroalgae such as *Ulva, Gracilaria*, and *Laminaria* are high in carbohydrates, especially in the form of complex polysaccharides like alginate and carrageenan.

The anaerobic digestion (AD) efficiency is often hindered by algal features such as complex cell wall structures and biochemical compositions. Pretreatment methods are used to open up the algal cell walls, make hydrolysis easier, and provide more available substrate. The techniques can be physical (e.g., ultrasound, thermal, autoclaving), chemical (e.g., alkaline or acid hydrolysis), biological (e.g., enzyme or microbial treatment), or in combination. Among these, physical pretreatments such as homogenization and ultrasound have shown significant promise. An example of this is that homogenization increased the methane yield by 189% at most, whereas ultrasound pretreatment increased yield by up to 164% [13, 19].

One of the factors influencing the methane yield is the algae species. Microalgae can give the highest methane yield of 540 mL CH_4/g volatile solids (VS) due to their high lipid and protein content, while carbohydrate-rich macroalgae such as *Ulva lactuca* provide yields of about 271 mL CH_4/g VS [4, 12]. Macroalgae have lower methane yields than microalgae, but they are much easier to grow in large quantities, particularly near the coasts, and they need very little fertilizer and other nutrients. Their availability during seasons also ensures the continuous provision of biomass for the production of biogas. One big disadvantage in the anaerobic digestion of microalgae is that the ammonia is emitted during the protein degradation which can be the result of methanogenic activity inhibition. In the same way, algae that are large in size can be the source of sulfur and halogenated substances which can become the cause of microbes' inability to digest them [19, 55]. Thus, readjusting the carbon-to-nitrogen ratio (C/N) and co-digestion with other organic wastes, for example, manure or food waste, are frequently adopted techniques for adjusting nutrient content and limiting negative effects.

6.3 *Bioethanol Production*

Algae, at both micro and macro levels, represent the third generation of feedstocks that are highly efficient for bioethanol production. They have fast growth rates, are high in carbohydrates, and require minimal resources. They do not need a lot of inputs such as fertilizers and water, which makes them a source of green energy [79]. This also makes them the best candidates for biofuel production. Unlike first-generation feedstocks that directly compete for food crops and second-generation lignocellulosic biomass that uses energy-intensive processing, algae offer a more sustainable and environmentally friendly route. A wide selection of algal species used in numerous studies proves that green macroalgae (e.g., *U. lactuca, Enteromorpha intestinalis*) and microalgae (e.g., *C. vulgaris, S. obliquus, Chlorococcum minutum*) have great performance in bioethanol yield when they are cultivated, pretreated, and fermented under optimized conditions [79, 84].

Green seaweeds, which are macroalgae, contain a lot of easily hydrolyzed carbohydrates that are the main source of energy for the cells and for the human diet in the form of cellulose and starch. As an example, *U. lactuca* exhibited the highest total carbohydrate content (62.15%) and cellulose content (14.03%) among several seaweeds examined, which means that it can easily be subjected to enzymatic hydrolysis followed by fermentation [37, 84]. If there is pretreatment with the cellulase enzyme, which is obtained from a fungus called *Cladosporium sphaerospermum, U. lactuca* can give ethanol that can be up to 0.47 g per gram of sugar with a theoretical conversion efficiency of 92.16% [93]. Likewise, *E. intestinalis* had a high cellulose content of 12.1%, and it was able to produce 1.63 g of ethanol from 6.32 g of sugar with an efficiency of 25.8% in separate hydrolysis and fermentation (SHF) and 77.4% in simultaneous saccharification and fermentation (SSF) using *Saccharomyces cerevisiae* [23].

Among microalgae, *C. vulgaris* is the most extensively researched species. A strain of *C. vulgaris* ESP-31 produced a staggering 98.11 g/L of ethanol under the optimal fermentation conditions [99]. The other strain *C. vulgaris* FSP-E obtained a yield of 43.9 g/L [27]. In the meantime, *C. sorokiniana* generated 422.44 mg/L in a medium enriched with nitrate, and the strain AK-1, on the other hand, gave 4.2 g/L when grown in 10% unsterilized swine wastewater [52]. Such outcomes emphasize the strain-specific dependence on culture media and fermentation optimization. One of the species from the genus *Chlorococcum*, such as *C. minutum*, has been proven to have bioethanol capabilities comparable to those of various other microalgae, with a 31.2–46.97 g/L range of bioethanol depending on the culture medium (TAP or Cr3 and Cm3 media) used [97, 98]. Correspondingly, some other strains such as *Chlorella* sp. ABC-001 and *Codium tomentosum* reached the ethanol titer of 2.8 g/L and 4 g/L, respectively, while, interestingly, those performances were lower than that of *Chlorella* or *Scenedesmus* strains [39, 88]. When supplemented with mepiquat chloride under wastewater conditions, *Scenedesmus acuminatus* gave 20.32 g/L of ethanol while nutrient-starved conditions allowed it to produce 17.2 mg/L/day, indicating the influence of external stressors on carbohydrate accumulation [14, 77].

Several factors such as fermentation process type, microbial strain, and pretreatment have the most significant impact on bioethanol outputs. Cellulases and amylases are the enzymes that help degrade polysaccharides in algal biomass during enzymatic hydrolysis and then fermentation is carried out with yeast cells like *S. cerevisiae* [97]. The fact that algal hydrolysates generally consist of a mixture of sugars (glucose, galactose, xylose, and mannitol) has led to an increased number of coculture or genetically modified organisms that can ferment both hexoses and pentoses. For instance, the hydrolysate of *Laminaria japonica* by *S. cerevisiae* generated 14 g/L of ethanol, and *Zymobacter palmae* as well as *E. coli* strains are reported to utilize mannitol and glucuronic acids, respectively, as sources for fermentation [46].

6.4 Other Value-Added Products

Besides biofuels such as biodiesel, bioethanol, and biogas, microalgae are a plentiful source of high-value-added products like pigments and bioplastics, which contribute to sustainable industrial applications. *A. platensis* (previously known as *Spirulina*) is a cyanobacterium that is extensively grown for its phycocyanin content, a blue phycobiliprotein that is used as a natural colorant and antioxidant in food and cosmetics. The phycocyanin in *A. platensis* may go up to 20% of its dry weight if the conditions are favorable [50]. In a similar manner, *Haematococcus pluvialis* is a leading source of astaxanthin, a powerful carotenoid that finds application in aquaculture, health food industry, and cosmetics. In stressful conditions (high light intensity, nutrient deprivation, etc.), *H. pluvialis* can stockpile astaxanthin to 4% of its dry weight [80, 102]. *D. salina*, a halotolerant green microalga, is also a strain that is a matter of industrial importance for β-carotene production with yields reportedly ranging between 8 and 14% of dry biomass [43, 44]. The freshwater microalga *Scenedesmus almeriensis* synthesizes lutein, a pigment of great interest due to its crucial role in eye health supplements, and the highest production corresponds to nitrogen limitation conditions, with up to 0.5% lutein content based on the dry weight of the cells [65]. On the other hand, *Porphyridium* spp., a species of red microalgae, is notable for phycoerythrin production, fluorescent-based applications, and as a food colorant. Nitrogen-based heterotrophic growth can yield 5–8% of phycoerythrin out of the total biomass [50]. The mentioned pigment quantities strongly determine that culture parameters such as light intensity, nutrient limitation, and stress induction play a significant role in pigment biosynthesis.

Microalgae including *C. vulgaris, S. platensis, Synechococcus* spp., and *Aulosira fertilissima* are the organisms on which scientists concentrate their research on the synthesis of intracellular biopolymers like polyhydroxyalkanoates (PHAs) and polyhydroxybutyrate (PHB) under nutrient-deprived conditions. PHAs are biodegradable polyesters that are compatible with the medical sector, agricultural sector, and packaging industry. Among cyanobacteria, *A. fertilissima* reveals great PHB buildup after feeding with acetate and under the condition of phosphate deprivation that was measured to be as high as 77% of dry cell weight. Similarly, *Synechococcus* sp. MA19 is also very capable, thereby producing around 55% PHB in phosphate-starved, autotrophic conditions. *Nostoc muscorum* can store as much as 43% PHB when it is under incubation in the dark with acetate [17, 87]. *Chlorella* sp., a species extensively utilized in food supplements, has also been found to possess qualities in PHB production, although generally, the yields are quite low (~20–30%) depending on the strain and the way of cultivation. *Spirulina maxima* and *Spirulina subsalsa* have comparably lower PHB yields (2–7%) but can be considered good species due to their resistance and capability of growing in seawater or alkaline media.

7 Conclusion

Algae-derived third-generation (3G) biorefineries are essentially a revolution in the sustainable energy market as well as in the world of bioproducts. By strategically combining the use of advanced cultivation methods, smart monitoring systems, and the efficient management of harvesting and processing, algal systems can break through current scalability and financial limitations. The process of selecting and upgrading productive algal strains that go hand in hand with the developments in photobioreactor design and the use of industrial effluents or wastewater goes a long way in giving the environmental and economic side of algal biorefineries a big boost. Further, the deep employment of algal matter through the multiproduct valorization—such as biofuels and high-value bioactives—leads to the zero-waste concept being put into practice and to the circular economy. The future of 3G biorefineries depends heavily on the interplay of disciplines like biotechnology, automation, and environmental engineering, which will take the process of turning algae into biofuel from a phase of potential to one of full performance. The discussed progressions are a solid base for sustainable industrial systems, and they make algae the main player in the future bioeconomy.

Acknowledgements Arumugam A greatly acknowledges the financial support from the DST/SERB, India (CRG/2022/003877).

References

1. Abdel-Raouf N, Al-Homaidan AA, Ibraheem IBM (2012) Microalgae and wastewater treatment. Saudi J Biol Sci 19:257–275. https://doi.org/10.1016/j.sjbs.2012.04.005
2. Abdur Razzak S, Bahar K, Islam KMO et al (2024) Microalgae cultivation in photobioreactors: sustainable solutions for a greener future. Green Chem Eng. https://doi.org/10.1016/j.gce.2023.10.004
3. Ahmad I, Abdullah N, Koji I et al (2022) Role of algae in built environment and green cities: a holistic approach towards sustainability. Int J Built Environ Sustain 9:69–80. https://doi.org/10.11113/ijbes.v9.n2-3.1039
4. Álvarez X, Arévalo O, Salvador M et al (2020) Cyanobacterial biomass produced in the wastewater of the dairy industry and its evaluation in anaerobic co-digestion with cattle manure for enhanced methane production. Processes 8:1290. https://doi.org/10.3390/pr8101290
5. Arunachalam Sivagurulingam AP, Sivanandi P, Pandian S, et al (2019) Optimization and kinetic studies on biodiesel production from microalgae (*Euglena sanguinea*) using calcium methoxide as catalyst. Energy Sources, Part A Recover Util Environ Eff 41:1497–1507. https://doi.org/10.1080/15567036.2018.1549124
6. Bagchi SK, Patnaik R, Sonkar S et al (2019) Qualitative biodiesel production from a locally isolated chlorophycean microalga *Scenedesmus obliquus* (Turpin) Kützing GA 45 under closed raceway pond cultivation. Renew Energy 139:976–987. https://doi.org/10.1016/j.renene.2019.02.115
7. Bai X, Acharya K (2017) Algae-mediated removal of selected pharmaceutical and personal care products (PPCPs) from Lake Mead water. Sci Total Environ 581–582:734–740. https://doi.org/10.1016/j.scitotenv.2016.12.192

8. Banerjee S, Dasgupta S, Atta A et al (2023) Flow rate optimization in a flat-panel photobioreactor for the cultivation of microalgae for mitigating waste gas. Water 15:2824. https://doi.org/10.3390/w15152824
9. Bhatt P, Bhandari G, Turco RF et al (2022) Algae in wastewater treatment, mechanism, and application of biomass for production of value-added product. Environ Pollut 309:119688. https://doi.org/10.1016/j.envpol.2022.119688
10. Bilal M, Rasheed T, Sosa-Hernández JE et al (2018) Biosorption: an interplay between marine algae and potentially toxic elements—A review. Mar Drugs 16:65. https://doi.org/10.3390/md16020065
11. Brennan L, Owende P (2010) Biofuels from microalgae—A review of technologies for production, processing, and extractions of biofuels and co-products. Renew Sustain Energy Rev 14:557–577. https://doi.org/10.1016/j.rser.2009.10.009
12. Bruhn A, Dahl J, Nielsen HB et al (2011) Bioenergy potential of *Ulva lactuca*: Biomass yield, methane production and combustion. Bioresour Technol 102:2595–2604. https://doi.org/10.1016/j.biortech.2010.10.010
13. Carpenter J, Badve M, Rajoriya S et al (2017) Hydrodynamic cavitation: an emerging technology for the intensification of various chemical and physical processes in a chemical process industry. Rev Chem Eng 33. https://doi.org/10.1515/revce-2016-0032
14. Chandra N, Shukla P, Mallick N (2020) Role of cultural variables in augmenting carbohydrate accumulation in the green microalga *Scenedesmus acuminatus* for bioethanol production. Biocatal Agric Biotechnol 26:101632. https://doi.org/10.1016/j.bcab.2020.101632
15. Chandra R, Iqbal HMN, Vishal G et al (2019) Algal biorefinery: a sustainable approach to valorize algal-based biomass towards multiple product recovery. Bioresour Technol 278:346–359. https://doi.org/10.1016/j.biortech.2019.01.104
16. Chaudhary R, Nawaz K, Khan AK et al (2020) An overview of the algae-mediated biosynthesis of nanoparticles and their biomedical applications. Biomolecules 10:1498. https://doi.org/10.3390/biom10111498
17. Cheah WY, Er AC, Aiyub K et al (2023) Current status and perspectives of algae-based bioplastics: a reviewed potential for sustainability. Algal Res 71:103078. https://doi.org/10.1016/j.algal.2023.103078
18. Chen C-L, Huang C-C, Ho K-C et al (2015) Biodiesel production from wet microalgae feedstock using sequential wet extraction/transesterification and direct transesterification processes. Bioresour Technol 194:179–186. https://doi.org/10.1016/j.biortech.2015.07.021
19. Chen H, Zhou D, Luo G et al (2015) Macroalgae for biofuels production: progress and perspectives. Renew Sustain Energy Rev 47:427–437. https://doi.org/10.1016/j.rser.2015.03.086
20. Cheng J, Huang R, Li T et al (2014) Biodiesel from wet microalgae: Extraction with hexane after the microwave-assisted transesterification of lipids. Bioresour Technol 170:69–75. https://doi.org/10.1016/j.biortech.2014.07.089
21. Chin-On R, de Boer M, van de Voort C et al (2024) Outdoor cultivation of Picochlorum sp. in a novel V-shaped photobioreactor on the Caribbean island Bonaire. Front Bioeng Biotechnol 12. https://doi.org/10.3389/fbioe.2024.1347291
22. Chin-On RC, Barbosa MJ, Wijffels RH, Janssen M (2022) A novel V-shaped photobioreactor design for microalgae cultivation at low latitudes: Modelling biomass productivities of *Chlorella sorokiniana* on Bonaire. Chem Eng J 449:137793. https://doi.org/10.1016/j.cej.2022.137793
23. Cho Y, Kim M-J, Kim S-K (2013) Ethanol production from seaweed, *Enteromorpha intestinalis*, by Separate Hydrolysis and Fermentation (SHF) and simultaneous saccharification and fermentation (SSF) with *Saccharomyces cerevisiae*. KSBB J 28:366–371. https://doi.org/10.7841/ksbbj.2013.28.6.366
24. Chou H-H, Su H-Y, Song X-D et al (2019) Isolation and characterization of Chlorella sp. mutants with enhanced thermo- and CO_2 tolerances for CO_2 sequestration and utilization of flue gases. Biotechnol Biofuels 12:251. https://doi.org/10.1186/s13068-019-1590-9

25. Chovanček E, Salazar J, Şirin S, Allahverdiyeva Y (2023) Microalgae from Nordic collections demonstrate biostimulant effect by enhancing plant growth and photosynthetic performance. Physiol Plant 175. https://doi.org/10.1111/ppl.13911
26. Chowdury KH, Nahar N, Deb UK (2020) The growth factors involved in microalgae cultivation for biofuel production: a review. Comput Water, Energy, Environ Eng 09:185–215. https://doi.org/10.4236/cweee.2020.94012
27. Condor BE, de Luna MDG, Chang Y-H et al (2022) Bioethanol production from microalgae biomass at high-solids loadings. Bioresour Technol 363:128002. https://doi.org/10.1016/j.biortech.2022.128002
28. Costa JAV, de Morais MG (2014) An open pond system for microalgal cultivation. In: Biofuels from Algae. Elsevier, pp 1–22
29. da Silva TL, Reis A (2015) Scale-up problems for the large scale production of algae. Algal biorefinery: an integrated approach. Springer International Publishing, Cham, pp 125–149
30. Dassey AJ, Theegala CS (2013) Harvesting economics and strategies using centrifugation for cost effective separation of microalgae cells for biodiesel applications. Bioresour Technol 128:241–245. https://doi.org/10.1016/j.biortech.2012.10.061
31. de Luna MDG, Doliente LMT, Ido AL, Chung T-W (2017) In situ transesterification of Chlorella sp. microalgae using LiOH-pumice catalyst. J Environ Chem Eng 5:2830–2835. https://doi.org/10.1016/j.jece.2017.05.006
32. Dhup S, Dhawan V (2014) Effect of nitrogen concentration on lipid productivity and fatty acid composition of Monoraphidium sp. Bioresour Technol 152:572–575. https://doi.org/10.1016/j.biortech.2013.11.068
33. Dormido R, Sánchez J, Duro N et al (2014) An interactive tool for outdoor computer controlled cultivation of microalgae in a tubular photobioreactor system. Sensors 14:4466–4483. https://doi.org/10.3390/s140304466
34. El-Araby R (2024) Biofuel production: exploring renewable energy solutions for a greener future. Biotechnol Biofuels Bioprod 17:129. https://doi.org/10.1186/s13068-024-02571-9
35. Feng P, Deng Z, Hu Z et al (2014) Characterization of Chlorococcum pamirum as a potential biodiesel feedstock. Bioresour Technol 162:115–122. https://doi.org/10.1016/j.biortech.2014.03.076
36. Fernández I, Acién FG, Berenguel M, Guzmán JL (2014) First principles model of a tubular photobioreactor for microalgal production. Ind Eng Chem Res 53:11121–11136. https://doi.org/10.1021/ie501438r
37. Fleurence J (1999) Seaweed proteins. Trends Food Sci Technol 10:25–28. https://doi.org/10.1016/S0924-2244(99)00015-1
38. Gabrielyan DA, Sinetova MA, Gabel BV et al (2022) Cultivation of *Chlorella sorokiniana* IPPAS C-1 in flat-panel Photobioreactors: from a laboratory to a pilot scale. Life 12:1309. https://doi.org/10.3390/life12091309
39. Gengiah K, Moses GLP, Baskar G (2024) Bioethanol production from *Codium tomentosum* residue. Energy Sources, Part A Recover Util Environ Eff 46:8912–8921. https://doi.org/10.1080/15567036.2020.1771481
40. Goswami L, Kayalvizhi R, Dikshit PK et al (2022) A critical review on prospects of biorefinery products from second and third generation biomasses. Chem Eng J 448:137677. https://doi.org/10.1016/j.cej.2022.137677
41. Guldhe A, Moura CVR, Singh P et al (2017a) Conversion of microalgal lipids to biodiesel using chromium-aluminum mixed oxide as a heterogeneous solid acid catalyst. Renew Energy 105:175–182. https://doi.org/10.1016/j.renene.2016.12.053
42. Guldhe A, Singh P, Ansari FA et al (2017b) Biodiesel synthesis from microalgal lipids using tungstated zirconia as a heterogeneous acid catalyst and its comparison with homogeneous acid and enzyme catalysts. Fuel 187:180–188. https://doi.org/10.1016/j.fuel.2016.09.053
43. Hejazi MA, de Lamarliere C, Rocha JMS et al (2002) Selective extraction of carotenoids from the microalga *Dunaliella salina* with retention of viability. Biotechnol Bioeng 79:29–36. https://doi.org/10.1002/bit.10270

44. Herrero M, Jaime L, Martín-Álvarez PJ et al (2006) Optimization of the extraction of antioxidants from *Dunaliella salina* microalga by pressurized liquids. J Agric Food Chem 54:5597–5603. https://doi.org/10.1021/jf060546q
45. Hincapie E, Stuart BJ (2015) Design, construction, and validation of an internally lit air-lift photobioreactor for growing algae. Front Energy Res 2. https://doi.org/10.3389/fenrg.2014.00065
46. Horn SJ, Aasen IM, Østgaard K (2000) Production of ethanol from mannitol by *Zymobacter palmae*. J Ind Microbiol Biotechnol 24:51–57. https://doi.org/10.1038/sj.jim.2900771
47. Hwang T, Kotte MR, Han J-I et al (2015) Microalgae recovery by ultrafiltration using novel fouling-resistant PVDF membranes with in situ PEGylated polyethyleneimine particles. Water Res 73:181–192. https://doi.org/10.1016/j.watres.2014.12.002
48. Hwang T, Park S-J, Oh Y-K et al (2013) Harvesting of chlorella sp. KR-1 using a cross-flow membrane filtration system equipped with an anti-fouling membrane. Bioresour Technol 139:379–382. https://doi.org/10.1016/j.biortech.2013.03.149
49. John RP, Anisha GS, Nampoothiri KM, Pandey A (2011) Micro and macroalgal biomass: a renewable source for bioethanol. Bioresour Technol 102:186–193. https://doi.org/10.1016/j.biortech.2010.06.139
50. Juin C, Chérouvrier J-R, Thiéry V et al (2015) Microwave-assisted extraction of phycobiliproteins from porphyridium purpureum. Appl Biochem Biotechnol 175:1–15. https://doi.org/10.1007/s12010-014-1250-2
51. Kao C-Y, Chen T-Y, Chang Y-B et al (2014) Utilization of carbon dioxide in industrial flue gases for the cultivation of microalga Chlorella sp. Bioresour Technol 166:485–493. https://doi.org/10.1016/j.biortech.2014.05.094
52. Kaur A, Taggar MS, Kalia A, Singh M (2022) Nitrate-Induced carbohydrate accumulation in *Chlorella sorokiniana* and its potential for ethanol production. BioEnergy Res 15:253–263. https://doi.org/10.1007/s12155-021-10292-2
53. Kim B, Chang YK, Lee JW (2017) Efficient solvothermal wet in situ transesterification of *Nannochloropsis gaditana* for biodiesel production. Bioprocess Biosyst Eng 40:723–730. https://doi.org/10.1007/s00449-017-1738-6
54. Kim K, Shin H, Moon M et al (2015) Evaluation of various harvesting methods for high-density microalgae, Aurantiochytrium sp. KRS101. Bioresour Technol 198:828–835. https://doi.org/10.1016/j.biortech.2015.09.103
55. Kumar BR, Mathimani T, Sudhakar MP et al (2021) A state of the art review on the cultivation of algae for energy and other valuable products: Application, challenges, and opportunities. Renew Sustain Energy Rev 138:110649. https://doi.org/10.1016/j.rser.2020.110649
56. Kumar K, Banerjee D, Das D (2014) Carbon dioxide sequestration from industrial flue gas by *Chlorella sorokiniana*. Bioresour Technol 152:225–233. https://doi.org/10.1016/j.biortech.2013.10.098
57. Kurniawati HA, Ismadji S, Liu JC (2014) Microalgae harvesting by flotation using natural saponin and chitosan. Bioresour Technol 166:429–434. https://doi.org/10.1016/j.biortech.2014.05.079
58. Lee D-J, Liao G-Y, Chang Y-R, Chang J-S (2012) Coagulation-membrane filtration of Chlorella vulgaris. Bioresour Technol 108:184–189. https://doi.org/10.1016/j.biortech.2011.12.098
59. Lee JWY, Chia WY, Ong W-J et al (2022) Advances in catalytic transesterification routes for biodiesel production using microalgae. Sustain Energy Technol Assessments 52:102336. https://doi.org/10.1016/j.seta.2022.102336
60. Lee S-M, Choi H-J (2015) Harvesting of microalgae species using Mg–sericite flocculant. Bioprocess Biosyst Eng 38:2323–2330. https://doi.org/10.1007/s00449-015-1466-8
61. Li C-T, Eng R, Zuniga C, et al (2023a) Optimization of nutrient utilization efficiency and productivity for algal cultures under light and dark cycles using genome-scale model process control. NPJ Syst Biol Appl 9:7. https://doi.org/10.1038/s41540-022-00260-7
62. Li S, Chang H, Zhang S, Ho S-H (2023) Production of sustainable biofuels from microalgae with CO_2 bio-sequestration and life cycle assessment. Environ Res 227:115730. https://doi.org/10.1016/j.envres.2023.115730

63. Lindblad P, Fuente D, Borbe F et al (2019) CyanoFactory, a European consortium to develop technologies needed to advance cyanobacteria as chassis for production of chemicals and fuels. Algal Res 41:101510. https://doi.org/10.1016/j.algal.2019.101510
64. Liu Z, Rochfort S, Cocks BG (2016) Optimization of a single phase method for lipid extraction from milk. J Chromatogr A 1458:145–149. https://doi.org/10.1016/j.chroma.2016.06.055
65. Macías-Sánchez MD, Fernandez-Sevilla JM, Fernández FGA et al (2010) Supercritical fluid extraction of carotenoids from *Scenedesmus almeriensis*. Food Chem 123:928–935. https://doi.org/10.1016/j.foodchem.2010.04.076
66. Maltsev Y, Maltseva K, Kulikovskiy M, Maltseva S (2021) Influence of light conditions on microalgae growth and content of lipids, carotenoids, and fatty acid composition. Biology (Basel) 10:1060. https://doi.org/10.3390/biology10101060
67. Markou G, Nerantzis E (2013) Microalgae for high-value compounds and biofuels production: a review with focus on cultivation under stress conditions. Biotechnol Adv 31:1532–1542. https://doi.org/10.1016/j.biotechadv.2013.07.011
68. Mata TM, Martins AA, Caetano NS (2010) Microalgae for biodiesel production and other applications: a review. Renew Sustain Energy Rev 14:217–232. https://doi.org/10.1016/j.rser.2009.07.020
69. Michalak I, Chojnacka K (2015) Algae as production systems of bioactive compounds. Eng Life Sci 15:160–176. https://doi.org/10.1002/elsc.201400191
70. Mohanty B, Majedi SM, Pavagadhi S et al (2022) Effects of light and temperature on the metabolic profiling of two habitat-dependent bloom-forming cyanobacteria. Metabolites 12:406. https://doi.org/10.3390/metabo12050406
71. Mohseni F, Moosavi Zenooz A (2022) Flocculation of chlorella vulgaris with alum and pH adjustment. Biotechnol Appl Biochem 69:1112–1120. https://doi.org/10.1002/bab.2182
72. Molazadeh M, Ahmadzadeh H, Pourianfar HR et al (2019) The use of microalgae for coupling wastewater treatment with CO_2 biofixation. Front Bioeng Biotechnol 7. https://doi.org/10.3389/fbioe.2019.00042
73. Molitor HR, Schnoor JL (2020) Using simulated flue gas to rapidly grow nutritious microalgae with enhanced settleability. ACS Omega 5:27269–27277. https://doi.org/10.1021/acsomega.0c03492
74. Neviani M, Bagnerini P, Paladino O (2021) Gas bubble dynamics in airlift photo-bioreactors for microalgae cultivation by level set methods. Fuel 292:120402. https://doi.org/10.1016/j.fuel.2021.120402
75. Olivares-Carrillo P, Quesada-Medina J (2011) Synthesis of biodiesel from soybean oil using supercritical methanol in a one-step catalyst-free process in batch reactor. J Supercrit Fluids 58:378–384. https://doi.org/10.1016/j.supflu.2011.07.011
76. Olivieri G, Salatino P, Marzocchella A (2014) Advances in photobioreactors for intensive microalgal production: configurations, operating strategies and applications. J Chem Technol Biotechnol 89:178–195. https://doi.org/10.1002/jctb.4218
77. Onay M, Aladag E (2022) Production and use of *Scenedesmus acuminatus* biomass in synthetic municipal wastewater for integrated biorefineries. Environ Sci Pollut Res 30:15808–15820. https://doi.org/10.1007/s11356-022-23332-x
78. Paliwal C, Mitra M, Bhayani K et al (2017) Abiotic stresses as tools for metabolites in microalgae. Bioresour Technol 244:1216–1226. https://doi.org/10.1016/j.biortech.2017.05.058
79. Pandit C, Pandit S, Kuhad RC et al (2024) Microalgal bioethanol production for sustainable development: current status and future prospects. Indian J Microbiol. https://doi.org/10.1007/s12088-024-01364-w
80. Panis G, Carreon JR (2016) Commercial astaxanthin production derived by green alga *Haematococcus pluvialis* : a microalgae process model and a techno-economic assessment all through production line. Algal Res 18:175–190. https://doi.org/10.1016/j.algal.2016.06.007
81. Parmar P, Kumar R, Neha Y, Srivatsan V (2023) Microalgae as next generation plant growth additives: functions, applications, challenges and circular bioeconomy based solutions. Front Plant Sci 14:. https://doi.org/10.3389/fpls.2023.1073546

82. Penloglou G, Pavlou A, Kiparissides C (2024) Recent advancements in photo-bioreactors for microalgae cultivation: a brief overview. Processes 12:1104. https://doi.org/10.3390/pr12061104
83. Qu S, Chen C, Guo M et al (2020) Synthesis of MgO/ZSM-5 catalyst and optimization of process parameters for clean production of biodiesel from *Spirulina platensis*. J Clean Prod 276:123382. https://doi.org/10.1016/j.jclepro.2020.123382
84. Ramachandra TV, Hebbale D (2020) Bioethanol from macroalgae: prospects and challenges. Renew Sustain Energy Rev 117:109479. https://doi.org/10.1016/j.rser.2019.109479
85. Sakthi Vignesh N, Vimali E, Sangeetha R, et al (2020) Sustainable biofuel from microalgae: application of lignocellulosic wastes and bio-iron nanoparticle for biodiesel production. Fuel 278:118326. https://doi.org/10.1016/j.fuel.2020.118326
86. Sarker NK, Kaparaju P (2024) Microalgal bioeconomy: a green economy approach towards achieving sustainable development goals. Sustainability 16:11218. https://doi.org/10.3390/su162411218
87. Senko O, Stepanov N, Maslova O, Efremenko E (2023) Transformation of enzymatic hydrolysates of chlorella-fungus mixed biomass into Poly(hydroxyalkanoates). Catalysts 13:118. https://doi.org/10.3390/catal13010118
88. Seon G, Kim HS, Cho JM et al (2020) Effect of post-treatment process of microalgal hydrolysate on bioethanol production. Sci Rep 10:16698. https://doi.org/10.1038/s41598-020-73816-4
89. Shen Z, Li Y, Wen H et al (2018) Investigation on the role of surfactants in bubble-algae interaction in flotation harvesting of Chlorella vulgaris. Sci Rep 8:3303. https://doi.org/10.1038/s41598-018-21629-x
90. Sivaramakrishnan R, Incharoensakdi A (2017) Production of methyl ester from two microalgae by two-step transesterification and direct transesterification. Environ Sci Pollut Res 24:4950–4963. https://doi.org/10.1007/s11356-016-8217-5
91. Sreenikethanam A, Raj S, J RB et al (2022) Genetic engineering of microalgae for secondary metabolite production: recent developments, challenges, and future prospects. Front Bioeng Biotechnol 10. https://doi.org/10.3389/fbioe.2022.836056
92. Tariq M, Ali S, Khalid N (2012) Activity of homogeneous and heterogeneous catalysts, spectroscopic and chromatographic characterization of biodiesel: a review. Renew Sustain Energy Rev 16:6303–6316. https://doi.org/10.1016/j.rser.2012.07.005
93. Trivedi N, Gupta V, Kumar M et al (2011) An alkali-halotolerant cellulase from Bacillus flexus isolated from green seaweed *Ulva lactuca*. Carbohydr Polym 83:891–897. https://doi.org/10.1016/j.carbpol.2010.08.069
94. Umamaheswari J, Kavitha MS, Shanthakumar S (2020) Outdoor cultivation of *Chlorella pyrenoidosa* in paddy-soaked wastewater and a feasibility study on biodiesel production from wet algal biomass through in-situ transesterification. Biomass Bioenerg 143:105853. https://doi.org/10.1016/j.biombioe.2020.105853
95. Vandamme D, Pohl PI, Beuckels A et al (2015) Alkaline flocculation of Phaeodactylum tricornutum induced by brucite and calcite. Bioresour Technol 196:656–661. https://doi.org/10.1016/j.biortech.2015.08.042
96. Vangenechten B, De Coninck B, Ceusters J (2025) How to improve the potential of microalgal biostimulants for abiotic stress mitigation in plants? Front Plant Sci 16. https://doi.org/10.3389/fpls.2025.1568423
97. Varaprasad D, Narasimham D, Paramesh K et al (2021) Improvement of ethanol production using green alga *Chlorococcum minutum*. Environ Technol 42:1383–1391. https://doi.org/10.1080/09593330.2019.1669719
98. Varaprasad D, Raghavendra P, Sudha NR et al (2022) Bioethanol production from green alga *Chlorococcum minutum* through reduced graphene oxide-supported platinum-ruthenium (Pt-Ru/RGO) nanoparticles. BioEnergy Res 15:280–288. https://doi.org/10.1007/s12155-021-10282-4
99. Yu KL, Chen W-H, Sheen H-K et al (2020) Bioethanol production from acid pretreated microalgal hydrolysate using microwave-assisted heating wet torrefaction. Fuel 279:118435. https://doi.org/10.1016/j.fuel.2020.118435

100. Zakova T, Jirout T, Kratky L, Belohlav V (2019) Hydrodynamics as a tool to remove biofilm in tubular photobioreactor. Chem Eng Trans 76:451–456. https://doi.org/10.3303/CET1976076
101. Zamalloa C, Vulsteke E, Albrecht J, Verstraete W (2011) The techno-economic potential of renewable energy through the anaerobic digestion of microalgae. Bioresour Technol 102:1149–1158. https://doi.org/10.1016/j.biortech.2010.09.017
102. Zou T-B, Jia Q, Li H-W et al (2013) Response surface methodology for ultrasound-assisted extraction of Astaxanthin from *Haematococcus pluvialis*. Mar Drugs 11:1644–1655. https://doi.org/10.3390/md11051644

Process Parameters Governing Effective Biomass Pretreatment and Enzyme-Mediated Oligosaccharide Recovery

Isha Bodhe, Chandamita Goswami, and G. Velvizhi

Abstract The circular bioeconomy uses lignocellulosic biomass (LCB) as a renewable feedstock to meet rising global energy demand and reduce fossil fuel emissions. The valorization of LCB, primarily cellulose, hemicellulose, and lignin, presents challenges due to its recalcitrant composition, necessitating pretreatment techniques. Oligosaccharides can be generated from partially degrading polysaccharides like cellulose and hemicellulose, and their selective recovery has drawn attention due to their commercial value and industrial importance. Enzymatic hydrolysis is ideal for this application due to its moderate conditions, excellent selectivity, and low byproduct generation. This chapter discusses several pretreatment methods, like physical, chemical, physicochemical, and biological treatments, that are used to break down LCB and make it more accessible for downstream processing. The chapter focuses on using the enzymatic hydrolysis of pretreated biomass to extract functional oligosaccharides (FOS), emphasizing the advantages of enzymatic methods over conventional techniques in terms of selectivity, efficiency, and mild conditions. This chapter discusses the processes of major enzymes (e.g., cellulases, hemicellulases, pectinases, amylases, inulinases) in the synthesis of xylo-, cello-, and fructooligosaccharides. The effects of temperature, pH, residence time, catalyst concentration, and solid-to-liquid ratio on pretreatment efficacy are explored. Consolidated bioprocessing (CBP) and simultaneous saccharification and fermentation (SSF) are discussed in the chapter to increase sugar yields and production efficiency. Recovered oligosaccharides are economically important since they are utilized in food, medicines, cosmetics, animal feed, agrochemicals, etc. The chapter concludes with biomass valorization's economic and environmental impacts in a sustainable bioeconomy.

Keywords Lignocellulosic biomass · Pretreatment · Enzymatic hydrolysis · Oligosaccharides · Biorefinery · Circular bioeconomy

---

I. Bodhe
School of Biosciences and Technology, Vellore Institute of Technology, Vellore, India

I. Bodhe · C. Goswami · G. Velvizhi (✉)
CO2 Research and Green Technologies Centre, Vellore Institute of Technology, Vellore, India
e-mail: velvizhi.g@vit.ac.in

© The Author(s), under exclusive license to Springer Nature Singapore Pte Ltd. 2025

A. K. Chandel and S. Sundaramurthy (eds.), *Biorefineries in Circular Bioeconomy*, Green Energy and Technology, https://doi.org/10.1007/978-981-95-3913-0_4

1 Introduction

Global energy demand is projected to rise significantly, with estimates indicating a 28% increase by 2040 [1]. This surge is primarily driven by population growth and economic development, especially in non-Organization for Economic Co-operation and Development (OECD) countries. Currently, fossil fuels—coal, oil, and natural gas—supply over 80% of the world's energy needs [2]. However, their limited reserves, high carbon emissions, and environmental impacts, such as global warming and air pollution, underscore the urgent need for alternative energy sources. Renewable energy sources, including wind, solar, hydro, and biomass, are gaining traction, accounting for approximately 23.7% of global energy consumption [2]. The transition to renewables aligns with the principles of a circular economy (CE), emphasizing the recovery and reuse of bio-based materials over the traditional "take, make, dispose" model [3, 4]. Lignocellulosic biomass (LCB), derived from plant-based materials like agricultural residues and forestry waste, offers a promising renewable energy source. With an estimated annual availability of 200 billion tons, LCB is abundant, nonedible, and cost-effective [5]. Its utilization can mitigate environmental issues associated with fossil fuels and contribute to energy security. First-generation (1G) biofuels, produced from edible crops, have raised concerns over food security and land use. In response, second-generation (2G) biofuels utilize nonedible lignocellulosic feedstocks, addressing the food-versus-fuel debate. However, the complex and recalcitrant structure of LCB, composed mainly of cellulose, hemicellulose, and lignin, poses challenges for efficient biofuel production. To overcome these challenges, various pretreatment methods have been developed to break down the rigid structure of LCB, enhancing its digestibility for enzymatic hydrolysis and fermentation [7]. These methods include:

- Physical pretreatments: Techniques like milling and extrusion reduce particle size, increasing surface area for enzymatic action.
- Chemical pretreatments: Alkaline treatments, using agents like sodium hydroxide, effectively remove lignin and hemicellulose, improving cellulose accessibility.
- Biological pretreatments: Microorganisms or enzymes are employed to degrade lignin and hemicellulose under mild conditions, offering an environmentally friendly approach.

The economic viability of LCB-based biofuels depends on pretreatment technology advances. Efficient pretreatment increases sugar yields, lower processing costs, and eliminates inhibitory byproducts. Integrating pretreatment with downstream processes streamlines biofuel production, enabling biorefineries to scale up [8].

Functional oligosaccharides (FOS) are nondigestible carbohydrates recognized for their various bioactivities, such as antitumor, antioxidant, anti-inflammatory, anticoagulant, radioprotective, and immunomodulatory properties [9, 10]. These features have attracted considerable interest in food science and biological studies. FOS are predominantly generated via microbial fermentation, chemical degradation, and enzymatic hydrolysis. Enzymatic approaches are favored for their specificity, efficiency, and eco-friendliness, frequently employing enzymes such as β-galactosidase and fructosyltransferase to produce molecules like galactooligosaccharides (GOS) and fructooligosaccharides (FOS) [11]. Nonetheless, the production method may add impurities such as residual sugars (e.g., glucose, sucrose, lactose), proteins, colloids, and microbial cells, potentially reducing the physiological efficiency of FOS. Consequently, effective purification methods—such as ethanol precipitation, activated carbon adsorption, dialysis, and membrane filtration—are crucial for guaranteeing product purity and functionality [11–14]. The term 'prebiotics' actually evolved in 2010, the concept was expanded to include selective fermentation that alters gut microbiota. In 1995, nondigestible dietary components were found to positively influenced the host by selectively activating specific colon bacteria [12][4, 15]. Prominent fructooligosaccharides with recognized prebiotic properties encompass lactulose, inulin-type fructans, and trans-galactooligosaccharides. The genus *Bifidobacterium* selectively metabolizes fructans and galactans via β-fructanosidase and β-galactosidase enzymes, respectively. Emerging options such as xylooligosaccharides (XOS), isomaltooligosaccharides (IMO), and pectic oligosaccharides (POS) are now being examined. The sustainable production of FOS can be accomplished by employing hemicellulose-rich agricultural byproducts, including fruit and vegetable solid waste (FVSW) [16]. Processing techniques such as hydrothermal treatment, acid hydrolysis, and enzymatic hydrolysis are utilized, with enzymatic treatments providing benefits in selectivity and less environmental impact. The integration of physicochemical and enzymatic processes offers a sustainable and economical alternative to conventional chemical procedures by reducing unwanted byproducts and toxic waste, thereby decreasing post-process decontamination and disposal expenses. This method enhances the value of agro-industrial byproducts while meeting the increasing demand for functional food ingredients that support intestinal health and general wellness [11].

Despite scientific, economic, and logistical challenges, biomass-to-energy conversion systems have great potential. Traditional 1G processes, like fermentation and anaerobic digestion for sugarcane and maize have sustainability issues due to the food-versus-fuel debate. Even in sugarcane-producing countries like India and Brazil, commercial-scale 1G bioethanol facilities are limited. Due to feedstock availability, second-generation (2G) technologies using lignocellulosic and algal biomass have greater potential, but their complex and expensive pretreatment processes limit scale [12]. Algal biomass is energy-intensive and still developing, limiting its commercial viability. Thermochemical processes such as pyrolysis, gasification, and liquefaction provide substantial fuel yields; nevertheless, they require considerable energy input, frequently derived from fossil sources, which undermines sustainability objectives. Elevated pressure demands in supercritical processes, coupled with the absence of optimized parameters in emerging technologies such as hydrothermal carbonization and torrefaction, impede scale-up and commercialization. Furthermore, the collection, classification, storage, and transportation of biomass—particularly from forest residues and urban waste—continue to be financially burdensome and logistically complex. High-moisture feedstocks require drying, hence elevating storage and energy expenses.

This chapter offers an innovative and comprehensive study of the conversion of lignocellulosic biomass into oligosaccharides (OS), highlighting recent developments in pretreatment technologies, enzymatic hydrolysis, and process parameters. The focus is on sustainable and economically viable methods and cutting-edge biological approaches to address the challenging characteristics of biomass. This chapter tackles the intricate structures of cellulose, hemicellulose, and lignin, while also outlining the synergistic impacts of various pretreatment strategies, thereby offering a comprehensive framework for the effective recovery of oligosaccharides. Simultaneous saccharification and fermentation (SSF) and simultaneous saccharification and co-fermentation (SSCF) enhance oligosaccharide production by reducing enzyme inhibition and efficiently utilizing sugars, especially mixed hexose/pentose sugars in SSCF. Thus, the chapter examines the incorporation of these processes for oligosaccharides within biorefinery frameworks, in accordance with circular economy principles to improve the production of various value-added products. Furthermore, the emphasis is placed on the economic and environmental implications of the diverse applications of recovered oligosaccharides.

2 Composition and Structure of Biomass

Biomass is often defined as a complex mix of biological components that include proteins, lignin, lipids, and carbohydrates like starch, hemicellulose, and cellulose [1]. Biomass may be categorized into two types, based on its origin, function, and end products: natural biomass and biomass used as feedstock. Research indicates that primary sources of biomass include agricultural residues (e.g., peanut shells, rice hulls, cane waste, corn leaves, coconut shells), agro-industrial byproducts (e.g., waste products from processing cassava peels, rice husks, peanut shells, coffee husks) and forest leftovers (waste from the wood industry, such as sawdust and shavings, residual tree trunks and foliage following wood extraction), as well as animal waste, algae, and aquatic crops. It can also include municipal solid waste (MSW) and waste streams generated from human activities when reuse is not feasible [2, 3]. Energy crops, crop residues, algae, industrial and domestic wastes, forestry waste (mostly trees, shrubs, and sawdust remnants), forest biomass (oak, spruce, birch, and poplar), and other animal excrement are all included in this category [4].

Straws derived from grains, rapeseed (*Brassica napus*), and sugarcane waste are "second-generation" biofuels, with assessments of their life cycles consistently demonstrating significant reductions in greenhouse gas (GHG) emissions [5]. Softwoods, hardwoods, and grasses have the highest lignin concentration in their biomass. Microalgal biomasses generally consist of lipids, carbohydrates, and proteins, with contents differing across species based on developmental environment [6, 7]. Elevated lipid concentration in microalgal species is primarily advantageous for biodiesel production, while carbohydrate-rich microalgal biomass serves as attractive raw material for fermentative biofuels [8]. Municipal solid waste (MSW) comprises significant quantities of plant-derived sugars that may serve as a biomass resource. A considerable fraction of MSW, including metals, dry cardboard, paper, and plastics, is easily recyclable [9]. Ethanol derived from waste sources is expected to possess a reduced carbon footprint compared to ethanol sourced from simple sugars or energy crops. LCBs are classified according to their chemical characteristics, chiefly into lignin, hemicellulose, and cellulose. Lignin act as the glue that maintains the structural integrity of LCB, making it insoluble in water. The breakdown of these bigger polysaccharides such as cellulose and hemicellulose during biomass treatment can generate intermediate compounds such as oligosaccharides [10].

2.1 Cellulose

Cellulose $(C_6H_{10}O_5)n$, which is made up of 10,000 to 15,000 monomers, is the main structural and integral polymer of LCB [11]. The D-glucose subunits are linked by β-(1,4)-glycosidic bonds [12], which contain extensive hydrogen bonds (H-bonds) between and within molecules [11–14]. Hence, it is a stable polymer that is very rigid in an axial direction. Over 50% of the crystalline domain in cellulose is associated with the crystallinity of lignocellulosic biomass. There are highly ordered crystalline regions and amorphous or disordered regions of cellulose microfibrils [4]. Because it has both organized and disorganized components, it does not dissolve in water or most organic liquids, except at very low or high pH levels [15, 16]. Cellulose can be soluble in two types of liquids: ionic liquids (ILs) and N-methylmorpholine-N-oxide [12]. This purified cellulose has several applications in fabrics, films, composites, platform chemicals, and fuels because of its biocompatibility, hydrophilicity, reactive hydroxyl groups, and stereoregularity [17].

2.2 Hemicellulose

Hemicellulose is made up of short chains of 500 to 3000 sugar molecules per polymer and is a heteropolysaccharide with branches made up of different sugars (C-5 and C-6) [18]. Depending on the plant species, the composition of these sugars varies significantly. The structure of their cell walls divides hemicellulose into five main groups: xyloglucan, galactomannan, glucuronoxylan, arabinoxylan, and glucomannan [19]. These groups are named after different types of polysaccharide side chains. The most important part is xylan which has short bands of polysaccharides that are joined by β-(1,4)- and/or β-(1,3)-glycosidic links [20]. When compared to cellulose, hemicellulose does not form a solid structure because it has branches coming from its backbone, which make it irregular, random, and weak. Compared to cellulose, hemicellulose is less polymerized and more flexible, which makes it easy to break down and dissolve. It is widely used in industry for applications like drug carriers, cosmetics, hydrogels, xylitols, bioethanol, etc. [21, 22]

2.3 Lignin

Lignin is a cross-linked large molecule comprising over 10,000 units, characterized by its hydrophobic and aromatic properties [23]. The distinctive features of lignin and its biological or physicochemical properties are determined by monolignols, specifically sinapyl, coniferyl, and p-coumaryl alcohols, which feature alkyl-aryl, alkyl-alkyl, and aryl-aryl group linkages [24]. The functional groups –OH, CH_3–O, R–COOH, and > C– –O in lignin polymers modulate reactivity [23, 25]. In addition to variations in monolignol content, lignin exhibits differences in bonding, with softwoods demonstrating a higher prevalence of C–C bonds relative to hardwoods [25]. Predicting the precise chemical structure of isolated lignin is challenging due to its diverse nature in terms of chemical bonds and monolignol composition. Covalent bonds between lignin and hemicelluloses form the lignin carbohydrate complex, facilitating the interconnection of lignin and cellulose, with cellulose and hemicellulose being H bonded [26–29]. Cross-linking involving ether and carbon–carbon linkages occurs through the dehydrogenation of phenylpropane monomers facilitated by in vivo enzymes [26]. The sequences of ether and carbon–carbon linkages vary across different plant species. The composition of monolignols is influenced by the environment, the cell type, and the type of plant [30]. Lignin primarily provides structural support to plant biomass, forming a matrix of various polymeric components of alcohol [29]. Lignin has served as an energy source for decades and has been utilized as an additive in cement, adhesives, dispersing agents, antioxidants, anti-UV agents, and antibacterial materials [31].

Oligosaccharides obtained from different LCBs: Oligosaccharides are chemically linked monosaccharides consisting of two to ten units. They are often produced by cellulose and hemicellulose breakdown during biomass preparation. Hemicellulose comprises a diverse array of monosaccharides, predominantly d-glucose, d-mannose, d-xylose, and l-arabinose. Consequently, three primary types of oligosaccharides can be extracted from hemicellulose: xylooligosaccharides (XOS), mannan oligosaccharides (MOS), and arabinooligosaccharides (AOS) [32]. These oligosaccharides are classified as nondigestible oligosaccharides (NDOs) derived from the bioconversion of hemicellulose via chemical and/or biological methods, which encompass pretreatments and fermentation [33]. They are recognized as food components with prebiotic properties, as they selectively promote the growth of good bacteria in the gastrointestinal (GI) tract [32, 34] (as shown in Fig. 1).

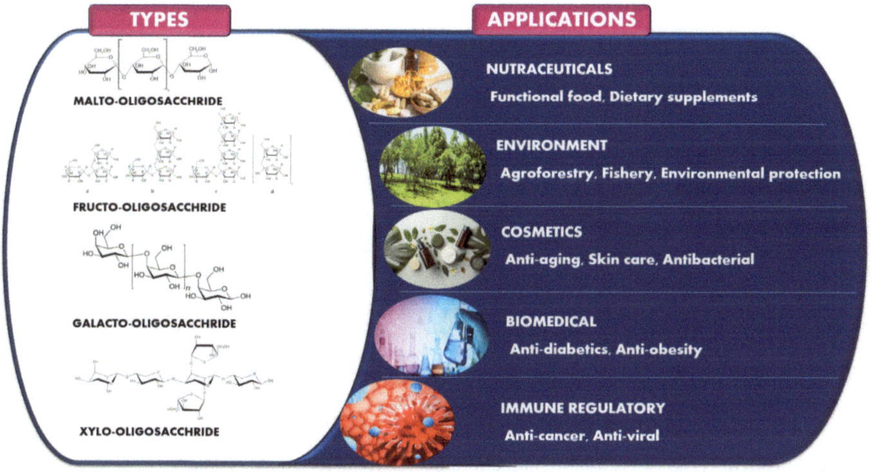

Fig. 1 Types and applications of oligosaccharides

Its renewable nature and extensive availability make it an attractive feedstock for biofuels and value-added chemicals. Chemical constituents are the primary factors in the recalcitrance of biomass, as they establish a spatial network that serves as a protective barrier. Both direct and indirect variables impact biomass cellulose accessibility. The direct factors correspond to the accessible surface area, while the indirect factors encompass biomass structure-relevant factors (pore size and volume, particle size, and specific surface area), chemical compositions (lignin, hemicellulose, and acetyl group), and cellulose structure-relevant factors (cellulose crystallinity and degree of polymerization [DP]). The process of pretreatment is essentially designed to enhance direct factors and modify indirect factors, thereby increasing the accessibility of cellulose [35]. The pretreatment process leads to lignin removal, degradation of cellulose's microcrystalline structure, and generation of oligosaccharides and monosaccharides. Liquid hot water (LHW) pretreatment, for instance, produces oligosaccharides and sugar, but the sugar content is relatively small [36].

3 Pretreatment Techniques

Pretreatment changes biomass from its natural crystalline state to one with less lignin and more noncrystalline cellulose and hemicellulose [37]. Another significant aspect to consider during pretreatment is depolymerization of biomass. Depolymerization and delignification happen when the ester and ether links between and within the units of the biomass heteromatrix are broken. The breaking of these bonds is also linked to the relaxation of the matrix structure and the decrease in cellulose crystallinity [38]. To select the optimal pretreatment method, the following criteria must be met: (a) The

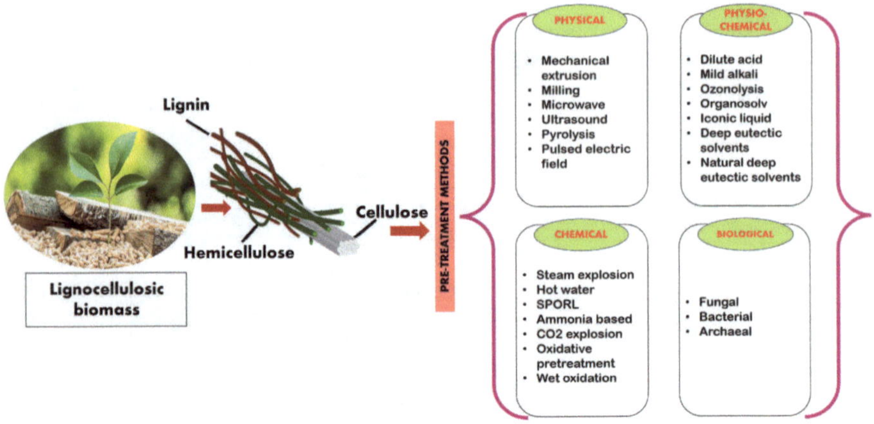

Fig. 2 Different types of pretreatment techniques for lignocellulosic biomass

chosen method should not diminish the size of biomass molecules; (b) the retention of hemicellulose fraction is essential; (c) the generation of degradation products must be minimized; (d) the energy requirements should be minimized; and (e) the catalyst used in pretreatment should be recyclable, with the regeneration of value-added lignin coproducts being cost-effective [39]. Evidence is required to demonstrate that the preparation influenced the costs of subsequent steps in the process, as well as the trade-offs among various costs, including operational costs, capital costs, and biomass costs [40]. Typically, pretreatment processes are categorized into two primary types: biological and non-biological. Non-biological methods exclude the use of microbes and can be categorized into three main groups: physical, chemical, and physicochemical [41] (Fig. 2).

3.1 Physical Methods

Mechanical extrusion is a common pretreatment method that involves pushing biomass into a heated barrel (>300 °C) with shear mixing conditions. Rotating screw blades provide shear pressures and high temperatures that shatter LCB's crystalline and amorphous cellulose. Disrupting the complex structure promotes enzyme accessibility and hydrolysis. This energy-intensive technology makes large-scale industrial use difficult owing to equipment costs and energy needs [42]. Another method, milling, lowers biomass particle size to improve surface area and minimize crystallinity. Depending on feedstock and intended outcomes, ball milling, attrition milling, planetary milling, and wet milling (including wet disk milling) are used. Studies suggest that ball milling improves enzymatic hydrolysis yields, yielding 78.7% glucose from treated bagasse [42]. Alkaline milling, which uses chemicals, improves enzymatic digestibility and sugar yields from plant biomass like oil palm frond fiber and run-down fruit bunch [23, 43].

Biomass may be pretreated quickly and efficiently using microwave irradiation. This approach degrades lignocellulosic structures with minimal inhibitor production by uniformly heating. Microwave pretreatment is easy to use, consumes low energy, has excellent heating capacity within short periods, and degrades cellulose and hemicellulose structures. These advantages make microwave treatment a good biomass digestibility option [23].

Ultrasound pretreatment causes biomass tissue microbubble implosion and physical disruption with 10–100 kHz sound pulses [44]. Breaking open cells improves enzyme access to carbohydrate chains and polymer breakdown. Ultrasound may improve biomass hydrolysis; however, high sonication energy or extended exposure can cause bubbles around the transducer tip, preventing energy transmission and possibly harming the biomass structure [41].

Another physical pretreatment technique, pyrolysis, is mostly employed in biorefineries to convert LCB into bio-oil, gaseous products, and charcoal, excluding sugars. Here, without oxidizing agents, biomass is heated to 500–800 °C [45]. Cellulose and other ingredients break down quickly during pyrolysis, producing a variety of end products with useful energy and industrial applications. Based on heat-up rate, pyrolysis is classified as slow and rapid, with reaction parameters and feedstock properties affecting product yield. Besides biofuel generation, pyrolysis is easy to transport, store, and burn, making it desirable for industrial use [23].

Pulsed electric field (PEF) pretreatment helps biomass breakdown by creating holes in cell membrane for sugar release. This approach uses brief bursts of high voltage (5–20 kV/cm) for nanoseconds to milliseconds and reduces energy usage, operating at ambient conditions owing to its simple design without moving components [23]. PEF has doubled or increased methane output from sewage sludge and pig manure by 80%. It also works on wood chips and switchgrass, depending on electric field intensity and pulse length. Rapeseed stem biomass optimization showed enhanced juice production, polyphenols, and protein levels at 8 kV/cm electric field, 10 bar pressure, 2 ms pulse duration [46].

Hydrodynamic cavitation occurs when mechanical constrictions like venturi pipes, orifice plates, and throttling valves change pressure enough to form vapor cavities that fall apart producing high energy and causing chemical and physical changes that interrupt the lignin–carbohydrate matrix [47, 48]. Cavitation occurs when liquid flow vapor microbubbles develop, grow, and collapse owing to rapid pressure drops. Acoustic, hydrodynamic, optic, and particle cavitation are categorized by generating modalities. Hydrodynamic cavitation helps in the oxidation of organic compounds which are low cost, simple to scale, power efficient, less polluting, and byproduct-free, making it more efficient than acoustic cavitation in many applications [47].

3.2 Chemical Methods

Most chemical pretreatments utilize acid and alkaline agents to break down lignocellulosic biomass. Acid pretreatment employs concentrated or diluted sulfuric (H_2SO_4) or hydrochloric acid (HCl) to break down hemicellulose and alter tissue structure, producing fermentable sugars [49]. Strong concentrated acids are efficient in cellulose hydrolysis but are caustic, poisonous, and need corrosion-resistant equipment. Acid pretreatment often produces fermentation inhibitors including furfural and hydroxymethylfurfural (HMF), which may impair subsequent fermentation. These difficulties may be addressed with gentler acid concentrations (below 4%) that are cheaper and less corrosive but still need cautious handling and neutralization [50, 51]. Alkaline pretreatment uses bases like sodium hydroxide (NaOH), calcium hydroxide ($Ca(OH)_2$), or others for treating biomass. This process breaks down lignin and hemicellulose to make cellulose easier to hydrolyze. Alkaline pretreatment is gentler and creates fewer inhibitory chemicals than acid treatment. Lignin is removed, while cellulose and hemicellulose are preserved, improving enzymatic digestibility. However, greater chemical costs and additional washing steps to eliminate remaining chemicals may reduce process viability [51, 52].

Other chemical processes are used to change LCB for biofuel generation besides acid and alkaline pretreatments. Salt pretreatment uses metal salts to form Lewis acid complexes that tear down lignin and enhance access to it. Organosolv pretreatment solubilizes lignin and hemicellulose in biomass using organic solvents, water, and acid catalysts for facile solvent recovery and excellent lignin purity [49, 53]. Wet oxidation delignifies and solubilizes lignin and hemicellulose with reduced inhibitor production than other procedures by oxidizing them with oxygen and water at high temperatures and pressures. Using hydrogen peroxide or peracetic acid to break down lignin and allow enzyme access improves cellulose digestibility but increases chemical costs [51]. Recent interest has focused on ionic liquid (IL) pretreatment, which dissolves lignin and carbohydrates simultaneously. ILs, organic salts with low melting temperatures, readily destroy biomass structures and are solvent recyclable. These technologies attempt to improve biomass digestibility and reduce chemical waste and environmental effect. Each chemical pretreatment approach has advantages, but cost, chemical recovery, and inhibitory byproducts need continued study to improve their performance in sustainable bioprocessing [51].

3.3 Physicochemical Methods

The physicochemical pretreatment technology integrates combined physical and chemical treatments to enhance the breakdown of lignocellulosic biomass. One such method, steam explosion (SE) pretreatment, disrupts lignocellulosic biomass structure chemically and physically. The process includes exposing biomass to high temperatures (160–260 °C) and pressure for a brief time, followed by fast depressurization, resulting in fibril breakdown and cellulose accessibility for enzymatic hydrolysis [51]. The efficacy relies on particle size, residence duration, and temperature. Although minimal chemicals and energy are required, the addition of acids or alkalis can enhance the results. It also stimulates the development of inhibitory chemicals like aromatic derivatives and furans which requires a lot of heat energy for breakdown of biomass [54, 55]. Another treatment technique like, liquid hot water (LHW) utilizes high-temperature and high-pressure water (170–230 °C; up to 5 MPa) to hydrolyze hemicellulose and partly remove lignin, enhancing enzyme accessibility. To reduce inhibitor development, this chemical-free procedure maintains pH between 4 and 7 [56, 57]. Although successful in breaking biomass components, downstream processing requires eliminating substantial amounts of effluent, which requires energy [51]. Ammonia fiber explosion (AFEX) is another physicochemical pretreatment technique where liquid ammonia disrupts LCB to improve enzymatic digestibility without removing sugars. Several minutes of high pressure (1.72–2.06 MPa) and moderate temperatures (60–120 °C) are followed by fast depressurization, causing ammonia expansion and rupture of biomass fibers [58, 59]. AFEX increases surface area and structure, resulting in > 90% ethanol yields in certain biomass types but less so in hardwoods. Despite being efficient and ecologically friendly, AFEX employs caustic ammonia, which is toxic and limits its usage [51].

In supercritical CO_2 explosion, CO_2 molecules, similar in size to those used in AFEX and steam explosion procedures, enter tiny biomass pores during supercritical CO_2 explosion pretreatments [60]. The nontoxic, nonflammable, and low-corrosion material works at lower temperatures than steam explosion and costs less than AFEX [60]. CO_2 pretreatment creates carbonic acid in water, permitting waste-free processes, while depressurization simplifies CO_2 removal. While supercritical CO_2 explosion is a green and sustainable technology for biomass processing, it faces hurdles before broad industrial implementation [51].

3.4 Biological Methods

Biological pretreatment is regarded as an efficient, environmentally friendly, and low-energy process when compared to traditional chemical and physical pretreatment methods. Nature contains a variety of cellulolytic and hemicellulolytic microbes that can be specifically utilized for efficient biomass pretreatment [61]. Biological pretreatments involve microorganisms, including brown-, white-, and soft-rot fungi, which primarily break down lignin and hemicellulose, with minimal impact on cellulose [62]. White-rot fungus degrades lignin by using peroxidases and laccases [63]. The white-rot fungi species frequently utilized for pretreatment include *Phanerochaete chrysosporium, Ceriporia lacerata, Cyathus stercoreus, Ceriporiopsis subvermispora, Pycnoporus cinnabarinus*, and *Pleurotus ostreatus.* Additionally, other Basidiomycete species were examined for their ability to degrade various lignocellulosic feedstocks. *Bjerkandera adusta, Fomes fomentarius, Ganoderma resinaceum, Irpex lacteus, Phanerochaete chrysosporium, Trametes versicolor*, and *Lepista nuda* have been extensively reported to exhibit high delignification efficiency [46]. The biological pretreatment presents significant interest; however, the hydrolysis rate of lignocellulosic fractions is slow, which limits its viability as a potential industrial-scale pretreatment method [64].

Although the lignin content of pretreated LCB is nearly equivalent to that of non-pretreated biomass, the digestibility of pretreated LCB exceeds that of untreated or raw biomass [35, 65]. LCB is robust and resilient in nature, but its rigid and compact structure makes biochemical conversion difficult, especially in enzymatic hydrolysis, where enzyme accessibility is limited and hydrolysis efficiency is reduced. Major structural challenges to enzymatic hydrolysis stem from the intrinsic physical and chemical resistance of biomass architecture to enzyme access and function.

The development of novel enzyme cocktails shows significant potential for improving the efficacy of pretreatment processes that convert lignocellulose into value-added products. Recent advancements in enzyme cocktails include tailored enzyme mixtures, enzyme engineering, genetic modifications, and synergistic combinations of enzymes [66]. Aishwarya et al. (2025) tailored enzyme cocktails to align with specific biomass characteristics and processing conditions [67]. This customization improves the efficiency of saccharification processes, leading to reduced enzyme usage and lower associated costs. Combining cellulases, hemicellulases, and accessory enzymes in optimal ratios can enhance the degradation of complex polysaccharides into fermentable sugars. The use of synergistic enzyme combinations generated from multiple microbial sources has resulted in higher saccharification yields. Co-cultivating filamentous fungus and bacteria produces potent LCB-degrading enzymatic blends [68]. Enzymes from extremophiles can improve biomass conversion efficiency under severe circumstances [68]. Genetic engineering has improved enzyme efficiency and stability. Engineered enzymes with increased activity and stability can withstand biomass pretreatment conditions. Enzyme engineering, including directed evolution and rational design, has improved catalytic characteristics [69].

4 Critical Process Parameters for Effective Pretreatment

LCB has immense potential; however, some transformation pathways encounter a serious challenge known as recalcitrance. Recalcitrance denotes the resistance to enzymatic and microbial degradation, arising from the intricate interplay of chemical, physical, and ultrastructural organization of LCB [70]. The challenge of transformation processes is to get past recalcitrance through the application of various treatments prior to enzymatic hydrolysis. The challenge arises from the multitude of parameters that must be addressed to overcome recalcitrance and subsequently develop a dependable process at the industrial scale [71]

4.1 Temperature

Pretreatment temperature is more important for lignin destabilization, allowing for process optimization. In biomass pretreatment, temperature affects lignin and hemicellulose breakdown and cellulose accessibility [72]. Higher temperatures accelerate reaction rates and deconstruct the lignocellulosic matrix. Thermal pretreatment increases reactivity by forming pores and releasing lignin from cell walls [73]. High temperatures can also degrade cellulose and create inhibitory byproducts like furfural and 5-hydroxymethylfurfural, which can impair enzymatic hydrolysis and fermentation [74]. Medium temperatures (10 °C) promote cell wall polymer mobility and disrupt cellulose. High temperature pretreatment helps create pores and release lignin [73]. This can minimize dilute acid pretreatment residence time, while lower temperatures necessitate longer processing durations to achieve the same yield. Time and temperature affects acid concentration nonlinearly. According to study reported by Pasquier et al. (2023), yields fall from 4% acid levels due to extremely severe conditions, but below 2%, favorable time–temperature combinations can be found. At acid levels below 2%, a wide range of time–temperature combinations can lead to good yields, but at acid concentrations beyond 4%, the conditions are too severe, reducing yields independent of time or temperature [71]. Biomass origin also affects pretreatment temperature. On average, forest biomass pretreatment requires 20–25 °C higher temperatures and shorter periods. Industrial biomass requires longer durations at moderate temperatures. Due to structural changes in biomass composition, biomass type impacts pretreatment parameter sensitivity [71]. Łukajtis et al. (2018) reported optimal conditions for alkaline pretreatment of *Salix viminalis* L., including 0.75 mm granulation, 7% catalyst concentration, 6-h pretreatment, and 65 °C temperature, based on saccharification efficiency. For maximum glucose yield in the tested range, 0.25 mm granulation, 7% catalyst concentration, 7-h pretreatment, and 80 °C temperature were ideal [75].

4.2 pH

pH is crucial in chemical pretreatment methods, as it regulates the solubility and breakdown of lignin and hemicellulose. At low pH, most hemicelluloses can be extracted from the solid component, thereby enhancing the accessibility of cellulose. Acidic conditions can result in the formation of inhibitors such as furfural and hydroxymethylfurfural. At elevated pH levels, lignin is efficiently dissolved, while cellulose and hemicelluloses form the solid fraction. Alkaline peroxide pretreatment necessitates an alkaline pH to generate oxidizing radicals that facilitate lignin breakdown. The effectiveness of lignin and hemicellulose breakdown is frequently highest around pH 11.5, where hydrogen peroxide stability is least, resulting in greater radical production [76]. pH optimization is essential for alkaline peroxide pretreatment. A pH near 11.5 is recognized as optimal for the decomposition of LCB. Maintaining a high pH requires additional adjustments prior to biological processing, which may result in elevated salt concentrations and the possible inhibition of microorganisms [77, 78]. The ideal range of pH for methanogens, which are responsible for biogas production in anaerobic digestion, is generally between 6.8 and 7.2. Changes from this range, particularly a decrease in pH resulting from the accumulation of volatile fatty acids, may inhibit methanogenesis [79, 80].

4.3 Residence Time

Residence time, also known as reaction time, denotes the period during which biomass undergoes pretreatment conditions. The extent of biomass deconstruction is directly influenced by this factor [81]. Extended residence times typically facilitate a more thorough decomposition of biomass constituents, resulting in increased sugar yields. Excessively prolonged residence times may cause over-degradation of carbohydrates, leading to the production of inhibitory compounds and diminished sugar recovery [82, 83]. In the alkaline pretreatment of hay biomass, an increase in process duration from 2 to 8 h at 22–80 °C typically resulted in elevated soluble chemical oxygen demand (COD) levels, signifying enhanced organic matter solubilization [84]. Optimal reaction times have been established for particular pretreatment methods. The optimal residence time used by Inbicon technology pretreatment process, utilizing a compressed thermal reactor was determined to be 15 min at a temperature of 195 °C [85]. Enzymatic hydrolysis is economically viable with short processing durations, although adequate hydrolysis may require longer residence times. For instance, pretreated spruce took 48 h for enzymatic hydrolysis and 24 h for separate hydrolysis and fermentation (SHF) [86].

4.4 Catalyst Concentration

Catalysts, such as acids, alkalis, and metal salts, are commonly employed in chemical pretreatment to enhance reaction rates and selectivity. Their concentration has a substantial impact on both the efficiency and the cost of the process. Elevating catalyst concentration typically accelerates the rate and degree of biomass deconstruction, resulting in enhanced delignification and hemicellulose solubilization [87]. Higher catalyst concentrations may elevate operational costs, exacerbate conditions, and potentially enhance the production of undesirable byproducts [88]. The alkaline-catalyzed organosolv pretreatment of rice straw was optimized using a reaction that included 5% (w/v) NaOH in an aqueous-acetone mixture, resulting in high glucose recovery. Research suggests that catalyst loading may be reduced by elevating solids concentrations during pretreatment, provided that the catalyst concentration remains unchanged [89]. The quantity of catalyst, such as zeolite, significantly influences conversion and selectivity in the catalytic upgrading of lignocellulosic biomass-derived compounds [90].

4.5 *Solid-To-Liquid Ratio*

Solids loading, or solid-to-liquid (S:L) ratio, is a crucial component in biomass pretreatment, affecting mass transfer, reaction kinetics, efficiency, and economic feasibility. The slurry gets viscous at large solids loading, making mixing difficult. Poor mixing limits mass transfer by preventing heat and chemicals (catalysts, water, or solvents) from spreading evenly among biomass particles. This can cause inconsistent pretreatment, poorer response speeds, and lower sugar yields [91]. High solids loading can cause "water constraint," meaning that there is insufficient free water to adequately hydrate biomass and dissolve and diffuse chemical reagents. This restricts enzyme and catalyst exposure to biomass components. [92]. The S:L ratio changes the number of reactants (like acids or bases) that are present per unit of biomass. When the S:L ratio is smaller (more liquid), there may be better contact and a faster reaction at first. Higher S:L ratios, on the other hand, compress the biomass portion, which can be good for processes further down the line if problems with mass transfer can be fixed [93].

5 Enzymatic Hydrolysis for Oligosaccharide Recovery

Oligosaccharide recovery involves the extraction or isolation of oligosaccharides from diverse biomass sources, especially LCB. Oligosaccharide frequently serve as intermediate products in the degradation of larger polysaccharides, such as cellulose and hemicellulose, during biomass processing. Hemicellulose comprises a variety of monosaccharides, predominantly d-glucose, d-mannose, d-xylose, and l-arabinose. As a result, three main categories of oligosaccharides can be derived from hemicellulose: xylooligosaccharides (XOS), mannan oligosaccharides (MOS), and arabinooligosaccharides (AOS) [32].

The mechanism of action for oligosaccharide recovery via enzymatic hydrolysis involves specific glycoside hydrolase (GH) enzymes that cleave glycosidic bonds in complex polysaccharides, such as hemicellulose (e.g., xylan) or cellulose, resulting in the formation of shorter-chain oligosaccharides [94–96]. The mechanism depends on the enzyme's active site binding to the substrate (polysaccharide) with high specificity. Upon binding, the enzyme promotes the hydrolysis reaction, typically via a catalytic mechanism that includes acid–base catalysis or covalent catalysis, resulting in the cleavage of the glycosidic bond and the liberation of oligosaccharide units [94, 97]. Xylanases catalyze the breakdown of xylan to yield xylooligosaccharides (XOS), whereas cellulases, which encompass endoglucanases and exoglucanases with diminished β-glucosidase activity, can be optimized to generate cellooligosaccharides (COS) from cellulose (Table1) [94, 95]. This specificity facilitates controlled depolymerization, targeting a specific degree of polymerization (DP) in the recovered oligosaccharides [94, 95].

The enzymes utilized for oligosaccharide recovery are contingent upon the polysaccharide undergoing hydrolysis and the targeted oligosaccharide product. Hemicellulases, such as xylanases (e.g., endo-1,4-β-xylanases), play a vital role in the degradation of xylan, leading to the formation of xylooligosaccharides (XOS) [98]. Cellulases, including endoglucanases, exoglucanases, and β-glucosidases, facilitate the production of cellooligosaccharides (COS) from cellulose, typically under controlled conditions to prevent complete hydrolysis to glucose [95]. Additional enzymes, including mannanases, generate mannan oligosaccharides (MOS), while pectinases facilitate the production of pectic oligosaccharides [95]. The choice of specific enzyme cocktails, which may include accessory enzymes, is crucial for achieving efficient and selective hydrolysis of complex biomass substrates, thereby ensuring the desired oligosaccharide profile [94, 95].

Advantages of Enzymatic Hydrolysis

Enzymatic hydrolysis presents multiple benefits compared to chemical hydrolysis techniques for the recovery of oligosaccharides. Initially, it functions under mild reaction conditions, specifically moderate temperature and pH, reducing the generation of unwanted byproducts and degradation products, typically linked to severe chemical treatments [94–96]. Enzymes demonstrate significant substrate specificity, facilitating the targeted cleavage of specific glycosidic bonds. This results in a controlled product profile characterized by a desired range of oligosaccharide lengths and purities [94, 97]. This specificity results in increased yields of particular oligosaccharides and a reduction in side reactions [94, 95]. Enzymatic processes are typically more environmentally sustainable, utilize less energy, and necessitate less corrosive equipment than acid or alkaline hydrolysis [95, 96].

Limitations of Enzymatic Hydrolysis

Enzymatic hydrolysis for oligosaccharide recovery, while beneficial, encounters specific constraints. A notable challenge is the high cost of enzymes, which can significantly impact the overall economics of the process [94, 95]. Enzymes may experience product inhibition, wherein accumulating oligosaccharides or monomeric sugars bind to the enzyme's active site, thereby diminishing its activity and hydrolysis rate [95, 96]. The diverse and resistant characteristics of lignocellulosic biomass typically require a pretreatment phase before enzymatic hydrolysis to improve enzyme accessibility, thereby increasing the complexity and cost of the overall process [95, 96]. Enzyme stability in industrial operating conditions, such as temperature, pH, and the presence of inhibitors, poses significant challenges that necessitate meticulous process design or creation of more resilient enzymes [94, 95]. The reaction rates may be slower than those of chemical methods, which could necessitate extended reaction times or increased enzyme concentrations to attain the desired yields [95]. However, the application of synergistic enzyme combinations derived from various microbial

Table 1 Enzyme types and their roles in the biosynthesis of functional oligosaccharides

Enzyme type	Substrate	Mechanism of action	Oligosaccharides produced	References
Cellulases	Crystalline/amorphous cellulose, lignocellulose	Endo-β-1,4-glucanases cleave internal bonds; exo-β-1,4-glucanases release cellobiose; β-glucosidases convert cellobiose to glucose; partial hydrolysis used for COS	Cellooligosaccharides (COS)	[137]
Hemicellulases	Xylans, mannans, arabinans	Endoxylanases cleave β-1,4 bonds; mannanases cleave β-1,4-mannosidic bonds; accessory enzymes remove side chains	Xylo-, mannan-, arabinooligosaccharides	[138, 139]
Pectinases	Pectins (homogalacturonan, RG-I/II)	Polygalacturonases hydrolyze α-1,4 linkages; lyases cleave via β-elimination; methyl esterases increase susceptibility	Pectic oligosaccharides (POS)	[140, 141]
Chitinases	Chitin (crustaceans, fungi)	Hydrolyze β-1,4 linkages in chitin to produce COS and N-acetylglucosamine	Chitin oligosaccharides (CHOS)	[142–144]
Alginate lyases	Alginate (brown seaweed)	Cleave β-1,4-D-mannuronate and α-1,4-L-guluronate via β-elimination	Alginate oligosaccharides (AOS)	[145, 146]
Dextranases	Dextran (*Leuconostoc* spp.)	Hydrolyze α-1,6-glucosidic bonds in dextran	Dextran oligosaccharides (DOS)	[147]
β-glucanases	β-glucans (cereals, yeast)	Hydrolyze β-1,3/β-1,4/β-1,6 glycosidic linkages. Enzyme specificity dictates oligomer type	β-glucan oligosaccharides	[148, 149]
Amylases	Starch	α-amylase cleaves internal α-1,4 bonds; β-amylase releases maltose; glucoamylase acts on α-1,4 and α-1,6 bonds	Malto-/isomaltooligosaccharides (MOS/IMO)	[150, 151]
Inulinases	Inulin (chicory, dahlia)	Cleaves β-2,1-fructosidic bonds to release fructose units and FOS	Fructooligosaccharides (FOS)	[152]

sources has resulted in increased saccharification yields. Co-cultivating filamentous fungi and bacteria yields highly active enzymatic cocktails that effectively degrade LCB. Enzymes derived from extremophiles, capable of operating under extreme conditions, have demonstrated potential for enhancing biomass conversion efficiency. Advancements in genetic engineering have enabled the production of more effective and stable enzymes [68]. Engineered enzymes exhibiting improved activity and stability can endure the severe conditions associated with biomass pretreatment. Techniques in enzyme engineering, including directed evolution and rational design, have resulted in enzymes exhibiting enhanced catalytic properties [69].

6 Integration of Pretreatment and Enzymatic Processes

Lignocellulosic biomass represents a significant and renewable resource with substantial potential to produce biofuels, biochemicals, and biomaterials. Nonetheless, its complex and resistant structure obstructs economical utilization. The intricate composition of cellulose, hemicellulose, and lignin renders fermentable carbohydrates inaccessible. Pretreatment is essential for the release of these significant sugars. Pretreatment deconstructs the lignocellulosic matrix, reduces cellulose crystallinity, depolymerizes hemicellulose, and removes or modifies lignin, thereby enhancing the hydrolysability of cellulose and hemicellulose [99, 100]. Following pretreatment, which enhances the biomass structure, enzymatic hydrolysis is essential. Under mild reaction conditions, cellulases and hemicellulases catalyze the depolymerization of cellulose and hemicellulose into fermentable sugars, such as glucose and xylose [101]. Integration of pretreatment and enzymatic hydrolysis is essential for economically viable and effective biomass conversion (Fig. 3).

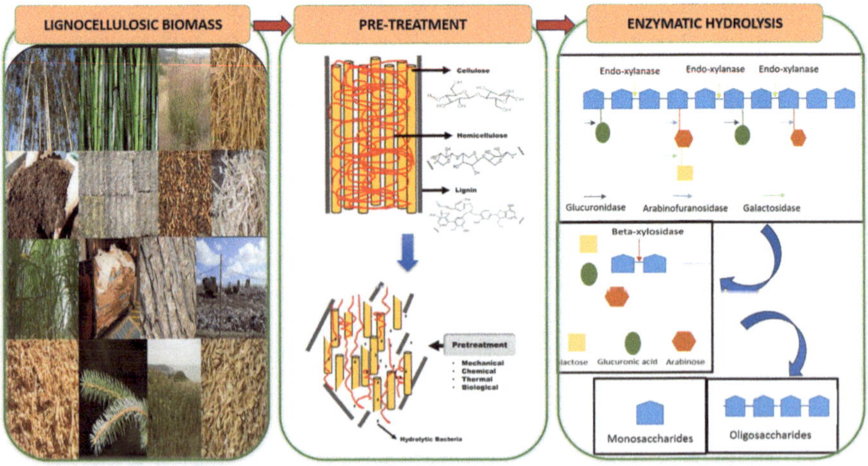

Fig. 3 Integration of pretreatment and enzymatic hydrolysis

Effective pretreatment enhances the enzymatic susceptibility of lignocellulosic material, thereby increasing sugar yields and decreasing enzyme requirements. Pretreatment methods should be compatible with subsequent enzymatic processes to reduce enzyme inhibitors and enhance substrate accessibility. The effectiveness of the interaction between these two stages is critical to the success of the biomass valorization pathway. These steps have typically been performed sequentially in distinct reactors. Segregated processes may elevate capital costs owing to the necessity for multiple reaction vessels, potential processing delays, and issues related to product inhibition or sugar degradation during intermediate handling. The emergence of these issues has catalyzed the advancement of integrated bioprocesses aimed at optimizing sugar synthesis and fermentation [102, 103].

6.1 Sequential Versus Simultaneous Saccharification and Fermentation (SSF)

Due to the resistance toward hydrolytic enzymes, microorganisms present a significant challenge in the production of bio-based value-added products from lignocellulosic biomass. In addressing these challenges, methods such as separate hydrolysis and fermentation, simultaneous saccharification and fermentation/co-fermentation (SSF/SSCF) and consolidated bioprocessing (CBP) have been employed. The processes of separate hydrolysis and fermentation (SHF) are conducted independently, allowing for the utilization of hydrolytic enzymes and fermentative microorganisms at their ideal temperatures to enhance enzyme activity and optimize byproduct yield [104, 105]. SHF presents multiple challenges, such as significant capital expenses associated with distinct reaction vessels, potential end-product inhibition of the enzyme, and likelihood of microbiological contamination arising from prolonged hydrolysis [105].

In the processes of SSF and SSCF, hydrolysis and (co-)fermentation can occur within a single vessel, allowing the fermentative bacteria to directly utilize the hydrolysis end product, which consists of simple sugars or monosaccharides. This setup effectively removes the end-product inhibition typically experienced by hydrolytic enzymes. The primary advantages of SSF include minimizing sugar waste and lowering capital costs through the utilization of fewer reaction vessels. The compromise is that hydrolytic enzymes function at a temperature that is below the optimal level. CBP is a breakthrough for cost-effective LC biofuel generation since it integrates saccharolytic enzyme synthesis (cellulases and hemicellulases), polysaccharide hydrolysis, and sugar fermentation by a single microbe or consortium [105, 106].

Simultaneous saccharification and fermentation (SSF) and simultaneous saccharification and co-fermentation (SSCF) present a compelling case study for the sustainable and efficient production of various oligosaccharides, highly sought after for their functional properties in food and pharmaceuticals. Traditional submerged fermentation often faces limitations in energy consumption and downstream processing, making SSF and SSCF attractive alternatives. By leveraging microorganisms grown on moist solid substrates, these methods mirror natural microbial habitats, fostering enhanced enzyme production crucial for oligosaccharide synthesis. For instance, the immobilization of *Aspergillus japonicus* on lignocellulosic materials like corn cobs significantly boosts β-fructofuranosidase (FFase) activity and fructooligosaccharide (FOS) yields and productivity, surpassing free-cell systems. A key advantage lies in their ability to utilize inexpensive agricultural byproducts as both substrates and immobilization carriers, thereby reducing costs and promoting waste valorization. The enhanced stability of immobilized cells and enzymes also facilitates repeated batch operations and simplifies product recovery, showcasing the immense potential of SSF and SSCF for industrial-scale oligosaccharide production [107].

6.2 Process Configurations for Enhanced Efficiency

To enhance efficiency in oligosaccharide processing, process configurations often involve a combination of extraction methods, hydrolysis techniques, and purification strategies. Enzymatic methods, hydrothermal pretreatment, and membrane reactors are commonly used to optimize extraction yields and purity while reducing energy consumption [108, 109]. The extraction methods used by Jovanovic-Malinovska et al. (2015) was ultrasound-assisted extraction (UAE), which markedly enhanced oligosaccharide output from diverse fruits and vegetables. Through response surface methods, researchers determined that optimal extraction was attained with 63% ethanol, a temperature of 40 °C, and a sonication duration of 10 min. The UAE enhanced oligosaccharide content by 2–4 times compared to conventional techniques, with Jerusalem artichoke (*Helianthus tuberosus*) exhibiting the most significant improvement, reaching 7.17 g/100 g FW. This underscores UAE as an effective method for improving oligosaccharide recovery [108]. Ultrasonic extraction doubled to quadrupled oligosaccharide concentration. Microwave-assisted extraction, superior to conventional methods, reduced chitosan extraction time and energy, making it more efficient and eco-friendly [110]. Biomass rich in hemicelluloses, like barley straw, is a promising source for oligosaccharides (OS) with prebiotic activity, aligning with biorefinery principles. Steam explosion pretreatment isolates hemicelluloses, which are then combined with enzymatic hydrolysis and other treatments. XOS1 (via ion exchange) and XOS2 (via membrane filtration, enzymatic hydrolysis, and ion exchange) demonstrate diverse linear and branched structures, some with acetyl or feruloyl substitutions. This process enhances hemicellulose value. Coupling fermentation and enzymatic hydrolysis offers a greener, continuous, and efficient method for OS production from cheap substrates [109, 111]. Along with acid and enzymatic hydrolysis, hydrothermal pretreatment is a type of process configuration where hydrothermal pretreatment is recognized as a particularly effective approach for biomass conversion, characterized by its straightforward process, economic viability, and limited interference with subsequent fermentation. This effective method produces both solid and liquid components. The material, abundant in cellulose and lignin, presents an excellent opportunity for ethanol production. The liquid hydrolysate, which contains dissolved hemicelluloses, presents considerable potential for the production of bio-based chemicals, fuels, and materials such as oligosaccharides and monosaccharides. Although a flawless pretreatment method is yet to be discovered, hydrothermal techniques emerge as a promising option for the biorefineries of the future [109] (Fig. 4).

Membrane filtration efficiency in oligosaccharide separation is significantly influenced by fluid properties, especially molecular size. The sequential batch reaction and filtration (SBRF) mode achieves higher permeate flux than the SRF mode because it generates smaller hydrolyzed molecules that pass through the membrane more easily. While xylan depolymerization by endo-1,4-β-xylanase and β-xylosidase produces small oligomers that enhance flux, these same small molecules can cause severe internal pore blocking, leading to significant membrane fouling and a lower product yield in SRF. Additionally, elevated temperature and pH cause membrane swelling, further increasing permeate flux and impacting the yield of target oligosaccharides [111].

In addressing the dual challenges of agricultural waste management and sustainable chemical production, researchers have explored the utilization of corncob residues—byproducts from xylooligosaccharide (XOS) manufacturing—as a cost-effective substrate for L-lactic acid production. Employing *Rhizopus oryzae* in a simultaneous saccharification and fermentation (SSF) process, the study achieved a notable L-lactic acid titer of 60.3 g/L and a yield of 0.60 g/g dry residue at 10% (w/v) substrate loading and 40 °C. This performance surpassed that of separate hydrolysis and fermentation (SHF), which yields 34.0 g/L and 0.34 g/g, respectively. The SSF approach not only streamlines the production process but also enhances efficiency, demonstrating the feasibility of converting lignocellulosic waste into valuable biochemicals. In order to separate oligosaccharides from other milk components, each approach uses different techniques that highlight different separation concepts

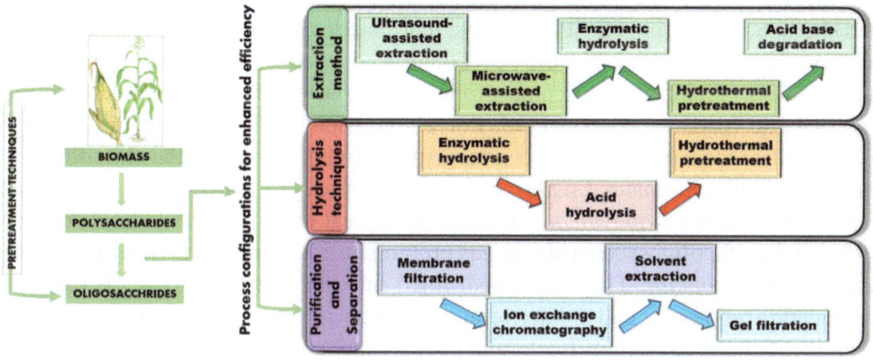

Fig. 4 Process configuration for enhanced efficiency of oligosaccharides

such as solvent extraction, ion exchange, gel filtration, and membrane-based filtration [112].This case underscores the potential of integrating waste valorization with biotechnological advancements to foster sustainable industrial practices [113].

Researchers explored the potential of apple pomace, a byproduct of juice processing, as a substrate for L-lactic acid production using simultaneous saccharification and fermentation (SSF). By optimizing parameters such as the cellulase-to-solid ratio (CSR), liquid-to-solid ratio (LSR), and β-glucosidase-to-cellulase ratio (BCR), they achieved a lactic acid concentration of 27.8 g/L within 10 h, corresponding to a productivity of 2.78 g/L·h. Processing 100 kg of dry apple pomace yielded 36.6 kg of lactic acid, 18.3 kg of oligomeric carbohydrates suitable for functional foods, 8.4 kg of microbial biomass, and 8 kg of uronic acids. This study demonstrates the feasibility of converting agro-industrial waste into valuable biochemicals, promoting sustainable waste management and resource utilization [114].

7 Applications

The extensive use of functional oligosaccharides (FOS) across various sectors such as food, feed, pharmaceuticals, cosmetics, and agrochemistry necessitates the advancement of scale-up synthesis techniques. Oligosaccharides may be obtained from natural sources or produced through various physical, chemical, or enzymatic techniques. Frequently utilized techniques include the hydrolysis of polysaccharides or the synthesis from disaccharide substrates through enzymatic and chemical processes [115, 116]. Raffinose oligosaccharides are obtained from plant materials using water or aqueous solutions of methanol or ethanol [116].

Many types of FOS are employed in food, prebiotic supplements, drug delivery, immunostimulators, cosmetics, animal feed, and agrochemicals (Table 2). FOS are frequently incorporated in food items as anticariogenic agents and low-sweetness humectants due to their many health benefits. Oligosaccharides are utilized in beverages (fruit drinks, coffee, chocolates, tea, soda, soya milk, soft drinks, health drinks, and alcoholic beverages) and milk products as nutraceuticals. Desserts (puddings and sherbets), table spreads (jellies, jams, marmalades, and honey products), confectionery (candies, cookies, biscuits, chocolate, and sweets), bakery products (breads and pastries), breakfast cereals, meat products (fish paste and tofu), and infant, elderly, and diabetic foods all contain FOS. Polyphenol-rich algal-oligosaccharide-lysates exhibit antioxidant properties. The synergistic action of lysozyme and chitooligosaccharides (COS) eliminates *Escherichia coli*, *Pseudomonas fluorescens*, and *Bacillus cereus* and reduces *Staphylococcus aureus* in a meat model system [117, 118].

In gut health, oligosaccharides act as prebiotics, nourishing beneficial *Bifidobacteria* and *Lactobacilli*. They are transformed into short-chain fatty acids (SCFAs) like acetic, propionic, and n-butyric acids, which improve gastrointestinal health, exhibit anticholesterolemic effects, and possess anticarcinogenic properties. Specific oligosaccharides, such as fructooligosaccharides and glucooligosaccharides, also

Table 2 Various application areas, uses, and types of oligosaccharides

Application area	Uses	Types of oligosaccharides	Reference
Food industry	**Prebiotics**: Oligosaccharides are used as prebiotics, which provide dietary supplement with antioxidant and possible immune-modulating characteristics, promoting the growth of beneficial gut bacteria	Galactooligosaccharides (GOS), xylooligosaccharides (XOS), fructooligosaccharides (FOS), mannan oligosaccharides (MOS), cellooligosaccharides (COS), and isomaltoligosaccharides (IOS)	[137, 153, 154]
	Low-calorie sweeteners: They add sweetness without calories to fermented dairy, sweets, cereals, and beverages	Galactooligosaccharides (GOS)	[155]
	Dietary fiber: It contributes to the overall fiber content of the diet, facilitating digestion and enhancing gut health	Algal-derived glycans and oligosaccharides	[156]
Pharmaceuticals and cosmetics	**Biomaterials**: Oligosaccharides can be used as biomaterials in medical applications (including antibacterial, anti-inflammatory, and hemostatic properties, thus promote and accelerate wound healing), offering properties like biocompatibility and biodegradability		[157]
	Cosmetics: They can be used in cosmetics for moisturizing, skin barrier improvement, and antimicrobial properties, potentially helping to prevent dermatitis	Cellooligosaccharides	[137]
	Immunomodulatory and anti-inflammatory agents: Certain oligosaccharides have demonstrated the ability to lower inflammation and strengthen the immune system	Fructooligosaccharides	[158]

(continued)

Table 2 (continued)

Application area	Uses	Types of oligosaccharides	Reference
Other applications	**Leather tanning**: Oligosaccharides include a lot of oxygen-containing groups; they can also be employed as masking agents for chrome-free metal tanning (such as Al, Zr, and Ti salts). In this case, oligosaccharides pre-coordinate with **metal** ions devoid of chrome, facilitating their entry into the leather matrix for further cross-linking processes	Starch-based oligosaccharides	[159]
	Lipid production: ***Cryptococcus curvatus*** is capable of utilizing oligocelluloses or oligoxyloses as its primary source of carbon to produce microbial	Oligocelluloses and oligoxyloses	[160]
	Ethanol production: Recently, researchers have proposed cellooligosaccharides as new substrates for fermentation and ethanol production due to their potential advantages over glucose (conventional process), including a reduced risk of contamination due to the use of engineered and selective yeast strains for prebiotic consumption and ethanol production, faster substrate consumption, and shorter fermentation times	Cellooligosaccharides (COS)	[137]

help to prevent urogenital infections. Xylooligosaccharides can suppress pathogenic bacteria and alleviate intestinal disorders [119, 120].

Pharmaceutical applications of oligosaccharides are diverse. They help control constipation, hyperlipidemia, and encephalopathy. Xylooligosaccharides exhibit anti-allergic, antimicrobial, anti-infective, and anti-inflammatory properties, alongside selective cytotoxicity and immunomodulatory action. Feruloyl xylooligosaccharides show antioxidative capacity. Oligosaccharides promote bone growth, while certain novel oligosaccharides suppress leukemia cell proliferation. They can also enhance the stability and bioavailability of peptides for drug delivery and act as carriers for hydrophobic drugs. Heparan sulfate mimetics and polysulfated oligosaccharide glycosides show promise as antiangiogenic and anticancer agents [121, 122].

In cosmetics, oligosaccharides serve as stabilizers and bulking agents. They provide skin emollient efficacy, protect cells from environmental damage, exhibit anti-inflammatory effects, regulate sebum, and chelate with metals for antioxidant and UV protection. Trehalose, an oligosaccharide, improves skin moisturization. Cyclodextrins are used to solubilize fragrances, neutralize odors, and stabilize antibacterial agents in skincare products [123]. For animal feeds, oligosaccharides improve nutrient digestibility, positively modulate gut microbial flora, and enhance immunity in various animals, offering a potential alternative to antibiotics, the examples include chitooligosaccharides and mannan oligosaccharides [124, 125]. In agriculture and environmental protection, cyclodextrins form inclusion complexes with herbicides, pheromones, and growth hormones, improving their solubility and reducing volatility. They also aid in degrading hydrocarbons. Pectin-derived acidic oligosaccharides promote plant growth, and mannan oligosaccharides can enhance laccase levels in fungi [123, 125].

8 Economic and Environmental Considerations

The valorization of each renewable carbon source is essential for the successful completion of the bioeconomy (BE) transition. The lignocellulosic biorefineries are the primary contributors to this framework, as they are abundant and contain specific components (cellulose, hemicellulose, and lignin) that are specifically utilized in the conversion processes [126]. The bioeconomy (BE) has attracted attention for its potential to reduce environmental impact and boost the economy. The circular economy (CE) and circular bioeconomy (CBE) aligns with this paradigm. CE uses reduction, reuse, and recycling to challenge the linear economic paradigm by reducing environmental impact, waste, and resource consumption.

Based on this, CBE recommends progressive biomass valorization to extract usable products before composting wastes or recovering energy [127, 128]. The optimization of biomass value over time is a key principle of CBE, seeking to prolong product, service, and resource usefulness. This idea is related to the "cascading" principle, which requires biomass to be used for products or services before composting or energy recovery [129]. Each stage in this cascade may degrade biomass quality but increase volume. Stegmann et al. (2020) found that biomass should be used in high-value applications like fine and specialty chemicals rather than energy conversion to maximize value. Studies on European BE clusters imply biomass will be used more in high-value chemical applications than bulk chemicals or energy in the future [130]. Louw et al. (2020) recommend extracting the most economically valuable items first, followed by those of lower value [131]. Biorefineries are crucial to BE concepts' sustainability goals, connecting CE and BE agendas. By producing biochemicals and bioenergy from biomass in biorefineries, CBE integrates the CE framework. A complete transition from a fossil-based economy to a CBE requires substantial scientific research and innovation [127, 129, 132].

Renewable energy and carbon use are evaluated using techno-economic analysis (TEA) where production, design optimization, and price estimation are its priorities. Emerging technologies must meet Department of Energy (DOE) environmental performance standards such as minimum fuel selling price (MFSP), greenhouse gas (GHG) effects, and energy return on investment (EROI). Process design, modeling, economic indicators, and cash flow analysis are used to drive early-stage R&D and assess a technology's commercial viability by evaluating theoretical scaled-up processes using lab or pilot data [133, 134]. TEA has evaluated the economic viability of various pretreatment methods for producing high-value oligosaccharides from *Miscanthus* Mx2779. They examined both individual and combined steam explosion (SE) and ionic liquid (IL) pretreatments, specifically using [TEA][HSO_4] and [C2mim][OAc]. The TEA focused on how these pretreatments altered biomass composition, directly impacting the efficiency of enzymatic hydrolysis for glucose or glucooligosaccharide production. Economically, standalone SE or [C2mim][OAc] pretreatments appear more cost-effective for producing xylo- and glucooligosaccharides than [TEA][HSO_4]. Notably, [C2mim][OAc] offers a significant economic advantage by recovering xylooligosaccharide (XOS) while leaving a highly digestible cellulose pulp for biofuels or glucooligosaccharide products. Conversely, combining SE with [TEA][HSO_4] showed promise for improved techno-economics through XOS coproduction. This study highlights that different pretreatment methods yield distinct outputs, making them suitable for various downstream bioprocesses. Long-term economic feasibility requires further bench- and pilot-scale experiments and comprehensive techno-economic evaluations [135].

To enhance value chains and production flexibility, producers are exploring coproducts like cellooligosaccharides (COS), functional oligosaccharides with high market value. This study investigated the economic and environmental feasibility of producing cellopentaose from sugarcane straw, a process rarely detailed in the literature. Analysis across six scenarios estimated cellopentaose unit production costs ranging from USD 0.40 to 1.15/mg. Results demonstrated that optimizing the upstream sector significantly reduces total plant direct costs and overall lifecycle impact. Environmental impact reductions between 16.2% and 19.9% were observed. The study also highlighted the critical need for a better understanding of the hydrolysis solid/liquid proportion to minimize bulk material costs. Sugarcane straw fermentation emerged as a prospective technology for COS production within an integrated biorefinery framework, showcasing a promising avenue for sustainable coproduct generation [136].

9 Conclusions

In a circular bioeconomy, valorizing lignocellulosic biomass into oligosaccharides produces value-added products sustainably. Using agricultural and forestry residues decreases pollution and boosts waste value. Recent advances in pretreatment, enzymatic hydrolysis, and process parameters have increased biomass conversion efficiency. These processes in biorefinery models enable the creation of low-carbon, profitable alternatives to fossil fuels. For environmental and carbon–neutral reasons, bio-based compounds from lignocellulosic feedstocks may be competitive, according to economic assessments. Biomass conversion systems are becoming more feasible and scalable due to technological advances, making them essential for sustainable resource use and energy production. Applications of recovered oligosaccharides are examined for economic and environmental impacts. This method cuts fossil fuel use and greenhouse gas emissions. Scaling up technological advances can improve resource use, health, and sustainable industrial practices. Oligosaccharides have broad applications as prebiotics in functional foods and beverages, promoting gut health. Opportunities lie in personalized nutrition, pharmaceutical drug delivery, cosmetics, and animal feed, driven by increasing health consciousness and demand for natural ingredients.

References

1. Onokwai AO, Ajisegiri ESA, Okokpujie IP, Ibikunle RA, Oki M, Dirisu JO (2022) Characterization of lignocellulose biomass based on proximate, ultimate, structural composition, and thermal analysis. Mater Today Proc 65:2156–2162. https://doi.org/10.1016/j.matpr.2022.05.313
2. Kalak T (2023) Potential use of industrial biomass waste as a sustainable energy source in the future. Energies (Basel) 16:1783. https://doi.org/10.3390/en16041783
3. Gupta J, Kumari M, Mishra A, Akram SM, Thakur IS (2022) Agro-forestry waste management–a review. Chemosphere 287:132321. https://doi.org/10.1016/j.chemosphere.2021.132321
4. Limayem A, Ricke SC (2012) Lignocellulosic biomass for bioethanol production: current perspectives, potential issues and future prospects. Prog Energy Combust Sci 38:449–467. https://doi.org/10.1016/j.pecs.2012.03.002
5. Lindner M, Dees MG, Anttila P, Verkerk PJ, Fitzgerald J, Datta P, Glavonjic B, Prinz R, Zudin S (2017) Assessing lignocellulosic biomass potentials from forests and industry. In: Modeling and optimization of biomass supply chains, Elsevier, pp 127–167. https://doi.org/10.1016/B978-0-12-812303-4.00006-9
6. Sooch BS, Lugani Y, Singh RS (2019) 2. Agro-industrial lignocellulosic residues for the production of industrial enzymes. In: Industrial biotechnology, de gruyter, pp 31–50. https://doi.org/10.1515/9783110563337-002
7. Prajapati SK, Kaushik P, Malik A, Vijay VK (2013) Phycoremediation coupled production of algal biomass, harvesting and anaerobic digestion: possibilities and challenges. Biotechnol Adv 31:1408–1425. https://doi.org/10.1016/j.biotechadv.2013.06.005
8. González-Fernández C, Sialve B, Bernet N, Steyer J (2012) Impact of microalgae characteristics on their conversion to biofuel. Part II: Focus Biomethane Prod, Biofuels, Bioprod Biorefining 6:205–218. https://doi.org/10.1002/bbb.337

9. Singh J, Gu S (2010) Commercialization potential of microalgae for biofuels production. Renew Sustain Energy Rev 14:2596–2610. https://doi.org/10.1016/j.rser.2010.06.014
10. Velvizhi G, Goswami C, Shetti NP, Ahmad E, Kishore Pant K, Aminabhavi TM (2022) Valorisation of lignocellulosic biomass to value-added products: paving the pathway towards low-carbon footprint. Fuel 313:122678. https://doi.org/10.1016/j.fuel.2021.122678
11. Sawatdeenarunat C, Surendra KC, Takara D, Oechsner H, Khanal SK (2015) Anaerobic digestion of lignocellulosic biomass: challenges and opportunities. Bioresour Technol 178:178–186. https://doi.org/10.1016/j.biortech.2014.09.103
12. Kumar R, Sharma RK, Singh AP (2017) Cellulose based grafted biosorbents–Journey from lignocellulose biomass to toxic metal ions sorption applications–A review. J Mol Liq 232:62–93. https://doi.org/10.1016/j.molliq.2017.02.050
13. Vanneste J, Ennaert T, Vanhulsel A, Sels B (2017) Unconventional pretreatment of lignocellulose with low-temperature plasma. Chemsuschem 10:14–31. https://doi.org/10.1002/cssc.201601381
14. Lee HV, Hamid SBA, Zain SK (2014) Conversion of lignocellulosic biomass to nanocellulose: structure and chemical process. Sci World J 2014:1–20. https://doi.org/10.1155/2014/631013
15. Liu Y-R, Thomsen K, Nie Y, Zhang S-J, Meyer AS (2016) Predictive screening of ionic liquids for dissolving cellulose and experimental verification. Green Chem 18:6246–6254. https://doi.org/10.1039/C6GC01827K
16. Baruah J, Nath BK, Sharma R, Kumar S, Deka RC, Baruah DC, Kalita E (2018) Recent trends in the pretreatment of lignocellulosic biomass for value-added products. Front Energy Res 6. https://doi.org/10.3389/fenrg.2018.00141
17. Ohno H, Fukaya Y (2009) Task specific ionic liquids for cellulose technology. Chem Lett 38:2–7. https://doi.org/10.1246/cl.2009.2
18. Haldar D, Purkait MK (2020) Micro and nanocrystalline cellulose derivatives of lignocellulosic biomass: a review on synthesis, applications and advancements. Carbohydr Polym 250:116937. https://doi.org/10.1016/j.carbpol.2020.116937
19. Spiridon I, Popa VI (2008) Hemicelluloses: major sources, properties and applications. In: Monomers, polymers and composites from renewable resources, Elsevier, pp 289–304
20. Hu L, Du M, Zhang J (2018) Hemicellulose-based hydrogels present status and application prospects: a brief review. Open J For 08:15–28. https://doi.org/10.4236/ojf.2018.81002
21. Huang L-Z, Ma M-G, Ji X-X, Choi S-E, Si C (2021) Recent developments and applications of hemicellulose from wheat straw: a review. Front Bioeng Biotechnol 9. https://doi.org/10.3389/fbioe.2021.690773
22. dIsabelle Herpoel Gimbert (2017) Efficient biomass pre-treatment using the white-rot fungus Polyporus Brumalis. Fungal Genom Biol 07. https://doi.org/10.4172/2165-8056.1000150
23. Phan PT, Nguyen B-S, Nguyen T-A, Kumar A, Nguyen V-H (2023) Lignocellulose-derived monosugars: a review of biomass pre-treating techniques and post-methods to produce sustainable biohydrogen. Biomass Convers Biorefin 13:8425–8439. https://doi.org/10.1007/s13399-020-01161-7
24. Farhat W, Venditti RA, Hubbe M, Taha M, Becquart F, Ayoub A (2017) A review of water-resistant hemicellulose-based materials: processing and applications. Chemsuschem 10:305–323. https://doi.org/10.1002/cssc.201601047
25. Zakzeski J, Bruijnincx PCA, Jongerius AL, Weckhuysen BM (2010) The catalytic valorization of lignin for the production of renewable chemicals. Chem Rev 110:3552–3599. https://doi.org/10.1021/cr900354u
26. Upton BM, Kasko AM (2016) Strategies for the conversion of lignin to high-value polymeric materials: review and perspective. Chem Rev 116:2275–2306. https://doi.org/10.1021/acs.chemrev.5b00345
27. Aldous L, Argyropoulos D, Budarin V, Cabrera-Rodríguez CI, Clark J, Coma M, Constable D, Fan J, Gschwend F, Hayes D, Hu C, Huber G, Hughes T, Hunt A, Lapkin A, Lye G, Mascal M, Matharu A, Mu X, Pant D, Papaioannou E, Rinaldi R, Rothenberg G, Samec J, Sarkanen S, Stankiewicz A, Stevens C, Titirici M, Waldron K, Wilson K, Zhenova A, Zhou L, Zuin V (2017) Feedstocks and analysis: general discussion. Faraday Discuss 202:497–519. https://doi.org/10.1039/C7FD90050C

28. Akhtar N, Gupta K, Goyal D, Goyal A (2016) Recent advances in pretreatment technologies for efficient hydrolysis of lignocellulosic biomass. Environ Prog Sustain Energy 35:489–511. https://doi.org/10.1002/ep.12257
29. Tarasov D, Leitch M, Fatehi P (2018) Lignin–carbohydrate complexes: properties, applications, analyses, and methods of extraction: a review. Biotechnol Biofuels 11:269. https://doi.org/10.1186/s13068-018-1262-1
30. Xu C, Ferdosian F (2017) Conversion of lignin into bio-based chemicals and materials, Springer Berlin Heidelberg, Berlin, Heidelberg. https://doi.org/10.1007/978-3-662-54959-9
31. Gaspar R, Fardim P (2023) Lignin-based materials for emerging advanced applications. Curr Opin Green Sustain Chem 41:100834. https://doi.org/10.1016/j.cogsc.2023.100834
32. de Freitas C, Terrone CC, Forsan CF, Milagres AMF, Brienzo M (2022) Oligosaccharides from lignocellulosic biomass and their biological and physicochemical properties, pp 275–309. https://doi.org/10.1007/978-981-16-3682-0_9
33. Mussatto SI, Mancilha IM (2007) Non-digestible oligosaccharides: a review. Carbohydr Polym 68:587–597. https://doi.org/10.1016/j.carbpol.2006.12.011
34. Sophonputtanaphoca S, Pridam C, Chinnak J, Nathong M, Juntipwong P (2018) Production of non-digestible oligosaccharides as value-added by-products from rice straw. Agric Nat Resour 52:169–175. https://doi.org/10.1016/j.anres.2018.06.013
35. Zhao X, Zhang L, Liu D (2012) Biomass recalcitrance. Part I: the chemical compositions and physical structures affecting the enzymatic hydrolysis of lignocellulose. Biofuels, Bioprod Biorefining 6:465–482. https://doi.org/10.1002/bbb.1331
36. Huang R, Su R, Qi W, He Z (2010) Understanding the key factors for enzymatic conversion of pretreated lignocellulose by partial least square analysis. Biotechnol Prog 26:384–392. https://doi.org/10.1002/btpr.324
37. Agbor VB, Cicek N, Sparling R, Berlin A, Levin DB (2011) Biomass pretreatment: fundamentals toward application. Biotechnol Adv 29:675–685. https://doi.org/10.1016/j.biotechadv.2011.05.005
38. Baksi S, Saha D, Saha S, Sarkar U, Basu D, Kuniyal JC (2023) Pre-treatment of lignocellulosic biomass: review of various physico-chemical and biological methods influencing the extent of biomass depolymerization. Int J Environ Sci Technol 20:13895–13922. https://doi.org/10.1007/s13762-023-04838-4
39. Wyman CE (1999) Biomass ethanol: technical progress, opportunities, and commercial challenges. Annu Rev Energy Env 24:189–226. https://doi.org/10.1146/annurev.energy.24.1.189
40. Lynd LR, Elander RT, Wyman CE (1996) Likely features and costs of mature biomass ethanol technology. In: Seventeenth symposium on biotechnology for fuels and chemicals, Humana Press, Totowa, NJ, pp 741–761. https://doi.org/10.1007/978-1-4612-0223-3_70
41. Kumar AK, Sharma S (2017) Recent updates on different methods of pretreatment of lignocellulosic feedstocks: a review. Bioresour Bioprocess 4:7. https://doi.org/10.1186/s40643-017-0137-9
42. Bhatia L, Johri S, Ahmad R (2012) An economic and ecological perspective of ethanol production from renewable agro waste: a review. AMB Express 2:65. https://doi.org/10.1186/2191-0855-2-65
43. da Silva AS, Inoue H, Endo T, Yano S, Bon EPS (2010) Milling pretreatment of sugarcane bagasse and straw for enzymatic hydrolysis and ethanol fermentation. Bioresour Technol 101:7402–7409. https://doi.org/10.1016/j.biortech.2010.05.008
44. Gogate PR, Sutkar VS, Pandit AB (2011) Sonochemical reactors: important design and scale up considerations with a special emphasis on heterogeneous systems. Chem Eng J 166:1066–1082
45. Kilzer FJ, Broido A (1965) Speculations on the nature of cellulose pyrolysis. Pyrodynamics 2:151–163 2
46. Kumar AK, Sharma S (2017) Recent updates on different methods of pretreatment of lignocellulosic feedstocks: a review. Bioresour Bioprocess 4:7

47. Bimestre TA, Júnior JAM, Canettieri EV, Tuna CE (2022) Hydrodynamic cavitation for lignocellulosic biomass pretreatment: a review of recent developments and future perspectives. Bioresour Bioprocess 9:7. https://doi.org/10.1186/s40643-022-00499-2
48. Ren X, Li C, Yang F, Huang Y, Huang C, Zhang K, Yan L (2020) Comparison of hydrodynamic and ultrasonic cavitation effects on soy protein isolate functionality. J Food Eng 265:109697
49. Banu JR, Kavitha S, Tyagi VK, Gunasekaran M, Karthikeyan OP, Kumar G (2021) Lignocellulosic biomass based biorefinery: a successful platform towards circular bioeconomy. Fuel 302:121086
50. Kininge MM, Gogate PR (2022) Intensification of alkaline delignification of sugarcane bagasse using ultrasound assisted approach. Ultrason Sonochem 82:105870
51. Kululo WW, Habtu NG, Abera MK, Sendekie ZB, Fanta SW, Yemata TA (2025) Advances in various pretreatment strategies of lignocellulosic substrates for the production of bioethanol: a comprehensive review. Discov Appl Sci 7:476. https://doi.org/10.1007/s42452-025-06748-1
52. Raina N, Boonmee R, Kirdponpattara S, Narasingha M, Sriariyanun M, Phitsuwan P, Chuetor S (2024) Process performance evaluation of different chemical pretreatments of lignocellulosic biomass for bioethanol production. Ind Crops Prod 211:118207
53. Borand MN, Karaosmanoğlu F (2018) Effects of organosolv pretreatment conditions for lignocellulosic biomass in biorefinery applications: a review. J Renew Sustain Energy 10
54. Sarker TR, Pattnaik F, Nanda S, Dalai AK, Meda V, Naik S (2021) Hydrothermal pretreatment technologies for lignocellulosic biomass: a review of steam explosion and subcritical water hydrolysis. Chemosphere 284:131372
55. Zhang B, Li H, Chen L, Fu T, Tang B, Hao Y, Li J, Li Z, Zhang B, Chen Q (2022) Recent advances in the bioconversion of waste straw biomass with steam explosion technique: a comprehensive review. Processes 10:1959
56. Champadang O, Boonsombuti A, Luengnaruemitchai A (2022) Enhanced enzymatic digestibility of water lettuce by liquid hot water pretreatment. Bioresour Technol Rep 18:101100
57. Prasad BR, Padhi RK, Ghosh G (2023) A review on key pretreatment approaches for lignocellulosic biomass to produce biofuel and value-added products. Int J Environ Sci Technol 20:6929–6944
58. Zhao C, Shao Q, Chundawat SPS (2020) Recent advances on ammonia-based pretreatments of lignocellulosic biomass. Bioresour Technol 298:122446
59. Mohammadi M, Alian M, Dale B, Ubanwa B, Balan V (2024) Multifaced application of AFEX-pretreated biomass in producing second-generation biofuels, ruminant animal feed, and value-added bioproducts. Biotechnol Adv, p 108341
60. Ramos LP, De León Mayorga FM, Silveira MHL, Galvão C, Corazza ML (2017) Pre-treatment of biomass using CO_2-based methods
61. Vats S, Maurya DP, Shaimoon M, Agarwal A, Negi S (2013) Development of a microbial consortium for production of blend of enzymes for hydrolysis of agricultural wastes into sugars
62. Sánchez C (2009) Lignocellulosic residues: biodegradation and bioconversion by fungi. Biotechnol Adv 27:185–194
63. Kumar P, Barrett DM, Delwiche MJ, Stroeve P (2009) Methods for pretreatment of lignocellulosic biomass for efficient hydrolysis and biofuel production. Ind Eng Chem Res 48:3713–3729
64. Sun Y, Cheng J (2002) Hydrolysis of lignocellulosic materials for ethanol production: a review. Bioresour Technol 83:1–11
65. Hendriks ATWM, Zeeman G (2009) Pretreatments to enhance the digestibility of lignocellulosic biomass. Bioresour Technol 100:10–18. https://doi.org/10.1016/j.biortech.2008.05.027
66. Adsul M, Sandhu SK, Singhania RR, Gupta R, Puri SK, Mathur A (2020) Designing a cellulolytic enzyme cocktail for the efficient and economical conversion of lignocellulosic biomass to biofuels. Enzyme Microb Technol 133:109442

67. Aishwarya A, Goyal A (2025) Development of a novel tailor-made cocktail from recombinant crude enzymes for efficient saccharification of pretreated elephant grass. Int J Biol Macromol 144645. https://doi.org/10.1016/j.ijbiomac.2025.144645
68. Zhu D, Qaria MA, Zhu B, Sun J, Yang B (2022) Extremophiles and extremozymes in lignin bioprocessing. Renew Sustain Energy Rev 157:112069. https://doi.org/10.1016/j.rser.2021.112069
69. Chica RA, Doucet N, Pelletier JN (2005) Semi-rational approaches to engineering enzyme activity: combining the benefits of directed evolution and rational design. Curr Opin Biotechnol 16:378–384. https://doi.org/10.1016/j.copbio.2005.06.004
70. Pu Y, Hu F, Huang F, Davison BH, Ragauskas AJ (2013) Assessing the molecular structure basis for biomass recalcitrance during dilute acid and hydrothermal pretreatments. Biotechnol Biofuels 6:1–13
71. du Pasquier J, Paës G, Perré P (2023) Principal factors affecting the yield of dilute acid pretreatment of lignocellulosic biomass: a critical review. Bioresour Technol 369:128439. https://doi.org/10.1016/j.biortech.2022.128439
72. Kovačić Đ, Radočaj D, Samac D, Jurišić M (2024) Influence of thermal pretreatment on lignin destabilization in harvest residues: an ensemble machine learning approach. AgriEngineering 6:171–184. https://doi.org/10.3390/agriengineering6010011
73. Podgorbunskikh EM, Bychkov AL, Ryabchikova EI, Lomovsky OI (2020) The effect of thermomechanical pretreatment on the structure and properties of lignin-rich plant biomass. Molecules 25:995. https://doi.org/10.3390/molecules25040995
74. Wang P, Howard B (2017) Impact of thermal pretreatment temperatures on woody biomass chemical composition, physical properties and microstructure. Energies (Basel) 11:25. https://doi.org/10.3390/en11010025
75. Łukajtis R, Rybarczyk P, Kucharska K, Konopacka-Łyskawa D, Słupek E, Wychodnik K, Kamiński M (2018) Optimization of saccharification conditions of lignocellulosic biomass under alkaline pre-treatment and enzymatic hydrolysis. Energies (Basel) 11:886. https://doi.org/10.3390/en11040886
76. Ummalyma SB, Herojit N, Sukumaran RK (2024) Alkaline hydrogen peroxide pretreatment of bamboo residues and its influence on physiochemical properties and enzymatic digestibility for bioethanol production. Front Energy Res 12. https://doi.org/10.3389/fenrg.2024.1444813
77. Kucharska K, Rybarczyk P, Hołowacz I, Łukajtis R, Glinka M, Kamiński M (2018) Pretreatment of lignocellulosic materials as substrates for fermentation processes. Molecules 23:2937. https://doi.org/10.3390/molecules23112937
78. Banerjee G, Car S, Scott-Craig JS, Hodge DB, Walton JD (2011) Alkaline peroxide pretreatment of corn stover: effects of biomass, peroxide, and enzyme loading and composition on yields of glucose and xylose. Biotechnol Biofuels 4:16. https://doi.org/10.1186/1754-6834-4-16
79. Cioabla AE, Ionel I, Dumitrel G-A, Popescu F (2012) Comparative study on factors affecting anaerobic digestion of agricultural vegetal residues. Biotechnol Biofuels 5:39. https://doi.org/10.1186/1754-6834-5-39
80. Beschkov VN, Angelov IK (2025) Volatile fatty acid production versus methane and hydrogen in anaerobic digestion. Fermentation 11:172. https://doi.org/10.3390/fermentation11040172
81. Brodeur G, Yau E, Badal K, Collier J, Ramachandran KB, Ramakrishnan S (2011) Chemical and physicochemical pretreatment of lignocellulosic biomass: a review. Enzyme Res 2011:1–17. https://doi.org/10.4061/2011/787532
82. Chen J, Ma X, Liang M, Guo Z, Cai Y, Zhu C, Wang Z, Wang S, Xu J, Ying H (2024) Physical–chemical–biological pretreatment for biomass degradation and industrial applications: a review. Waste 2:451–473. https://doi.org/10.3390/waste2040024
83. Shobana S, Kumar G, Bakonyi P, Saratale GD, Al-Muhtaseb AH, Nemestóthy N, Bélafi-Bakó K, Xia A, Chang J-S (2017) A review on the biomass pretreatment and inhibitor removal methods as key-steps towards efficient macroalgae-based biohydrogen production. Bioresour Technol 244:1341–1348. https://doi.org/10.1016/j.biortech.2017.05.172

84. Zdeb M, Pawłowska M, Skóra A (2021) Influence of temperature and reaction time on the efficiency of alkaline pretreatment of hay biomass. J Eco Eng 22:120–127. https://doi.org/10.12911/22998993/130882
85. Prunescu R, Blanke M, Jensen J, Sin G (2012) Temperature modelling of the biomass pretreatment process. https://doi.org/10.13140/2.1.1512.1287
86. Zhang L, Peng H, Yu Q, Zhang Y, Wang Z, Tang X (2015) Optimization of liquid ammonia treatment for enzymatic hydrolysis of miscanthus sinensis anderss. J Power Energy Eng 03:26–32. https://doi.org/10.4236/jpee.2015.37005
87. Cai J, Wei L, Wang J, Lin N, Li Y, Li F, Zha X, Li W (2024) Application of catalysts in the conversion of biomass and its derivatives. Catalysts 14:499. https://doi.org/10.3390/catal14080499
88. Li Z, Chen CH, Hegg EL, Hodge DB (2013) Rapid and effective oxidative pretreatment of woody biomass at mild reaction conditions and low oxidant loadings. Biotechnol Biofuels 6:119. https://doi.org/10.1186/1754-6834-6-119
89. Wongkaew M, Tinpovong B, Sringarm K, Leksawasdi N, Jantanasakulwong K, Rachtanapun P, Hanmoungjai P, Sommano SR (2021) Crude pectic oligosaccharide recovery from thai chok anan mango peel using pectinolytic enzyme hydrolysis. Foods 10:627. https://doi.org/10.3390/foods10030627
90. Vasconcelos SC, Rossa V, Quattrociocchi DGS, Madriaga VGC, Ribeiro MCO, Pinheiro SB, dos Santos RD, Archanjo BS, Urquieta-González EA, Soares AVH, Passos FB, Varma RS, Lima TM (2024) Catalytic upgrading of lignocellulosic biomass-derived compounds using hierarchical ZSM-5 zeolites. Energy Fuels 38:23823–23835. https://doi.org/10.1021/acs.energyfuels.4c04000
91. Zhang B, Liu X, Bao J (2023) High solids loading pretreatment: the core of lignocellulose biorefinery as an industrial technology—an overview. Bioresour Technol 369:128334. https://doi.org/10.1016/j.biortech.2022.128334
92. Chen H, Liu Z (2017) Enzymatic hydrolysis of lignocellulosic biomass from low to high solids loading. Eng Life Sci 17:489–499. https://doi.org/10.1002/elsc.201600102
93. Tao L, Ren J, Yu FK, Ni TR (2013) Effects of liquid-to-solid ratio and reaction temperature on NaOH pretreatment of achnatherum splendens. Asian J Chem 25:3545–3548. https://doi.org/10.14233/ajchem.2013.13424
94. Huang C, Yu Y, Li Z, Yan B, Pei W, Wu H (2022) The preparation technology and application of xylo-oligosaccharide as prebiotics in different fields: a review. Front Nutr 9. https://doi.org/10.3389/fnut.2022.996811
95. Saini R, Patel AK, Saini JK, Chen C-W, Varjani S, Singhania RR, Di Dong C (2022) Recent advancements in prebiotic oligomers synthesis via enzymatic hydrolysis of lignocellulosic biomass. Bioengineered 13:2139–2172. https://doi.org/10.1080/21655979.2021.2023801
96. Cano ME, García-Martin A, Comendador Morales P, Wojtusik M, Santos VE, Kovensky J, Ladero M (2020) Production of oligosaccharides from agrofood wastes. Fermentation 6:31. https://doi.org/10.3390/fermentation6010031
97. Yang S, Wu C, Yan Q, Li X, Jiang Z (2023) Nondigestible functional oligosaccharides: enzymatic production and food applications for intestinal health. Annu Rev Food Sci Technol 14:297–322. https://doi.org/10.1146/annurev-food-052720-114503
98. Yuansah SC, Laga A, Pirman (2023) Production strategy of functional oligosaccharides from lignocellulosic biomass using enzymatic process: a review. Food Bioproc Tech 16:2359–2381. https://doi.org/10.1007/s11947-023-03063-8
99. Blasi A, Verardi A, Lopresto CG, Siciliano S, Sangiorgio P (2023) Lignocellulosic agricultural waste valorization to obtain valuable products: an overview. Recycling 8. https://doi.org/10.3390/recycling8040061
100. Borrero-López AM, Valencia C, Franco JM (2022) Lignocellulosic materials for the production of biofuels, biochemicals and biomaterials and applications of lignocellulose-based polyurethanes: a review. Polymers (Basel) 14:881. https://doi.org/10.3390/polym14050881
101. Mittal A, Katahira R, Himmel ME, Johnson DK (2011) Effects of alkaline or liquid-ammonia treatment on crystalline cellulose: changes in crystalline structure and effects on enzymatic digestibility. Biotechnol Biofuels 4:41. https://doi.org/10.1186/1754-6834-4-41

102. Taherzadeh MJ, Karimi K (2008) Pretreatment of lignocellulosic wastes to improve ethanol and biogas production: a review. Int J Mol Sci 9:1621–1651. https://doi.org/10.3390/ijms9091621
103. Broda M, Yelle DJ, Serwańska K (2022) Bioethanol production from lignocellulosic biomass—challenges and solutions. Molecules 27:8717. https://doi.org/10.3390/molecules27248717
104. Nagarajan D, Lee D-J, Chang J-S (2019) Recent insights into consolidated bioprocessing for lignocellulosic biohydrogen production. Int J Hydrogen Energy 44:14362–14379. https://doi.org/10.1016/j.ijhydene.2019.03.066
105. Rodionova MV, Bozieva AM, Zharmukhamedov SK, Leong YK, Chi-Wei Lan J, Veziroglu A, Veziroglu TN, Tomo T, Chang J-S, Allakhverdiev SI (2022) A comprehensive review on lignocellulosic biomass biorefinery for sustainable biofuel production. Int J Hydrogen Energy 47:1481–1498. https://doi.org/10.1016/j.ijhydene.2021.10.122
106. Cripwell RA, Favaro L, Viljoen-Bloom M, van Zyl WH (2020) Consolidated bioprocessing of raw starch to ethanol by Saccharomyces cerevisiae: achievements and challenges. Biotechnol Adv 42:107579. https://doi.org/10.1016/j.biotechadv.2020.107579
107. Mussatto SI, Aguilar CN, Rodrigues LR, Teixeira JA (2009) Fructooligosaccharides and β-fructofuranosidase production by Aspergillus japonicus immobilized on lignocellulosic materials. J Mol Catal B Enzym 59:76–81. https://doi.org/10.1016/j.molcatb.2009.01.005
108. Jovanovic-Malinovska R, Kuzmanova S, Winkelhausen E (2015) Application of ultrasound for enhanced extraction of prebiotic oligosaccharides from selected fruits and vegetables. Ultrason Sonochem 22:446–453. https://doi.org/10.1016/j.ultsonch.2014.07.016
109. Yue P, Hu Y, Tian R, Bian J, Peng F (2022) Hydrothermal pretreatment for the production of oligosaccharides: a review. Bioresour Technol 343:126075. https://doi.org/10.1016/j.biortech.2021.126075
110. Chen Y, Sui X, Wang Y, Zhao Z, Han T, Liu Y, Zhang J, Zhou P, Yang K, Ye Z (2024) Preparation, structural characterization, biological activity, and nutritional applications of oligosaccharides. Food Chem X 22:101289. https://doi.org/10.1016/j.fochx.2024.101289
111. Su Z, Luo J, Li X, Pinelo M (2020) Enzyme membrane reactors for production of oligosaccharides: a review on the interdependence between enzyme reaction and membrane separation. Sep Purif Technol 243:116840. https://doi.org/10.1016/j.seppur.2020.116840
112. Chauhan DDAPS, Usmani SA, Chauhan S, Singh A, Deepak D (2024) Optimized purification processes for isolation and modification of oligosaccharides from rathi cow's milk. Bio Life Sci Forum 40. https://doi.org/10.3390/blsf2024040021
113. Zhang L, Li X, Yong Q, Yang S-T, Ouyang T, Yu S (2015) Simultaneous saccharification and fermentation of xylo-oligosaccharides manufacturing waste residue for l-lactic acid production by Rhizopus oryzae, Biochem Eng J 94:92–99. https://doi.org/10.1016/j.bej.2014.11.020
114. Gullón B, Garrote G, Alonso JL, Parajó JC (2007) Production of l-lactic acid and oligomeric compounds from apple pomace by simultaneous saccharification and fermentation: a response surface methodology assessment. J Agric Food Chem 55:5580–5587. https://doi.org/10.1021/jf070442v
115. Courtois J (2009) Oligosaccharides from land plants and algae: production and applications in therapeutics and biotechnology. Curr Opin Microbiol 12:261–273. https://doi.org/10.1016/j.mib.2009.04.007
116. Mussatto SI, Mancilha IM (2007) Non-digestible oligosaccharides: a review, Carbohydr Polym 68:587–597. https://doi.org/10.1016/j.carbpol.2006.12.011
117. Qiang X, YongLie C, QianBing W (2009) Health benefit application of functional oligosaccharides. Carbohydr Polym 77:435–441. https://doi.org/10.1016/j.carbpol.2009.03.016
118. Wu S-C, Wen T-N, Pan C-L (2005) Algal-oligosaccharide-lysates prepared by two bacterial agarases stepwise hydrolyzed and their anti-oxidative properties. Fish Sci 71:1149–1159. https://doi.org/10.1111/j.1444-2906.2005.01075.x
119. Tuohy KM, Rouzaud GCM, Bruck WM, Gibson GR (2005) Modulation of the human gut microflora towards improved health using prebiotics–assessment of efficacy. Curr Pharm Des 11:75–90. https://doi.org/10.2174/1381612053382331

120. Sánchez O, Guio F, Garcia D, Silva E, Caicedo L (2008) Fructooligosaccharides production by Aspergillus sp. N74 in a mechanically agitated airlift reactor. Food Bioproducts Process 86:109–115. https://doi.org/10.1016/j.fbp.2008.02.003
121. Johnstone KD, Karoli T, Liu L, Dredge K, Copeman E, Li CP, Davis K, Hammond E, Bytheway I, Kostewicz E, Chiu FCK, Shackleford DM, Charman SA, Charman WN, Harenberg J, Gonda TJ, Ferro V (2010) Synthesis and biological evaluation of polysulfated oligosaccharide glycosides as inhibitors of angiogenesis and tumor growth. J Med Chem 53:1686–1699. https://doi.org/10.1021/jm901449m
122. Hosomi O, Misawa Y, Takeya A, Matahira Y, Sugahara K, Kubohara Y, Yamakura F, Kudo S (2009) Novel oligosaccharide has suppressive activity against human leukemia cell proliferation. Glycoconj J 26:189–198. https://doi.org/10.1007/s10719-008-9175-z
123. Patel S, Goyal A (2011) Functional oligosaccharides: production, properties and applications. World J Microbiol Biotechnol 27:1119–1128. https://doi.org/10.1007/s11274-010-0558-5
124. Mahious AS, Gatesoupe FJ, Hervi M, Metailler R, Ollevier F (2006) Effect of dietary inulin and oligosaccharides as prebiotics for weaning turbot, Psetta maxima (Linnaeus, C. 1758). Aquaculture Int 14:219–229. https://doi.org/10.1007/s10499-005-9003-4
125. Li XJ, Piao XS, Kim SW, Liu P, Wang L, Shen YB, Jung SC, Lee HS (2007) Effects of chito-oligosaccharide supplementation on performance, nutrient digestibility, and serum composition in broiler chickens. Poult Sci 86:1107–1114. https://doi.org/10.1093/ps/86.6.1107
126. Sadhukhan J, Martinez-Hernandez E, Amezcua-Allieri MA, Aburto J, Honorato SJA (2019) Economic and environmental impact evaluation of various biomass feedstock for bioethanol production and correlations to lignocellulosic composition. Bioresour Technol Rep 7:100230. https://doi.org/10.1016/j.biteb.2019.100230
127. Ubando AT, Felix CB, Chen W-H (2020) Biorefineries in circular bioeconomy: a comprehensive review. Bioresour Technol 299:122585. https://doi.org/10.1016/j.biortech.2019.122585
128. Dhamodharan K, Ahlawat S, Kaushal M, Rajendran K (2020) 25–economics and cost analysis of waste biorefineries. In: Kumar RP, Gnansounou E, Raman JK, Baskar G (Eds.), Refining biomass residues for sustainable energy and bioproducts, Academic Press, pp 545–565. https://doi.org/10.1016/B978-0-12-818996-2.00025-9
129. Patel R, Rajaraman TS, Rana PH, Ambegaonkar NJ, Patel S (2025) A review on techno-economic analysis of lignocellulosic biorefinery producing biofuels and high-value products. Results Chem 13:102052. https://doi.org/10.1016/j.rechem.2025.102052
130. Stegmann P, Londo M, Junginger M (2020) The circular bioeconomy: its elements and role in European bioeconomy clusters. Resour, Conserv & Recycl: X 6:100029. https://doi.org/10.1016/j.rcrx.2019.100029
131. Louw J, Dogbe ES, Yang B, Görgens JF (2023) Prioritisation of biomass-derived products for biorefineries based on economic feasibility: a review on the comparability of techno-economic assessment results. Renew Sustain Energy Rev 188:113840. https://doi.org/10.1016/j.rser.2023.113840
132. Hadley Kershaw E, Hartley S, McLeod C, Polson P (2021) The sustainable path to a circular bioeconomy. Trends Biotechnol 39:542–545. https://doi.org/10.1016/j.tibtech.2020.10.015
133. Velvizhi G, Nair R, Goswami C, Arumugam SK, Shetti NP, Aminabhavi TM (2023) Carbon credit reduction: a techno-economic analysis of "drop-in" fuel production. Environ Pollut 316:120507. https://doi.org/10.1016/j.envpol.2022.120507
134. Zimmermann AW, Wunderlich J, Müller L, Buchner GA, Marxen A, Michailos S, Armstrong K, Naims H, McCord S, Styring P, Sick V, Schomäcker R (2020) Techno-economic assessment guidelines for CO_2 utilization. Front Energy Res 8. https://doi.org/10.3389/fenrg.2020.00005
135. Bhatia R, Lad JB, Bosch M, Bryant DN, Leak D, Hallett JP, Franco TT, Gallagher JA (2021) Production of oligosaccharides and biofuels from Miscanthus using combinatorial steam explosion and ionic liquid pretreatment. Bioresour Technol 323:124625. https://doi.org/10.1016/j.biortech.2020.124625
136. Barbosa FC, Nogueira GP, Kendrick E, Franco TT, Leak D, Dias MOS, Cavaliero CKN, Goldbeck R (2021) Production of cello-oligosaccharides through the biorefinery concept: a

technical-economic and life-cycle assessment. Biofuels, Bioprod Biorefining 15:1763–1774. https://doi.org/10.1002/bbb.2276
137. Ávila PF, Silva MF, Martins M, Goldbeck R (2021) Cello-oligosaccharides production from lignocellulosic biomass and their emerging prebiotic applications. World J Microbiol Biotechnol 37:73. https://doi.org/10.1007/s11274-021-03041-2
138. Gullón B, Dávila I, García-Torreiro M, Yáñez R, Labidi J, Gullón P (2017) Production and emerging applications of bioactive oligosaccharides from biomass hemicelluloses by hydrothermal processing. Hydrothermal Process Biorefineries: Prod Bioethanol High Added-Value Compd Second Third Gener Biomass, pp 253–283
139. Polizeli MD, Rizzatti ACS, Monti R, Terenzi HF, Jorge JA, Amorim DS (2005) Xylanases from fungi: properties and industrial applications. Appl Microbiol Biotechnol 67:577–591
140. Rebello S, Anju M, Aneesh EM, Sindhu R, Binod P, Pandey A (2017) Recent advancements in the production and application of microbial pectinases: an overview. Rev Environ Sci Biotechnol 16:381–394
141. Chuenkaek T, Vo TMT, Nakajima K, Kobayashi T (2025) Pectin materials sourced from agriculture waste: extraction, purification, properties, and applications. Build Low-Carbon Soc Appl Environ Mater Sci, pp 285–320
142. Kaczmarek MB, Struszczyk-Swita K, Li X, Szczęsna-Antczak M, Daroch M (2019) Enzymatic modifications of chitin, chitosan, and chitooligosaccharides. Front Bioeng Biotechnol 7:243
143. Kim S-K, Rajapakse N (2005) Enzymatic production and biological activities of chitosan oligosaccharides (COS): a review. Carbohydr Polym 62:357–368
144. Stoykov YM, Pavlov AI, Krastanov AI (2015) Chitinase biotechnology: production, purification, and application. Eng Life Sci 15:30–38
145. Kim HS, Lee C-G, Lee EY (2011) Alginate lyase: structure, property, and application. Biotechnol Bioprocess Eng 16:843–851
146. Liu J, Yang S, Li X, Yan Q, Reaney MJT, Jiang Z (2019) Alginate oligosaccharides: Production, biological activities, and potential applications. Compr Rev Food Sci Food Saf 18:1859–1881
147. Dhaneshwar S, Bhilare N, Roy S (2021) Dextran pharmaceutical applications. In: Polysaccharides of microbial origin: biomedical applications, Springer, pp 1–28
148. Lazaridou A, Biliaderis CG, Izydorczyk MS (2007) Cereal beta-glucans: structures, physical properties, and physiological functions. Funct Food Carbohydr, pp 1–72
149. Jin X, Wang J-K, Wang Q (2023) Microbial β-glucanases: production, properties, and engineering. World J Microbiol Biotechnol 39:106
150. Van Hijum S, FT A, Van Geel-Schutten GH, Rahaoui H. Van der Maarel MJEC, Dijkhuizen L (2002) Characterization of a novel fructosyltransferase from Lactobacillus reuteri that synthesizes high-molecular-weight inulin and inulin oligosaccharides. Appl Environ Microbiol 68:4390–4398
151. Saman P, Kuancha C, Chaiongkarn A, Moonmangmee S, Fungsin B (2019) Isomalto-oligosaccharides production from rice flour and cassava starch. J Food Sci Agric Technol (JFAT) 5:188–192
152. Singh RS, Singh RP, Kennedy JF (2016) Recent insights in enzymatic synthesis of fructooligosaccharides from inulin. Int J Biol Macromol 85:565–572
153. Narisetty V, Parhi P, Mohan B, Hakkim Hazeena S, Naresh Kumar A, Gullón B, Srivastava A, Nair LM, Paul Alphy M, Sindhu R, Kumar V, Castro E, Kumar Awasthi M, Binod P (2022) Valorization of renewable resources to functional oligosaccharides: recent trends and future prospective. Bioresour Technol 346:126590. https://doi.org/10.1016/j.biortech.2021.126590
154. Martínez-Herrera RE, Alemán-Huerta ME, Almaguer-Cantú V, Rosas-Flores W, Martínez-Gómez VJ, Quintero-Zapata I, Rivera G, Rutiaga-Quiñones OM (2020) Efficient recovery of thermostable polyhydroxybutyrate (PHB) by a rapid and solvent-free extraction protocol assisted by ultrasound. Int J Biol Macromol 164:771–782. https://doi.org/10.1016/j.ijbiomac.2020.07.101

155. De Gregorio PR, Gennari A, Nied CV, Volpato G, Volken de Souza CF (2023) Production of oligosaccharides, a prebiotic from lactose, using β-galactosidase. In: Enzymes beyond traditional applications in dairy science and technology, Elsevier, pp 383–401. https://doi.org/10.1016/B978-0-323-96010-6.00015-1
156. Verma P, Kaushik R, Sirohi R (2024) A review on the preparation, characterization, and applications of agro-waste-derived oligosaccharides. Food Biosci 62:105221. https://doi.org/10.1016/j.fbio.2024.105221
157. Gullón B, Dávila I, García-Torreiro M, Yáñez R, Labidi J, Gullón P (2017) Production and emerging applications of bioactive oligosaccharides from biomass hemicelluloses by hydrothermal processing. In: Hydrothermal processing in biorefineries, Springer International Publishing, Cham, pp 253–283. https://doi.org/10.1007/978-3-319-56457-9_10
158. Thurl S, Munzert M, Boehm G, Matthews C, Stahl B (2017) Systematic review of the concentrations of oligosaccharides in human milk. Nutr Rev 75:920
159. Jiang Z, Ding W, Fan J, Liao Y, Remón J, Shi B (2023) Biomass derived oligosaccharides for potential leather tanning. Collagen Leather 5:7. https://doi.org/10.1186/s42825-023-00113-x
160. Gong Z, Wang Q, Shen H, Wang L, Xie H, Zhao ZK (2014) Conversion of biomass-derived oligosaccharides into lipids. Biotechnol Biofuels 7:13. https://doi.org/10.1186/1754-6834-7-13

Key Analysis of Process Parameters for an Effective Enzymatic Hydrolysis for Cellulosic Sugar Production

James Villar, Matheus I. B. Aragão, Pedro R. Faria, Lucas Ramos, Susan H. Soriano Morales, Astrid Corrales Centeno, Rita C. L. B. Rodrigues, and Anuj K. Chandel

Abstract As a cornerstone of circular bioeconomy strategies, enzymatic hydrolysis transforms lignocellulosic biomass into fermentable monosaccharides (primarily glucose and xylose). This transformation is carried out using enzymatic systems (cocktails) that include cellulases, hemicellulases, and lytic polysaccharide monooxygenases (LPMOs). The derived sugars act as versatile precursors (platforms) for bio-based products, including fuels, plastics, and chemicals, enabling a transition from fossil resources to renewable alternatives. Nevertheless, several technical challenges limit the efficiency of enzymatic hydrolysis, as the inherent recalcitrance of the lignocellulosic biomass matrix. High enzyme costs, product inhibition, and mass transfer limitations, particularly at solids loadings above 15%, further restrict the process. Therefore, pretreatment methods are essential to alter biomass structure and improve enzyme accessibility. To overcome such limitations, optimization strategies target multiple fronts. Recent advancements in enzyme engineering include the development of stable, inhibition-resistant enzymes and optimized multienzyme cocktails, with LPMO incorporation proving particularly effective in addressing biomass accessibility challenges.

Keywords Cellulase · Biomass hydrolysis · 2G Sugar · Reactor configurations

J. Villar · M. I. B. Aragão · P. R. Faria · L. Ramos · A. C. Centeno · R. C. L. B. Rodrigues · A. K. Chandel (✉)
Department of Biotechnology, Engineering School of Lorena, University of São Paulo (EEL-USP), Lorena, SP, Brazil
e-mail: anuj10@usp.br

S. H. S. Morales
Department of Food Engineering (FEA), University of Campinas—UNICAMP, Campinas, Brazil

© The Author(s), under exclusive license to Springer Nature Singapore Pte Ltd. 2025
A. K. Chandel and S. Sundaramurthy (eds.), *Biorefineries in Circular Bioeconomy*, Green Energy and Technology, https://doi.org/10.1007/978-981-95-3913-0_5

1 Introduction

Enzymatic hydrolysis of lignocellulosic biomass is a key process in the circular bioeconomy, allowing the sustainable conversion of agricultural, forestry, and food residues into fermentable sugars such as glucose and xylose. This process primarily uses enzymes such as cellulases and hemicellulases to efficiently break down cellulose and hemicellulose, yielding high-value products with minimal formation of toxic byproducts [91, 105].

The sugars produced through enzymatic hydrolysis are essential platform molecules for the production of biofuels, bioplastics, biochemicals, and other high-value industrial compounds. Valorization of residual biomass streams by enzymatic hydrolysis supports waste reduction, resource circularity, and environmental sustainability-core principles of the circular bioeconomy[115].

Integrating enzymatic hydrolysis into biorefineries enhances biomass utilization efficiency. It is commonly combined with upstream pretreatment and downstream fermentation steps to maximize sugar yields. For example, feedstocks such as eucalyptus bark and agave bagasse, when pretreated through methods like kraft pulping or autohydrolysis, can reach glucose conversion efficiencies of 76–83% [5, 116]. Subsequently, enzymatic hydrolysis using tailor-made enzyme cocktails—such as Cellic® CTec3 HS in combination with hemicellulases (NS22244)—can enable the efficient depolymerization of lignocellulosic biomass into fermentable sugars, including glucose and xylose. Moreover, the addition of compounds such as hydrogen peroxide (H_2O_2) and Tween® 80 further enhances enzymatic activity and improves substrate accessibility (Fig. 1).

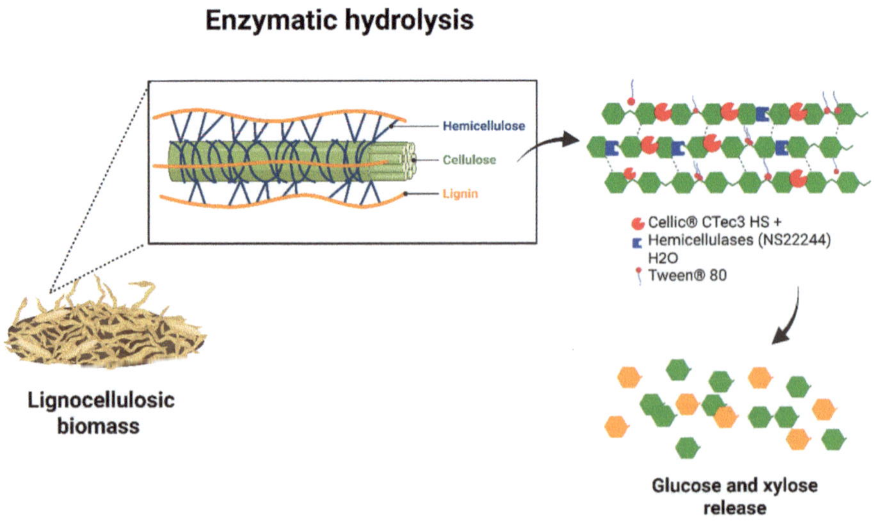

Fig. 1 Enzymatic attack from cocktails, such as Cellic® CTec3 HS, and hemicellulases (NS22244)

Nevertheless, enzymatic hydrolysis faces constraints such as the natural recalcitrance of lignocellulosic biomass to enzymatic breakdown and the high cost of the enzymes required [67, 99]. Such factors limit the efficiency and scalability of the process, despite advances in enzyme engineering and biomass pretreatment strategies [150].

2 Fundamentals of Enzymatic Hydrolysis

Plant biomass exhibits significant structural complexity due to its tripartite composition—cellulose (40–60%), hemicellulose (20–40%), and lignin (20–35%)—which form a highly recalcitrant matrix resistant to enzymatic hydrolysis [166]. Hemicellulose and lignin act as cross-linking agents, conferring rigidity and limiting enzyme accessibility. Cellulose, a linear polysaccharide of β-D-glucose units linked by β-1,4-glycosidic bonds, is organized into crystalline and amorphous regions [40, 120, 169]. Due to their crystalline packing, cellulose microfibrils present a major barrier to enzymatic breakdown, reducing hydrolysis efficiency.

Crystalline cellulose forms tightly packed microfibrils stabilized by extensive intrachain and interchain hydrogen bonding, making it less accessible to hydrolytic enzymes. Nonetheless, recent studies suggest that crystallinity alone is not the dominant factor in recalcitrance. To illustrate, chemical modification of cellulose using CO_2 and organic bases was shown to disrupt its hydrogen bonding network, transforming cellulose I into cellulose II and significantly improving enzymatic digestibility, achieving 100% glucose yield under optimal conditions [44]. Similarly, irradiation and oxidative cleavage pretreatment were found to enhance enzyme access by introducing new chain ends and weakening glycosidic bonds in crystalline cellulose [168].

Efficient deconstruction of this matrix requires multifunctional enzyme cocktails composed of synergistic biocatalysts. Cellulases are central to this process and are typically classified into three main enzyme groups:

- **Endoglucanases (EGs)**: Endoglucanases cleave internal β-1,4-glycosidic bonds in amorphous cellulose, generating new chain ends and oligosaccharides of various lengths [3, 59, 95].
- **Cellobiohydrolases (CBHs)**: Cellobiohydrolases types I and II (CBH I and II) are exoglucanases that degrade cellulose processively from the chain termini—CBH I from the reducing end and CBH II from the nonreducing end—generating cellobiose. Accumulation of cellobiose inhibits both CBH and endoglucanase (EG) activity, creating a rate-limiting feedback loop [8, 136].
- ***β*-glucosidases (BGs)**: β-glucosidases convert cellobiose and other oligosaccharides into glucose, mitigating product inhibition and providing fermentable sugars [48, 128, 137].

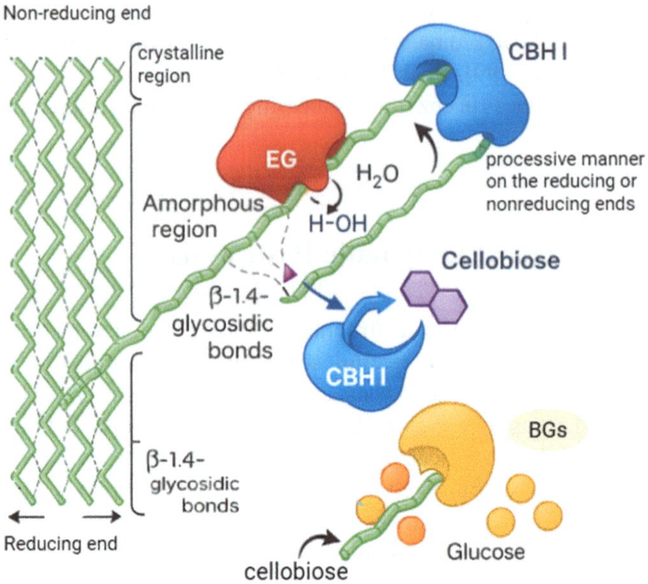

Fig. 2 Catalytic mechanisms of the three key cellulolytic enzyme classes. *CBH* cellobiohydrolases, *CBH I* cellobiohydrolases type I, *EG* endoglucanases, *BGs* β-glucosidases

Although their mechanisms differ, these enzymes act synergistically to achieve efficient cellulose degradation (Fig. 2). Exoglucanases are generally processive, remaining attached to the cellulose chain and sequentially cleaving off cellobiose units, whereas most endoglucanases are nonprocessive. However, processive endoglucanases have been identified in both bacteria and fungi [10, 152]. β-glucosidases, either free or membrane-bound, convert cellobiose and cellodextrins into glucose, completing the saccharification process [22].

Modern formulations increasingly include lytic polysaccharide monooxygenases (LPMOs), which catalyze the oxidative cleavage of glycosidic bonds in crystalline polysaccharides, enhancing cellulose accessibility and improving saccharification efficiency [50, 63, 66, 109, 134, 137].

Hemicellulose, the second most abundant biomass component, is structurally diverse and highly branched. Its breakdown requires a suite of enzymes, including xylanases, mannanases, and debranching enzymes such as arabinofuranosidases, galactosidases, deacetylases, and glucuronoyl esterases [27, 90, 96, 107]. Notably, some endoglucanases (EGs) and LPMOs exhibit broad substrate specificity, allowing them to act on various hemicelluloses such as xyloglucans, xylans, and glucomannans. This enzymatic versatility supports the development of universal cocktails for biomass, facilitating industrial scalability and cost reduction [12, 17, 41, 87, 96, 142].

Enzymatic hydrolysis plays a pivotal role in the circular bioeconomy by converting low-value lignocellulosic residues into fermentable sugars, thereby reducing landfill burden and greenhouse gas (GHG) emissions [88, 163]. Compared to chemical hydrolysis, enzymatic approaches consume less energy and avoid the formation of toxic by-products, aligning well with green chemistry principles [115].

The effectiveness of enzymatic hydrolysis hinges on substrate accessibility, enzyme affinity, and specificity. Physical characteristics such as crystallinity, surface area, and the presence of lignin and hemicellulose significantly influence enzyme–substrate interactions [153]. For processive cellulases like Cel7A, productive binding and surface mobility are crucial and are modulated by substrate structure and surface charge [57, 158]. Carbohydrate-binding modules (CBMs) further enhance performance by increasing enzyme concentration at the biomass surface and improving targeting efficiency [49].

To respond to industrial demands, commercial enzyme blends such as Accellerase and CTec have been developed to optimize cellulase and hemicellulase activity across various feedstocks. These cocktails combine core enzymes with accessory components like xylanases and mannanases, which improve cellulose accessibility by removing hemicellulose barriers and enhance sugar yields [91, 97]. Their formulation is often tailored to specific biomass types and pretreatment strategies, enabling cost-effective and scalable hydrolysis in industrial settings.

3 Key Process Parameters Affecting Hydrolysis Efficiency

The enzymatic hydrolysis of lignocellulosic biomass into fermentable sugars is influenced by a complex interaction of biochemical, physical, and operational parameters. The most important of them are enzyme and solids loading, substrate properties, reactor configuration, and environmental factors such as pH and temperature. Those parameters affect enzyme activity, substrate accessibility, and overall sugar yield.

3.1 Kinetic and Thermodynamic Considerations

Enzyme loading, substrate concentration, and the enzyme-to-substrate (E/S) ratio critically influence hydrolysis efficiency, following Michaelis–Menten-type kinetics [111]. Initial increases in enzyme concentration enhance hydrolysis rates, but beyond a threshold, product inhibition by glucose and cellobiose can lead to diminishing returns [51]. Similarly, substrate concentrations exceeding optimal levels can induce substrate inhibition, reducing hydrolysis efficiency over time [18]. The kinetic behavior of enzymes, characterized by parameters such as $V_{\max}$ and K_m, varies significantly, for instance, bromelain and papain exhibit higher catalytic efficiencies and lower K_m values than neutrase, indicating superior substrate affinity [110]. Models incorporating enzyme adsorption dynamics and inhibition kinetics—such as Langmuir-type or modified exponential decay equations—offer more accurate predictions of enzymatic performance on solid substrates [9].

Environmental conditions also play an essential role. Most commercial cellulases perform optimally between 45 °C and 55 °C and at pH 4.8–5.5; deviations may reduce catalytic activity or cause enzyme denaturation [164]. Integrating pH and temperature into kinetic models improves the reliability of sugar yield predictions and aids in process optimization [26].

3.2 *Enzyme, Solids Loading, and Substrate Structure and Pretreatment*

Balancing enzyme and substrate concentrations is essential for maximizing conversion while minimizing costs. Although higher enzyme dosages accelerate hydrolysis, they increase production costs. Optimized enzyme loading—such as 8.1 filter paper unit (FPU)/g dry substrate for wheat straw—strikes a balance between efficiency and cost-effectiveness [164]. High-solids hydrolysis (> 15% w/w) improves sugar concentration and economic viability, nevertheless, it introduces challenges like increased viscosity, poor mixing, and enzyme crowding, which limit mass transfer and reduce digestibility [118].

Substrate crystallinity, degree of polymerization, and lignin content significantly affect enzymatic accessibility. Highly crystalline cellulose resists enzymatic attack, while effective pretreatments that disrupt crystallinity and remove lignin enhance digestibility. For instance, alkaline hydrogen peroxide and subcritical water pretreatments improve cellulose accessibility by removing lignin and hemicellulose, enabling shorter reaction times [25, 162]. Under optimized conditions, peak sugar release from wheat straw occurred within 33 h [164]. Nonetheless, in some pretreatment systems such as those involving ionic liquids, cellulose phase transitions may exert a greater influence than crystallinity [155]. Additionally, lignin-derived inhibitors such as phenolics and furans can promote nonproductive enzyme adsorption, reducing hydrolysis efficiency. The degree of inhibition depends on the chemical composition of residual lignins, particularly in cholinium ionic liquid–pretreated biomass [7].

3.3 *Agitation and Reactor Configuration*

Mixing and agitation are essential, especially under high-solids conditions, to maintain uniform enzyme distribution and overcome mass transfer limitations. Controlled agitation (250–300 min^{-1}) significantly enhances enzymatic hydrolysis of microcrystalline and pretreated substrates [36, 147, 149]. Nevertheless, excessive shear can disrupt enzyme–substrate interactions or denature proteins, reducing hydrolysis efficiency [171]. Use of nitrogen atmospheres can mitigate oxidative damage during high-shear mixing, though specific studies on enzymatic hydrolysis under nitrogen are limited in this dataset. Enzymes with carbohydrate-binding modules (CBMs) have shown greater resistance to mechanical stress, helping retain activity under harsh processing conditions [23]. Hydrolysis in stirred-tank reactors (STRs) may require higher agitation power due to the increased viscosity post-pretreatment, which can compromise enzyme stability [122]. In contrast, plug-flow reactors (PFRs) experience viscosity gradients that affect fluid dynamics and impose mechanical stress on enzymes differently, sometimes reducing enzyme deactivation and improving productivity [139].

3.4 Enzyme Synergy and Cost Considerations

Enzymatic cocktails comprising endoglucanases (EGs), cellobiohydrolases (CBHs), β-glucosidases (BGs), and lytic polysaccharide monooxygenases (LPMOs) exhibit a synergistic activity that maximizes glucose yields while mitigating inhibition. Proper formulation of those cocktails enhances hydrolysis rates across diverse biomass feedstocks [101, 143]. However, enzymes remain a major cost component in bioethanol production, contributing approximately USD 1.05 per gallon compared to a market price of USD 1.752, underscoring the need for further cost optimization [64]. Efficient enzymatic hydrolysis depends on an integrated optimization of kinetic, structural, and operational parameters. Understanding enzyme kinetics, optimizing loading strategies, employing effective pretreatments, and designing suitable reactor configurations are all critical to enhancing sugar yield and ensuring economic viability in biomass conversion processes.

4 Inhibitory Factors and Mitigation Strategies

In the enzymatic hydrolysis of lignocellulosic biomass, inhibitors derived from lignin and hemicellulose significantly impair enzyme efficiency. Compounds such as phenolics, furans (e.g., 5-hydroxymethylfurfural [5-HMF]), and organic acids (e.g., acetic, formic, and levulinic acids) interfere with enzyme function by altering protein conformation, competing for catalytic sites, or inducing nonproductive adsorption [20, 74]. Residual lignin acts both as a physical barrier and a sink for nonspecific enzyme binding, reducing catalytic activity through deactivation.

Environmental stressors exacerbate enzyme instability during hydrolysis. Agitation, especially under low enzyme loading, increases exposure to the air–liquid interface, destabilizing proteins [16]. Additionally, high shear forces from mixing can unfold enzymes, leading to irreversible functional loss [80].

To mitigate those limitations, surfactants and protein stabilizers have been widely used. Additives such as polyethylene glycol (PEG), Tween, and bovine serum albumin (BSA) help preserve enzyme integrity by reducing lignin adsorption and interfacial deactivation. PEG, for example, enhances cellobiohydrolase I activity and increases glucose yields by minimizing nonproductive binding [56]. Synergistic combinations including PEG–Tween or PEG–casein have been shown to restore enzyme secondary structures, such as α-helices, further improving hydrolysis efficiency in lignin-rich systems [39].

Beyond stabilizing enzymes, surfactants can alter substrate properties. Surfactant-assisted ionic liquid pretreatments enhance cellulose accessibility while reducing lignin content, achieving enzymatic digestibilities of up to 91% [47]. BSA additionally blocks lignin adsorption sites and supports hemicellulose removal, reducing enzyme demand by up to 33% [35].

Nevertheless, inhibitory byproducts from pretreatment remain a major barrier. Processes such as steam explosion, acid, alkali, and sulfite-based treatments improve biomass digestibility but generate toxic compounds, including volatile acids, furfurals, and phenolics [34, 38, 165]. Such molecules inhibit carbohydrate-active enzymes (CAZymes) via competitive or allosteric mechanisms and also hinder downstream microbial fermentation [157]. To illustrate, just 0.4% (w/v) furfural can suppress microbial activity by 50% [65], while phenolics such as tannic acid inhibit *β*-glucosidase and cellobiohydrolase activities in a concentration- and time-dependent manner [119, 86].

To face this challenge, several detoxification strategies have been explored. Traditional approaches include pH neutralization, sulfite addition [92], enzymatic oxidation using laccases [138], ion exchange, and adsorption with activated charcoal [113]. Polymeric flocculants like polyethyleneimine have also demonstrated up to 50% removal efficiency for toxic compounds [28]. A promising recent development involves hydrophobic magnetic deep eutectic solvents (HMDES) composed of menthol, thymol, and ferric chloride ($FeCl_3$), which achieved extraction efficiencies of 98.5% for furfural and 78.8% for 5-HMF from real hydrolysates. Nonetheless, economic viability remains limited due to the high solvent volumes required [82].

On the whole, overcoming enzyme inhibition and deactivation is essential for the success of enzymatic hydrolysis. A multifaceted strategy—combining surfactants, enzyme stabilizers, effective detoxification, and optimized pretreatment—is necessary to enhance enzymatic performance. Improving enzyme robustness against inhibitors and environmental stress is equally critical for cost-effective and scalable bioconversion processes [64].

5 Synergistic Effects and Enzyme Engineering

Recent advances in enzyme discovery and engineering are shaping the future of enzymatic hydrolysis by enabling more active, specific, and recyclable enzyme systems. A notable breakthrough came from researchers at the Brazilian Center for Research in Energy and Materials (CNPEM), who discovered CelOCE, a novel copper-dependent oxidative metalloenzyme composed of only 115 amino acids. When co-expressed in

Trichoderma reesei, CelOCE enhanced glucose release by approximately 20% under high-solids conditions through synergistic interaction with conventional cellulases and hemicellulases, highlighting its potential to boost industrial hydrolysis efficiency [112].

Tailored cellulases from genetically engineered or newly discovered fungal and bacterial strains offer superior catalytic efficiency and increased tolerance to lignocellulose-derived inhibitors, enabling lower enzyme dosages and reducing costs [31, 123, 151]. Lignin-compatible enzymes are also being developed to prevent nonproductive binding, potentially converting lignin into a co-catalyst or enzyme carrier. In parallel, immobilization strategies using magnetic or lignin-based supports have proven effective in enzyme reuse and process stabilization, decreasing enzyme consumption across multiple hydrolysis cycles [100].

A core principle of effective enzymatic hydrolysis is synergy between hydrolases. The coordinated action of endoglucanases (EGs), which randomly cleave internal β-1,4-glycosidic bonds; cellobiohydrolases (CBHs), which liberate cellobiose units from chain ends; and β-glucosidases (BGs), which convert cellobiose to glucose while relieving product inhibition, is critical for efficient cellulose breakdown. Lytic polysaccharide monooxygenases (LPMOs) further enhance this synergy by oxidatively disrupting crystalline cellulose, improving enzyme access and glucose yields [13, 127, 159].

The importance of enzyme balance is underscored by studies showing that optimized cocktails with specific ratios of EGs, CBHs, and BGs—such as Cel45A, Cel6A, and Cel3A—achieve higher sugar yields from substrates like phosphoric acid-swollen cellulose, with synergistic effects becoming more pronounced over time [152]. Similarly, engineered yeast consortia co-expressing these enzymes effectively converted lignocellulose to fermentable sugars [72].

Enzyme engineering, particularly through directed evolution and protein design, has enabled the development of enhanced enzyme variants with improved synergy, stability, and activity. For example, [78] used high-throughput screening to evolve an exoglucanase in *Escherichia coli*, which, when combined with engineered EG and BG variants, yielded an 8.2-fold increase in glucose release over wild-type enzymes. Further studies using directed co-evolution of enzyme pairs confirmed that simultaneously optimizing synergistic enzymes can accelerate the development of highly efficient enzyme systems [79].

Rational design approaches, such as random mutagenesis, constraint network analysis, and computational modeling, have also been instrumental in improving enzyme performance under industrial conditions. A mutant *Aspergillus niger* β-glucosidase engineered for higher thermal stability and catalytic efficiency showed significantly enhanced k_{cat}/K_m values and greater resilience to harsh processing environments [79]. Similarly, stabilizing mutations identified through network-based predictions led to improved thermostability in endoglucanases with reduced experimental effort [30].

In summary, the integration of synergistic enzyme mechanisms, tailored protein engineering, and intelligent enzyme reuse strategies is central to optimizing enzymatic hydrolysis for diverse biomass feedstocks. These innovations are driving the development of more efficient, robust, and cost-effective bioconversion platforms for biofuels and biochemicals.

6 Reactor Configurations and Analytical Techniques for Monitoring Hydrolysis

Figure 3 presents an integrated strategy to improve enzymatic hydrolysis of cellulose into fermentable sugars by combining high-throughput screening (HTS), advanced reactor design, and analytical optimization. HTS enables rapid, parallel testing of over 1000 enzyme variants per hour using light-scattering and fluorometric sensors to monitor enzyme–substrate interactions in real time. This platform allows for efficient enzyme variant screening, kinetic modeling, and eliminates the need for separation steps during sugar release [10]. Once high-performing enzymes are selected, they are implemented in an evolving reactor system—from batch to semi-batch and fed-batch configurations—culminating in two-stage systems, which provide superior cost-efficiency and productivity [11].

Reactor performance is further enhanced by continuous membrane systems that improve enzyme reuse and process stability. Capillary and submerged membrane reactors, particularly those with fouling-resistant polyethersulfone (PES) membranes, maintain high enzymatic activity and enable up to 86% enzyme reuse [152]. Additionally, ceramic membrane systems allow for enzyme immobilization and continuous product removal, while submerged bioreactors minimize enzyme loss by retaining proteins within the reaction zone. These configurations outperform batch systems in sugar yield and cost-effectiveness, especially when implemented as two-stage continuous operations [29].

HTS technologies complement these reactors by enabling rapid, microscale evaluation of enzyme performance under varying pretreatment conditions, pH, and temperature [45, 60, 71, 130]. Methods such as co-hydrolysis assays—where enzymes are directly added to biomass slurries—eliminate the need for separation steps and deliver sugar yields comparable to traditional assays [130]. Scattered-light detection and fluorometric monitoring allow real-time tracking of enzymatic activity on insoluble substrates, improving throughput and reducing operational costs [145, 172].

Analytical methods remain foundational to optimizing enzymatic hydrolysis. High-performance liquid chromatography (HPLC) remains the standard for quantifying glucose and xylose during hydrolysis, particularly in membrane-integrated and continuous systems [102]. Supporting assays, such as 3,5-dinitrosalicylic acid (DNS) for reducing sugars and p-nitrophenyl-β-D-glucopyranoside (pNPG) for β-glucosidase activity, provide valuable insights into enzyme kinetics and inform process control decisions.

However, scaling enzymatic hydrolysis remains challenging under high-solids conditions. Increased viscosity hinders mass transfer, requiring greater mixing energy in stirred tank reactors (STRs), which can damage enzymes [171]. Fed-batch STRs

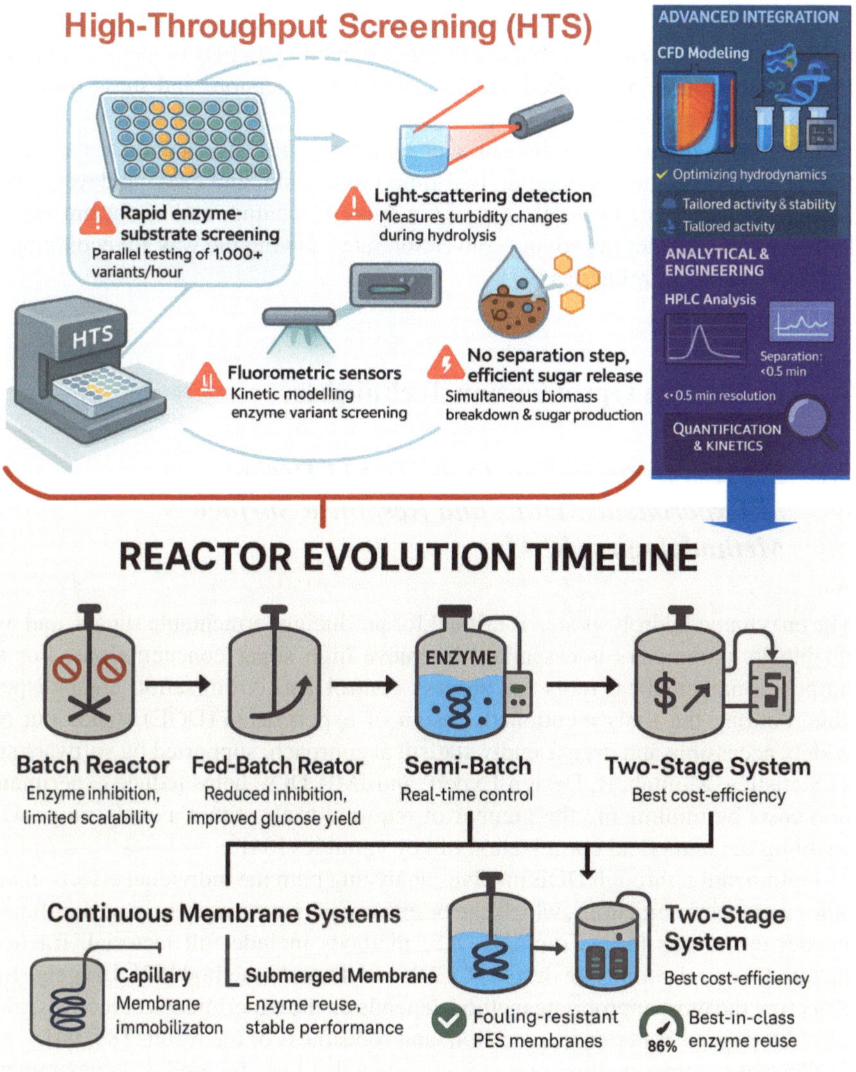

Fig. 3 Evolution of reactor configuration, enzyme engineering, and challenges

alleviate some of these issues by adjusting feed rates based on power input or viscosity to enhance mixing and reduce energy demands [129]. Plug-flow reactors (PFRs) face similar limitations due to shear stress and nonuniform flow, which disrupts enzyme–substrate interactions [70, 132].

Membrane-based alternatives help mitigate these problems. Submerged STRs with PES membranes allow continuous operation and enzyme retention, while newer radial-flow tubular designs enable efficient sugar diffusion and achieve higher hydrolysis yields—up to 53% within 8 h—compared to conventional STRs [73]. To optimize these systems, computational fluid dynamics (CFD) modeling is increasingly used to simulate hydrodynamics, enzyme dispersion, and mass transfer in high-viscosity media [19, 55, 106].

Ultimately, the success of industrial enzymatic hydrolysis depends on the integration of smart reactor engineering, real-time monitoring, and enzyme development. Advances in enzyme cocktail design, CAZyme screening, and surfactant-assisted hydrolysis are further improving conversion rates, paving the way for economically viable, scalable biorefineries [77].

7 Advances in Optimization Techniques

7.1 *Process Optimization Tools: Use of Design of Experiments (DoE) and Response Surface Methodologies (RSM)*

The enzymatic hydrolysis step is critical for producing fermentable sugars, and optimizing its parameters is essential to achieve high sugar concentrations. For this purpose, mathematical tools for process control and optimization are indispensable. Among the tools mentioned, design of experiments (DOE) stands out as a widely accessible and user-friendly statistical approach, supported by software such as Statistica, Minitab, R, Design-Expert, and JMP. DOE helps reduce experimentation costs by minimizing the number of required trials within a study range while enabling the numerical optimization of key variables [84].

Optimization through DOE involves analyzing both the individual effects of variables and their interactions, which can be either positive or negative, and significantly impact the results [85]. Common DOE methods include full factorial, fractional factorial, central composite design (CCD), and Taguchi method [53], Drégelyi-[37]. Selecting the most appropriate method depends on the experiment's objectives, along with the desired confidence, precision, and robustness of the results [84, 94].

Currently, scientific literature presents several studies focused on key parameters influencing enzymatic hydrolysis, such as solids content, hydrolysis time, enzyme loading, product inhibition, and the use of additives like ionic liquids. Those factors represent a crucial role in determining process efficiency and are vital for accurate modeling and optimization [68, 170]. In this context, Table 1 presents selected studies that have used DOE.

Response surface methodology (RSM) is a valuable tool for interpreting DOE results, particularly when using CCD, which captures nonlinear relationships between variables [61, 167]. RSM enables the graphical visualization of how independent variables influence the response, making it easier to identify optimal process conditions through 3D surfaces and contour plots [135].

Table 1 Practical studies for bioprocess optimization

Study (Citation)	Process evaluated and key parameters studied	DOE method and response variable	Model confirmed?	Optimal conditions
[62]	Dilute acid pretreatment + hydrolysis; particle size, pH, time	Taguchi method; glucose and hemicellulose yield	Yes	55 °C, 48 h, small particles (53–106 μm)
[114]	Enzyme mix optimization; cellulase blends, temperature, time	Central composite design; reducing sugars	Yes	10% substrate, 55 °C, 48 h
[89]	Sugarcane bagasse; pretreatment methods	Plackett–Burman; glucose/xylose yield	Yes	10% solids; kinetic model-based
[108]	Pretreatment of elephant grass; temperature, chemicals	2^{5-1} Fractional factorial design; reducing sugars	Yes	NaOH 4.5%, 85 °C, 100 min

Although response surfaces are visual in nature, they are based on mathematical models fitted to experimental data. Such models help clarify variable interactions, predict outcomes in unexplored regions, and support more accurate optimization decisions [126].

7.2 *Mathematical Modeling and Simulations: Development of Kinetic Models to Predict Process Outcomes*

Mathematical modeling and process simulation are essential tools in engineering and applied sciences, enabling the description, analysis, and prediction of complex system behavior through mathematical relationships. By formulating equations that represent physical, chemical, or biological phenomena, these models allow for the simulation of various operational scenarios, optimization of process variables, and a reduction in the need for large-scale experimentation. This leads to greater efficiency, resource savings, and more informed decision-making in industrial applications [33, 160].

One key advantage of mathematical modeling is its ability to support real-time decision-making by continuously monitoring process dynamics. In enzymatic hydrolysis, modeling is particularly valuable for estimating the optimal reaction time. As glucose—the primary product—accumulates, it can inhibit cellulolytic enzymes through product feedback [104]. Modeling helps determine whether extending the reaction duration is beneficial, thereby improving yields while avoiding unnecessary reagent use and time expenditure [76].

Given the high costs associated with extensive experimental trials, mathematical models are critical for understanding enzymatic hydrolysis and optimizing its conditions. Such models reduce experimental demands by predicting process behavior based on key parameters [33]. However, accurately modeling enzymatic hydrolysis is challenging due to the process's heterogeneous nature, requiring advanced models that capture complex phenomena such as enzyme kinetics and product inhibition [148].

Numerous studies have proposed models for enzymatic hydrolysis across various process configurations, including batch and fed-batch systems, and using diverse biomass substrates such as sugarcane bagasse and corn stover [161].

7.3 *Integration with Machine Learning: Advanced Methods for Parameter Tuning and Scale-Up*

The use of computational tools has expanded rapidly in recent years, significantly advancing data analysis and predictive capabilities. Traditional methods such as the one-factor-at-a-time (OFAT) approach—which evaluates variables independently—have notable limitations. These shortcomings have been largely addressed by modern techniques that analyze the combined effects of multiple variables. As a result, researchers are increasingly adopting holistic methods such as machine learning (ML) to study complex interactions within systems.

In the field of lignocellulosic biomass (LCB) conversion, ML has proven especially valuable. It enhances prediction accuracy and supports technological innovation by enabling the integrated evaluation of various factors, including substrate concentration, enzyme dosage, and the presence of surfactants or ionic liquids, alongside operational parameters [69, 125]. This approach streamlines analysis, reduces experimental time and costs, and improves process efficiency. ML also aids in feature selection by identifying and ranking the most influential variables affecting pretreatment and hydrolysis efficiency, enabling more targeted and effective optimization strategies [93]. For instance, artificial neural networks (ANN) have been used to model the effects of cellulase dosage, substrate concentration, and reaction time on sugar yield, outperforming traditional methods like response surface methodology [154].

Further, ML-assisted optimization has been used to refine enzymatic hydrolysis of pretreated wood waste, identifying optimal concentrations of cellulase, β-glucosidase, and surfactants like Tween 80 to maximize sugar production [103]. ML-supported kinetic modeling has also been used to simulate complex hydrolysis mechanisms, facilitating better reactor design and scalability for industrial applications

[21]. Together, these applications demonstrate how ML enables data-driven optimization of enzymatic hydrolysis, improving efficiency, reducing costs, and accelerating the development of biofuels and bioproducts from lignocellulosic biomass.

8 Process Intensification Approaches, Case Studies, Industrial Insights, and Circular Bioeconomy

Enzymatic hydrolysis of lignocellulosic biomass is relevant for the circular bioeconomy by converting waste into fermentable sugars for biofuels, biochemicals, and biomaterials. Nonetheless, industrial scale-up remains limited due to low reaction rates and high enzyme costs. To manage those challenges, process intensification (PI) strategies—such as integrated bioreactors, advanced pretreatments, and novel energy-assisted techniques—have been developed to enhance sugar yields, resource circularity, and overall process efficiency.

Integrated membrane bioreactors enable simultaneous hydrolysis and sugar removal, improving enzyme retention and reducing inhibition [43]. Ultrasound-assisted hydrolysis increases glucose yields by enhancing cellulose fragmentation and surface area, improving mass transfer [1]. Combining pretreatments like cold plasma or alkaline delignification with ultrasound further breaks down lignin barriers, increasing enzyme accessibility and boosting sugar yields by up to 100% compared to conventional methods [117, 131]. They have been successfully applied at scale. For example, fed-batch enzymatic hydrolysis of eucalyptus bark at 20% solids content achieved 161 g/L sugar and 76% hydrolysis efficiency, demonstrating scalability for biorefineries [4].

8.1 Integration in Biorefinery Operations

Coupling hydrolysis with upstream and downstream processes is key for efficient biorefinery design. High-solids hydrolysis (up to 20%) integrates smoothly with fermentation, boosting sugar concentrations and enabling large-scale biofuel production [121]. Simultaneous saccharification and fermentation (SSF) combines hydrolysis and fermentation in one step, reducing product inhibition and capital costs. Thermotolerant strains and enzyme-compatible conditions enhance SSF performance under high solids [52, 121]. Fed-batch SSF further improves enzyme utilization and reduces viscosity, achieving up to 64% ethanol yield without additional enzymes or yeast [144] and can be optimized using dynamic modeling [140].

Consolidated bioprocessing (CBP), the most integrated strategy, merges enzyme production, hydrolysis, and fermentation into a single step using engineered microbes. Though promising, CBP still faces limitations in enzyme activity and strain robustness [52]. Recent advancements in synthetic biology and metabolic engineering have enabled the development of CBP strains capable of expressing complete cellulolytic systems and fermenting both hexoses and pentoses, further enhancing feasibility. For instance, engineered *Saccharomyces cerevisiae* strains have been developed with surface-displayed cellulases and hemicellulases along with integrated pathways for xylose and cellobiose fermentation, allowing simultaneous conversion of mixed sugars in lignocellulosic hydrolysates [146].

In a broader context, integrating enzymatic hydrolysis into biorefinery operations requires harmonization with pretreatment, detoxification, and fermentation steps to maximize yield and cost-effectiveness. For example, coupling hydrolysis with in situ product recovery (ISPR) can alleviate end-product inhibition and improve enzyme turnover, while integration with process water and heat recovery systems contributes to energy efficiency and sustainability [124]. These approaches support continuous biorefining, which is essential for the economic viability of cellulosic sugar production on an industrial scale [46].

8.2 Enzyme Engineering and Process Control

Recent enzyme engineering efforts have produced tailored cocktails with enhanced catalytic efficiency and thermal stability, significantly improving lignocellulose conversion [83]. Strategies like on-site enzyme production and recycling can reduce enzyme-related costs by up to 30% [2]. Immobilized enzymes, such as those supported on bacterial cellulose, maintain stability across multiple cycles, improving economic feasibility [156].

Among PI techniques, fed-batch and semi-batch systems maintain stable biomass concentrations and reactor viscosity, supporting more consistent performance and lower energy input [32, 129]. Continuous systems offer stability, but are less suited for high-solid biomass due to shear-sensitive enzyme deactivation [14].

To address this, integrated fed-batch systems using submerged membranes have been developed to enable real-time control, stream recycling, and enzyme/sugar recovery—further enhancing scalability and operational reliability.

8.3 Economic and Environmental Sustainability

Techno-economic and lifecycle assessments consistently show that enzyme reuse and process optimization significantly reduce costs and emissions. Immobilized enzyme systems and in situ hydrolysis lower enzyme demand and energy input [42, 81], while fed-batch and semi-batch operations improve glucose concentrations and cut water and energy use [54]. Life cycle assessments (LCAs) show that enzymatic hydrolysis can reduce GHG emissions by 50–70% compared to fossil-based sugar production [98], reinforcing its environmental benefits. Public initiatives, like the

European Union (EU) Circular Economy Action Plan, are accelerating the adoption of such sustainable technologies [6].

Beyond glucose, hydrolysis recovers hexose and pentose sugars, critical for producing organic acids, biopesticides, and biopolymers. Residual lignin can also be transformed into adhesives or carbon-based materials, supporting full biomass utilization and sustainability goals [105, 115].

8.4 Real-World Applications and Waste Valorization

Enzymatic hydrolysis is a promising strategy for converting diverse lignocellulosic waste streams into fermentable sugars. High sugar recoveries have been reported across various substrates: cassava pulp pretreated with liquid hot water achieved 99% recovery [141], eucalyptus bark produced 161 g/L sugars with 76% conversion under high-solids fed-batch conditions [5], and two-stage hydrolysis of food waste yielded up to 204.2 g/L of sugar, highlighting strong economic viability in circular bioeconomy models [163].

Table 2 compares several studies that highlight diverse process intensification strategies aimed at enhancing enzymatic efficiency, including high-solids loading with surfactants, magnetic nanoparticle-immobilized enzymes, nanostructure modification, ultrasound-assisted hydrolysis, and fungal enzyme production. Although sugar yields varied—from 19.5 g/L to 161 g/L—and conversion efficiencies reached up to 86%, most studies remained at lab scale, with only one advancing to pilot scale [58]. Major operational constraints included enzyme instability, poor mixing at high solids, crystallinity-related inhibition, and substrate heterogeneity. Notably, none of the studies incorporated life cycle assessment (LCA) or techno-economic analysis (TEA), limiting the assessment of scalability and commercial feasibility.

9 Conclusion and Future Directions

Advancements in enzyme engineering—such as the development of LPMOs and customized enzyme cocktails—have significantly improved the efficiency of enzymatic hydrolysis. However, major challenges remain, including high enzyme costs, biomass recalcitrance, and difficulties in scaling up the process. Overcoming these barriers will require process intensification, with artificial intelligence (AI) and machine learning playing key roles. These technologies can predict optimal enzyme–substrate interactions, enable real-time adjustment of reaction conditions, and minimize reliance on trial-and-error experimentation.

Table 2 Case studies for sugar production from lignocellulose biomass

Study	Feedstock	Intensification strategy	Enzymatic cocktail and key parameters	Scale	Challenges	Performance outcomes	LCA/TEA
Amândio et al. [5]	Kraft-pretreated eucalyptus bark	High solids loading (20%) fed-batch + surfactants	Commercial cellulase blend; 25 FPU/g carbs; 24 h; pH 4.8	5 L	High viscosity, poor mixing, additive inefficacy	161 g/L sugar; 76% conversion	No
Libardi et al. [75]	Cellulose from wastewater	Waste-grown fungal enzyme production	Endoglucanase + CBH + β-glucosidase; 107.6 mg sugar/g	Lab	Substrate heterogeneity; enzyme consistency	62% higher efficiency than commercial enzyme	No
Gaikwad et al. [42]	Pure cellulose	Magnetic nanoparticle-immobilized enzyme	Immobilized cellulase; 48 h per reuse cycle	Lab	Decline in activity after 2nd use	19.5 g/L glucose (1st cycle); reusable 3x	No
Chávez-guerrero et al. [24]	Cellulose nanoplatelets	Nanostructure/crystallinity reduction	*Trichoderma* sp. cellulase; 37 °C; 60 h	Lab	High crystallinity inhibits access	47% glucose yield (vs. 15% in untreated)	No
Hu et al. [58]	Wood pulp	Optimization of temperature, mixing, enzyme ratio	Cellulase + xylanase; 45 °C; manual mixing mode	Pilot	Mixing limitations, lower performance at low temperatures	Improved sugar yield; process viable for scale-up	No

(continued)

Table 2 (continued)

Study	Feedstock	Intensification strategy	Enzymatic cocktail and key parameters	Scale	Challenges	Performance outcomes	LCA/TEA
Szabo and Csiszar [133]	Bleached cotton cellulose	Ultrasound-assisted hydrolysis	Cellulase; 40–80% amplitude; up to 43.4 W/cm^2	Lab	Local overheating, probe distance sensitivity	54% yield improvement over the best mechanical stirring	No
Beyene et al. [15]	Wood pulp	CNC + sugar co-production	Commercial cellulase; 2–10 h + acid hydrolysis	Lab	Product separation, enzyme recovery	20–40% sugar; CNC yield ↑ by 86%	No

LCA life cycle assessment, *TEA* techno-economic analysis, *FPU* filter paper units, *CBH* cellobiohydrolases, *CNC* cellulose nanocrystals

To further reduce costs, new strategies are emerging, including enzyme recycling via magnetic supports, in situ enzyme production, and immobilization on fouling-resistant membranes. Next-generation bioreactors will increasingly utilize continuous membrane systems and CFD-optimized designs to mitigate limitations associated with high-solids viscosity, poor mass transfer, and shear stress. Novel formats such as radial-flow tubular membranes and modular microreactors promise greater sugar recovery and scalability.

Evaluating the true potential of these technologies requires integration with TEA and LCA to ensure process improvements are both cost-effective and environmentally sustainable. Ultimately, the industrial success of enzymatic hydrolysis will hinge on the convergence of enzyme innovation, intelligent reactor design, AI-driven process control, and holistic system integration. Together, these advances pave the way for scalable, economically viable, and sustainable next-generation biorefineries.

Acknowledgements A.K. Chandel thanks The Brazilian National Council for Scientific and Technological Development (CNPq), Brazil for the Scientific Productivity Project (Process number: 304451/2025-8).

References

1. Adewuyi YG, Deshmane VG (2015) Intensification of enzymatic hydrolysis of cellulose using high-frequency ultrasound: an investigation of the effects of process parameters on glucose yield. Energy Fuels 29:4998–5006. https://doi.org/10.1021/acs.energyfuels.5b00661
2. Adsul MG (2024) Cellulolytic enzymes recycling strategies for the economic conversion of lignocellulosic biomass to fuels. Process Biochem 147:62–74. https://doi.org/10.1016/j.procbio.2024.08.009
3. Alnoch RC, Salgado JCS, Alves GS, et al (2023) Biochemical characterization of an endoglucanase GH7 from thermophile thermothielavioides terrestris expressed on *Aspergillus nidulans*. Catalysts 13. https://doi.org/10.3390/catal13030582
4. Amândio MST, Rocha JMS, Serafim LS, Xavier AMRB (2021) Cellulosic bioethanol from industrial eucalyptus globulus bark residues using kraft pulping as a pretreatment. Energies 14:1–18. https://doi.org/10.3390/en14082185
5. Amândio MST, Rocha JMS, Xavier AMRB (2023) Enzymatic hydrolysis strategies for cellulosic sugars production to obtain bioethanol from *Eucalyptus globulus* bark. Fermentation 9:241. https://doi.org/10.3390/fermentation9030241
6. Amândio MST, Rocha JMS, Serafim LS, Xavier AMRB (2022) Bioethanol from wastes for mobility: Europe on the road to sustainability. Di Blasio G, Agarwal AK, Belgiorno G, Shukla PC (eds). Springer Singapore, Singapore, pp 97–123
7. An YX, Zong MH, Hu SQ, Li N (2017) Effect of residual lignins present in cholinium ionic liquid-pretreated rice straw on the enzymatic hydrolysis of cellulose. Chem Eng Sci 161:48–56. https://doi.org/10.1016/j.ces.2016.12.014
8. Annamalai N, Rajeswari MV, Sivakumar N (2016) Cellobiohydrolases: role, mechanism, and recent developments, Gupta VK (ed) Springer International Publishing, Cham, pp 29–35
9. Anuganti M, Fu H, Ekatan S et al (2021) Kinetic study on enzymatic hydrolysis of cellulose in an open, inhibition-free system. Langmuir 37:5180–5192. https://doi.org/10.1021/acs.langmuir.1c00115

10. Asha P, Divya J, Bright Singh IS (2016) Purification and characterisation of processive-type endoglucanase and β-glucosidase from *Aspergillus ochraceus* MTCC 1810 through saccharification of delignified coir pith to glucose. Bioresour Technol 213:245–248. https://doi.org/10.1016/j.biortech.2016.03.013
11. Asztalos A, Daniels M, Sethi A et al (2012) A coarse-grained model for synergistic action of multiple enzymes on cellulose. Biotechnol Biofuels 5:1–15. https://doi.org/10.1186/1754-6834-5-55
12. Bailey M, Siika-aho M, Valkeajarvi A, Penttila M (1993) Hydrolytic properties of two cellulases of *Trichoderma reesei* expressed in yeast. Biotechnol Appl Biochem 17:65–76. https://doi.org/10.1111/j.1470-8744.1993.tb00233.x
13. Barbosa FC, Silvello MA, Goldbeck R (2020) Cellulase and oxidative enzymes: new approaches, challenges and perspectives on cellulose degradation for bioethanol production. Biotechnol Lett 42:875–884. https://doi.org/10.1007/s10529-020-02875-4
14. Battista F, Bolzonella D (2018) Some critical aspects of the enzymatic hydrolysis at high dry-matter content: a review. Biofuels Bioprod Biorefining 12:711–723. https://doi.org/10.1002/bbb.1883
15. Beyene D, Chae M, Dai J et al (2017) Enzymatically-mediated co-production of cellulose nanocrystals and fermentable sugars. 1–13. https://doi.org/10.3390/catal7110322
16. Bhagia S, Dhir R, Kumar R, Wyman CE (2018) Deactivation of cellulase at the air-liquid interface is the main cause of incomplete cellulose conversion at low enzyme loadings. Sci Rep 8:1–12. https://doi.org/10.1038/s41598-018-19848-3
17. Borisova AS, Isaksen T, Dimarogona M et al (2015) Structural and functional characterization of a lytic polysaccharide monooxygenase with broad substrate specificity. J Biol Chem 290:22955–22969. https://doi.org/10.1074/jbc.M115.660183
18. Bravo V, Paez MP, Aoulad M, Reyes A (2000) The influence of temperature upon the hydrolysis of cellobiose by β-1,4-glucosidases from *Aspergillus niger*. Enzyme Microb Technol 26:614–620. https://doi.org/10.1016/S0141-0229(00)00136-8
19. Brüning S, Weuster-Botz D (2014) CFD analysis of interphase mass transfer and energy dissipation in a milliliter-scale stirred-tank reactor for filamentous microorganisms. Chem Eng Res Des 92:240–248. https://doi.org/10.1016/j.cherd.2013.07.024
20. Cai C, Zhang C, Li N et al (2023) Changing the role of lignin in enzymatic hydrolysis for a sustainable and efficient sugar platform. Renew Sustain Energy Rev 183:113445. https://doi.org/10.1016/j.rser.2023.113445
21. Caminal G, López-Santín J, Solà C (1985) Kinetic modeling of the enzymatic hydrolysis of pretreated cellulose. Biotechnol Bioeng 27:1282–1290. https://doi.org/10.1002/bit.260270903
22. Chandel AK, Chandrasekhar G, Silva MB, Silva SS (2012) The realm of cellulases in biorefinery development. Crit Rev Biotechnol 32:187–202. https://doi.org/10.3109/07388551.2011.595385
23. Chang C-M, Lu WJ, Own KS, Hwang S-J (1994) Comparison of airlift and stirred tank reactors for immobilized enzyme reactions. Process Biochem 29:133–138. https://doi.org/10.1016/0032-9592(94)80006-5
24. Chávez-guerrero L, Silva-mendoza J, Sepúlveda-guzmán S et al (2019) Enzymatic hydrolysis of cellulose nanoplatelets as a source of sugars with the concomitant production of cellulose nano fi brils. 210:85–91. https://doi.org/10.1016/j.carbpol.2019.01.055
25. Chen J, Wang X, Zhang B et al (2021) Integrating enzymatic hydrolysis into subcritical water pretreatment optimization for bioethanol production from wheat straw. Sci Total Environ 770:145321. https://doi.org/10.1016/j.scitotenv.2021.145321
26. Chen Y-T, Wang F-S (2012) Determination of kinetic parameters for enzymatic cellulose hydrolysis using hybrid differential evolution. Int J Chem React Eng 9. https://doi.org/10.2202/1542-6580.2498
27. Chibrikov V, Pieczywek PM, Cybulska J, Zdunek A (2024) The effect of hemicelluloses on biosynthesis, structure and mechanical performance of bacterial cellulose—hemicellulose hydrogels. Sci Rep 1–15. https://doi.org/10.1038/s41598-024-72513-w

28. Cho D-H, Kim Y-H (2009) Evaluation of biological and physico-chemical detoxification methods for the removal of inhibitors in lignocellulose hydrolysate. KSBB J 24:415–419
29. Climent Barba F, Grasham O, Puri DJ, Blacker AJ (2022) A simple techno-economic assessment for scaling-up the enzymatic hydrolysis of MSW pulp. Front Energy Res 10:1–13. https://doi.org/10.3389/fenrg.2022.788534
30. Contreras F, Nutschel C, Beust L et al (2021) Can constraint network analysis guide the identification phase of KnowVolution? A case study on improved thermostability of an endo-β-glucanase. Comput Struct Biotechnol J 19:743–751. https://doi.org/10.1016/j.csbj.2020.12.034
31. Contreras F, Pramanik S, Rozhkova AM et al (2020) Engineering robust cellulases for tailored lignocellulosic degradation cocktails. Int J Mol Sci 21. https://doi.org/10.3390/ijms21051589
32. Corrêa LJ, Badino AC, Cruz AJG (2016) Power consumption evaluation of different fed-batch strategies for enzymatic hydrolysis of sugarcane bagasse. Bioprocess Biosyst Eng 39:825–833. https://doi.org/10.1007/s00449-016-1562-4
33. de Godoy CM, Machado DL, da Costa AC (2019) Batch and fed-batch enzymatic hydrolysis of pretreated sugarcane bagasse—assays and modeling. Fuel 253:392–399. https://doi.org/10.1016/j.fuel.2019.05.038
34. Dessie W, Xin F, Zhang W et al (2019) Inhibitory effects of lignocellulose pretreatment degradation products (hydroxymethylfurfural and furfural) on succinic acid producing *Actinobacillus succinogenes*. Biochem Eng J 150:107263. https://doi.org/10.1016/j.bej.2019.107263
35. Ding D, Li P, Zhang X et al (2019) Synergy of hemicelluloses removal and bovine serum albumin blocking of lignin for enhanced enzymatic hydrolysis. Bioresour Technol 273:231–236. https://doi.org/10.1016/j.biortech.2018.11.024
36. dos Santos ACF, Ximenes E, N. Thompson D, et al (2020) Effect of using a nitrogen atmosphere on enzyme hydrolysis at high corn stover loadings in an agitated reactor. Biotechnol Prog 36:1–10. https://doi.org/10.1002/btpr.3059
37. Drégelyi-Kiss Á (2024) Application of experimental design-based predictive models and optimization in additive manufacturing—a review. Hungarian J Ind Chem 52:55–70. https://doi.org/10.33927/hjic-2024-08
38. Dziekońska-Kubczak U, Berłowska J, Dziugan P et al (2018) Comparison of steam explosion, dilute acid, and alkali pretreatments on enzymatic saccharification and fermentation of hardwood sawdust. BioResources 13:6970–6984. https://doi.org/10.15376/biores.13.3.6970-6984
39. Eckard AD, Muthukumarappan K, Gibbons W (2014) The role of polymeric micelles on chemical changes of pretreated corn stover, cellulase structure, and adsorption. Bio Energy Res 7:389–407. https://doi.org/10.1007/s12155-013-9379-3
40. Etale A, Onyianta AJ, Turner SR, Eichhorn SJ (2023) Cellulose: a review of water interactions, applications in composites, and water treatment. Chem Rev 123:2016–2048. https://doi.org/10.1021/acs.chemrev.2c00477
41. Filiatrault-Chastel C, Navarro D, Haon M et al (2019) AA16, a new lytic polysaccharide monooxygenase family identified in fungal secretomes. Biotechnol Biofuels 12:1–15. https://doi.org/10.1186/s13068-019-1394-y
42. Gaikwad S, Ingle AP, da Silva SS, Rai M (2018) Immobilized nanoparticles-mediated enzymatic hydrolysis of cellulose for clean sugar production: a novel approach. Curr Nanosci 15:296–303. https://doi.org/10.2174/1573413714666180611081759
43. Gan Q, Allen SJ, Taylor G (2005) Analysis of process integration and intensification of enzymatic cellulose hydrolysis in a membrane bioreactor. J Chem Technol Biotechnol 80:688–698. https://doi.org/10.1002/jctb.1195
44. Gan J, Peng Y, Chen Q et al (2020) Reversible covalent chemistry of carbon dioxide unlocks the recalcitrance of cellulose for its enzymatic saccharification. Bioresour Technol 295:122230. https://doi.org/10.1016/j.biortech.2019.122230
45. Gao X, Kumar R, Demartini JD et al (2013) Application of high throughput pretreatment and co-hydrolysis system to thermochemical pretreatment. Part 1: Dilute acid. Biotechnol Bioeng 110:754–762. https://doi.org/10.1002/bit.24751

46. García-Granados R, Lerma-Escalera JA, Morones-Ramírez JR (2019) Metabolic engineering and synthetic biology: synergies, future, and challenges. Front Bioeng Biotechnol 7:1–4. https://doi.org/10.3389/fbioe.2019.00036
47. Goshadrou A, Lefsrud M (2017) Synergistic surfactant-assisted [EMIM]OAc pretreatment of lignocellulosic waste for enhanced cellulose accessibility to cellulase. Carbohydr Polym 166:104–113. https://doi.org/10.1016/j.carbpol.2017.02.076
48. Goswami S, Gupta N, Datta S (2016) Using the B-glucosidase catalyzed reaction product glucose to improve the ionic liquid tolerance of B-glucosidases. Biotechnol Biofuels 9:1–12. https://doi.org/10.1186/s13068-016-0484-3
49. Gourlay K, Arantes V, Saddler JN (2012) Use of substructure-specific carbohydrate binding modules to track changes in cellulose accessibility and surface morphology during the amorphogenesis step of enzymatic hydrolysis. Biotechnol Biofuels 5:1–14. https://doi.org/10.1186/1754-6834-5-51
50. Hagemann MM, Hedegård ED (2023) Molecular mechanism of substrate oxidation in lytic polysaccharide monooxygenases: insight from theoretical investigations. Chem—A Eur J 29. https://doi.org/10.1002/chem.202202379
51. Haldar D, Gayen K, Sen D (2018) Enumeration of monosugars' inhibition characteristics on the kinetics of enzymatic hydrolysis of cellulose. Process Biochem 72:130–136. https://doi.org/10.1016/j.procbio.2018.06.008
52. Hasunuma T, Kondo A (2012) Consolidated bioprocessing and simultaneous saccharification and fermentation of lignocellulose to ethanol with thermotolerant yeast strains. Process Biochem 47:1287–1294. https://doi.org/10.1016/j.procbio.2012.05.004
53. Heinisch J, Lockner Y, Hopmann C (2021) Comparison of design of experiment methods for modeling injection molding experiments using artificial neural networks. J Manuf Process 61:357–368. https://doi.org/10.1016/j.jmapro.2020.11.011
54. Hernández-Escoto H, Torres Zúñiga I, López-Caamal F (2023) Controlled semi-batch process to enhance glucose concentration in an enzymatic hydrolysis stirred tank reactor. Memorias del Congr Nac Control Automático 6:139–144. https://doi.org/10.58571/CNCA.AMCA.2023.009
55. Hou W, Bao J (2018) Rheology characterization of lignocellulose feedstock during high solids content pretreatment and hydrolysis BT—fungal cellulolytic enzymes: microbial production and application, Fang X, Qu Y (eds). Springer Singapore, Singapore, pp 257–266
56. Hsieh CC, Cannella D, Jørgensen H et al (2015) Cellobiohydrolase and endoglucanase respond differently to surfactants during the hydrolysis of cellulose. Biotechnol Biofuels 8:52. https://doi.org/10.1186/s13068-015-0242-y
57. Hu J, Gourlay K, Arantes V et al (2015) The accessible cellulose surface influences cellulase synergism during the hydrolysis of lignocellulosic substrates. Chemsuschem 8:901–907. https://doi.org/10.1002/cssc.201403335
58. Hu B, Zhu S, Fang S et al (2022) Optimization and scale-up of enzymatic hydrolysis of wood pulp for cellulosic sugar production. 11:7242–7257
59. Ibrahim AM, Hamouda RA, El-Naggar NEA, Al-Shakankery FM (2021) Bioprocess development for enhanced endoglucanase production by newly isolated bacteria, purification, characterization and in-vitro efficacy as anti-biofilm of *Pseudomonas aeruginosa*. Sci Rep 11:1–24. https://doi.org/10.1038/s41598-021-87901-9
60. Jäger G, Wulfhorst H, Zeithammel EU et al (2011) Screening of cellulases for biofuel production: online monitoring of the enzymatic hydrolysis of insoluble cellulose using high-throughput scattered light detection. Biotechnol J 6:74–85. https://doi.org/10.1002/biot.201000387
61. Jang S, Lee AY, Lee AR et al (2017) Optimization of ultrasound-assisted extraction of glycyrrhizic acid from licorice using response surface methodology. Integr Med Res 6:388–394. https://doi.org/10.1016/j.imr.2017.08.003
62. Karapatsia A, Pappas I, Penloglou G et al (2017) Optimization of dilute acid pretreatment and enzymatic hydrolysis of *Phalaris aquatica* L. lignocellulosic biomass in batch and fed-batch processes. Bioenergy Res 10:225–236. https://doi.org/10.1007/s12155-016-9793-4

63. Karnaouri A, Chorozian K, Zouraris D et al (2022) Lytic polysaccharide monooxygenases as powerful tools in enzymatically assisted preparation of nano-scaled cellulose from lignocellulose: a review. Bioresour Technol 345:126491. https://doi.org/10.1016/j.biortech.2021.126491
64. Klein-Marcuschamer D, Oleskowicz-Popiel P, Simmons BA, Blanch HW (2012) The challenge of enzyme cost in the production of lignocellulosic biofuels. Biotechnol Bioeng 109:1083–1087. https://doi.org/10.1002/bit.24370
65. Ko JK, Um Y, Park Y-C et al (2015) Compounds inhibiting the bioconversion of hydrothermally pretreated lignocellulose. Appl Microbiol Biotechnol 99:4201–4212. https://doi.org/10.1007/s00253-015-6595-0
66. Kumar A, Singh A, Sharma VK et al (2024) The upsurge of lytic polysaccharide monooxygenases in biomass deconstruction: characteristic functions and sustainable applications. FEBS J 291:5081–5101. https://doi.org/10.1111/febs.17063
67. Kumar D, Murthy GS (2016) Enzymatic hydrolysis of cellulose for ethanol production: fundamentals, optimal enzyme ratio, and hydrolysis modeling. Elsevier B.V
68. Larnaudie V, Ferrari MD, Lareo C (2019) Techno-economic analysis of a liquid hot water pretreated switchgrass biorefinery: Effect of solids loading and enzyme dosage on enzymatic hydrolysis. Biomass Bioenergy 130:105394. https://doi.org/10.1016/j.biombioe.2019.105394
69. Lee SH, Li J, Wang X, Yang KL (2022) Online-learning-aided optimization and interpretation of sugar production from oil palm mesocarp fibers with analytics for industrial applications. Resour Conserv Recycl 180:106206. https://doi.org/10.1016/j.resconrec.2022.106206
70. Li Y, Cao X, Geng Z, Zhang M (2018) A novel quasi plug-flow reactor design for enzymatic hydrolysis of cellulose using rheology experiment and CFD simulation. Can J Chem Eng 96:770–778. https://doi.org/10.1002/cjce.22963
71. Li H, Gao X, Demartini JD et al (2013) Application of high throughput pretreatment and co-hydrolysis system to thermochemical pretreatment. Part 2: Dilute alkali. Biotechnol Bioeng 110:2894–2901. https://doi.org/10.1002/bit.24951
72. Liang Y, Si T, Ang EL, Zhao H (2014) Engineered pentafunctional minicellulosome for simultaneous saccharification and ethanol fermentation in *Saccharomyces cerevisiae*. Appl Environ Microbiol 80:6677–6684. https://doi.org/10.1128/AEM.02070-14
73. Liao Q, Zhu T, Zhang C, et al (2024) Numerical simulation of enzymatic hydrolysis in a bionic intestinal segmentation reactor. Phys Fluids 36. https://doi.org/10.1063/5.0186353
74. Liao W, Wen Z, Hurley S et al (2005) Effects of hemicellulose and lignin on enzymatic hydrolysis of cellulose from dairy manure BT—twenty-sixth symposium on biotechnology for fuels and chemicals, Davison BH, Evans BR, Finkelstein M, McMillan JD (eds). Humana Press, Totowa, NJ, pp 1017–1030
75. Libardi N, Porto L, Vandenberghe DS et al (2022) A non-waste strategy for enzymatic hydrolysis of cellulose recovered from domestic wastewater. Environ Technol 43:1503–1512. https://doi.org/10.1080/09593330.2020.1840635
76. Lischeske JJ, Stickel JJ (2019) A two-phase substrate model for enzymatic hydrolysis of lignocellulose: application to batch and continuous reactors. Biotechnol Biofuels 12:1–15. https://doi.org/10.1186/s13068-019-1633-2
77. Liu Q, Sun C, Madadi M et al (2025) Elucidation of mixed surfactants reducing mass transfer resistance during high-solids enzymatic hydrolysis from lignocellulosic biomass. Chem Eng J 507:160332. https://doi.org/10.1016/j.cej.2025.160332
78. Liu M, Xie W, Xu H et al (2014) Directed evolution of an exoglucanase facilitated by a co-expressed β-glucosidase and construction of a whole engineered cellulase system in *Escherichia coli*. Biotechnol Lett 36:1801–1807. https://doi.org/10.1007/s10529-014-1541-2
79. Liu M, Ye L, Yu H (2018) Directed coevolution of two cellulosic enzymes in *Escherichia coli* based on their synergistic reactions, Bornscheuer UT, Höhne M (eds). Springer New York, New York, NY, pp 255–267
80. Lou H, Zeng M, Hu Q et al (2018) Nonionic surfactants enhanced enzymatic hydrolysis of cellulose by reducing cellulase deactivation caused by shear force and air-liquid interface. Bioresour Technol 249:1–8. https://doi.org/10.1016/j.biortech.2017.07.066

81. Lu S, Wang Q, Wang X et al (2023) Preparation of reducing sugars from corncob by solid acid catalytic pretreatment combined with in situ enzymatic hydrolysis. Biomass Convers Biorefinery 13:12619–12629. https://doi.org/10.1007/s13399-021-01943-7
82. Makoś-Chełstowska P, Kucharska K, Słupek E et al (2023) Magnetic deep eutectic solvents as efficient media for extraction of furfural and 5-hydroxymethylfurfural from aqueous samples. J Mol Liq 390. https://doi.org/10.1016/j.molliq.2023.122945
83. Maleki M, Shahraki MF, Kavousi K et al (2020) A novel thermostable cellulase cocktail enhances lignocellulosic bioconversion and biorefining in a broad range of pH. Int J Biol Macromol 154:349–360. https://doi.org/10.1016/j.ijbiomac.2020.03.100
84. Mandenius C-F, Brundin A (2008) Bioprocess optimization using design-of-experiments methodology. Biotechnol Prog 24:1191–1203. https://doi.org/10.1002/btpr.67
85. Marrubini G, Dugheri S, Cappelli G et al (2020) Experimental designs for solid-phase microextraction method development in bioanalysis: a review. Anal Chim Acta 1119:77–100. https://doi.org/10.1016/j.aca.2020.04.012
86. Mhlongo SI, den Haan R, Viljoen-Bloom M, van Zyl WH (2015) Lignocellulosic hydrolysate inhibitors selectively inhibit/deactivate cellulase performance. Enzyme Microb Technol 81:16–22. https://doi.org/10.1016/j.enzmictec.2015.07.005
87. Mikkelson A, Maaheimo H, Hakala TK (2013) Hydrolysis of konjac glucomannan by *Trichoderma reesei* mannanase and endoglucanases Cel7B and Cel5A for the production of glucomannooligosaccharides. Carbohydr Res 372:60–68. https://doi.org/10.1016/j.carres.2013.02.012
88. Molina-Peñate E, Artola A, Sánchez A (2022) Organic municipal waste as feedstock for biorefineries: bioconversion technologies integration and challenges. Rev Environ Sci Bio/Technol 211(21):247–267. https://doi.org/10.1007/S11157-021-09605-W
89. Moreira Neto J, Costa JM, Bonomi A, Costa AC (2023) A novel kinetic modeling of enzymatic hydrolysis of sugarcane bagasse pretreated by hydrothermal and organosolv processes. Molecules 28:5617. https://doi.org/10.3390/molecules28145617
90. Mosbech C, Holck J, Meyer AS, Agger JW (2018) The natural catalytic function of CuGE glucuronoyl esterase in hydrolysis of genuine lignin-carbohydrate complexes from birch. Biotechnol Biofuels 11:1–9. https://doi.org/10.1186/s13068-018-1075-2
91. Moya EB, Syhler B, Dragone G, Mussatto SI (2024) Tailoring a cellulolytic enzyme cocktail for efficient hydrolysis of mildly pretreated lignocellulosic biomass. Enzyme Microb Technol 175:110403. https://doi.org/10.1016/j.enzmictec.2024.110403
92. Mpabanga TP, Chandel AK, da Silva SS, Singh OV (2012) Detoxification strategies applied to lignocellulosic hydrolysates for improved xylitol production BT—D-xylitol: fermentative production, application and commercialization, da Silva SS, Chandel AK (eds). Springer Berlin Heidelberg, Berlin, Heidelberg, pp 63–82
93. Mu L, Wang Z, Wu D et al (2022) Prediction and evaluation of fuel properties of hydrochar from waste solid biomass: machine learning algorithm based on proposed PSO–NN model. Fuel 318. https://doi.org/10.1016/j.fuel.2022.123644
94. Müller ALH, De Oliveira JA, Prestes OD et al (2019) Design of experiments and method development. Solid-Phase Extr 589–608. https://doi.org/10.1016/B978-0-12-816906-3.00022-4
95. Nagl M, Haske-Cornelius O, Skopek L et al (2021) Biorefining: the role of endoglucanases in refining of cellulose fibers. Cellulose 28:7633–7650. https://doi.org/10.1007/s10570-021-04022-2
96. Østby H, Hansen LD, Horn SJ et al (2020) Enzymatic processing of lignocellulosic biomass: principles, recent advances and perspectives. Springer International Publishing
97. Patel H, Divecha J, Shah A (2017) Microwave assisted alkali treated wheat straw as a substrate for co-production of (hemi)cellulolytic enzymes and development of balanced enzyme cocktail for its enhanced saccharification. J Taiwan Inst Chem Eng 71:298–306. https://doi.org/10.1016/j.jtice.2016.12.032
98. Patel K, Singh SK (2024) Sustainable waste management: a comprehensive life cycle assessment of bioethanol production from agricultural and municipal waste. Environ Sci Pollut Res 31:51431–51446. https://doi.org/10.1007/s11356-024-34612-z

99. Patel AK, Pandey A, Singhania RR (2019) Production of celluloytic enzymes for lignocellulosic biomass hydrolysis, 2nd edn. Elsevier Inc.
100. Paul M, Mohapatra S, Kumar Das Mohapatra P, Thatoi H (2021) Microbial cellulases—an update towards its surface chemistry, genetic engineering and recovery for its biotechnological potential. Bioresour Technol 340:125710. https://doi.org/10.1016/j.biortech.2021.125710
101. Perea-Moreno M-A, Samerón-Manzano E, Perea-Moreno A-J (2019) Biomass as renewable energy: worldwide research trends. Sustainability 11:863. https://doi.org/10.3390/su11030863
102. Petzelbauer I, Splechtna B, Nidetzky B (2002) Development of an ultrahigh-temperature process for the enzymatic hydrolysis of lactose. III. Utilization of two thermostable β-glycosidases in a continuous ultrafiltration membrane reactor and galacto-oligosaccharide formation under steady-state conditio. Biotechnol Bioeng 77:394–404
103. Phummala K, Imai T, Reungsang A et al (2015) Optimization of enzymatic hydrolysis for pretreated wood waste by response surface methodology in fermentative hydrogen production. J Water Environ Technol 13:153–166. https://doi.org/10.2965/jwet.2015.153
104. Pihlajaniemi V, Sipponen MH, Kallioinen A et al (2016) Rate-constraining changes in surface properties, porosity and hydrolysis kinetics of lignocellulose in the course of enzymatic saccharification. Biotechnol Biofuels 9:1–12. https://doi.org/10.1186/s13068-016-0431-3
105. Pranaw K, Drewniak L, Nain L, Singh S (2022) Editorial: waste to wealth: a sustainable circular bioeconomy approach. Front Bioeng Biotechnol 10:1–2. https://doi.org/10.3389/fbioe.2022.1010811
106. Ranganathan P, Sivaraman S (2011) Investigations on hydrodynamics and mass transfer in gas-liquid stirred reactor using computational fluid dynamics. Chem Eng Sci 66:3108–3124. https://doi.org/10.1016/j.ces.2011.03.007
107. Rao J, Lv Z, Chen G, Peng F (2023) Hemicellulose: structure, chemical modification, and application. Prog Polym Sci 140:101675. https://doi.org/10.1016/j.progpolymsci.2023.101675
108. Rezende CA, Atta BW, Breitkreitz MC et al (2018) Optimization of biomass pretreatments using fractional factorial experimental design. Biotechnol Biofuels 11:1–15. https://doi.org/10.1186/s13068-018-1200-2
109. Rieder L, Sørlie M (2023) Recent advances in understanding LPMO catalysis. Biochemistry 62:3170–3172. https://doi.org/10.1021/acs.biochem.3c00458
110. Roskoski R (2015) Michaelis-Menten Kinetics☆. In: Reference module in biomedical sciences. Elsevier, pp 1–9
111. Ryu DDY, Lee SB (1986) Enzymatic hydrolysis of cellulose: determination of kinetic parameters. Chem Eng Commun 45:119–134. https://doi.org/10.1080/00986448608911377
112. Santos CA, Morais MAB, Mandelli F et al (2025) A metagenomic 'dark matter' enzyme catalyses oxidative cellulose conversion. Nature 639. https://doi.org/10.1038/s41586-024-08553-z
113. Sarawan C, Suinyuy TN, Sewsynker-Sukai Y, Gueguim Kana EB (2019) Optimized activated charcoal detoxification of acid-pretreated lignocellulosic substrate and assessment for bioethanol production. Bioresour Technol 286:121403. https://doi.org/10.1016/j.biortech.2019.121403
114. Sharma S, Kuila A, Sharma V (2017) Enzymatic hydrolysis of thermochemically pretreated biomass using a mixture of cellulolytic enzymes produced from different fungal sources. Clean Technol Environ Policy 19:1577–1584. https://doi.org/10.1007/s10098-017-1346-9
115. Sharma D, Saini A (2020) Lignocellulosic waste valorization and biorefineries concept. In: Sharma D, Saini A (eds) Lignocellulosic Ethanol production from a biorefinery perspective. Springer Singapore, Singapore, pp 195–215
116. Shiva, Rodríguez-Jasso RM, Rosero-Chasoy G et al (2023) Enzymatic hydrolysis, kinetic modeling of hemicellulose fraction, and energy efficiency of autohydrolysis pretreatment using agave bagasse. Bioenergy Res 16:75–87. https://doi.org/10.1007/s12155-022-10442-0
117. Sidana A, Yadav SK (2022) Recent developments in lignocellulosic biomass pretreatment with a focus on eco-friendly, non-conventional methods. J Clean Prod 335:130286. https://doi.org/10.1016/j.jclepro.2021.130286

118. da Silva AS, Espinheira RP, Teixeira RSS et al (2020) Constraints and advances in high-solids enzymatic hydrolysis of lignocellulosic biomass: a critical review. Biotechnol Biofuels 13:58. https://doi.org/10.1186/s13068-020-01697-w
119. da Silva VM, Sato JAP, Araujo JN et al (2017) Systematic studies of the interactions between a model polyphenol compound and microbial β-glucosidases. PLoS ONE 12:e0181629. https://doi.org/10.1371/journal.pone.0181629
120. Silva TAL, Varão LHR, Pasquini D (2024) Lignocellulosic biomass BT—handbook of biomass. In: Hosur M, Pasquini D, Jose Chirayil C (eds) Thomas S. Springer Nature Singapore, Singapore, pp 105–142
121. Silva-Fernandes T, Marques S, Rodrigues RCLB et al (2016) Enzymatic hydrolyses of pretreated eucalyptus residues, wheat straw or olive tree pruning, and their mixtures towards flexible sugar-based biorefineries. Biomass Convers Biorefinery 6:385–396. https://doi.org/10.1007/s13399-016-0201-2
122. Simon E, Augier F, Morchain J et al (2024) Investigation of the local apparent viscosity in a stirred tank with a shear-thinning fluid through particle image velocimetry. Chem Eng Res Des 206:378–385. https://doi.org/10.1016/j.cherd.2024.05.008
123. Singh H, Janiyani K, Gangawane A et al (2024) Engineering cellulolytic fungi for efficient lignocellulosic biomass hydrolysis: advances in mutagenesis, gene editing, and nanotechnology with CRISPR-Cas innovations. Springer International Publishing
124. Singhvi MS, Gokhale DV (2019) Lignocellulosic biomass: hurdles and challenges in its valorization. Appl Microbiol Biotechnol 103:9305–9320. https://doi.org/10.1007/s00253-019-10212-7
125. Smuga-Kogut M, Kogut T, Markiewicz R, Słowik A (2021) Use of machine learning methods for predicting amount of bioethanol obtained from lignocellulosic biomass with the use of ionic liquids for pretreatment. Energies 14. https://doi.org/10.3390/en14010243
126. Song C, Kitamura Y, Li S (2014) Optimization of a novel cryogenic CO2 capture process by response surface methodology (RSM). J Taiwan Inst Chem Eng 45:1666–1676. https://doi.org/10.1016/j.jtice.2013.12.009
127. Song B, Li B, Wang X et al (2018) Real-time imaging reveals that lytic polysaccharide monooxygenase promotes cellulase activity by increasing cellulose accessibility. Biotechnol Biofuels 11:1–11. https://doi.org/10.1186/s13068-018-1023-1
128. Sørensen A, Lübeck M, Lübeck PS, Ahring BK (2013) Fungal beta-glucosidases: a bottleneck in industrial use of lignocellulosic materials. Biomolecules 3:612–631. https://doi.org/10.3390/biom3030612
129. Sotaniemi VH, Taskila S, Ojamo H, Tanskanen J (2016) Controlled feeding of lignocellulosic substrate enhances the performance of fed-batch enzymatic hydrolysis in a stirred tank reactor. Biomass Bioenerg 91:271–277. https://doi.org/10.1016/j.biombioe.2016.05.037
130. Studer MH, DeMartini JD, Brethauer S et al (2010) Engineering of a high-throughput screening system to identify cellulosic biomass, pretreatments, and enzyme formulations that enhance sugar release. Biotechnol Bioeng 105:231–238. https://doi.org/10.1002/bit.22527
131. Subhedar PB, Ray P, Gogate PR (2018) Intensification of delignification and subsequent hydrolysis for the fermentable sugar production from lignocellulosic biomass using ultrasonic irradiation. Ultrason Sonochem 40:140–150. https://doi.org/10.1016/j.ultsonch.2017.01.030
132. Sun C, Meng X, Sun F et al (2023) Advances and perspectives on mass transfer and enzymatic hydrolysis in the enzyme-mediated lignocellulosic biorefinery: a review. Biotechnol Adv 62:108059. https://doi.org/10.1016/j.biotechadv.2022.108059
133. Szabo OE, Csiszar E (2017) Some factors affecting efficiency of the ultrasound-aided enzymatic hydrolysis of cotton cellulose. Carbohydr Polym 156:357–363. https://doi.org/10.1016/j.carbpol.2016.09.039
134. Tandrup T, Frandsen KEH, Johansen KS et al (2018) Recent insights into lytic polysaccharide monooxygenases (LPMOs). Biochem Soc Trans 46:1431–1447
135. Taran M, Aghaie E (2015) Designing and optimization of separation process of iron impurities from kaolin by oxalic acid in bench-scale stirred-tank reactor. Appl Clay Sci 107:109–116. https://doi.org/10.1016/j.clay.2015.01.010

136. Taylor LE, Knott BC, Baker JO et al (2018) Engineering enhanced cellobiohydrolase activity. Nat Commun 9:1–10. https://doi.org/10.1038/s41467-018-03501-8
137. Teugjas H, Väljamäe P (2013) Selecting β-glucosidases to support cellulases in cellulose saccharification. Biotechnol Biofuels 6:1–13. https://doi.org/10.1186/1754-6834-6-105
138. Tramontina R, Brenelli LB, Sodré V et al (2020) Enzymatic removal of inhibitory compounds from lignocellulosic hydrolysates for biomass to bioproducts applications. World J Microbiol Biotechnol 36:166. https://doi.org/10.1007/s11274-020-02942-y
139. Truppo MD, Pollard DJ, Moore JC, Devine PN (2008) Production of (S)-γ-fluoroleucine ethyl ester by enzyme mediated dynamic kinetic resolution: comparison of batch and fed batch stirred tank processes to a packed bed column reactor. Chem Eng Sci 63:122–130. https://doi.org/10.1016/j.ces.2007.09.022
140. Unrean P, Khajeeram S, Laoteng K (2016) Systematic optimization of fed-batch simultaneous saccharification and fermentation at high-solid loading based on enzymatic hydrolysis and dynamic metabolic modeling of Saccharomyces cerevisiae. Appl Microbiol Biotechnol 100:2459–2470. https://doi.org/10.1007/s00253-015-7173-1
141. Varongchayakul S, Jomkamsing S, Chaiprasert P (2024) Maximizing sugars production from cassava pulp using efficient hydrothermal pretreatment coupled with cellulase. Biomass Convers Biorefinery 12163–12174. https://doi.org/10.1007/s13399-024-06041-y
142. Vlasenko E, Schülein M, Cherry J, Xu F (2010) Substrate specificity of family 5, 6, 7, 9, 12, and 45 endoglucanases. Bioresour Technol 101:2405–2411. https://doi.org/10.1016/j.biortech.2009.11.057
143. Wang R, He Z, Chen H et al (2024) Enhancing biomass conversion to bioenergy with machine learning: gains and problems. Sci Total Environ 927:172310. https://doi.org/10.1016/j.scitotenv.2024.172310
144. Wang Z, Lv Z, Yang X, Tian S (2013) Fed-batch mode optimization of SSF for cellulosic ethanol production from steam-exploded corn stover. BioResources 8:5773–5782. https://doi.org/10.15376/biores.8.4.5773-5782
145. Wei R, Oeser T, Billig S, Zimmermann W (2012) A high-throughput assay for enzymatic polyester hydrolysis activity by fluorimetric detection. Biotechnol J 7:1517–1521. https://doi.org/10.1002/biot.201200119
146. Wei N, Oh EJ, Million G et al (2015) Simultaneous utilization of cellobiose, xylose, and acetic acid from lignocellulosic biomass for biofuel production by an engineered yeast platform. ACS Synth Biol 4:707–713. https://doi.org/10.1021/sb500364q
147. Wojtusik M, Zurita M, Villar JC et al (2016) Influence of fluid dynamic conditions on enzymatic hydrolysis of lignocellulosic biomass: effect of mass transfer rate. Bioresour Technol 216:28–35. https://doi.org/10.1016/j.biortech.2016.05.042
148. Xu C, Zhang J, Zhang Y et al (2019) Enhancement of high-solids enzymatic hydrolysis efficiency of alkali pretreated sugarcane bagasse at low cellulase dosage by fed-batch strategy based on optimized accessory enzymes and additives. Bioresour Technol 292:121993. https://doi.org/10.1016/j.biortech.2019.121993
149. Xu C, Zhu Y, Wang K et al (2025) Substrate-specific responses to mixing conditions in high-solids enzymatic hydrolysis: insights from microcrystalline cellulose and dilute-acid pretreated corncob. Int J Biol Macromol 294:139431. https://doi.org/10.1016/j.ijbiomac.2024.139431
150. Yang B, Dai Z, Ding SY, Wyman CE (2011) Enzymatic hydrolysis of cellulosic biomass. Biofuels 2:421–449. https://doi.org/10.4155/bfs.11.116
151. Yang J, Yue HR, Pan LY et al (2023) Fungal strain improvement for efficient cellulase production and lignocellulosic biorefinery: current status and future prospects. Bioresour Technol 385:129449. https://doi.org/10.1016/j.biortech.2023.129449
152. Yang M, Zhang K-D, Zhang P-Y et al (2016) Synergistic cellulose hydrolysis dominated by a multi-modular processive endoglucanase from *Clostridium cellulosi*. Front Microbiol 7:1–8. https://doi.org/10.3389/fmicb.2016.00932
153. Yoshida M, Liu Y, Uchida S et al (2008) Effects of cellulose crystallinity, hemicellulose, and lignin on the enzymatic hydrolysis of *Miscanthus sinensis* to monosaccharides. Biosci Biotechnol Biochem 72:805–810. https://doi.org/10.1271/bbb.70689

154. Yu Z, Jingliang X, Zhenhong Y, Xinshu Z (2009) Kinetic model study on enzymatic hydrolysis of cellulose using artificial neural networks. Chinese J Catal 30:355
155. Yuan TQ, Wang W, Zhang LM et al (2013) Reconstitution of cellulose and lignin after [C2mim][OAc] pretreatment and its relation to enzymatic hydrolysis. Biotechnol Bioeng 110:729–736. https://doi.org/10.1002/bit.24743
156. Zanuso E, Ruiz HA, Domingues L, Teixeira JA (2022) Magnetic nanoparticles as support for cellulase immobilization strategy for enzymatic hydrolysis using hydrothermally pretreated corn cob biomass. BioEnergy Res 15:1946–1957. https://doi.org/10.1007/s12155-021-10384-z
157. Zhai R, Hu J, Jin M (2022) Towards efficient enzymatic saccharification of pretreated lignocellulose: enzyme inhibition by lignin-derived phenolics and recent trends in mitigation strategies. Biotechnol Adv 61:108044. https://doi.org/10.1016/j.biotechadv.2022.108044
158. Zhai R, Hu J, Saddler JN (2018) The inhibition of hemicellulosic sugars on cellulose hydrolysis are highly dependant on the cellulase productive binding, processivity, and substrate surface charges. Bioresour Technol 258:79–87. https://doi.org/10.1016/j.biortech.2017.12.006
159. Zhang X, Chen K, Long L, Ding S (2021) Two C1-oxidizing AA9 lytic polysaccharide monooxygenases from *Sordaria brevicollis* differ in thermostability, activity, and synergy with cellulase. Appl Microbiol Biotechnol 105:8739–8759. https://doi.org/10.1007/s00253-021-11677-1
160. Zhang H, Han L, Dong H (2021) An insight to pretreatment, enzyme adsorption and enzymatic hydrolysis of lignocellulosic biomass: experimental and modeling studies. Renew Sustain Energy Rev 140:110758. https://doi.org/10.1016/j.rser.2021.110758
161. Zhang X, Han X, Xie J, Zhang H (2024) Composite additives enhance high-solids enzymatic hydrolysis of CuCl2-catalyzed organosolv pretreated sugarcane bagasse. Cellulose 31:6681–6692. https://doi.org/10.1007/s10570-024-06004-6
162. Zhang H, Huang S, Wei W et al (2019) Investigation of alkaline hydrogen peroxide pretreatment and Tween 80 to enhance enzymatic hydrolysis of sugarcane bagasse. Biotechnol Biofuels 12:1–9. https://doi.org/10.1186/s13068-019-1454-3
163. Zhang C, Kang X, Wang F et al (2020) Valorization of food waste for cost-effective reducing sugar recovery in a two-stage enzymatic hydrolysis platform. Energy 208:118379. https://doi.org/10.1016/j.energy.2020.118379
164. Zhang H, Wu J (2023) Optimization of potassium hydroxide combined urea pretreatment and enzymatic hydrolysis of wheat straw using response surface methodology for improving sugar production. BioResources 19:1079–1106. https://doi.org/10.15376/biores.19.1.1079-1106
165. Zhao Q, Wang L, Chen H (2019) Effect of novel pretreatment of steam explosion associated with ammonium sulfite process on enzymatic hydrolysis of corn straw. Appl Biochem Biotechnol 189:485–497. https://doi.org/10.1007/s12010-019-03018-w
166. Zheng Y, Tao L, Yang X et al (2019) Comparative study on pyrolysis and catalytic pyrolysis upgrading of biomass model compounds: Thermochemical behaviors, kinetics, and aromatic hydrocarbon formation. J Energy Inst 92:1348–1363. https://doi.org/10.1016/j.joei.2018.09.006
167. Zhou B, Wang T, Li C et al (2020) Multi-objective optimization of the preparation parameters of the powdered activated coke for SO2 adsorption using response surface methodology. J Anal Appl Pyrolysis 146:104776. https://doi.org/10.1016/j.jaap.2020.104776
168. Zhou H, Wang L, Liu Y (2018) Physico-chemical oxidative cleavage strategy facilitates the degradation of recalcitrant crystalline cellulose by cellulases hydrolysis. Biotechnol Biofuels 11:1–13. https://doi.org/10.1186/s13068-018-1016-0
169. Zhu JY, Pan X (2022) Efficient sugar production from plant biomass: current status, challenges, and future directions. Renew Sustain Energy Rev 164:112583. https://doi.org/10.1016/j.rser.2022.112583
170. Zhu Y, Wang K, Xu C et al (2024) Enhancing high-solid corn cob hydrolysis via mixing optimization strategies. Bioenergy Res 17:1460–1468. https://doi.org/10.1007/s12155-024-10740-9

171. Zhu T, Xia A, Lin K et al (2022) Numerical investigation of bio-inspired mixing enhancement for enzymatic hydrolysis. Chem Eng Sci 260:117950. https://doi.org/10.1016/j.ces.2022.117950
172. Zumstein MT, Kohler HPE, McNeill K, Sander M (2017) High-throughput analysis of enzymatic hydrolysis of biodegradable polyesters by monitoring cohydrolysis of a polyester-embedded fluorogenic probe. Environ Sci Technol 51:4358–4367. https://doi.org/10.1021/acs.est.6b06060

Role of Counter-Current, Percolation, Plug-Flow, and Shrinking-Bed Reactor in Lignocellulosic Biomass Pretreatment and Enzymatic Hydrolysis

Kiara A. García-Bustos, Kevin Tejada-Meza, Miguel A. D. Flores-Alarcón, and Ruly Terán-Hilares

Abstract Lignocellulosic biomass is a widely available, renewable resource with significant potential for sustainable bioconversion. However, its effective utilization requires a sequence of processing steps, primarily pretreatment followed by enzymatic hydrolysis, to release fermentable sugars from structural carbohydrates. These processes are commonly implemented in various reactor configurations, including counter-current, percolation, plug-flow, and shrinking-bed reactors operating under batch, semicontinuous, or continuous modes. Such systems offer distinct advantages, including the capacity to handle high solid loadings, generate sugar-rich hydrolysates, and minimize water usage and wastewater production. This chapter presents a concise review of current literature on the application of these reactor technologies in the pretreatment and hydrolysis of lignocellulosic biomass. The insights provided aim to inform the development of more efficient and environmentally sustainable bioconversion strategies for the advancement of bio-based industries.

Keywords Lignocellulosic biomass · Pretreatment · Enzymatic hydrolysis · Reactors

1 Introduction

The effective conversion of lignocellulosic biomass (LCB) into fermentable sugars represents a pivotal step in the development of sustainable biofuels and high-value bioproducts. However, this process is significantly hindered by the intrinsic structural

K. A. García-Bustos · K. Tejada-Meza · R. Terán-Hilares (✉)
Laboratorio de Bioprocesos, Facultad de Ciencias FarmacéuticasBioquímicas y Biotecnológicas, Universidad Católica de Santa María—UCSM, Urb. San José s/n—Umacollo, Arequipa, Peru
e-mail: rteran@ucsm.edu.pe

M. A. D. Flores-Alarcón
Department of Biotechnology, Engineering School of Lorena, University of São Paulo, Lorena, Brazil

© The Author(s), under exclusive license to Springer Nature Singapore Pte Ltd. 2025

A. K. Chandel and S. Sundaramurthy (eds.), *Biorefineries in Circular Bioeconomy*, Green Energy and Technology, https://doi.org/10.1007/978-981-95-3913-0_6

complexity and recalcitrance of lignocellulosic materials, which resist enzymatic degradation and limit sugar release [1]. As a result, substantial research efforts have been directed toward optimizing both the pretreatment and enzymatic hydrolysis stages to enhance the efficiency, yield, and scalability.

Among the numerous factors that influence the success of these operations, reactor configuration plays a central and strategic role. It not only determines sugar yields and energy efficiency but also affects the feasibility of process integration within modern biorefinery systems. Reactor design and optimization have therefore emerged as critical elements in achieving effective feedstock utilization. Various reactor configurations have been investigated for their capacity to facilitate high solids loading, improve mass and heat transfer, and maintain favorable conditions for hydrolytic and microbial activities [1].

In recent years, innovative reactor systems, including percolation reactors (PCRs), plug-flow reactors (PFRs), and shrinking-bed reactors (SBRs), have gained increasing attention for their effectiveness under high-solids conditions. PCRs enable continuous enzymatic hydrolysis by allowing liquid percolation through static solid beds, thereby eliminating the need for solid–liquid separation steps. However, they face operational limitations such as bed compaction and pressure buildup. PFRs, on the other hand, offer precise control over residence time and thermal-hydrolytic conditions, making them highly suitable for large-scale pretreatment and hydrolysis, with successful applications in the production of fermentable sugars, bioethanol, and furfural. Meanwhile, SBRs utilize dynamic bed compaction mechanisms that enhance enzyme–substrate interaction and mass transfer, resulting in improved hydrolysis kinetics and sugar recovery with reduced degradation losses.

This chapter provides a comprehensive analysis of these advanced reactor designs, focusing on their operational principles, engineering characteristics, and performance outcomes in the context of lignocellulosic biomass deconstruction. Furthermore, various reactor types, including counter-current, percolation, and simulated counter-current systems, have been evaluated at the laboratory scale to optimize biomass breakdown and sugar yield. By comparing recent technological advancements and performance data, this text highlights the potential of PCRs, PFRs, and SBRs for integration into next-generation biorefineries, aiming to enhance both the environmental sustainability and economic viability of lignocellulosic biofuel production.

2 Lignocellulosic Biomass, Pretreatment, and Enzymatic Hydrolysis

Biomass refers to naturally occurring organic matter derived from diverse sources, including agriculture, forestry, households, urban settings, and industry [2]. These materials, often categorized as second-generation waste, are heterogeneous and typically rich in moisture. Agricultural and food processing industries alone generate approximately 1.3 billion tons of lignocellulosic waste annually. This lignocellulosic biomass (LCB) represents the most abundant, yet underutilized, renewable biomaterial on earth [3]. Valorizing plant-based lignocellulosic waste presents a promising approach to not only mitigate environmental impacts but also to generate economic benefits through the conversion of waste into bio-based products and bioenergy [4, 5]. This strategy supports the circular economy by reducing raw material consumption and enhancing the share of renewable resources [6, 7].

LCB is mainly composed of 30–70% cellulose, 15–35% hemicellulose, and 10–20% lignin, with trace amounts of other minerals [8, 9]. The type of plant, species, and even within species, depending on the environment, determines the ratio of these components [5, 10–12]. The three polymers are mutually crosslinked, forming a complex and stable structure. Cellulose is a glucose-based polymer; it is biodegradable but insoluble in water and most organic solvents. It is usually broken down and solubilized by acid or enzyme hydrolysis. Hemicellulose is composed of xylan and glucomannan; however, due to the amorphous structure and hydrophilicity, hemicellulose can be more easily broken down. Lignin is a nonsugar-based polymer with a three-dimensional network structure; it is neither soluble nor biodegradable. Being rigid and resistant, lignin serves as the main composition protecting LCB from biochemical decomposition. Thus, lignin is the key factor hindering the biodegradation of LCB [13].

Pretreatment is essential to overcome the natural resistance to biodegradation. It facilitates the breakdown of lignin, reduces the crystallinity of cellulose into a more accessible form, and enhances the accessibility of biodegradable oligosaccharides and the partial solubilization of hemicelluloses [5, 12, 14].

The selection of the pretreatment method and its conditions are highly dependent on the type and the chemical composition of the biomass. In the last few decades, many pretreatment methods have been developed and validated for different types of LCB. The biological, physical, and chemical methods are often used to separate lignin, cellulose, and hemicellulose from LCB [14].

Bioreactors performing the enzymatic hydrolysis require a certain level of agitation to promote the mass transfer and mixing of the phases. The timescale of the operation is quite long (more than 48 h) and can be shortened by increasing the enzyme loading of the reactor. However, as the enzyme is one of the costliest components, this would obviously affect the economies of the operation [15]. Currently, cellulase costs range between 3 and 300 Euro/kg. For example, in the case of ethanol production from lignocellulose biomass, this counts for 0.05–1 Euro/kg of the total production cost [15, 16]. Thus, the operation asks for a careful design of the bioreactor and an efficient use of enzymes as biocatalysts [17].

The benchmark technology allows a solid–liquid reaction step performed batchwise in stirred vessels at an optimized temperature, sacrificing the enzyme at each batch. Since the main cost of the enzymatic hydrolysis lies in the cost of the cellulolytic enzyme cocktail [17], the continuous operation has also been investigated together with strategies for enzyme recycling. This aims, in theory, at significantly reducing the net loading of fresh enzyme per unit of cellulose below 20 mg/g.

However, various bioreactor configurations have been evaluated not only for enzymatic hydrolysis but also for the pretreatment of lignocellulosic biomass. For example, a percolation reactor involves the flow of an acid liquid through a biomass-packed bed to hydrolyze the polysaccharides, and the hydrolysate liquid is collected at the outlet of the bed. PCR shows some merits as: (1) high sugar yield because of immediate discharge of sugar products with a minimum sugar decomposition; (2) high sugar concentration by a high solid/liquid ratio; and (3) easy separation of liquid products at the outlet of the reactor [18]. This reactor is tubular, and it is filled with solid catalyst particles, most often used to catalyze gas reactions. The chemical reaction takes place on the surface of the catalyst [19]. Another interesting reactor is the shrinking-bed reactor (SBR), which has a fixed and a movable end supported by a compressed spring. As the hydrolysis proceeds, the biomass bed becomes less dense. The spring-attached end then takes effect, pressing the loose biomass particles to maintain the bulk packing density constant. Due to the compressing action, the liquid throughput per solid mass as well as mass and heat transfer in the reactor are enhanced, and thus the sugar yield can be increased [18, 20]. Results indicate that bed shrinking provides a positive effect on both hemicellulose and cellulose hydrolysis, resulting in high yield and product concentration. The effect of bed shrinking is more profound on cellulose hydrolysis than on hemicellulose hydrolysis [21].

The plug flow reactor is the second most common ideal reactor and is similar to the continuous stirred-tank bioreactor. Here, polymerization and conversion reactions are performed in a noncatalytic mode [22]. In this type of reactor, a high concentration of solid or semisolid fermentation raw material enters from one end of the feed inlet and is then discharged from the other end after supplementation. The advantages of this type of reactor are that it does not require mixing, has low energy consumption, a simple structure, and good stability. Its limitations include the fact that the reactants are easily precipitated, affecting the effective volume of the reactor, reducing the hydraulic retention time (HRT) and solid retention time (SRT), and the efficiency is low [23].

3 Reactors for Pretreatment and Enzymatic Hydrolysis of Lignocellulosic Biomass

Lignocellulosic materials undergo pretreatment followed by enzymatic hydrolysis, essential steps in the conversion process. These operations are typically conducted in a variety of reactor configurations, including counter-current, percolation, plug flow, and a shrinking-bed reactor at the bench scale. This section provides a concise overview of each reactor, covering both the pretreatment and enzymatic hydrolysis stages.

3.1 Counter-Current Reactor

A counter-current reactor utilizes opposing flows to enhance reaction conversion and transport properties, making it generally more efficient than batch reactors. Its advantages include improved substrate utilization, continuous processing that eliminates idle times from loading and unloading, and reduced product inhibition [24]. This type of reactor has been used in the pretreatment and enzymatic hydrolysis of lignocellulosic biomass. For instance, counter-current liquid hot water flow-through pretreatment has been used for corn stover hydrolysis, resulting in higher solubilized xylan concentrations and 3–6 times less degradation compared to a fixed reactor [25]. Moreover, [26] performed counter-current extraction of water-soluble sugars, followed by mild NaOH (1 h at 80 °C with 0.10 g NaOH/g biomass) or liquid hot water (1 h at 120 °C) pretreatment of the extracted bagasse. Their study demonstrated that this process could extract 97% of the water-extractable sugars within five extraction stages, yielding extractive-free bagasse and an extraction juice containing 80.1 g/L of sucrose, fructose, and glucose. Additionally, glucose hydrolysis yields varied from 64.9% for low-severity liquid hot water pretreatment to 87.3% for the most severe NaOH pretreatment condition.

In the enzymatic process, counter-current is a novel approach of a continuous process, consisting of a moving-bed reactor in which the direction of the lignocellulosic biomass and liquid (enzymes, buffer, and antimicrobials) are reversed (Fig. 1a) [27]. Therefore, the counter-current is presented as an alternative system to batch reactors commonly used for enzymatic hydrolysis, such as stirred tank reactors (STR). In STR, significant time (days) and high enzyme loadings are usually required to achieve reasonable conversion yields, and the leftover active enzymes are usually discarded [28]. In contrast, the counter-current system provides an internal recycling of enzymes, since the enzymes released from the enzyme–substrate complex can hydrolyze fresh biomass, reducing the enzyme loadings and thus reducing the impact of enzyme cost on the sugar production process from biomass [29, 30]. Also, sugar concentrations increase over the length of the counter-current reactor, thus the least reactive biomass contacts the lowest sugar concentrations, and sugars are continuously removed from the system; while in STR, a high sugar concentration is maintained, especially when operated in continuous mode, triggering product inhibition [31].

The scalability (bench to large-scale production) of a counter-current system can be performed in a packed column where fresh lignocellulosic biomass is added at the top and flows downward, while fresh liquid is pumped through the bottom and flows counter-currently to the solids [32]. The hydrolysate is harvested from the top of the column, and the enzyme is injected at the appropriate depth [33]. It is worth mentioning that antibiotic booster doses must be added to prevent the growth of contaminating microorganisms, since it is performed in continuous mode [33]. Other alternative antimicrobial agents have also been studied [33], including essential oils of thyme and oregano, which can even be added as whole plants to the biomass feedstock [34]. However, at bench-scale, the counter-current flow pattern is usually

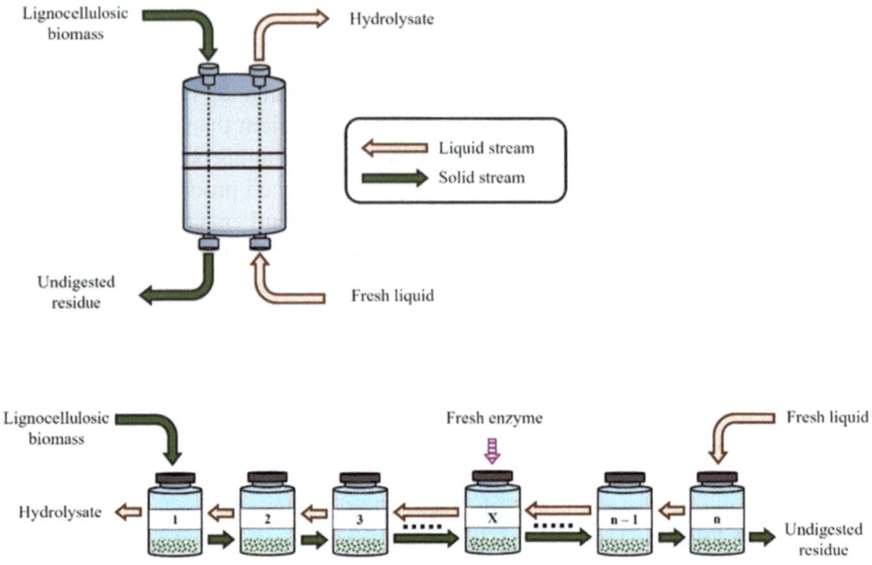

Fig. 1 Schematic diagram of the enzymatic hydrolysis of lignocellulosic biomass in the following systems: **a** Counter-current and **b** Simulated counter-current at bench scale

simulated using multiple reactors (centrifuge bottles) containing biomass (Fig. 1b), such as percolation reactors, connected in series (train), moving in one direction, and the flow of liquid moving in the opposite direction [28]. As can be seen in Fig. 1b, this process involves separating the bottle contents, contacting the supernatant solution (hydrolysate) from one batch reactor (stage) with a second batch of biomass, and adding fresh enzyme to a fixed, intermediate batch (stage "X") or a moving terminal batch (stage "1" or "n"). The process can continue until all batches in the train have been contacted with fresh or supernatant enzyme solutions [35].

In this counter-current reactor scheme (Fig. 1b), the highest concentration of sugars is produced in the vicinity of the lignocellulosic biomass inlet point (stage "1") and the variations in sugar concentrations are greater near the liquid inlet point (stage "n"), since there is more enzyme moving downstream with the biomass (from reactor "1" to "n") than moving upstream with the liquid (from stage "n" to "1″). This is because most cellulolytic enzymes remain tightly bound to biomass rather than remaining suspended in the liquid phase. Consequently, these enzymes retained on the biomass can assist any fresh incoming enzyme with hydrolysis [32, 36].

As observed in the studies summarized in Table 1, the enzymes were mostly added to stage "4" instead of stage "1," because it is necessary to use the enzymes that remain in the liquid phase and are not adsorbed on the biomass before leaving the system. At the fresh liquid inlet point, digested biomass is washed to recover spent enzymes and product sugars, thus improving process efficiency [37]. When the sugar concentrations are still changing from stage to stage, more stages are needed to fully utilize enzyme activity; while no increase in sugar concentration indicates that enzymes should be added closer to the biomass inlet point [27, 37]. Moreover, it is true that sugar conversion increases with increasing enzyme loading and the number of stages (Table 1). However, with optimized load and point of enzyme addition, the conversion could be maximized by more fully utilizing the leftover active enzymes. Thus, more sugar is produced without the need for a higher enzyme load. In addition, a reduced number of stages is preferable, aiming to reduce enzyme inhibition, raising conversion yield, and sugar concentration [28]. In this sense, Zentay et al. [32] demonstrated that, at optimum conditions of operation, it is possible to efficiently use the enzyme, reducing the enzyme loading by 8–20 times compared to the batch mode to achieve the same glucan conversion yields when using α-cellulose as substrate (Table 1). However, when the counter-current enzymatic hydrolysis of pretreated corn stover was performed by Liang et al. [33], the enzyme requirements were similar to those of batch mode, probably due to the unproductive binding of enzymes to the lignin and the weak binding of hemicellulases to the substrate as compared to cellulases. Therefore, pretreatments that can remove most of the lignin from biomass are desirable to improve the benefits of counter-current enzymatic hydrolysis. Recently, supplementation with hemicellulases was used as a strategy to improve both cellulose and hemicellulose conversion yield, since the digestion of hemicellulose makes cellulose more accessible to enzymes [37].

Besides the enzyme addition location, the total number of stages, and enzyme loading, there are other variables, such as liquid residence time (LRT) and solids loading rate (SLR), which affect the sugar conversion yield and concentration in a counter-current system[33]. A longer LRT allows higher sugar concentration, whereas conversion decreases. As SLR decreases, conversion increases significantly, whereas sugar concentration decreases. Thus, a lower SLR allows longer LRT. Therefore, in the Counter-current system, conversion yields depend on the conditions and feedstock used [32]. Counter-current process was also performed, aiming to reduce the water consumption, allowing high extraction yield, high concentration, low solvent consumption, and small space requirements [38]

3.2 Percolation Reactor

A percolation reactor (PCR) is a specialized system engineered for the pretreatment and enzymatic hydrolysis of lignocellulosic biomass. Its primary function is to convert agricultural residues and other lignocellulosic feedstocks into fermentable sugars to produce biofuels and value-added bioproducts [39]. PCRs play a critical role in biomass pretreatment by promoting the dissolution of lignin and hemicellulose, thereby improving the accessibility of cellulose to hydrolytic enzymes. Figure 1b illustrates a conceptual diagram of a PCR, which operates as a packed-bed, flow-through reactor [18, 20]. In this system, the biomass remains stationary within percolation baskets, while the hydrolysate is continuously recirculated through the reactor, ensuring intimate contact with the solid substrate [39].

Table 1 Studies of counter-current enzymatic hydrolysis of lignocellulosic biomass compared to the batch process

Mode of operation	Substrate	Number of stages	Enzyme addition point/Solid inlet point	Enzyme load (mg protein/g biomass)	Solids content (% w/w)	Conversion yield (%)		Factor improvement** (Batch: counter-current)		References
						For glucose	For xylose	For glucose	For xylose	
Batch	Alkali-treated sugarcane bagasse	1	–	–	11	44.3[a]		–	–	[31]
PFR with a counter-current flow		3	1, 2, and 3/3	–	11	56.4[a]				
Batch	α-cellulose	1	–	5 (CTec2)	10	35.6[b]	35.9[b]	–	–	[32]
		1	–	2 (CTec2)	10	18.6[b]	17.6[b]	–	–	
Multistage semi-continuous counter-current		8	5/1	5 (CTec2)	10	89.06	73.61	16.8	6.8	
		8	4/1	2 (CTec2)	10	55.42	52.56	8	6	
		16	8/1	2 (CTec2)	10	74.66	71.75	20.5	15.5	
Batch	Lime-pretreated corn stover	1	–	1 (CTec3)	~ 10		~ 41[b]	–	–	[24]
		1	–	1 (CTec3) + 1 (HTec3)	~ 10			–	–	
Pseudo-continuous counter-current		16	4/1	1 (CTec3)	~ 10	61	41	1.6	0.7	
		16	4/1	1 (CTec3) + 1 (HTec3)	~ 10	67	53	1.4	0.5	
	Lime + shock pretreated corn stover	16	4/1	1 (CTec3)	~ 10	64	49	1.9	0.82	[35]

(continued)

Table 1 (continued)

Mode of operation	Substrate	Number of stages	Enzyme addition point/Solid inlet point	Enzyme load (mg protein/g biomass)	Solids content (% w/w)	Conversion yield (%)		Factor improvement** (Batch: counter-current)		References
						For glucose	For xylose	For glucose	For xylose	
		16	4/1	1 (CTec3) + 1 (HTec3)	~ 10	72	62	1.4	0.8	
		16	4/1	2 (CTec3) + 2 (HTec3)	~ 10	82	79	0.8	0.73	

* Factor of enzyme requirements to achieve the same conversion yield as was obtained in counter-current enzymatic hydrolysis
[a]Only reported the overall conversion yield
[b]Calculated conversion yield from available equation curve, graph, or text of batch enzymatic hydrolysis

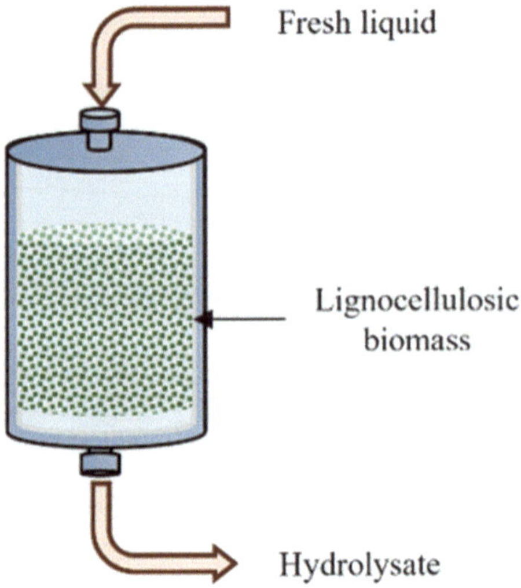

Fig. 2 Schematic diagram of the enzymatic hydrolysis of LCB in a percolated reactor

Structurally, a PCR comprises a fixed or packed bed containing stationary lignocellulosic biomass (Fig. 2). An acid or alkaline solution is continuously introduced, percolating through the bed either in a top-down or bottom-up configuration. This design promotes efficient fluid–solid interactions, facilitating the removal of lignin and hemicellulose and improving the overall pretreatment efficacy [40]. PCRs are valued for their compact design, operational simplicity, and energy efficiency. The use of recirculation pumps and integrated heat exchange systems minimizes energy consumption, while hot streams can further enhance thermodynamic efficiency by reducing heat losses and enabling energy recovery [41]. Some reactor designs incorporate thermal jackets or built-in heating coils, providing precise internal temperature control and eliminating the need for external heat exchangers [42].

By selectively removing or modifying lignin—an inherent structural barrier—PCRs enhance enzymatic hydrolysis efficiency. Hemicellulose removal also contributes to improved fermentable sugar release, a key step in bioethanol and bioproduct production. A notable example is the organosolv pretreatment with acetone in PCRs, which has shown high effectiveness in recovering cellulose while solubilizing lignin and C5 sugars. Pilot-scale trials have yielded results consistent with laboratory data, highlighting the method's potential for industrial application [43].

Table 2 compares the performance of various lignocellulosic biomass pretreatment processes using PCRs, including operational parameters and metrics such as lignin removal, cellulose recovery, and fermentable sugar release. Key factors influencing pretreatment efficiency, temperature, chemical concentration, and residence time are also outlined. PCRs demonstrate significant advantages in lignin removal and cellulose accessibility.

While PCRs are primarily used for biomass pretreatment, Kirsch et al. [44] demonstrated that thermal pretreatment and enzymatic hydrolysis of rye straw (10% biomass

Table 2 Efficiency of pretreatments in percolation reactors for different types of lignocellulosic biomass

Lignocellulosic biomass	Scope	Operation mode	Solvent and operational conditions	Yield or efficiency	References
Sugarcane bagasse	Lignin and hemicellulose removal	Batch (recirculation)	Alkali (0,3 M NaOH); 70 °C; 4 h,	61%	[49]
Softwood (Norway spruce)	Lignin removal	Semi-continuous	Mix of ethanol, water, and sulfuric acid; 140–193 °C;	86% lignin yield, 67% cellulose recovery	[70]
Poplar wood	Detoxification	Batch (Percolation)	Saturated steam at 170 °C for 120 min; resin treatment postprocess	97% removal of ASL, 95.3% sugar utilization	[71]
Sawdust (***Platycladus orientalis***)	Oxidative degradation	Continuous	NaOH solution with O_2 at 2.0 MPa, 140 °C; flow rate 5 mL/min	70% carbon degradation	[72]
Beech wood	Lignin removal	Semi-continuous	Aqueous acetone (60%) with sulfuric acid at 140 °C; L/S ratio 3:1	86% lignin removal, 97% glucan hydrolysis	[73]
Birch wood	Lignin removal	Semi-continuous	Mix of aqueous acetone (50%) with 20 mM H_2SO_4; 140 °C	92% (C5 sugar yield)	[74]
Beech wood	Lignin precipitation	Semi-continuous	Aqueous acetone with direct solvent evaporation (LigniSep)	86% (lignin yield)	

loading) can be carried out in a single fixed-bed reactor. Using Celluclast 1.5 L at 11 filter paper units (FPU)/g glucan, they achieved high glucan conversion and glucose concentrations. In that study, the enzyme solution was pumped through the reactor at elevated temperature (60 °C) and pressure (100 bar), increasing reaction rates and reducing residence time from 14 h to 5.5 h compared to conditions at 1 bar and 50 °C. Reynolds et al. [45], from the same research group, used a high-pressure fixed-bed reactor to conduct sequential hydrothermal and enzymatic hydrolysis of approximately 22 wt% pelletized wheat straw using Cellic CTec2 at 15 FPU/g cellulose. Optimized conditions (50 °C, 75 bar, 450 mL/min circulation) resulted in a 51% glucose yield after 72 h, demonstrating that hydrolysis efficiency can be improved by adjusting operational parameters and reactor configuration.

One advantage of enzymatic hydrolysis in PCRs is the elimination of solid–liquid separation, which is typically required in batch reactors. PCRs also support higher solids loading [29, 46]. However, a key operational challenge is bed collapse or increased voidage due to biomass solubilization during hydrolysis [47]. Consequently, relatively few studies have explored PCRs for enzymatic hydrolysis, as the compressibility of both untreated and pretreated biomass can lead to excessive pressure buildup, restricted flow, or bed clogging—factors that constrain operational throughput [28, 48].

For example, Terán-Hilares et al. [49] used a packed-bed, flow-through reactor to hydrolyze alkaline-pretreated sugarcane bagasse at a solid-to-liquid ratio of 17.1% with Celluclast at 20 FPU/g dry biomass, achieving glucan and hemicellulose conversion yields of approximately 49% and 57%, respectively. Parallel experiments in Erlenmeyer flasks at a 5% solids ratio showed liquid absorption by the biomass, limiting enzyme activity. In contrast, the PCR ensured effective mixing and enzyme distribution via continuous liquid recirculation. Hydrolysis efficiency can be further improved through hemicellulase supplementation, as demonstrated in countercurrent systems.

Similarly, Ramos et al. [50] conducted enzymatic hydrolysis of alkaline sulfite-pretreated sugarcane bagasse in a flow-through packed-bed reactor at various solids loadings, achieving glucan hydrolysis yields of 87%, 77%, and 65% at 9.0%, 12.5%, and 16.0% solids, respectively. The decline in hydrolysis at higher solids was attributed to mass transfer limitations, enzyme product inhibition, and unproductive enzyme adsorption. Nonetheless, sugar concentrations in the PCR were twice as high as those achieved in agitated flasks with $\leq 7\%$ solids. The favorable structural properties of the pretreated bagasse allowed efficient bed formation and liquid flow, facilitating a simple and scalable process design.

In conclusion, percolation reactors offer a promising and straightforward configuration for enzymatic hydrolysis of lignocellulosic biomass at high solids content. They enable the efficient production of sugar-rich hydrolysates, thereby enhancing the economic viability of downstream processes in integrated biorefineries.

3.3 Plug Flow Reactor

A plug flow reactor (PFR), also known as a continuous tubular reactor (CTR), is a cylindrical reactor where reactants flow unidirectionally through a tube with minimal axial mixing, ensuring that each "plug" of fluid maintains its composition as it moves along the reactor length. This design allows for uniform residence time across all reactants, optimizing conversion rates and reaction efficiency [51]. PFRs are particularly effective in large-scale industrial applications due to their continuous operation and precise control over reaction conditions [52]. They are well-suited for rapid or highly exothermic reactions, as the tubular design facilitates efficient heat management throughout the process [53].

Catalysts can be applied to the reactor walls or packed inside the tube, enabling reactions such as ethylene polymerization or ammonia synthesis [51]. Additionally, energy efficiency is enhanced by incorporating recirculation pumps and heat exchangers, which reduce energy consumption by recovering heat from the process itself [54]. Commonly used in biofuel production, PFRs offer high conversion rates and scalability but can face challenges such as clogging with viscous materials or uneven temperature distribution in large reactors [55]. To address these limitations, multiple PFRs can be arranged in parallel or series configurations, maintaining consistent space–time and minimizing pressure drops [55].

Structurally, a plug-flow reactor (PFR) consists of a tubular system where lignocellulosic biomass is continuously introduced at one end, while the processed material exits at the other. This configuration facilitates a unidirectional flow of reactants, effectively minimizing back-mixing and enabling precise control over the residence time of the biomass throughout the reactor [52]. Key operational parameters affecting its performance include flow rate, temperature, type and concentration of the pretreatment agent, and pressure. These factors contribute to the efficient breakdown of lignocellulosic structures, facilitating the removal of lignin and hemicellulose while enhancing the enzymatic accessibility of cellulose [56].

Energy efficiency in plug flow reactors (PFRs) can be significantly improved by integrating recirculation pumps and heat exchange systems, which help minimize overall energy consumption by recovering and reusing process heat [57]. Additionally, recent advancements in solvolysis and catalyst development have shown potential for improving the efficiency of lignocellulosic biomass conversion, particularly in the production of value-added chemicals such as furfural [54].

The efficiency of PFRs in the pretreatment of lignocellulosic biomass is influenced by factors such as temperature, concentration of the pretreatment agent, and residence time [42]. Operating the reactor at higher temperatures can improve the solubilization of lignin and hemicellulose, but excessive heat may lead to the degradation of sugars, thereby reducing overall yields [58]. The plug-flow design ensures minimal back-mixing and consistent exposure of biomass to pretreatment agents, which results in more efficient conversion processes and improved recovery of fermentable sugars [59].

Hydrothermal liquefaction (HTL) has emerged as a promising method for converting wet biomass into biofuels and valuable chemicals. However, challenges such as corrosion, coke formation, and the precipitation of inorganic salts have limited its industrial scalability. Tran [60] emphasizes the need for PFRs in addressing these issues due to their ability to facilitate rapid heating, short residence times, and minimize fouling during continuous operation.

By enhancing heat transfer rates and reducing reaction time, PFRs can significantly improve the efficiency and selectivity of biocrude production from wet biomass, offering a viable pathway for scaling up HTL processes in biofuel industries. PFRs have demonstrated significant potential in enhancing the efficiency of bioethanol production from lignocellulosic biomass. As highlighted by Solarte-Toro et al. [61], the uniform flow profile of PFRs ensures consistent exposure of the biomass to pretreatment agents, improving mass transfer and reaction kinetics. This results in more efficient hydrolysis of hemicellulose and cellulose, leading to higher sugar recovery and conversion rates. Additionally, the continuous operation of PFRs allows for better process scalability and energy efficiency, making them particularly suitable for industrial applications in biofuel production. Their ability to maintain controlled residence times and minimize back-mixing contributes to higher yields and improved process stability.

Table 3 provides a comparative analysis of the performance of PFRs in the hydrolysis of lignocellulosic biomass, showcasing their efficiency across various feedstocks and operational conditions. It highlights key factors such as biomass type, process objectives, catalyst type, and critical operational parameters like temperature, pressure, acid concentration, and residence time. The data reveal that PFRs consistently achieve high yields of sugars and other value-added products, such as furfural and bioethanol, while offering enhanced scalability and process control. This versatility makes PFRs particularly suitable for industrial applications, optimizing the conversion of biomass into biofuels and other bioproducts.

3.4 Shrinking-Bed Reactor

The shrinking-bed reactor (SBR) is a specialized configuration designed for the hydrolysis of lignocellulosic biomass, particularly under acid-catalyzed conditions [18]. This reactor allows for efficient hydrolysis of both cellulose and hemicellulose by exploiting the natural reduction in bed volume as biomass is solubilized during the reaction. Unlike conventional reactors, SBRs improve reaction kinetics and minimize sugar degradation by controlling the residence time of sugars within the system [21, 62].

In an SBR, the biomass bed gradually shrinks due to the solubilization of structural components, primarily hemicellulose and cellulose. A spring-loaded mechanism or a moving screw system helps maintain compaction of the bed, reducing liquid throughput and increasing sugar concentration. This results in higher reaction efficiency compared to nonshrinking systems [63]. The design ensures that liquid

Table 3 Performance of plug-flow reactors (PFRs) in lignocellulosic biomass hydrolysis

Lignocellulosic biomass	Scope	Catalyst type	Solvent and operational conditions	Yield or efficiency	References
Corn fiber	Glucose and xylose production	Cation Exchange resin	Water at 160 °C, pH 4–7; pressurized; residence time: 105 min	88% hydrolysis at 130 °C; 90% glucose yield at 160 °C	[75]
Lignocellulosic biomass	Monosaccharide production and acid recycling	Concentrated sulfuric acid	Continuous PFR; 1.0–1.5 mol/L H_2SO_4; optimized hydrolysis and separation	Up to 90% reduction in reactor volume; high-purity sugar yield	[76]
Olive tree pruning biomass	Bioethanol production using acid pretreatment	Sulfuric acid (dilute)	160–164 °C; 4.9–5.9% acid concentration; solids loading: 15–35%	97.3–110.8 kg bioethanol/t raw material	[61]
Various lignocellulosic residues	Furfural production via continuous process	Sulfuric acid (H_2SO_4)	Continuous plug-flow reactor; 160–165 °C; 10 atm pressure	35–50% furfural yield; 98.5% furfural purity	[77]
Lignocellulosic biomass residues	Hydrothermal liquefaction for bio-crude	No catalyst (thermal process)	Temperature: 280–370 °C; Pressure: 10–25 MPa	High-energy content bio-crude yield	[55]
Agro-forestry waste	Furfural production by solvolysis	Homogeneous acid catalyst	C6-rich cellulose feedstock, organic solvents, 160–180 °C	Up to 12 wt% furfural yield	[54]
Corn cobs and sugarcane bagasse	Furfural production and process intensification	Acid catalyst	Continuous PFR; heat integration at 170–190 °C	Enhanced energy efficiency; improved furfural yield	[78]

flow and solid compaction occur simultaneously, allowing for high sugar yields while minimizing the decomposition of sugars under severe hydrolysis conditions.

SBRs are particularly effective in treating hardwood residues, wheat straw, and olive pruning waste. The shrinking mechanism enhances the hydrolysis of hemicellulose, resulting in sugar yields exceeding 90% for xylose and 85% for glucose under optimized conditions [63, 64]. Compared to plug-flow reactors (PFRs), SBRs offer superior performance in maintaining high sugar concentrations while reducing degradation losses, particularly for hemicellulose sugars such as xylose.

The configuration of SBRs, as detailed by Ortíz and Quintero [65], involves a dynamic bed compaction system that progressively reduces the volume of biomass during the reaction process. This shrinkage is typically controlled by mechanical compression systems or spring-loaded plungers, which maintain bed density and improve contact between the solid and liquid phases. This configuration increases mass transfer efficiency and enhances sugar concentration in the liquid phase. Additionally, newer designs incorporate countercurrent flow mechanisms that improve the removal of solubilized sugars and reduce the residence time of sugars in the reactor, thus preventing degradation. These design improvements make SBRs particularly effective for high-solid biomass feedstocks, allowing for consistent operation under high-pressure and temperature conditions.

The efficiency of SBRs is influenced by factors such as acid concentration, temperature, and flow rate. Increasing the temperature (up to 240 °C) and optimizing acid concentration (typically around 1% H_2SO_4) has been shown to improve glucose and xylose yields significantly [66]. When compared to PFRs, SBRs can achieve up to 16% higher yields due to better sugar recovery and reduced decomposition times [63]. Shrinking-bed reactors represent a promising solution for the efficient hydrolysis of lignocellulosic biomass, offering high sugar yields, better control over residence times, and enhanced process scalability. Their ability to maintain high product concentrations while minimizing decomposition losses makes them a viable option for integration into modern biorefineries focused on sustainable biofuel production.

Recent studies by Javanmard et al. [67] emphasize the importance of optimizing operational parameters in shrinking-bed reactors (SBRs) for enhancing the efficiency of lignocellulosic biomass hydrolysis. Critical factors such as temperature, acid concentration, and flow rate significantly influence the reactor's performance. Optimal hydrolysis conditions are typically achieved at temperatures ranging from 190 °C to 240 °C, where cellulose and hemicellulose break down effectively without causing excessive sugar degradation. Acid concentrations between 0.5% and 1.5% have been shown to increase the yield of fermentable sugars while avoiding the formation of inhibitory byproducts. Furthermore, adjusting the liquid flow rate ensures effective solubilization of biomass, enhancing sugar recovery while minimizing operational costs, particularly in systems designed for continuous processing.

Giuliano et al. [64] highlight the superior efficiency of shrinking-bed reactors (SBRs) in lignocellulosic biomass hydrolysis compared to conventional reactors. Their study demonstrates that SBRs enhance sugar recovery by maintaining higher solid–liquid contact and reducing the residence time of hydrolyzed sugars, which minimizes degradation. SBRs enhance the pretreatment of lignocellulosic biomass by improving solid–liquid contact and increasing sugar concentration through bed volume reduction. This leads to higher conversion rates of hemicellulose and cellulose, reduced sugar degradation, and greater efficiency, making SBRs ideal for industrial bioethanol production [68].

Wongsirichot [69] emphasizes the potential of computational fluid dynamics (CFD) in optimizing SBRs for biomass hydrolysis. CFD simulations can model complex mass and heat transfer processes, allowing researchers to fine-tune reactor configurations and operational parameters for higher efficiency. Notably, SBRs, when simulated with transient multiphase models, show improvements in mixing homogeneity and sugar yields while minimizing energy consumption. This makes SBRs particularly promising for industrial applications where maximizing conversion rates while reducing operational costs is critical.

Table 4 presents a comparative analysis of the performance of SBRs in the hydrolysis of lignocellulosic biomass. The table summarizes key parameters from various studies, highlighting the reactor's efficiency in enhancing sugar recovery and improving biomass conversion rates. Each entry details the type of biomass used, the objective of the process, the catalyst applied, operational conditions (such as temperature, acid concentration, and flow rate), the specific shrinking mechanism employed, and the overall yield or efficiency achieved. These data illustrate how SBRs optimize solid–liquid contact, minimize sugar degradation through reduced residence times, and maintain high product concentrations.

4 Future Perspectives

The development of bioreactor design is crucial to realizing the complete potential of lignocellulosic biomass for sustainable biofuel and bioproduct production. Research in the future must be directed toward optimization and integration of reactor systems for optimal conversion efficiency, scalability, and economic viability. Novel configurations like shrinking-bed reactors (SBRs), plug-flow reactors (PFRs), percolation reactors (PCRs), and counter-current systems have been demonstrated to be effective in sugar recovery and process integration. However, technical problems remain, requiring more work to allow their industrial-scale deployment.

To overcome these limitations, next-generation bioreactors must be versatile in terms of accepting multiple biomass feedstocks and accommodating high-solids operation and efficient mass and heat transfer. SBRs, for instance, are favored by their self-compaction properties, which could be optimized using CFD. PFRs are energy-conserving and highly integrable, particularly when integrated with recirculation loops and heat recovery systems. On the other hand, the development of bed design and flow dynamics is critical to the maximization of PCR performance. Similarly, counter-current systems require the development of automated multistage setups that incorporate enzyme recovery and reuse strategies to maximize conversion rates and minimize enzyme costs.

Table 4 Performance of shrinking-bed reactors (SBRs) in lignocellulosic biomass hydrolysis

Lignocellulosic biomass	Objective/scope	Catalyst type	Operational conditions	Shrinking mechanism	Yield or efficiency (%)	References
Yellow poplar sawdust	Hydrolysis of hemicellulose and cellulose	Dilute sulfuric acid	190 °C; 0.08 wt% H_2SO_4; liquid flow: 2.0 cm/min	Spring-loaded plunger	95% hemicellulose, 85% cellulose	[63]
Mixed agricultural residues	Hydrolysis optimization for glucose recovery	Dilute sulfuric acid	240 °C; 1% acid concentration; Rws = 1	Radial compaction with crossflow	88% glucose yield, 91% xylose yield	[63]
Corn stover	Removal of hemicellulose and lignin	No catalyst (hot water only)	180–220 °C; flow rates: 1–10 mL/min; 2.1–2.4 MPa	Natural shrinking during flow-through	Up to 97% xylose recovery; 75% lignin removal	[79]
Lignocellulosic biomass residues	Structural modification for enhanced digestibility	Biological pretreatment (fungi)	150–180 °C; atmospheric pressure; enzymatic saccharification	Natural shrinkage during biodegradation	72% cellulose accessibility improvement	[65]
Olive pruning biomass	Bioethanol production via acid pretreatment	Dilute sulfuric acid	160–164 °C; 4.9–5.9% acid; solids loading: 15–35%	Mechanical compaction screw	97.3–110.8 kg bioethanol/t raw material	[61]
Empty fruit bunches (EFB)	Enhanced sugar recovery and energy yield	Dilute sulfuric acid	190–220 °C; 1% H_2SO_4; enzymatic hydrolysis post-treatment	Mechanical bed compaction	85% cellulose conversion; 78% hemicellulose recovery	[67]
Wheat straw	Acid hydrolysis for sugar recovery	Dilute sulfuric acid	180 °C; 1% acid concentration; residence time: 30 min	Continuous bed compaction	87% xylose, 82% glucose	[64]

References

1. Li C, Chen K, Wang B et al (2025) Bioreactor design for efficient biofuels production from lignocellulosic biomass. Biofuels Prod Lignocellul Mater 181–217
2. Antar M, Lyu D, Nazari M et al (2021) Biomass for a sustainable bioeconomy: an overview of world biomass production and utilization. Renew Sustain Energy Rev 139:110691
3. Chaudhary G, Chaudhary N, Saini S, et al (2023) Assessment of pretreatment strategies for valorization of lignocellulosic biomass: path forwarding towards lignocellulosic biorefinery. Waste Biomass Valorization 15:1–36
4. Magalhães AI, de Carvalho JC, de Melo Pereira GV et al (2019) Lignocellulosic biomass from agro-industrial residues in South America: current developments and perspectives. Biofuels Bioprod Biorefining 13:1505–1519
5. Manyatshe A, Sibali LL (2025) A review on the recovery of cellulose, lignin, and hemicellulose biopolymers from the same source of lignocellulosic biomass—methodology, characterization and applications. J Water Process Eng 70:107037
6. Ischia G, Fiori L (2020) Hydrothermal carbonization of organic waste and biomass: a review on process, reactor, and plant modeling. Waste and Biomass Valorization 12:2797–2824
7. Prasannamedha G, Kumar PS, Balasubramani SR et al (2025) Engineering properties of hydrochar fabricated from hydrothermal carbonization of lignocellulosic biomass: practice as adsorbent and catalyst in water treatment. J Water Process Eng 74:107739
8. Yang J, Ching YC, Chuah CH (2019) Applications of lignocellulosic fibers and lignin in bioplastics: a review. Polymers 11:751
9. Mankar AR, Pandey A, Modak A et al (2021) Pretreatment of lignocellulosic biomass: a review on recent advances. Bioresour Technol 334:125235
10. Manyatshe A, Cele ZED, Balogun MO et al (2022) Lignocellulosic derivative-chitosan biocomposite adsorbents for the removal of soluble contaminants in aqueous solutions—preparation, characterization and applications. J Water Process Eng 47:102654
11. Tu WC, Hallett JP (2019) Recent advances in the pretreatment of lignocellulosic biomass. Curr Opin Green Sustain Chem 20:11–17
12. Yin Y, Wang J (2022) Production of biohydrogen. In: Biofuels and biorefining: volume 1: current technologies for biomass conversion, pp 283–337
13. Dharmaraja J, Shobana S, Arvindnarayan S et al (2023) Lignocellulosic biomass conversion via greener pretreatment methods towards biorefinery applications. Bioresour Technol 369:128328
14. Baksi S, Saha D, Saha S et al (2023) Pre-treatment of lignocellulosic biomass: review of various physico-chemical and biological methods influencing the extent of biomass depolymerization. Int J Environ Sci Technol 20:13895–13922
15. Olson DG, McBride JE, Joe Shaw A et al (2012) Recent progress in consolidated bioprocessing. Curr Opin Biotechnol 23:396–405
16. Olivieri G, Wijffels RH, Marzocchella A et al (2021) Bioreactor and bioprocess design issues in enzymatic hydrolysis of lignocellulosic biomass. Catalysts 11:680
17. Ferreira RG, Azzoni AR, Freitas S (2021) On the production cost of lignocellulose-degrading enzymes. Biofuels Bioprod Biorefining 15:85–99
18. Zhou Z, Liu D, Zhao X (2021) Conversion of lignocellulose to biofuels and chemicals via sugar platform: an updated review on chemistry and mechanisms of acid hydrolysis of lignocellulose. Renew Sustain Energy Rev 146:111169
19. Titiladunayo IF, Ibitowa OA, Titiladunayo IF et al (2018) Design of a percolation reactor for the hydrolysis of lignocellulosic biomass. World J Eng Technol 6:874–884
20. Lee YY, Iyer P, Torget RW (1999) Dilute-acid hydrolysis of lignocellulosic biomass 93–115
21. Lee YY, Wu Z, Torget RW (2000) Modeling of countercurrent shrinking-bed reactor in dilute-acid total-hydrolysis of lignocellulosic biomass. Bioresour Technol 71:29–39
22. Naga Vignesh S (2020) Working principle of typical bioreactors. Bioreact Sustain Des Ind Appl Mitig GHG Emiss 145–173
23. Youcai Z, Ran W. Anaerobic fermentation process for biomethane production from vegetable waste. Biomethane Prod Veg Water Hyacinth Waste 1–62

24. Lonkar S, Liang C, Bond A et al (2017) Countercurrent saccharification of lime-pretreated corn stover–Part 1. Biomass Bioenergy 96:28–37
25. Shao X, Lynd L (2013) Kinetic modeling of xylan hydrolysis in co-and countercurrent liquid hot water flow-through pretreatments. Bioresour Technol 130:117–124
26. Williams DL, Ong RG, Mullet JE et al (2019) Integration of pretreatment with simultaneous counter-current extraction of energy sorghum for high-titer mixed sugar production. Front Energy Res 6:133
27. Shafiei M, Kumar R, Karimi K (2015) Pretreatment of lignocellulosic biomass. Lignocellul-Based Bioprod 85–154
28. Pihlajaniemi V Pretreatment categories, process alternatives and material characteristics in enzymatic hydrolysis of lignocellulose
29. Modenbach AA, Nokes SE (2013) Enzymatic hydrolysis of biomass at high-solids loadings–a review. Biomass Bioenergy 56:526–544
30. Vázquez-Villegas P, Aguilar O, Rito-Palomares M (2015) Continuous enzyme aqueous two-phase extraction using a novel tubular mixer-settler in multi-step counter-current arrangement. Sep Purif Technol 141:263–268
31. Fox DJ, Gray PP, Dunn NW et al (1989) Comparison of alkali and steam (acid) pretreatments of lignocellulosic materials to increase enzymic susceptibility: evaluation under optimised pretreatment conditions. J Chem Technol Biotechnol 44:135–146
32. Zentay AN, Liang C, Lonkar S et al (2016) Countercurrent enzymatic saccharification of cellulosic biomass. Biomass Bioenergy 90:122–130
33. Liang C (2015) Batch enzymatic hydrolysis of pretreated corn stover and improvements with countercurrent saccharification. Master Thesis, https://oaktrust.library.tamu.edu/items/2f065bdb-3d48-4720-8026-3bd51f1ee73b. Accessed on 28 May 2025
34. Çetin B, Çakmakçı S, Cakmakci R (2011) The investigation of antimicrobial activity of thyme and oregano essential oils. Turk J Agric For 35:145–154
35. Liang C, Gu C, Karim MN et al (2019) Kinetic modeling of countercurrent saccharification. Biotechnol Biofuels 12:1–13
36. Yadav SK (2017) Technological advances and applications of hydrolytic enzymes for valorization of lignocellulosic biomass. Bioresour Technol 245:1727–1739
37. Goldbeck R, Gonçalves TA, Damásio ARL et al (2016) Effect of hemicellulolytic enzymes to improve sugarcane bagasse saccharification and xylooligosaccharides production. J Mol Catal B Enzym 131:36–46
38. Conrad M, Smirnova I (2021) Counter-current suspension extraction process of lignocellulose in biorefineries to reach low water consumption, high extraction yields, and extract concentrations. Processes 9:1585
39. Titiladunayo IF, Ibitowa OA (2018) Design of a percolation reactor for the hydrolysis of lignocellulosic biomass. World J Eng Technol 06:874–884
40. Franca LS, Ornelas-Ferreira B, Pereira CP et al (2022) Performance of a percolation reactor integrated to solid-state anaerobic garage-type digesters with leachate recirculation for organic fraction of municipal solid waste treatment. Bioresour Technol Rep 20. Epub ahead of print 1 December 2022. https://doi.org/10.1016/j.biteb.2022.101215
41. Kim JS, Lee YY, Kim TH (2016) A review on alkaline pretreatment technology for bioconversion of lignocellulosic biomass. Bioresource Technology 199:42–48
42. Chandel AK, Antunes FAF, de Arruda PV et al (2012) Dilute acid hydrolysis of agro-residues for the depolymerization of hemicellulose: state-of the art. D-Xylitol Ferment Prod Appl Commer 39–61
43. Smit AT, Verges M, Schulze P et al (2022) Laboratory-to pilot-scale fractionation of lignocellulosic biomass using an acetone organosolv process. ACS Sustain Chem Eng 10:10503–10513
44. Kirsch C, Zetzl C, Smirnova I (2010) Development of an integrated thermal and enzymatic hydrolysis for lignocellulosic biomass in fixed-bed reactors: 11th EWLP, Hamburg, Germany, August 16–19

45. Reynolds W, Kirsch C, Smirnova I (2015) Thermal-enzymatic hydrolysis of wheat straw in a single high pressure fixed bed. Chemie Ingenieur Technik 87:1305–1312
46. Pino MS, Rodríguez-Jasso RM, Michelin M et al (2018) Bioreactor design for enzymatic hydrolysis of biomass under the biorefinery concept. Chem Eng J 347:119–136
47. Reynolds W, Conrad M, Mbeukem S et al (2019) Pressure drop, mechanic deformation, stabilization and scale-up of wheat straw fixed-beds during hydrothermal pretreatment: experiments and modeling. Chem Eng J 360:1587–1600
48. Kinnarinen T Pressure filtration characteristics of enzymatically hydrolyzed biomass suspensions
49. Terán-Hilares R, Reséndiz AL, Martínez RT et al (2016) Successive pretreatment and enzymatic saccharification of sugarcane bagasse in a packed bed flow-through column reactor aiming to support biorefineries. Bioresour Technol 203:42–49
50. Ramos L, Vasconcelos MH, Milagres AMF et al (2021) High-solid enzymatic hydrolysis of sugarcane bagasse and ethanol production in repeated batch process using column reactors. 3 Biotech 11:1–11
51. Foutch GL, Johannes AH (2001) Reactors in process engineering
52. Perman E, Karlsson A, Westerholm M et al (2024) High-solid digestion—a comparison of completely stirred and plug-flow reactor systems. Waste Manag 189:265–275
53. Naga Vignesh S (2020) Working principle of typical bioreactors. In: Bioreactors: sustainable design and industrial applications in mitigation of GHG emissions. Elsevier, pp 145–173
54. Zhou Q, Ding A, Zhang L et al (2024) Furfural production from the lignocellulosic agro-forestry waste by solvolysis method—a technical review. Fuel Process Technol 255. Epub ahead of print 1 May 2024. https://doi.org/10.1016/j.fuproc.2024.108063
55. El Bast M, Allam N, Abou Msallem Y et al (2023) A review on continuous biomass hydrothermal liquefaction systems: process design and operating parameters effects on biocrude. J Energy Inst 108. Epub ahead of print 1 June 2023. https://doi.org/10.1016/j.joei.2023.101260
56. Veluchamy C, Kalamdhad AS, Gilroyed BH (2021) Evaluating and modelling of plug flow reactor digesting lignocellulosic corn silage. Fuel 287. Epub ahead of print 1 March 2021. https://doi.org/10.1016/j.fuel.2020.119498
57. Shoshaa R, Ghanimeh S, Almomani F (2024) Recent advances in plug flow reactors for anaerobic digestion and in-depth evaluation of mixing approaches: a review. Fuel 377:132711
58. Yang B, Tao L, Wyman CE (2018) Strengths, challenges, and opportunities for hydrothermal pretreatment in lignocellulosic biorefineries. Biofuels Bioprod Biorefining 12:125–138
59. Abatzoglou N, Koeberle PG, Chornet E et al (1990) Dilute acid hydrolysis of lignocellulosics: an application to medium consistency suspensions of hardwoods using a plug flow reactor. Can J Chem Eng 68:627–638
60. Tran KQ (2016) Fast hydrothermal liquefaction for production of chemicals and biofuels from wet biomass—the need to develop a plug-flow reactor. Bioresour Technol 213:327–332
61. Solarte-Toro JC, Romero-García JM, Martínez-Patiño JC et al (2019) Acid pretreatment of lignocellulosic biomass for energy vectors production: a review focused on operational conditions and techno-economic assessment for bioethanol production. Renew Sustain Energy Rev 107:587–601
62. Zhang J, Bao J (2018) Lignocellulose pretreatment using acid as catalyst. In: Handbook of biorefinery research and technology. Springer Netherlands, pp 1–14
63. Converse AO (2002) Simulation of a cross-flow shrinking-bed reactor for the hydrolysis of lignocellulosics. Bioresour Technol 81:109–116
64. Giuliano A, Robinson Medici A, Barletta D (2025) Integrating techno-economic and environmental criteria in the design and optimization of a full utilization multiproduct lignocellulosic biorefinery. ACS Sustain Chem Eng 13(5):1961–1973
65. Ortíz I, Quintero R (2014) Recent advancements in pretreatment technologies of biomass to produce bioenergy. In: Bioenergy research: advances and applications. Elsevier Inc., pp 57–69
66. Zhou Z, Liu D, Zhao X (2021) Conversion of lignocellulose to biofuels and chemicals via sugar platform: an updated review on chemistry and mechanisms of acid hydrolysis of lignocellulose.

Renew Sustain Energy Rev 146. Epub ahead of print 1 August 2021. https://doi.org/10.1016/j.rser.2021.111169
67. Javanmard A, Wan Daud WMA, Patah MFA et al (2024) Breaking barriers for a green future: a comprehensive study on pre-treatment techniques for empty fruit bunches in the bio-based economy. Process Saf Environ Prot 182:535–558
68. Jofre FM, Prado CA, Shibukawa VP et al (2025) Critical analysis of process parameters towards smart bioreactors development in biorefinery for biorenewables production. Int J Biol Macromol 305. Epub ahead of print 1 May 2025. https://doi.org/10.1016/j.ijbiomac.2025.140957
69. Wongsirichot P (2024) Development and future potential of computation fluid dynamics for improved biomass hydrolysis. Chem Eng J 482. Epub ahead of print 15 February 2024. https://doi.org/10.1016/j.cej.2024.149032
70. Løhre C, Kleinert M, Barth T (2017) Organosolv extraction of softwood combined with lignin-to-liquid-solvolysis as a semi-continuous percolation reactor. Biomass Bioenergy 99:147–155
71. Yu Y, Christopher LP (2017) Detoxification of hemicellulose-rich poplar hydrolysate by polymeric resins for improved ethanol fermentability. Fuel 203:187–196
72. Wang JX, Wang DM, Xu WL et al (2024) Investigation of the degradation process of biomass during oxidative alkaline treatment by a flow-through reactor. J Anal Appl Pyrolysis 183. Epub ahead of print 1 October 2024. https://doi.org/10.1016/j.jaap.2024.106734
73. de Vrije T, Dussan K, van de Vondervoort RHAM et al (2024) Bioethanol production from organosolv treated beech wood chips obtained at pilot scale. Biomass Bioenergy 181. Epub ahead of print 1 February 2024. https://doi.org/10.1016/j.biombioe.2023.107003
74. Smit AT, Hoek M, Bonouvrie PA et al (2024) Semicontinuous aqueous acetone organosolv fractionation of lignocellulosic biomass: improved biorefinery processing and output. ACS Sustain Chem Eng 12:4731–4742
75. Kim Y, Hendrickson R, Mosier N et al (2005) Plug-flow reactor for continuous hydrolysis of glucans and xylans from pretreated corn fiber. Energy Fuels 19:2189–2200
76. Heinonen J, Sainio T (2015) Performance evaluation of a recycle-integrated process for the production and purification of monosaccharides from lignocellulosic biomass. Sep Purif Technol 156:561–571
77. Yong KJ, Wu TY, Lee CBTL et al (2022) Furfural production from biomass residues: current technologies, challenges and future prospects. Biomass Bioenergy 161. Epub ahead of print 1 June 2022. https://doi.org/10.1016/j.biombioe.2022.106458
78. Pierrat L, García-Triñanes P (2024) Optimising furfural production from lignocellulosic biomass: feedstock selection, process enhancement, and techno-economic and environmental viability. Chem Eng Res Des 212:261–280
79. Liu C, Wyman CE (2003) The effect of flow rate of compressed hot water on xylan, lignin, and total mass removal from corn stover. Ind Eng Chem Res 42:5409–5416

Comparative Assessment of Batch, Fed-Batch, and Continuous Feeding of Lignocellulosic Biomass on Sugar Recovery

Ana Karine Furtado de Carvalho, Bruno Chaboli Gambarato, Murilo de Paula Garcia Leal, Milena Lorenzi da Silva, Júlio Gabriel Oliveira de Lima, Pedro Garcia Pereira Silva, José Eduardo Holler Branco, and Heitor Buzetti Simões Bento

Abstract The transition to sustainable bioeconomy models has intensified interest in lignocellulosic biomass hydrolysis as a pathway for renewable energy and bioproduct generation. This chapter offers a comprehensive comparative analysis of batch, fed-batch, and continuous reactor configurations in the enzymatic hydrolysis of lignocellulosic feedstocks, emphasizing how reactor design, pretreatment strategies, and feedstock characteristics collectively influence process performance. Batch systems, while prevalent in laboratory-scale research due to their operational simplicity, face significant scalability constraints. Fed-batch configurations mitigate these challenges by enabling gradual substrate addition, which enhances hydrolysis efficiency and process flexibility. Continuous reactors, though complex and capital-intensive, offer the greatest potential for industrial integration and alignment with circular bioeconomy principles due to their high productivity and steady-state operation. The chapter further explores the interplay between pretreatment methods—such as acid, alkaline, and organosolv—and reactor design, showing the necessity of integrated system optimization. By synthesizing recent literature and evaluating industrial

A. K. F. de Carvalho (✉) · M. de Paula Garcia Leal · J. E. H. Branco
Department of Basic and Environmental Sciences, Engineering School of Lorena, University of Sao Paulo, Lorena, Brazil
e-mail: anacarvalho@usp.br

B. C. Gambarato
Department of Material Science and Technology, University Center of Volta Redonda (UniFOA), Volta Redonda, Rio de Janeiro, Brazil

M. L. da Silva · J. G. O. de Lima · P. G. P. Silva · H. B. S. Bento
Department of Engineering of Bioprocesses and Biotechnology,, School of Pharmaceutical Sciences, São Paulo State University - UNESP, Araraquara, Brazil

J. E. H. Branco
Department of Chemical Engineering, Engineering School of Lorena, University of São Paulo, Lorena, Brasil

© The Author(s), under exclusive license to Springer Nature Singapore Pte Ltd. 2025
A. K. Chandel and S. Sundaramurthy (eds.), *Biorefineries in Circular Bioeconomy*, Green Energy and Technology, https://doi.org/10.1007/978-981-95-3913-0_7

trends, the chapter advances the understanding of reactor-preprocessing compatibilities and provides strategic insights for the development of scalable and sustainable biorefineries.

Keywords Lignocellulosic biomass · Enzymatic hydrolysis · Reactor configurations · Circular bioeconomy · Biorefinery optimization

1 Introduction

The global transition toward sustainable energy systems has intensified the focus on lignocellulosic biomass as a renewable and abundant feedstock for biorefineries. Central to these processes is the efficient release of fermentable sugars through hydrolysis, a stage profoundly influenced by the design and operation of reactors. Among these, batch, fed-batch, and continuous configurations are the most studied, each with distinct operational characteristics and implications for sugar recovery [6, 27]. Selecting the most suitable reactor configuration requires a comprehensive understanding of biomass properties, pretreatment methods, and enzymatic behavior.

Historically, batch reactors have dominated laboratory-scale hydrolysis experiments due to their operational simplicity and controlled environment, allowing for detailed kinetic studies and parameter optimization [23, 28]. However, their limitations become apparent at industrial scales, where product inhibition, viscosity issues, and mass transfer limitations hinder performance [1, 7]. To overcome these challenges, fed-batch and continuous reactors have been developed. Fed-batch systems enable gradual substrate addition, which mitigates inhibitory effects and improves process control [20, 26]. Meanwhile, continuous reactors function under steady-state conditions, providing high productivity and seamless integration into industrial workflows [25, 30].

The development of these reactor configurations occurs within the broader context of the circular bioeconomy, which emphasizes resource efficiency, waste minimization, and sustainable production [18, 19]. In this framework, the choice of reactor system is not merely a technical decision but one that impacts environmental and economic sustainability. Continuous reactors, due to their operational stability and integration potential, are particularly well-suited to circular bioeconomy principles [12, 31]. However, their complexity and cost may limit their applicability in certain contexts, making batch and fed-batch systems valuable alternatives for specific scales and feedstocks [9, 29].

Despite the growing body of research on lignocellulosic hydrolysis, comparative analyses that integrate findings across different reactor configurations remain scarce. Most studies focus on optimizing a single mode of operation for a specific biomass and pretreatment combination [14, 16]. However, understanding the trade-offs and synergies between batch, fed-batch, and continuous reactors is essential for advancing biorefinery design and maximizing sugar recovery [22, 23]. This chapter aims to address this gap by synthesizing data from recent literature, comparing the performance of these reactor configurations across diverse feedstocks and pretreatment strategies.

Pretreatment methods are another fundamental variable in this analysis. The structural complexity of lignocellulosic biomass necessitates pretreatment to enhance enzymatic accessibility, with the effectiveness of these methods dependent on feedstock properties and reactor design [5, 15]. For instance, alkaline pretreatment is often paired with batch systems due to its simplicity and cost-effectiveness, whereas advanced methods such as organosolv are better suited for continuous systems because of their integration potential and reduced inhibitor formation [2, 25]. Understanding these interdependencies is crucial for optimizing the entire hydrolysis process.

Therefore, this chapter provides a comprehensive comparative assessment of batch, fed-batch, and continuous reactor configurations in the context of lignocellulosic biomass hydrolysis. By reviewing a wide range of recent studies, we explore how reactor design, pretreatment strategies, and feedstock characteristics interact to influence sugar recovery. Our objective is to provide a critical evaluation of these systems, highlighting their advantages, limitations, and potential for integration into sustainable biorefineries aligned with the principles of circular bioeconomy [17, 21].

2 Lignocellulosic Feedstocks and Pretreatment Strategies

Lignocellulosic biomass comprises a complex matrix of cellulose, hemicellulose, and lignin, with structural variations that depend on the source material, as presented in Fig. 1. Agricultural residues such as sugarcane bagasse, corn stover, and pineapple leaves have been extensively investigated due to their abundance and relatively high carbohydrate content [7, 11]. Forestry residues and dedicated energy crops also present viable feedstocks, though their higher lignin content often requires more intensive pretreatment strategies [1]. Each biomass type presents unique challenges during hydrolysis, necessitating careful consideration of both pretreatment and reactor design to optimize sugar release and ensure process efficiency [5].

Pretreatment serves as the critical initial step in lignocellulosic biomass valorization, breaking down the complex plant structure to improve enzymatic access to polysaccharides. Hydrothermal pretreatment, as demonstrated by Wang et al. [30] and Saini et al. [24], is particularly effective for hemicellulose solubilization and partial delignification, enhancing enzymatic hydrolysis without the use of additional chemicals. However, high energy requirements and potential equipment scaling issues have led researchers to explore alternative methods, such as acid and alkaline pretreatments, which can be more cost-effective and adaptable for diverse feedstocks [23, 28]. These methods, however, come with their own limitations, including inhibitor formation and increased wastewater generation [21].

Advanced pretreatment technologies, such as organosolv and ionic liquid systems, offer promising alternatives by efficiently fractionating biomass components while

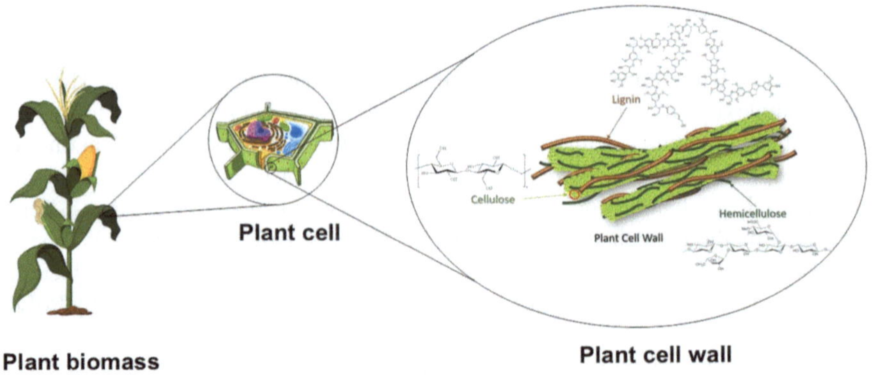

Fig. 1 Cell wall of a plant biomass, highlighting its structural components and role in biomass processing

reducing inhibitor production [2, 25]. The OrganoCat process, for example, integrates well into continuous reactor configurations, allowing for seamless transitions between pretreatment, hydrolysis, and fermentation stages [12]. While these technologies demonstrate high sugar recovery yields, their implementation requires careful economic and environmental analysis, particularly considering solvent recovery and recycling requirements [18, 19]. The scalability of these advanced systems remains a subject of ongoing research, with pilot-scale demonstrations providing valuable insights into their practical application [3].

Biomass properties, including particle size, crystallinity, and lignin content, influence pretreatment efficiency and downstream performance. Smaller particle sizes generally enhance enzyme accessibility but increase energy consumption during milling and handling [8]. Highly crystalline cellulose, common in certain grasses and hardwoods, poses additional challenges, requiring more aggressive pretreatment conditions to disrupt the ordered structures [14]. The choice of pretreatment, therefore, is closely linked to biomass characterization, to maximize sugar release while minimizing degradation products and operational costs [16].

Feedstock variability is another critical factor affecting pretreatment and reactor integration. Seasonal fluctuations in biomass composition may lead to inconsistencies in process performance, particularly in continuous systems where steady-state operation is essential [17]. Strategies such as blending different biomass sources and adjusting pretreatment severity have been explored to ensure consistent feedstock quality and optimize sugar yields [15]. Such adaptive approaches require flexible reactor designs and real-time monitoring to dynamically adjust process parameters in response to feedstock variation [6].

Given the wide range of lignocellulosic feedstocks and pretreatment methods available, integrating these elements with appropriate reactor configurations is essential to achieve efficient hydrolysis and high sugar recovery. The interplay between biomass characteristics, pretreatment severity, and reactor operation must be managed through comprehensive process optimization, ensuring that each component supports the performance of the overall system [9, 22]. The following sections delve deeper into how reactor configurations interact with these upstream factors, shaping the outcomes of lignocellulosic hydrolysis.

3 Batch Reactors in Lignocellulosic Hydrolysis

Batch reactors have long served as the foundation of laboratory-scale research on lignocellulosic biomass hydrolysis due to their straightforward operation and precise control over experimental conditions. In these systems, all reactants, including pretreated biomass, enzymes, and water, are introduced at the start, and the reaction proceeds without further intervention. This static environment facilitates comprehensive kinetic studies, allowing researchers to evaluate the effects of varying pretreatment methods, enzyme loadings, and temperatures on sugar release [6, 23]. For instance, experiments with elephant foot yam in batch systems demonstrated the significance of pretreatment severity on hydrolysis efficiency, providing valuable insights for optimizing process parameters [23].

However, as biomass loading increases, batch reactor configurations encounter notable challenges that limit their scalability. High solids content leads to increased slurry viscosity, complicating mixing and reducing mass transfer efficiency. This problem is exacerbated in lignocellulosic hydrolysis due to the heterogeneous nature of pretreated biomass, which may form clumps or sediment during reaction [10]. Product inhibition is another critical barrier, as the accumulation of sugars like glucose and cellobiose may reduce enzymatic activity, leading to incomplete hydrolysis [7]. These limitations create operational bottlenecks, particularly in efforts to translate laboratory-scale successes to industrial-scale applications.

The effects of pretreatment on batch system performance have been widely studied. Alkaline pretreatment, as explored by Thornburg et al. [28], is frequently paired with batch operations due to its compatibility with short-duration, high-intensity reactions. However, it often requires subsequent neutralization and generates wastewater streams that must be carefully managed [4]. Acid pretreatment, while effective in solubilizing hemicellulose and enhancing cellulose accessibility, introduces additional challenges, such as equipment corrosion and inhibitor formation, necessitating complex detoxification steps before hydrolysis [23]. These trade-offs underscore the importance of aligning pretreatment methods with reactor capabilities to maximize sugar yields and process efficiency.

Recent innovations in batch reactor design aim to address these inherent limitations. Studies have explored advanced agitation systems to improve mixing, as well as staged enzyme addition strategies to maintain catalytic activity over extended reaction periods [13]. Additionally, the incorporation of real-time monitoring tools to track sugar release and viscosity changes has enhanced process control, enabling more dynamic responses to shifting reaction conditions [12]. Nevertheless, these improvements add complexity and costs to what is otherwise considered a simple and accessible reactor configuration, raising concerns about the long-term viability of batch systems for large-scale biorefineries.

Despite these operational challenges, batch reactors remain indispensable for testing new pretreatment protocols and enzyme formulations. Their controlled environment enables researchers to systematically evaluate the effects of different variables, generating valuable datasets that inform the development of fed-batch and continuous systems [15]. Furthermore, batch systems play a vital role in producing baseline data for techno-economic and life cycle assessments, helping quantify the costs and environmental impacts associated with lignocellulosic biofuel production [14]. In this context, batch reactors serve as a critical stepping stone in the broader progression of biorefinery design.

As research advances, the continued use of batch reactors will be essential for understanding fundamental reaction mechanisms and testing novel feedstocks under controlled conditions. However, scaling up batch processes remains an ongoing challenge, particularly as biorefineries seek to operate within circular bioeconomy principles that demand high efficiency and minimal waste [16]. The transition from batch to more dynamic systems, such as fed-batch and continuous operation modes, will require careful consideration of the knowledge generated from batch experiments, ensuring that insights gained at small scales are effectively translated into industrial practice.

4 Fed-Batch Operations: Advantages and Challenges

Fed-batch reactors have gained prominence in lignocellulosic biomass hydrolysis due to their ability to address some of the key operational challenges encountered in batch systems, particularly regarding high solids loading and viscosity management. By gradually introducing biomass and enzymes during the reaction, fed-batch systems reduce the viscosity challenges inherent in batch systems, improving mass transfer [20, 26]. This strategy enables higher total solids concentrations compared to batch systems, ultimately increasing sugar concentrations in the hydrolysate while avoiding process failure due to mechanical limitations. Studies by Zanuso et al. [32] demonstrate that controlled substrate addition significantly improves hydrolysis efficiency, particularly during the processing of heterogeneous feedstocks such as mixed agricultural residues.

The effectiveness of fed-batch systems is also influenced by the compatibility of pretreatment methods. Sequential pretreatments, such as those combining alkaline and enzymatic steps, have shown particular promise in fed-batch configurations by reducing lignin content and improving cellulose accessibility without excessive inhibitor formation [20]. However, the complexity of managing multistage pretreatment and gradual feeding schedules requires advanced process control, making fed-batch systems more operationally demanding than traditional batch processes [3]. Consequently, the success of fed-batch systems relies heavily on precise timing and dosing strategies tailored to the characteristics of both the feedstock and pretreatment conditions.

One of the primary advantages of fed-batch operations lies in their flexibility and adaptability, particularly when handling biomass with variable compositions or moisture contents. Seasonal fluctuations and supply inconsistencies often affect feedstock quality, leading to potential disruptions in fully continuous systems [17]. Fed-batch reactors mitigate these issues by enabling incremental adjustments during operation, allowing operators to respond dynamically to feedstock variability without halting the process [9]. This operational resilience is particularly valuable in decentralized biorefinery models, where local feedstock supplies may vary significantly over time, necessitating adaptable processing technologies that can maintain consistent performance under fluctuating conditions.

Despite these advantages, fed-batch systems present specific challenges related to enzyme stability and recycling. Over prolonged hydrolysis periods, enzymes are subjected to mechanical shear, thermal stress, and fluctuating substrate concentrations, all of which can reduce catalytic activity [26]. Maintaining optimal enzyme performance often requires supplemental dosing or enzyme recycling strategies, which add operational complexity and increase costs [22]. Furthermore, as more solids are introduced incrementally, there is a risk of localized substrate buildup, which can reduce the efficiency of enzymatic action and create uneven hydrolysis zones within the reactor. These operational nuances highlight the need for careful reactor design and process optimization when deploying fed-batch systems for industrial-scale applications.

In the broader context of biorefinery development, the fed-batch system serves as a critical intermediary between batch simplicity and continuous productivity. They provide an operational model that supports higher sugar yields while preserving some of the control and flexibility of batch systems [32]. This balance makes fed-batch configurations particularly attractive for pilot and demonstration-scale facilities, where process conditions are still being refined, and scalability remains under evaluation. By leveraging the strengths of gradual substrate addition and adaptive feeding strategies, fed-batch reactors have proven to be a versatile tool in advancing lignocellulosic biomass hydrolysis, especially in contexts where feedstock variability and operational flexibility are critical. In this way, the swot matrix (Fig. 2) summarizes the strengths, weaknesses, opportunities, and threats of the fed-batch reactors discussed in this session.

5 Continuous Feeding Systems and Their Optimization

Continuous feeding systems have become a focal point in the advancement of lignocellulosic biomass hydrolysis due to their potential to support high-throughput, steady-state operations essential for industrial-scale biorefineries. Unlike batch and fed-batch operation modes, continuous systems maintain a constant flow of substrate and enzymes while simultaneously removing hydrolysate, which helps

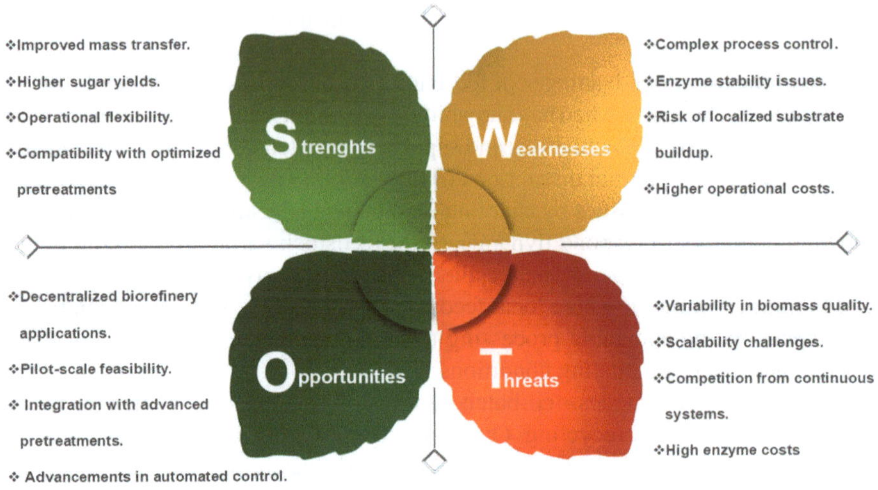

Fig. 2 SWOT analysis of fed-batch reactors for lignocellulosic biomass hydrolysis, highlighting key strengths, weaknesses, opportunities, and threats in industrial applications

sustain optimal reaction conditions over extended periods [30, 31]. This steady-state operation minimizes downtime and ensures more uniform product quality, making continuous reactors particularly attractive for integrating pretreatment, hydrolysis, and fermentation stages within a seamless process flow [25]. The ability to maintain consistent sugar release is critical in reducing operational disruptions and maximizing overall process efficiency.

Optimization of continuous systems requires meticulous balancing of residence time, flow rates, and enzyme loading to prevent process imbalances. Inappropriate flow rates can result in insufficient hydrolysis, leaving unconverted polysaccharides in the effluent or allowing inhibitory compounds to accumulate, negatively affecting enzyme activity [27]. Studies by González-Ríos [12] and Olivieri et al. [19] emphasize the importance of reactor design, such as the implementation of flow-through and oscillatory systems, which enhance mass transfer and mixing. These designs help alleviate the risk of channeling and dead zones, common problems in continuous operations using lignocellulosic slurries, where particle heterogeneity poses a persistent challenge to maintaining homogeneous reaction conditions.

Another critical consideration in continuous systems is enzyme recycling and stability. Extended operation times place significant stress on enzymatic catalysts, which may experience gradual deactivation due to prolonged exposure to shear forces, temperature fluctuations, and substrate depletion zones [16]. Various approaches, including membrane-based retention systems and immobilization techniques, have been explored to prolong enzyme activity and reduce overall consumption [25]. However, these strategies must be carefully integrated into reactor design to avoid introducing additional pressure drops or fouling issues, particularly in systems handling high-viscosity lignocellulosic materials [17]. The economic viability of continuous hydrolysis often hinges on the effectiveness of these enzyme management solutions [27].

As discussed in Sect. 2, feedstock variability adds complexity to continuous operations, requiring robust control strategies. Unlike batch and fed-batch systems, which may be adjusted between cycles or during feeding intervals, continuous reactors require a consistent and predictable input stream to maintain steady-state performance [3, 7]. Seasonal and geographic variations in biomass composition, moisture content, and particle size can disrupt the delicate balance of continuous hydrolysis, leading to reduced yields or operational instability [9]. Blending feedstocks, real-time monitoring of feedstock properties, and adaptive control systems have been proposed as strategies to mitigate these risks, but implementing these solutions adds to the overall complexity and cost of continuous reactor management.

Continuous feeding systems, despite their operational demands, remain the most promising configuration for scaling up lignocellulosic biorefineries within the principles of the circular bioeconomy. Their inherent efficiency, integration potential, and productivity gains make them ideal for large-scale applications focused on maximizing resource use and minimizing waste [18]. As research continues to refine reactor designs and enzyme stabilization techniques, continuous reactors are poised to play a central role in the commercialization of lignocellulosic biofuels. However, achieving this requires a comprehensive understanding of the interactions between feedstock variability, pretreatment strategies, and operational parameters to fully realize the benefits of continuous hydrolysis at an industrial scale.

6 Comparative Performance Analysis

The comparative performance of batch, fed-batch, and continuous reactors in lignocellulosic biomass hydrolysis has been extensively investigated to understand the trade-offs associated with each system. Batch reactors, as highlighted by Rakshitha et al. [23], Thornburg et al. [28], and Chatterjee et al. [6], offer simplicity and ease of operation, making them ideal for fundamental studies and early-stage optimization. However, their scalability is limited when compared to continuous reactors, which demonstrate superior productivity and sustained sugar release [25, 30]. Fed-batch reactors, exemplified by the work of Zanuso et al. [32] and Pérez et al. [22], provide an operational bridge between these two extremes, allowing gradual substrate addition to balance enzyme activity and reduce viscosity issues.

Beyond overall sugar yields, the kinetic behavior of each reactor type reveals critical distinctions. Batch reactors often exhibit rapid initial hydrolysis rates, which decline as product inhibition escalates, whereas fed-batch systems sustain moderate, stable rates through controlled substrate addition. Continuous reactors, by contrast, maintain relatively constant hydrolysis rates over extended periods, facilitated by steady-state conditions that minimize fluctuations in enzyme activity and substrate concentration [30, 31]. Incorporating kinetic performance metrics into reactor evaluation frameworks enhances the capacity to predict long-term operational behavior under varying loading conditions.

A key factor influencing reactor performance is the pretreatment method employed. Pretreatment methods such as organosolv, steam explosion, ammonia fiber expansion (AFEX), and liquid hot water (LHW) have been successfully integrated into continuous systems, each offering advantages in sugar yield and inhibitor management. For instance, organosolv facilitates lignin removal with solvent recovery, while AFEX and steam explosion enhance cellulose accessibility with reduced chemical input [3, 25]. In contrast, acid pretreatment, frequently used in batch operations [23], often necessitates detoxification steps to remove fermentation inhibitors. Sequential pretreatments, explored by Paini et al. [20] and Zanuso et al. [32], complement fed-batch processes by reducing recalcitrance while minimizing viscosity buildup. These pretreatment-reactor alignments underscore the importance of an integrated approach to process design, where the choice of pretreatment and reactor type is made in tandem to maximize overall system performance [14].

Operational stability is another distinguishing feature across reactor types. Batch processes provide stability within individual runs but lack the throughput demanded by industrial biorefineries. Fed-batch systems enhance operational stability through controlled feeding, while continuous systems maintain productivity under steady-state conditions, albeit requiring advanced management strategies to handle feedstock variability and mechanical challenges [3, 9]. Monitoring technologies and adaptive control algorithms are essential in continuous reactors to maintain optimal conditions despite fluctuations in biomass quality or environmental parameters [17].

Environmental and economic considerations further differentiate reactor configurations. Batch reactors often result in higher water and energy usage due to their cyclical operation and downtime between runs [13]. Fed-batch reactors improve resource efficiency by extending reaction periods and reducing the number of startups and shutdowns required. Continuous reactors, however, maximize energy and water use efficiency by operating in a steady state, allowing for heat integration and waste minimization strategies [11]. Nevertheless, the capital costs and operational complexities of continuous systems remain significant barriers to widespread adoption, especially in regions with variable biomass availability or limited infrastructure [19].

Although continuous reactors exhibit advantages in theoretical productivity and integration potential, industrial deployment remains constrained by factors such as high capital investment, operational complexity, and sensitivity to biomass variability [17]. In practice, many commercial-scale biorefineries continue to favor fed-batch systems due to their operational flexibility, lower maintenance demands, and capacity to handle inconsistent feedstock qualities [7]. Thus, broad statements regarding the dominance of continuous reactors must be contextualized within economic, logistical, and regional realities.

Comparing reactor configurations across performance metrics reveals that no single system is universally optimal for all contexts. Batch reactors are indispensable in research and small-scale operations, fed-batch reactors excel in balancing operational control with enhanced sugar yields, and continuous systems dominate in large-scale, high-throughput biorefineries seeking to adhere to circular bioeconomy principles [21]. The selection of the most appropriate reactor type must account for factors such as feedstock characteristics, pretreatment compatibility, resource availability, and economic constraints. Consequently, ongoing research aims to develop hybrid systems that combine the strengths of multiple configurations, leveraging batch flexibility, fed-batch control, and continuous productivity to build resilient, efficient biorefinery processes.

7 Influence of Pretreatment on Reactor Performance

Pretreatment plays a fundamental role in the hydrolysis of lignocellulosic biomass, as it significantly alters the structural complexity of raw materials, facilitating enzymatic access to cellulose and hemicellulose. The effectiveness of batch, fed-batch, and continuous reactors is heavily influenced by the pretreatment strategy applied to the feedstock. Studies such as those by Wang et al. [30] and Saini et al. [24] have shown that hydrothermal pretreatment, particularly when integrated into continuous processes, enhances biomass digestibility while minimizing chemical inputs. However, Argeiti et al. [2] and Conrad et al. [8] highlight that the high energy demand associated with hydrothermal methods can increase operational costs, requiring careful evaluation of process parameters, especially in large-scale settings with fluctuating feedstock properties.

Acid pretreatment remains a widely employed method, especially in batch reactor operations, due to its efficiency in solubilizing hemicellulose and exposing cellulose fibers [23]. Despite these advantages, acid pretreatment produces significant amounts of inhibitory compounds, such as furfural and hydroxymethylfurfural (HMF), which negatively affect enzymatic activity and downstream fermentation [28]. In batch processes, these inhibitors can be managed through additional detoxification steps, but their accumulation presents greater challenges in continuous operations, where real-time mitigation strategies are more difficult to implement. Therefore, the use of acid pretreatment must be carefully matched with reactor configurations that can accommodate its operational demands and manage its limitations effectively [3].

Organosolv pretreatment, particularly through the OrganoCat process, has gained significant attention for continuous reactor systems due to its selective lignin removal and minimal inhibitor generation [25]. This pretreatment technique enables efficient hydrolysis of hardwood biomass, such as beech wood, and facilitates high sugar recovery while supporting the circular bioeconomy through solvent recovery and reuse [12, 19]. However, implementing organosolv pretreatment requires substantial capital investment and sophisticated process control, making it more viable for large-scale, integrated continuous systems where these costs can be justified. The effectiveness of organosolv in sustaining long-term continuous operations highlights its growing relevance in commercial biorefineries [18].

Alkaline pretreatment is frequently applied to agricultural residues such as corn stover and sugarcane bagasse, offering moderate delignification and improved enzymatic hydrolysis performance [22, 28]. This method is well-suited to batch and fed-batch reactors, where it supports manageable viscosity and enhances process flow. However, the environmental impacts of alkaline pretreatment, including high water usage and the generation of alkaline waste streams, necessitate effective effluent treatment strategies to mitigate ecological harm [21]. As summarized in Table 1, the comparison of hydrothermal, acid, organosolv, alkaline, and sequential pretreatments reveals varying levels of sugar recovery, operational complexity, environmental impact, and reactor compatibility. This comparison underscores the importance of selecting pretreatment methods that align with the operational characteristics of each reactor type to maximize efficiency while managing sustainability concerns.

Although organosolv pretreatment has demonstrated high compatibility with continuous reactor systems due to its efficient lignin removal and integration potential, it is important to acknowledge that other pretreatment methods, such as steam explosion, ammonia fiber expansion (AFEX), and hydrothermal processes, also exhibit robust performance across continuous, fed-batch, and batch configurations. The association between specific pretreatment methods and reactor types is not inherently rigid, as operational feasibility depends on factors like biomass composition, regional feedstock variability, economic considerations, and infrastructure availability [1, 3, 12, 21]. Industrial practice has shown that flexible integration of diverse pretreatments across multiple reactor configurations can optimize process stability and sugar recovery, highlighting the need for adaptable biorefinery designs rather than predetermined technological pairings.

Adaptive pretreatment strategies are increasingly necessary to accommodate variations in feedstock composition, particularly in continuous systems that demand consistent input quality to maintain steady-state performance. Research by Deshavath et al. [9] and Mokaram and Esfahanian [17] highlights the value of real-time process adjustments in response to seasonal and geographical differences in biomass characteristics. These approaches allow operators to modify pretreatment severity dynamically, stabilizing downstream hydrolysis and sugar recovery. Such adaptability is crucial in modern biorefineries, where feedstock variability is unavoidable and process resilience depends on the seamless integration of pretreatment with reactor design and operational strategies [15, 16].

8 Future Perspectives

The continuous development of lignocellulosic biorefineries is increasingly dependent on the refinement of reactor configurations and their alignment with specific feedstocks and pretreatment strategies. Future research is expected to explore hybrid systems that merge the flexibility of fed-batch operations with the high productivity of continuous reactors, aiming to mitigate the operational challenges associated with each mode. Studies such as those by Mokaram and Esfahanian [17] and Deshavath

Table 1 Influence of pretreatment methods on reactor performance across batch, fed-batch, and continuous systems

Pretreatment method	Example studies	Biomass source	Reactor mode (s)	Impact on sugar recovery	Key operational challenges	Environmental impact	Integration feasibility	Inhibitor formation	Pretreatment severity (mild/moderate/severe)
Hydrothermal	Wang et al. [30], Saini et al. [24]	Pineapple leaves, mixed residues	Continuous, Batch	High	High energy demand, scaling issues	Moderate	High (continuous systems)	Low to moderate	Moderate
Acid pretreatment	Rakshitha et al. [23]	Elephant foot yam	Batch	Moderate	Equipment corrosion, acid recovery, neutralization steps	High	Low (due to acid recovery needs)	High	Severe
Organosolv (organoCat)	Schoofs et al. [25]	Beech wood	Continuous	Very high	Solvent recovery, process cost	Low	Very high (integrated systems)	Low	Moderate
Alkaline	Thornburg et al. [28], Pérez et al. [22]	Corn stover, Sugarcane bagasse	Batch, fed-batch	Variable (moderate to high)	Wastewater management, high chemical use	Moderate	Moderate	Moderate	Mild
Sequential (combination)	Paini et al. [20], Zanuso et al. [32]	Mixed agricultural residues	Fed-batch	High	Complex control, extended time	Moderate	High (adaptable to fed-batch)	Low	Moderate

et al. [9] suggest that modular reactor designs capable of adjusting feeding strategies dynamically based on real-time biomass characterization will play a critical role in enhancing process stability and sugar recovery across variable input streams. These flexible systems may provide a more resilient platform for regions where biomass heterogeneity and supply fluctuations challenge continuous operation.

Advancements in pretreatment technologies are expected to directly impact reactor optimization, as novel approaches focusing on lower energy demand, minimal chemical use, and reduced inhibitor formation continue to emerge. Processes like organosolv and advanced hydrothermal techniques will likely evolve to integrate seamlessly into continuous systems, enhancing their sustainability and economic viability [12, 25]. In parallel, enzyme recycling and stabilization strategies are anticipated to become standard features in future reactor designs, improving catalytic longevity and reducing overall costs [16]. These enhancements will require deeper integration between reactor control systems and bioprocess monitoring technologies, enabling adaptive responses to changes in enzymatic activity and substrate composition.

The increasing adoption of digitalization tools, such as artificial intelligence and machine learning, is poised to transform reactor management through predictive analytics and automated adjustments in operational parameters. Research by Martins et al. [15] and Manandhar and Shah [14] indicates that intelligent control systems can optimize reaction conditions, feeding rates, and pretreatment severity based on historical and real-time data, minimizing human intervention and maximizing efficiency. These technologies will be essential for large-scale biorefineries that operate across diverse feedstocks and geographic regions, ensuring consistent performance despite external variability and reducing resource consumption through precise process control.

Scaling lignocellulosic biorefineries while maintaining alignment with circular bioeconomy principles will require integrated design approaches that consider the entire value chain, from biomass sourcing and pretreatment to hydrolysis, fermentation, and waste valorization. Research priorities are shifting toward holistic system modeling, where reactor configuration decisions are made in tandem with life cycle assessments, techno-economic analyses, and regional biomass mapping [18, 21]. By focusing on the systemic interdependencies of biorefinery components, future developments will enable optimized resource use and reduce environmental impacts, reinforcing the strategic role of lignocellulosic biofuels in global sustainability efforts.

9 Conclusion

The comparative analysis of batch, fed-batch, and continuous reactors reveals that no single configuration universally outperforms the others across all operational, economic, and environmental metrics. Batch reactors remain essential in experimental and small-scale settings due to their controllability, whereas fed-batch systems show a practical balance between flexibility and throughput, making them suitable for pilot or decentralized operations. Continuous reactors, while offering superior productivity and integration potential, demand high levels of operational precision, consistent biomass quality, and elevated capital investment. Furthermore, the synergy between reactor type and pretreatment method plays a key role in determining efficiency and sustainability. As the field moves toward large-scale implementation, future research should prioritize hybrid reactor models, digital control systems, and adaptive pretreatment protocols to accommodate feedstock variability and maximize sugar recovery. Finally, the advancement of lignocellulosic hydrolysis within biorefineries will depend not only on technological innovations but also on systemic, integrative approaches that align with circular economy principles and global sustainability goals.

References

1. Antunes FAF, Freitas JBF, Prado CA, Castro-Alonso MJ, Diaz-Ruiz E, Mera AE, Santos JC, da Silva SS (2023) Valorization of corn cobs for xylitol and bioethanol production through column reactor process. Energies 16:4841. https://doi.org/10.3390/en16134841
2. Argeiti C, Psaki O, Filippi K, Ladakis D, Scally L, Cullen PJ, Koutinas A, Stylianou E (2024) Biorefinery electrification of brewers' spent grains using plasma bubbles for sustainable production of poly(3-hydroxybutyrate). Chem Eng J 496:153548. https://doi.org/10.1016/j.cej.2024.153548
3. Barron C, Devaux MF, Foucat L, Falourd X, Looten R, Joseph-Aime M, Durand S, Bonnin E, Lapierre C, Saulnier L, Rouau X, Guillon F (2021) Enzymatic degradation of maize shoots: monitoring of chemical and physical changes reveals different saccharification behaviors. Biotechnol Biofuels 14:1. https://doi.org/10.1186/s13068-020-01854-1
4. Barreto ES, Fonseca YA, Adarme OFH, Silva DF, Brandão RL, Baêta BEL, Guimarães VM, Gurgel LVA (2025) Optimization of 2G ethanol production from sugarcane bagasse: upscaling of soda pretreatment with redox mediator followed by fed-batch enzymatic hydrolysis and co-fermentation. Energy Convers Manage 323(A):119225. https://doi.org/10.1016/j.enconman.2024.119225
5. Cai C, Wang G, Wang L, Wang L, Zhang X (2024) Rethinking the role of steam explosion on biomass composition, structure, and saccharification with new-designed pilot-operated pneumatic decompression mode. Biomass Conv. Bioref 14:16899–16910. https://doi.org/10.1007/s13399-023-03789-7
6. Chatterjee S, Mohan SV (2021) Simultaneous production of green hydrogen and bioethanol from segregated sugarcane bagasse hydrolysate streams with circular biorefinery design. Chem Eng J 425:130386. https://doi.org/10.1016/j.cej.2021.130386
7. Choudhury SR, Dias A, Chakraborty R (2024) Review on progress in cellulase catalyzed saccharification of agricultural lignocellulosic biomass towards fermentable sugar and bioethanol: Kinetics & reactor configurations. Biocatal Agric Biotechnol 58:103142. https://doi.org/10.1016/j.bcab.2024.103142
8. Conrad M, Häring H, Smirnova I (2021) Design of an industrial autohydrolysis pretreatment plant for annual lignocellulose. Biomass Conv. Bioref. 11:2293–2310. https://doi.org/10.1007/s13399-019-00479-1
9. Deshavath NN, Dien BS, Slininger PJ, Jin Y-S, Singh VA (2023) A chemical-free pretreatment for biosynthesis of bioethanol and lipids from lignocellulosic biomass: an industrially relevant 2G biorefinery approach. Fermentation 9:5. https://doi.org/10.3390/fermentation9010005

10. Gan J, Iqbal HMN, Show PL, Rahdar A, Bilal M (2024) Upgrading recalcitrant lignocellulosic biomass hydrolysis by immobilized cellulolytic enzyme–based nanobiocatalytic systems: a review. Biomass Conv. Bioref 14:4485–4509. https://doi.org/10.1007/s13399-022-02642-7
11. Gomes GR, de Jesus EG, Jacintho JCC, García DLG, Alencar BRA, Gabetto FP, Gomes JJ, Carvalho JLN, Strauss M, Driemeier C (2025) Peculiarities of bio-oil and biochar obtained from the lignin-rich residue of the enzymatic hydrolysis of sugarcane bagasse. Renewable Energy 241:122282. https://doi.org/10.1016/j.renene.2024.122282
12. González-Ríos JA, Valle-Pérez AU, Amaya-Delgado L, Sanchez A (2023) A quick fed-batch saccharification strategy of wheat straw at high solid loadings improving lignocellulosic ethanol productivity. Biomass Convers Biorefinery 13:4991–5003. https://doi.org/10.1007/s13399-021-01580-0
13. Hanstein S (2024) Wheat bran extract as a source of amino compounds and sugar for biotech processes. Chem Ing Tec 96(4):440–445. https://doi.org/10.1002/cite.202300228
14. Manandhar A, Shah A (2023) Techno-economic analysis of the production of lactic acid from lignocellulosic biomass. Fermentation 9:641. https://doi.org/10.3390/fermentation9070641
15. Martins PL, Reis A, Duarte LC, Carvalheiro F (2022) Effective fractionation of microalgae biomass as an initial step for its utilization as a bioenergy feedstock. Energy Convers Manage: X 16:100317. https://doi.org/10.1016/j.ecmx.2022.100317
16. Meoli CM, Iervolino G, Procentese A (2023) Non-thermal plasma as a biomass pretreatment in biorefining processes. Processes 11:536. https://doi.org/10.3390/pr11020536
17. Mokaram MM, Esfahanian M (2025) Bioethanol fermentation integrated with PDMS membrane modified by carbon nanotube and graphyene oxide from sugarcane bagasse. Biomass Convers Biorefinery. https://doi.org/10.1007/s13399-024-05990-8
18. Monção M, Anukam AI, Hrůzová K, Rova U, Christakopoulos P, Matsakas LA (2024) A parametric study of the Organosolv Fractionation of Norway Spruce Sawdust. Energies 17:3276. https://doi.org/10.3390/en17133276
19. Olivieri G, Wijffels RH, Marzocchella A, Russo ME (2021) Bioreactor and bioprocess design issues in enzymatic hydrolysis of lignocellulosic biomass. Catalysts 11:680. https://doi.org/10.3390/catal11060680
20. Paini J, Benedetti V, Menin L, Baratieri M, Patuzzi F (2021) Subcritical water hydrolysis coupled with hydrothermal carbonization for apple pomace integrated cascade valorization. Biores Technol 342:125956. https://doi.org/10.1016/j.biortech.2021.125956
21. Pasquier J, Paës G, Perré P (2023) Principal factors affecting the yield of dilute acid pretreatment of lignocellulosic biomass: a critical review. Biores Technol 369:128439. https://doi.org/10.1016/j.biortech.2022.128439
22. Pérez ASL, Mahecha VR, Suarez LEC, Fajardo CAG (2024) Improved hydrothermal conversion of pea pod biomass for production of platform chemicals with organic acid catalysts. ChemEngineering 8:110. https://doi.org/10.3390/chemengineering8060110
23. Rakshitha SP, Keerthana KV, Anjuna P, Sangam SG, Sandesh K, Shet VB, Neetha JN, Ujwal P, Mubarak NM (2023) Statistical optimization of saccharification process using *Amorphophallus paeoniifolius* tubers into fermentable sugars for bioethanol production in stirred tank batch reactor (STBR). Biomass Conversion and Biorefinery 13:9315–9323. https://doi.org/10.1007/s13399-021-01748-8
24. Saini R, Chen C-W, Patel AK, Saini JK, Dong C-D, Singhania RR (2022) Valorization of pineapple leaves waste for the production of bioethanol. Bioengineering 9:557. https://doi.org/10.3390/bioengineering9100557
25. Schoofs L, Weidener D, Leitner W, Klose H, Grande PM (2024) Lignocellulose treatment using a flow-through variant of OrganoCat process. Chem Sustain Energy Mater 18:e202401063. https://doi.org/10.1002/cssc.202401063
26. Shi X, Wang L, Sun L, Chen H (2022) Sufficient premixing enhances enzymatic hydrolysis efficiency of lignocellulose at high-solids loading. Chem Eng J 444:136612. https://doi.org/10.1016/j.cej.2022.136612

27. Sun C, Meng X, Sun F, Zhang J, Tu M, Chang J-S, Reungsang A, Xia A, Ragauskas AJ (2023) Advances and perspectives on mass transfer and enzymatic hydrolysis in the enzyme-mediated lignocellulosic biorefinery: a review. Biotechnol Adv 62:108059. https://doi.org/10.1016/j.biotechadv.2022.108059
28. Thornburg NE, Ness RM, Crowley MF, Bu L, Pecha MB, Usseglio-Viretta FLE, Bharadwaj VS, Li Y, Chen X, Sievers DA, Wolfrum EJ, Resch MG, Ciesielski PN (2022) Mass transport limitations and kinetic consequences of corn stover deacetylation. Front Energy Res 10:841169. https://doi.org/10.3389/fenrg.2022.841169
29. Valladares-Diestra KK, Vandenberghe LPS, Torres LAZ, NishidaVS, Zandoná Filho A, Woiciechowski AL, Soccol CR (2021) Imidazole green solvent pre-treatment as a strategy for second-generation bioethanol production from sugarcane bagasse. Chem Eng J 420(2):127708. https://doi.org/10.1016/j.cej.2020.127708
30. Wang Y, Yang H, Zhang B, Liu X, Bao J (2022) Continuous enzymatic saccharification and its rheology profiling under high solids loading of lignocellulose biomass. Biochem Eng J 186:108543. https://doi.org/10.1016/j.bej.2022.108543
31. Xu J, Shao Z, Li Y, Dai L, Wang Z, Si C (2021) A flow-through reactor for fast fractionation and production of structure-preserved lignin. Ind Crops Prod 164:113350. https://doi.org/10.1016/j.indcrop.2021.113350
32. Zanuso E, Ruiz HA, Domingues L, Teixeira JA (2022) Oscillatory flow bioreactor operating at high solids loading for enzymatic hydrolysis of lignocellulosic biomass. Biochem Eng J 187:108632. https://doi.org/10.1016/j.bej.2022.108632

Bioreactors Configuration in the Development of Process Consolidation for Biofuels and Biochemicals in Lignocellulosic Biorefineries

C. A. Prado, M. T. F. R. Raymundo, V. P. Shibukawa, F. M. Jofre, Mónica M. Cruz-Santos, G. L. Arruda, D. J. F. dos Santos, N. O. Silos, W. S. M. Reis, S. S. da Silva, and J. C. Santos

Abstract The sustainable development of biofuels and biochemicals in lignocellulosic biorefineries is a critical focus in the search for alternatives to fossil fuels, which are major contributors to greenhouse gas emissions. In this context, bioreactor configuration plays a pivotal role in establishing efficient and scalable production processes. Recent studies have explored key strategies for optimizing bioreactor configurations, with particular emphasis on submerged fermentation systems and enzymatic hydrolysis step. Critical parameters such as temperature control, pH, aeration, and agitation have been extensively analyzed for their impact on productivity in the conversion of renewable raw materials into biomolecules of industrial interest. Among the established approaches for obtaining biochemicals, second-generation ethanol (2G) production relies on strategies such as Simultaneous Saccharification and Fermentation (SSF) and Simultaneous Saccharification and Co-Fermentation (SSCF). These strategies may employ either single or separate reactors, sometimes integrated with membrane systems to enhance process efficiency. Simultaneous conversion of sugars into ethanol or other products minimizes glucose accumulation, which could otherwise inhibit cellulolytic enzyme activity. Furthermore, combining saccharification and fermentation or co-fermentation in a single reactor eliminates the need for the sequential addition of microorganisms and enzymes in separate equipment, thereby improving overall process efficiency. Additionally, the ability to utilize both five-carbon (C5) and six-carbon (C6) sugars significantly enhances the efficiency of biomass conversion into biofuels in co-fermentation processes. The integration of simultaneous processing steps also reduces total production time, leading to increased productivity. An alternative strategy is the consolidated bioprocess (CBP), which leverages microbial strains, natural or genetically engineered, which possess the

C. A. Prado · M. T. F. R. Raymundo · V. P. Shibukawa · F. M. Jofre · M. M. Cruz-Santos · G. L. Arruda · D. J. F. dos Santos · N. O. Silos · W. S. M. Reis · S. S. da Silva · J. C. Santos (✉)
Department of Biotechnology, Engineering School of Lorena, University of São Paulo, Lorena, Brazil
e-mail: jsant200@usp.br

© The Author(s), under exclusive license to Springer Nature Singapore Pte Ltd. 2025
A. K. Chandel and S. Sundaramurthy (eds.), *Biorefineries in Circular Bioeconomy*, Green Energy and Technology, https://doi.org/10.1007/978-981-95-3913-0_8

ability to produce the enzymes required for biomass hydrolysis while simultaneously carrying out fermentation within the same system, streamlining the entire conversion process. Various reactor configurations, including membrane-integrated ones and interconnected columns, can be implemented to optimize SSF/SSCF and CBP, depending on specific process requirements. Furthermore, development of these bioprocesses in different bioreactor systems can improve yields and productivities in biorefineries. Actually, tailoring bioreactor configurations to the specific characteristics of the process configuration is required to ensure maximum efficiency and economic feasibility. This chapter provides a discussion of the main aspects of bioreactor design for lignocellulosic biorefineries, highlighting the crucial role of well-engineered and properly operated equipment with optimized process parameters. Special attention is given to key strategies, including SSF, SSCF, and CBP, with a focus on their contributions to enabling the efficient and sustainable production of biofuels and biochemicals.

1 Introduction

Currently, economic and environmental crises associated with dependence on fossil fuels and derived products have driven the search for more sustainable alternatives. In this context, the transition from a linear fossil-based economy to a circular bioeconomy has emerged as a viable solution from economic, social, and environmental perspectives [33]. Within this framework, biorefineries play a crucial role by enabling the integrated and sustainable processing of lignocellulosic biomass to produce chemicals, fuels, and energy. These facilities apply technologies from various fields to fractionate biomass components and convert them into biofuels and high-value-added biochemicals [38].

Biochemical conversion pathways have been widely employed for the production of biofuels and biochemicals. These processes rely on the action of enzymes and microorganisms (bacteria, yeasts, and filamentous fungi) to convert renewable raw materials into commercially valuable bioproducts [66]. Among these bioproducts, biofuels have gained prominence as sustainable alternatives to fossil fuels, offering a means to mitigate environmental impacts. The main classes of biofuels include bioethanol, biodiesel, biohydrogen, and biogas, with bioethanol and biodiesel standing out nowadays due to their integration into the energy matrices of several countries [69]. In addition to biofuels, various biochemicals—including enzymes, biopolymers, biopigments, pharmaceuticals, organic acids, and sweeteners—can be derived from renewable sources. These bioproducts have applications in diverse sectors such as agriculture, healthcare, the food industry, and the chemical industry [92].

The use of appropriately designed bioreactors is a key factor in ensuring the technical feasibility of bioprocesses. Bioreactors must provide optimal conditions to support the specific requirements of biological systems (microorganisms or enzymes) while enabling efficient bioproduct production. Various bioreactor types and configurations exist, classified according to their operating mode as well as their agitation and aeration strategies [22].

Bioreactors can be categorized into three main groups based on their operating mode: batch, fed-batch, and continuous. In batch bioreactors, all nutrients and microorganisms are introduced at the beginning of cultivation, with no further additions or products removal until the process is complete. In fed-batch systems, substrates are incrementally fed into the bioreactor during cultivation, allowing for prolonged processing times and improved productivity. Similarly to batch reactors, products are only removed at the end of the cultivation cycle. In contrast, continuous bioreactors operate with a constant substrate supply and simultaneous product removal, reducing substrate inhibition and enhancing overall efficiency [22].

Beyond operating mode, agitation and aeration are critical parameters in bioprocesses, ensuring medium homogeneity, mass transfer, and oxygen availability. Bioreactors are also classified according to these characteristics, and their selection must align with the specific requirements of each process. Mechanically agitated bioreactors, such as stirred-tank reactors, are the most employed ones due to their effectiveness in mixing and oxygenation. However, pneumatically agitated bioreactors, such as bubble columns and airlift bioreactors, provide alternative aeration and agitation strategies while minimizing shear stress on cells [12].

Several bioprocess configurations related to the biological steps of biorefineries have been evaluated to increase the efficiency of biofuel and bioproduct production using raw materials from renewable sources. There is a usual strategy of separate hydrolysis and fermentation (SHF), in which enzymatic hydrolysis and fermentation are performed in sequence in different systems. Additionally, there are three strategies in which biological steps occur simultaneously in the process: simultaneous saccharification and fermentation (SSF), simultaneous saccharification and co-fermentation (SSCF), and consolidated bioprocess (CBP) [39]. Among these three configurations, the simplest is simultaneous saccharification and fermentation (SSF), in which enzymatic hydrolysis and fermentation of hexoses from the cellulolytic hydrolysate occur simultaneously in a single step. The simultaneous saccharification and co-fermentation (SSCF) process is like SSF, but in this configuration, enzymatic hydrolysis and fermentation of hydrolysates containing both hexoses and pentoses occur [39].

The CBP represents a promising alternative for converting lignocellulosic raw materials into bioproducts. This strategy adds the enzyme production to the saccharification and fermentation, all together performed in a single step of the process. The main challenge related to the implementation of CBP is the need for microbial strains capable of producing enzymes to hydrolyze the raw material and of converting the hydrolysate into the products of interest (biofuels or biochemicals). Despite this, CBP represents a promising strategy in terms of cost reduction, product yield, environmental impact reduction, and simplification of operational processes [33, 50].

Given the significant role of bioreactors in optimizing biochemical conversion processes, this chapter explores the different configurations and operational strategies that support the development of process consolidation in lignocellulosic biorefineries. The discussion considered different aspects of bioreactor design, operation mode, and parameters that can influence the efficiency of bioprocesses. There was a focus on discussing integrated enzymatic hydrolysis and fermentation steps, and key approaches such as SSF, SSCF, and CBP were examined, highlighting their advantages, challenges, and technological advancements. Additionally, bioreactor configurations such as SSCF systems with interconnected columns and membrane-integrated reactors were presented. By addressing these aspects, this chapter aimed to provide a comprehensive understanding of bioreactor selection and optimization as a fundamental step toward efficient and sustainable biorefineries.

2 Bioproducts and Bioprocesses in Biorefineries: General Aspects and Bioreactors

2.1 *Main Approaches for Obtaining Biofuels and Biochemicals in Biorefineries*

Biorefineries are conversion technology platforms capable of producing multiple products from a single feedstock based on integrated processes, using raw materials, by-products, or waste, either locally available or underutilized from other production lines [5, 46].

One of the main product categories in biorefineries is biofuels, which serve as renewable energy sources. Biofuels can be obtained through physicochemical processing of biomass, resulting in products such as charcoal, bio-oil, and volatile compounds (e.g., methanol, acetone, and volatile acids) [70]. Alternatively, they can be produced through biological processes mediated by microbial metabolism or biocatalysis, such as enzymatic conversion. The most well-known biofuels include bioethanol, biogas, and biodiesel, among others [7, 80].

Biofuels are generally classified into two main categories: those produced from primary agricultural products (edible crops) and advanced biofuels, which are derived from secondary raw materials generated during food processing or consumption (e.g., bagasse, husks, and waste) or from nonfood sources such as microalgae, wood, and other biomasses [29]. This classification forms the basis for the widely recognized nomenclature of first-generation (produced from food crops), second-generation (derived from non-edible sources), third-generation (obtained from microbial sources), and fourth-generation biofuels (developed using genetically modified organisms). Each successive generation has introduced technological advancements, innovations, and improvements in sustainability, many of which are intricately linked to biorefinery production chains [34].

Currently, advances in biofuel production are largely driven by the availability of waste and by-products, as well as the efficiency of key processing steps such as pretreatment and hydrolysis of raw materials, particularly lignocellulosic biomass. These advancements also contribute to the valorization of carbohydrate fractions through intensified fermentation processes, enabling the production of a wide range of industrially relevant biochemicals. These include biopigments (e.g., carotenoids and polyketides), biopolymers, biosurfactants, lipids, sugar alcohols (e.g., xylitol), enzymes, organic acids (e.g., lactic, succinic, acetic, and propionic acids), amino acids, and vitamins [6, 21, 47, 48, 59].

Many of the aforementioned biochemical products can also be synthesized through chemical routes, some of which currently dominate the market. In most cases, these synthetic processes offer greater economic viability, simplified purification steps, and high yields [19]. However, conventional synthesis methods often rely on raw materials and production processes that pose significant environmental challenges, including waste generation, high-energy consumption, and a lack of sustainability [2].

The production of biomolecules through bioprocesses has gained prominence due to the growing demand from the consumer market increasingly conscious of clean and sustainable technologies. In this context, the selection of microbial species and process design plays a crucial role, supported by cultivation optimization, the integration of processing steps, and efficient bioreactor design and operation [14, 44].

Submerged cultures for the production of biofuels and biochemicals typically are used for microorganisms such as bacteria, yeasts, and microalgae, whereas solid-state cultures are generally employed for filamentous fungi species. Figure 1 outlines the main process steps and shows some examples of microbial species used to obtain various bioproducts and biomolecules in a lignocellulosic biorefinery.

Among the different types of biofuels, ethanol stands out for its widespread production and use in various countries, particularly Brazil and the USA. The most well-known ethanol-producing microorganism is the yeast *Saccharomyces cerevisiae*, due to its high industrial production capacity. However, several other microbial species are also capable of converting sugars into ethanol, such as alcohol-producing bacteria, including *Zymomonas mobilis* and *Clostridium thermocellum* [87]. Although ethanol can also be derived from petrochemical sources, the majority of commercial ethanol is obtained via fermentation, using sugar-based raw materials such as starch, sucrose, or hydrolysates from lignocellulosic biomass. The production potential is significant, given the broad range of substrates that can serve as nutrient sources for ethanol biosynthesis. Additionally, the ability to leverage existing infrastructure, supply chains, and plant locations strengthens the robustness of ethanol biorefineries [30, 43, 54].

In some cases, biofuel production is driven by a complete microbial consortium. For instance, in biogas production through anaerobic digestion, a highly diverse microbial community is responsible for the decomposition of organic matter. This microbial diversity includes groups such as *Methanothermobacter, Methanosarcina, Methanoculleus, Saccharofermentans, Proteobacteria, Acidobacteria*, and *Verrucomicrobia*. Key factors such as the biodegradability of the raw material, pretreatment

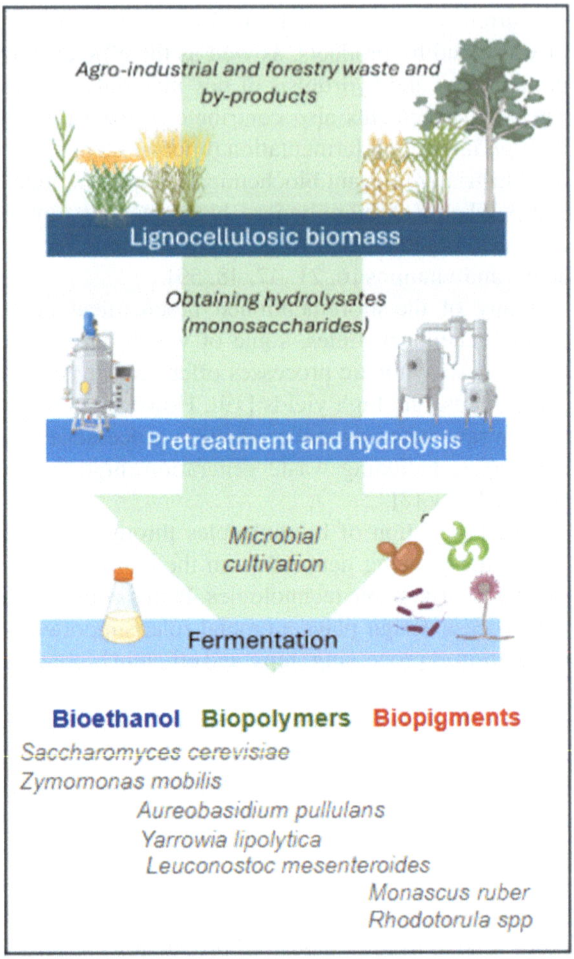

Fig. 1 A general illustrative example of a process and some microorganisms producing biofuels and biochemicals of interest within the context of biorefineries

technologies, pH, and temperature are crucial parameters that influence both biogas production and variations in the microbial community [15, 55, 57].

Many other bioproducts of interest can be obtained in lignocellulosic biorefineries for applications in various industrial sectors, including, for example, biopolymers and biopigments.

Biopolymers include xanthan, produced by *Xanthomonas campestris*; dextran, produced by *Leuconostoc mesenteroides*; pullulan, produced by *Aureobasidium pullulans*; alginate, produced by *Pseudomonas*; and cellulose, produced by *Acetobacter* bacteria [89, 90]. These polymers have great potential for applications, particularly as substitutes for synthetic polymer materials, especially when obtained through fermentation processes based on sustainable methods and formulations [11, 78].

Biopigments also have significant industrial value, especially in the food, cosmetics, and pharmaceutical sectors, either due to the demand for new coloring molecules or because of the nutraceutical and biological properties associated with microbial pigments. Several important microbial groups are capable of producing biopigments, including yeasts (*Rhodotorula glutinis, Pichia guilliermondii,* and *Yarrowia lipolytica*) [9], filamentous fungi (*Monascus ruber, Penicillium purpurogenum,* and *Talaromyces atroroseus*) [6], and microalgae (*Chlorella sp., Dunaliella salina,* and *Chlamydomonas reinhardtii*) [84].

Beyond these examples, lignocellulosic biorefineries can also produce a wide range of other valuable biomolecules, including xylitol, biosurfactants, organic acids, vitamins, lipids, and secondary metabolites. This diversification not only expands the portfolio of sustainable alternatives to petrochemical products but also enhances the economic and technological feasibility of biorefineries. To further optimize biomolecule production, advanced technologies and innovative bioreactors can be employed.

2.2 *Bioreactors' Operation Modes*

A bioreactor is a system designed to cultivate microorganisms or enzymes under controlled conditions. In microbial cultivation, it allows and favors biological reactions that lead to the synthesis of target products, such as biofuels and other valuable biomolecules [22]. To accommodate the specific requirements of different processes, several types of bioreactors have been developed.

Bioreactors can be operated in different modes, most commonly as batch, fed-batch, or continuous systems [83]. The choice of operation mode fundamentally influences process cycles, particularly in terms of cell and nutrient addition, and in product harvesting, as illustrated in Fig. 2. Each mode has distinct advantages and limitations. For example, in biochemical production, batch processes may lead to the accumulation of inhibitory compounds, potentially reducing yields. In contrast, fed-batch and continuous systems can mitigate this issue, minimizing inhibitory effects and improving overall productivity [85].

Batch reactors operate with a specific initial concentration of cells and nutrients under controlled conditions, without the continuous addition of fresh nutrients or the removal of products during operation. However, certain interventions can occur without altering the batch nature of the process. These include aeration, which is essential for microbial metabolism; sample collection for monitoring purposes; and the addition of antifoam agents or acid/basic solutions for pH control [18, 53, 91]. This mode of operation is relatively simple to control and holds significant industrial relevance, particularly in laboratory-scale studies aimed at optimizing production conditions [10].

In contrast, fed-batch operation involves the gradual addition of nutrients during the cultivation process, while the culture remains in the reactor without continuous product removal. This strategy allows better control over nutrient concentration,

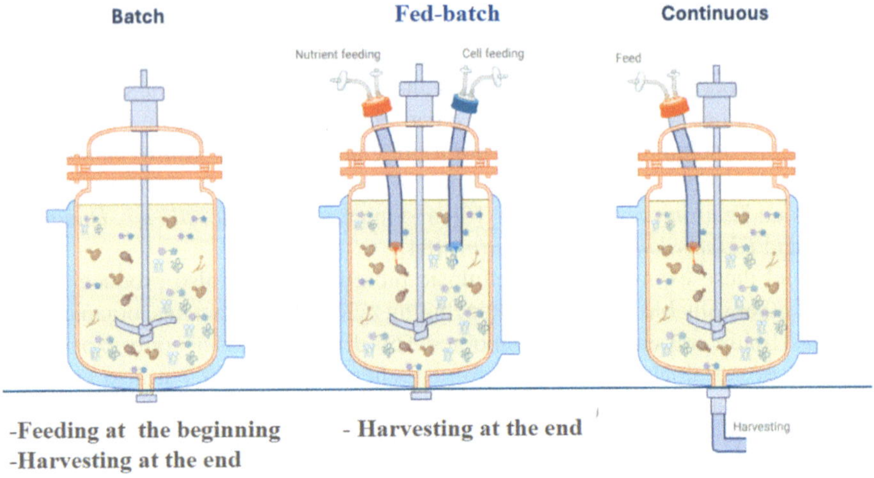

Fig. 2 Representation of different operation modes of bioreactors

reducing substrate inhibition and enhancing biomass or product formation. The feed rate is a critical factor in fed-batch processes, playing a key role in process optimization studies [53]. This approach has been widely explored for the industrial production of various bioproducts, such as polyhydroxyalkanoates (PHA). Fed-batch operation enhances cell growth and supports successful PHA production by allowing for controlled nutrient supply, while also enabling a higher dissolved oxygen availability, which favors PHA accumulation [71]. However, despite its extensive industrial application, fed-batch processing may present challenges in certain bioprocesses. For example, in the production of organic acids by some filamentous fungi, the accumulation of organic acids in the medium can lead to pH reduction and potential inhibition of microbial growth. Although fed-batch strategies allow for better control of substrate concentration and can help mitigate these effects, careful process monitoring is required to optimize production [18, 88].

Continuous system reactors are characterized by the continuous inflow of nutrients and the simultaneous collection of products at the reactor outlet, without interrupting operation [31]. Although this mode of operation requires more precise control, it offers several advantages. For example, continuous operation maintains constant nutrient concentration over time, and supports high cell densities without a lag phase. Additionally, it minimizes downtime associated with emptying, cleaning, and transferring media or products. These advantages make continuous processes attractive compared to other operation modes [13, 25].

2.3 Bioreactors According to the Mixing and Aeration Mode

Mixing and aeration are two key factors of utmost importance in bioprocess development. Ensuring efficient mixing is essential to prevent gradients and maintain homogeneity within the process vessel. However, it is equally important to ensure that mixing does not become detrimental to cells or the desired bioproduct due to mechanical stress [73, 75].

Providing an adequate amount of oxygen is crucial for successful bioprocesses, as oxygen is essential for aerobic processes such as microbial growth, and for microaerobic processes. However, an excess of oxygen can reduce bioprocess yield by diverting the carbon source toward microbial growth instead of bioproduct production. Additionally, aeration plays a key role in removing carbon dioxide generated during the process and dissipating the heat produced [73, 75].

A continuous oxygen supply is essential for bioprocesses since oxygen has low solubility in water. The concentration of dissolved oxygen is directly influenced by the oxygen consumption rate (e.g., for cell growth and maintenance), the oxygen transfer rate from the gas to the liquid phase, and the transport dynamics within the cultivation medium. The traditional aeration method in bioreactors involves sparging (bubbling) gas through the culture medium, where air or mixed gases are forced through small sparger pores or tips under slight overpressure. Smaller bubbles increase the surface area, enhancing oxygen diffusion at the gas–liquid interface [81].

Achieving a balance between mixing and aeration is crucial for successful and efficient bioprocess. These two factors are closely interconnected, as improved mixing generally enhances aeration. However, in certain cases, it may be necessary to reduce mixing intensity, even at the cost of less efficient aeration, to minimize shear stress that could lead to cell disruption or bioproduct losses [73]. Nowadays, there are several types of bioreactors categorized by their mixing and aeration modes, each with its own characteristics, making them suitable—or unsuitable—for specific processes [63].

Stirred Tank Reactors (STRs) are among the most widely used bioreactors for submerged fermentation due to their simple design, versatility, scalability, reduced risk of contamination, and excellent control systems. They also provide efficient mixing with high mass and heat transfer rates. However, there are some drawbacks associated with their use, such as pressure differences between the bottom and top layers (which can vary from 1 to 3 bar), high shear stress due to mechanical stirring, and a complex agitation system [63, 75, 81]. Nowadays, there are modified STR capable of reducing shear stress, however, there is an increase in their prices, making it sometimes difficult for an industrial scenario [63].

The agitation system in STRs is a crucial element to ensure homogenization inside the vessel. Common systems include Rushton turbines and pitched-blade impellers, which generate two flow patterns: radial and axial flow. Turbines and impellers can be combined with baffles to break vortices and create turbulence, enhancing mixing efficiency [63, 75]. A Rushton turbine is a disk-type impeller with 6–8 vertical blades that push the medium in a radial flow direction. Compared to propellers,

Rushton turbines are more suitable for aeration under high superficial air velocities. These turbines are ideal for cells that are resistant to shear stress. However, for cell cultivation under lower shear stress conditions, pitched-blade impellers (or even other types of impellers), which generate an axial flow pattern, are more appropriate and offer a higher performance-to-energy ratio [75]. Combinations of Rushton turbines and pitched-blade impellers can be used to prevent compartmentalization, enhance mixing, and reduce shear stress. The Rushton turbine provides strong radial mixing, while the pitched-blade impeller promotes axial flow, ensuring more uniform distribution of the medium and more effective oxygen transfer. This combination helps achieve a balance between efficient mixing and reduced mechanical stress on sensitive cells or bioproducts [63, 75].

Pneumatic bioreactors have gained attention in recent years due to their ability to cultivate sensitive shear-stress cells (such as plant and animal cells) while providing good mixing with low shear stress. They also offer advantages such as high mass transfer, low energy consumption, and simple design and operation. Currently, bubble column reactors and airlift reactors have been studied for different bioprocesses [63].

Bubble column reactors have a simple cylindrical structure with a height-to-diameter ratio typically exceeding 2:1, reaching values as 6:1. They utilize a sparger system for aeration and agitation is promoted by the air moving up in the system, without additional internal structures. At high aeration rates, the liquid flow becomes random, and turbulence from rising air bubbles ensures homogeneity. Perforated plates at the reactor's base help to distribute air bubbles, which should have a high surface-to-volume ratio to maximize mass-transfer efficiency, as there are no moving parts to break them down. Gas hold-up, a key dimensionless parameter for mass transfer, is influenced by superficial gas velocity, liquid viscosity, temperature, and system pressure [75]

Airlift bioreactors operate using a draft tube and consist of a cylindrical vessel divided into two sections: the riser, where gas is injected and released at the top, and the downcomer, were degassed liquid flows downward. Thus, airlift bioreactors create a continuous circulation of the medium, promoting aeration and homogeneity with minimal shear stress. This makes them ideal for plant and mammalian cell cultures, as well as immobilized cell or enzyme systems. Depending on the riser and downcomer placement, airlift bioreactors are classified into two main configurations: internal loop reactors have a centrally placed draft tube, while external loop reactors feature a separate tubular structure connected by a horizontal tube. External loop designs offer more efficient air bubble disengagement, leading to higher downcomer velocity, faster circulation, and improved mixing compared to internal loop systems [63, 75].

An interesting alternative to STR and pneumatic reactors is the wave bioreactor, especially for cells that are highly sensitive to shear stress. This bioreactor was developed for cells with anchorage-dependent growth and does not require submerged gasification. It consists of a cell bag (made of plastic and sterile, with varying volume) featuring a hydrophobic filter for gas exchange and a rocking table responsible for agitation (oscillatory movements) and temperature control. Wave bioreactors offer minimal shear stress and have the added advantage of eliminating the need for sterilization, as the bags are sterile and disposable. However, they face the disadvantage of limited scalability beyond 500 L of effective volume. Currently, these reactors are primarily used in the pharmaceutical industry for vaccine and biopharmaceutical production. They are particularly suitable for animal cell cultivation and the production of unstable products, such as bioconjugates [63].

3 Bioreactors and the Different Strategies for Process Consolidation

3.1 Simultaneous Saccharification and Fermentation (SSF) and Simultaneous Saccharification and Co-Fermentation (SSCF)

As previously explained, in the production of second-generation (2G) ethanol and other bioproducts, fermentation can be conducted separately or in combination with enzymatic hydrolysis. Furthermore, the co-fermentation of hexoses and pentoses, as well as enzyme production, can be integrated into various process configurations, as reported in the literature [4, 67, 76, 86].

Studies have shown that high concentrations of cellobiose and glucose can inhibit enzymatic hydrolysis. Therefore, the simultaneous conversion of glucose into ethanol during enzymatic hydrolysis helps to prevent sugar accumulation, thereby enhancing enzymatic activity. In this context, Simultaneous Saccharification and Fermentation (SSF) and Simultaneous Saccharification and Co-Fermentation (SSCF) offer several advantages over separate hydrolysis and fermentation, such as reduced processing times, simplified operations, and lower capital investment [39]. Table 1 provides an overview of some examples of SSF/SSCF processes applied to products of interest in biorefineries.

Implementing SSF/SSCF strategies enables microorganisms to rapidly consume sugars as they are released during hydrolysis, thereby reducing the inhibitory effects of hydrolysis byproducts on cellulolytic enzymes [39]. However, a major limitation of the most employed configuration—where hydrolysis and fermentation occur in the same reactor—is the compromise in operating temperature. This setup typically requires conditions that are suboptimal for both enzymatic hydrolysis and fermentation/co-fermentation, potentially limiting overall process efficiency [61].

3.1.1 Use of Single or Separate Reactors and Membrane Systems

The choice between using single or separate reactors, as well as the integration of membrane systems, can significantly influence process performance, product yield, and operational costs [16, 39]. In single-reactor systems, multiple bioprocess steps

Table 1 Recent examples of SSF/SSCF processes used for different biomass to produce bioproducts

Process	Microorganism	Biomass	Reaction system	Results	References
SSF	*S. cerevisiae Thermosacc*	Spruce	3.6 L STR	Ethanol production of 40 g/L	Koppram and Olsson [45]
SSF and SSCF	*S. cerevisiae* ATCC-200062 and *E. coli* ATCC-55124	Paper mill sludge	250 mL Erlenmeyer flask	Ethanol production of 45 g/L (SSF) and 42 g/L (SSCF)	Kang et al. [42]
SSCF	*C. tropicalis* Y-27290	Sorghum	250 mL Erlenmeyer flask	Production of 36 g/L and 1 g/L of ethanol and xylitol, respectively	Joy et al. [41]
SSCF	*S. cerevisiae* LFB-1	Corn stover and corn cob	50 mL flask	Production of 46.1 g/L (stover) and 72.8 g/L (cob) of ethanol	Huang et al. [32]
SSCF	*Bacillus coagulans*	Sweet sorghum bagasse	500 mL vessel	L-Lactic acid production of 210 nM	Ou et al. [62]
SSCF	*Pediococcus acidilactici* ZY271	Wheat straw	5 L bioreactor equipped with the helical impeller	L-Lactic acid production of 107 g/L	Zhang et al. [93]

are integrated within a single vessel. This approach is commonly applied in SSF and SSCF configurations, where hydrolysis and fermentation steps are carried out simultaneously [1].

Single-reactor systems offer several advantages, including reduced risk of contamination due to closed-system operation, enhanced product yields, and lower capital investment by eliminating the need for multiple reactors. Additionally, these systems simplify process control [1]. Despite these benefits, a major challenge lies in the difficulty of maintaining optimal conditions for the distinct biological reactions involved. Furthermore, the accumulation of inhibitory compounds—such as organic acids and ethanol—can negatively impact enzymatic activity and microbial metabolism [79].

Antunes et al. [4] investigated ethanol production from corn cob cellulignin using a semi-simultaneous saccharification and fermentation (SSSF) strategy. The process was conducted sequentially in a single boron silicate column reactor (90 cm height × 4 cm diameter, and 1 cm thickness jacket) operating in a fluidized bed configuration, homogenized by air insertion at 0.3 min^{-1}. The process began with a hydrolysis phase followed by fermentation, which continued alongside ongoing hydrolysis.

Initially, the column reactor was loaded with cellulignin, citrate buffer, and 20 FPU/g biomass of cellulolytic enzymes. The hydrolysis step was maintained for 48 h at 50 °C. Subsequently, the medium's pH was adjusted to 5.5 using a NaOH solution, and fermentation nutrients (yeast extract, ammonium sulfate, and malt extract) were added along with *Scheffersomyces parashehatae*. The temperature was then lowered to 30 °C, and the process was continued for another 48 h. This strategy resulted in a maximum ethanol production rate (Q_P) of 0.4 g/($L \cdot h$) and an ethanol yield ($Y_{p/s}$) of 0.43 g/g after 24 h.

Another study that investigated saccharification and fermentation in a single vessel was conducted by Sharma et al. [76], who evaluated ethanol production from rice straw using a simultaneous saccharification and co-fermentation (SSCF) system. The experiments were carried out in a stirred-tank bioreactor equipped with two elephant ear blade impellers (oriented upward and downward), which promoted efficient mixing while minimizing cellulase degradation at low energy consumption. Both batch and fed-batch modes were tested. Following a scheme similar to that used by Antunes et al. [4], saccharification was maintained for 6 h at 50 °C. Fermentation began with the inoculation of *Saccharomyces cerevisiae* DBTIOC and a temperature reduction to 42 °C. After 24 h, co-fermentation was initiated by introducing *S. cerevisiae* GSE1618, accompanied by a further temperature decrease to 30 °C. In the batch process, a final ethanol concentration of 56.38 g/L was achieved after 72 h, with a productivity of 0.78 g/($L \cdot h$). In a fed-batch process, a maximum ethanol concentration of 68.72 g/L was reached after 72 h, with a productivity of 0.94 g/($L \cdot h$). This performance was obtained with solid loadings of 29.5% (w/v) through sequential feeding of dilute acid-pretreated rice straw, without the need for detoxification, solid–liquid separation, or sterilization.

On the other hand, separate reactor configurations, which utilize distinct units for enzymatic hydrolysis and fermentation, allow for more precise control over each individual process. This approach is typically employed in separate hydrolysis and fermentation (SHF), but it can also be adapted for SSCF, in which hydrolysis is conducted under optimal conditions in one reactor and the liquid is recirculated to another reactor for fermentation. The primary advantages of using separate reactors include enhanced enzymatic and microbial performance due to the ability to apply optimal conditions for each step [17, 72]. However, this configuration poses challenges such as increased energy requirements and higher capital investment, as it necessitates the use of multiple reactors.

Terán-Hilares et al. [86] investigated ethanol production using sugarcane bagasse (SCB) pretreated by an alkaline hydrodynamic cavitation-assisted process (20 min, 0.3 mol/L NaOH, 60 °C). For the enzymatic hydrolysis and fermentation steps, the authors implemented an SSCF system using two bioreactors: a column reactor for enzymatic hydrolysis operated at 50 °C, and a bubble column reactor for fermentation maintained at 30 °C with immobilized *Scheffersomyces stipitis* cells. The hydrolysis reactor was loaded with 110 g of pretreated SCB, while the fermentation reactor contained alginate-encapsulated yeast cells. A peristaltic pump promoted medium recirculation between the two reactors at a flowrate of 26 mL min^{-1}, and the fermentation vessel was aerated at a rate of 0.26 vvm. The study reported an ethanol concentration of 8 g/L after 24 h, reaching 17 g/L after 96 h, with a yield (Y_p/s) of 0.51 g/g.

Prado et al. [67] investigated ethanol production using *Scheffersomyces parashehatae* UFMG-HM 52.2 (immobilized in alginate) from SCB pretreated with an oxidizing medium in a process assisted by hydrodynamic cavitation (Fenton process). The authors reported an SSCF system with two separate reactors. One reactor was used for enzymatic hydrolysis at 50 °C, and the other for fermentation at 30 °C, following a configuration similar to that described by Terán-Hilares et al. [86]. The system of Prado et al. [67] comprised a column reactor packed with pretreated SCB for hydrolysis, and a bubble column reactor containing alginate-immobilized *S. parashehatae* cells for fermentation. The medium was continuously recirculated between the reactors using a peristaltic pump set at a flow rate of 18 mL·min^{-1}, while the fermentation reactor was aerated at a rate of 30 mL·min^{-1}. In the fed-batch process, ethanol production reached 26 g/L after 48 h, while the batch process yielded up to 17 g/L of ethanol in the same period. In the fed-batch system, an additional 11 g of pretreated SCB was fed into the reactor at 10 h of operation.

To address some of the challenges associated with SSF and SSCF processes, membrane bioreactors (MBRs) have emerged as a promising alternative. MBRs can help mitigate inhibition issues while simultaneously reducing downstream processing and energy costs. Key advantages of MBRs include enhanced separation efficiency, low energy consumption, and the production of a high-quality final product [27].

Ishola et al. [49] demonstrated the use of a membrane-integrated system for simultaneous saccharification, filtration, and fermentation (SSFF). The authors adopted a strategy similar to that used by Terán-Hilares et al. [86] and Prado et al. [67], employing two separate vessels—one for saccharification and another for fermentation—but in this case, connected by a membrane to separate the two compartments. This configuration aimed to ensure a clear, sugar-rich stream for the fermentation vessel and facilitate the use of a flocculating yeast strain. Five sequential batches were conducted, each achieving more than 27 g/L of ethanol, with a maximum concentration of 34.6 g/L and an ethanol yield of 85% of the theoretical maximum. The authors also compared SSFF with conventional SSF, reporting ethanol productivities of 51.3 g ethanol/kg slurry for SSFF and 55.1 g ethanol/kg slurry for SSF.

3.2 *Consolidated Bioprocess for Biofuels and Biochemicals*

Industrial bioprocesses can be configured in different ways depending on the target product. The selection and integration of unit operations must be carefully planned to ensure the development of a robust process from both technical and economic standpoints [28]. Generally, increasing the number of steps in either upstream or downstream operations adds complexity and raises operational costs. In this context, consolidated bioprocessing (CBP) has emerged as a promising alternative to traditional multistep processes, offering advantages such as lower costs and reduced operational complexity [51, 60].

CBP is a strategy in biorefineries that integrates enzyme production, saccharification, and fermentation within a single reactor. This approach employs microorganisms—either isolated strains or microbial consortia, genetically modified or not—that are capable of degrading lignocellulosic biomass and directly converting the released sugars into biofuels or bioproducts such as ethanol, butanol, or organic acids [51, 82]. CBP offers significant advantages by streamlining the bioconversion process, reducing the number of unit operations, lowering operational costs, and enhancing overall process efficiency [51].

Among the main advantages, CBP reduces the number of individual processing steps, which significantly lowers initial capital investment in equipment, such as tanks, and reduces operational costs associated with labor, energy, and chemical or biochemical reagents. Furthermore, CBP minimizes handling and transfer of the fermentation medium, thereby reducing the risk of contamination. A disadvantage of CBP is that, so far, it is hard to find a naturally occurring microorganism that can perform all the required steps with sufficient efficiency for industrial-scale applications. As a result, the method can require genetically modified microorganisms, whose use faces regulatory and bureaucratic restrictions in many countries due to potential public health and environmental risks. Furthermore, another significant challenge in CBP application is the longer fermentation time often observed [28, 40].

Although CBP is not yet implemented on an industrial scale, it holds great potential to become economically viable, potentially replacing or complementing the methods currently used for lignocellulosic biomass valorization. To achieve this goal, numerous studies have been focused on both the identification of microorganisms capable of performing the necessary bioconversions and the use of synthetic biology to genetically modify model species widely applied in traditional processes [26].

Jin et al. [37] investigated a CBP approach for converting ammonia fiber expansion (AFEX)–pretreated corn stover (AFEX–CS) into ethanol using *Clostridium phytofermentans* ATCC 700394. The experiments were conducted in sealed serum vials, evaluating key parameters such as temperature (20–30 °C), inoculum size (5–40% v/v), nutrient supplementation (yeast extract at 6–12 g/L and YP medium—comprising 10 g/L yeast extract and 20 g/L tryptone–at 0.3–0.5 fractions), and initial pH (6–7.5). Under optimized conditions (30 °C, 5% inoculum, 0.4 YP fraction, and pH 7), the process achieved 76% glucan conversion and 88.6% xylan conversion within 10 days, yielding 174.1 g of ethanol per kg of AFEX–CS (equivalent to 2.8 g/L ethanol concentration). However, when the process was carried out at higher solid loadings, a notable decline in performance was observed, likely due to mass transfer limitations and microbial inhibition. Although the experiments were conducted in small-scale serum vials rather than conventional bioreactors, the findings offer valuable insights into the potential and limitations of CBP systems for lignocellulosic ethanol production. Further investigations using controlled bioreactor setups are necessary to assess scalability and optimize performance under industrial conditions [51].

Just as SSF and SSCF processes, which can be implemented using different configurations, CBP can also be conducted through various approaches. For instance, Matano et al. [52] employed a cell recycling strategy using a genetically modified *Saccharomyces cerevisiae* strain capable of producing cellulases to ferment rice straw into ethanol. The study compared the performance of this modified strain with that of a wild-type *S. cerevisiae*, which, lacking cellulase production, does not meet CBP criteria. Results showed that the CBP system achieved higher ethanol concentrations than the process using the wild-type strain. This difference was attributed to the sustained cellulase secretion by the engineered yeast, which may have mitigated the inhibitory effects of lignin on enzyme activity. After five fermentation cycles, the CBP system yielded an average ethanol concentration of 42.2 g/L, while the process using the wild-type strain resulted in 34.5 g/L.

CBP can be conducted in various bioreactor configurations, including MBRs. For example, Shahab et al. [74] evaluated CBP for lactic acid production in an MBR using a microbial consortium of *Trichoderma reesei* and *Lactobacillus pentosus*. The reactor included a quadrangular structure at its center, wrapped with a polydimethylsiloxane membrane for air dispersion—required to support the aerobic growth of *T. reesei*. The fungus formed a biofilm on the membrane surface, capturing and consuming all released oxygen, thereby maintaining anaerobic conditions in the bulk phase—favorable for *L. pentosus* and lactic acid fermentation. The system was continuously stirred at 250 rpm using either a Rushton turbine or a helical PEEK stirrer. Initially, the reactor was loaded with steam-pretreated beech wood and inoculated with *T. reesei* for cellulase production and biofilm formation. After 48 h, *L. pentosus* was introduced. Three initial solid loadings were tested (1.93%, 2.21%, and 3.86%), with the highest lactic acid concentration of 15.1 g L^{-1} obtained at 3.86%, and a maximum yield (Y_p/s) of 0.78 g g^{-1} at 2.21% solids.

CBP can also be applied to hydrogen production using anaerobic reactors. For instance, Álvarez-Guzmán [3] evaluated hydrogen production from various biomass mixtures—including agave bagasse, corn stover, wheat straw, and guinea grass—using 500 mL anaerobic reactors. The reactors were fed with 30 g of unsterilized total solids, maintained at 37 °C with agitation at 150 rpm. The study tested single biomass substrates, binary, and ternary mixtures, with the highest hydrogen yield of 9.3 mL H_2 g^{-1} TS observed for the combination of guinea grass and corn stover. In another study, Pérez-Rangel [65] explored hydrogen production from wheat straw at varying organic loading rates and temperatures (37 and 55 °C) in 1 L anaerobic bottles. These assays, conducted in a fed-batch mode, involved replacing 40–60% of the digestate with fresh wheat straw twice a week under anaerobic conditions. The authors reported that both organic loading rate and temperature influenced hydrogen yield, with the maximum production reaching 19.5 mL H_2 g^{-1} volatile solids at 10% organic loading and 55 °C. It is interesting to notice that CBP does not demand highly sophisticated bioreactors, even simple anaerobic reactors are able to hold consolidated bioprocesses.

4 Challenges and Perspectives

Biomanufacturing is a rapidly growing sector with an expanding range of applications in areas such as food, agricultural products, biopharmaceuticals, biofuels, and sustainable environmental solutions. To meet the high global demand for bio-based molecules, it is crucial that industry is equipped with advanced bioprocessing technologies. The transition from classical to digitized manufacturing, driven by the Fourth Industrial Revolution, brings new opportunities but also challenges due to the complexity of biological systems and the need to ensure reproducibility and process sustainability [35, 64].

The expansion of bioproduct and biofuel production at an industrial scale faces challenges such as process scalability, uniformity of operational conditions, and control of environmental variables. The transition to large volumes requires new designs and technologies to ensure efficiency and quality, as well as the adaptation of control and automation systems to keep costs low, ensuring the economic sustainability of biomanufacturing [20, 23].

Configurations such as SSF and SSCF are strategies that allow microorganisms to rapidly consume sugars as they are released during hydrolysis, minimizing the inhibitory effects of hydrolysis by-products on cellulolytic enzymes. Also, CBP offers significant benefits by unifying enzyme production, biomass hydrolysis, and fermentation within a single reactor, thereby simplifying the bioconversion pathway and enhancing overall process efficiency. Biorefineries are increasingly capable of reducing costs, simplifying operations, and improving yields while addressing environmental concerns. Although challenges remain, particularly regarding strain development and high-solids fermentation, technological innovations such as membrane-integrated systems and novel reactor designs continue to enhance performance. Therefore, the strategic selection and configuration of bioreactors—aligned with the nature of the biomass, microbial systems, and desired products—represent a decisive factor for the success of next-generation biofuel and biochemical production systems.

Currently, there is a noticeable gap in literature regarding the application of bioreactors for SSF, SSCF, and CBP processes. While some studies have addressed the use of bioreactors for SSF and SSCF, reports on scaling up to semi-pilot or pilot-scale systems remain limited in scientific databases. CBP research is even further behind in this aspect, with bioreactor-based implementations being scarce. Among the different bioreactor types, membrane bioreactors (MBRs) stand out as promising systems, particularly due to their potential to mitigate substrate and product inhibition, thereby improving overall process yields. However, MBRs are still underexplored for these bioconversion processes. Furthermore, integrated approaches that combine MBRs with SSF, SSCF, or CBP strategies are largely absent in the literature. This highlights a significant opportunity for future research aimed at advancing process development and scaling up toward economically viable industrial applications.

The large-scale feasibility of bioreactors, aimed at maximizing production and optimizing bioprocesses, relies on the development of new reactor types or the improvement of existing designs. These systems must also be economically viable, easy to handle, and suitable for industrial applications. However, several challenges persist, including high initial investment costs, regulatory complexities, and technical issues such as contamination and scalability [36, 68]. Specific scale-up challenges involve problems in solid waste recirculation, greater risk of inoculum and medium contamination, and significant downtime for maintenance and cleaning in batch operations, which could be reduced by implementing alternative operation modes [8].

The future of bioreactors requires significant technological advancements, particularly the integration of sensors and automation to enhance precision, efficiency, and real-time monitoring of production processes. These innovations not only support consistent product quality and regulatory compliance but also aid in early anomaly detection [56, 58]. Sustainability is another key focus, driving the development of eco-friendly, energy-efficient, and increasingly disposable or recyclable bioreactors. Despite progress, challenges such as integrating heterogeneous data and improving sensor accuracy remain and require continued research. Meanwhile, the growing interest in biotechnology is fostering a dynamic innovation ecosystem, with startups and small companies playing an active role in developing sustainable solutions to address global challenges [63].

Economic analysis is a crucial tool for ensuring the efficiency and viability of biofuel and bioproduct production and should be applied throughout all stages of plant development, from laboratory scale to commercial implementation [77]. This analysis encompasses the evaluation of process flowcharts, equipment sizing, material selection, operational expenses, and profit estimates, including costs related to supplies, labor, quality control, material handling, sanitation, taxes, and insurance [24, 77]. Moreover, bioreactor equipment costs vary based on factors such as manufacturing location, system size, and time, with key components—such as head volume, agitator, aeration systems, temperature and pH control, sterilization, and separation—significantly impacting capital investment. These costs can be updated using appropriate cost indices and adjusted based on parameters like fermenter capacity and membrane surface area [77].

5 Conclusion

The efficient and sustainable production of biofuels and biochemicals is vital for reducing dependence on fossil fuels and mitigating environmental impacts, particularly greenhouse gas emissions. The transition toward a circular bioeconomy depends significantly on the advancement of integrated, efficient, and sustainable bioprocesses for converting lignocellulosic biomass into valuable products. In this context, the role of bioreactors is fundamental not only as vessels for cultivation but as enablers of process optimization, scalability, and economic viability.

SSF, SSCF, and CBP represent promising bioprocess configurations capable of enhancing overall process performance. However, the current literature still lacks comprehensive studies involving these processes conducted in bioreactors, with most research limited to small-scale flasks and without considerations for scale-up. To advance these technologies toward industrial application, the integration of novel bioreactor designs—such as membrane bioreactors (MBRs)—should be explored. Additionally, combining such bioreactors with advances in synthetic biology, including the use of genetically modified organisms, or microbial consortia, presents a compelling strategy for optimizing process efficiency and robustness.

References

1. Afedzi AEK, Parakulsuksatid P (2023) Recent advances in process modifications of simultaneous saccharification and fermentation (SSF) of lignocellulosic biomass for bioethanol production. Biocatal Agric Biotechnol 54:102961. https://doi.org/10.1016/j.bcab.2023.102961
2. Ahmad S, Jaiswal R, Yadav R, Verma S (2024) Recent advances in green chemistry approaches for pharmaceutical synthesis. Sustain Chem One World 4:100029. https://doi.org/10.1016/j.scowo.2024.100029
3. Alvarez-Guzmán CL, Rodríguez-Hipólito F, Chávez-Reyes Y, Valdez-Vazquez I (2025) Lignocellulosic biomass mixtures improve hydrogen production by promoting microbial complementation in a consolidated bioprocess. J Clean Prod 489:144691. https://doi.org/10.1016/j.jclepro.2025.144691
4. Antunes FA, Freitas JB, Prado CA, Castro-Alonso MJ, Diaz-Ruiz E, Mera AE, Santos JC, da Silva SS (2023) Valorization of corn cobs for xylitol and bioethanol production through column reactor process. Energies 16(13):4841. https://doi.org/10.3390/en16134841
5. Aristizábal-Marulanda V, Alzate CAC (2019) Methods for designing and assessing biorefineries: review. Biofuels Bioprod Biorefin 13(3):789–808. https://doi.org/10.1002/bbb.1961
6. Arruda GL, Raymundo MTFR, Cruz-Santos MM, Shibukawa VP, Jofre FM, Prado CA, Da Silva SS, Mussatto SI, Santos JC (2024) Lignocellulosic materials valorization in second generation biorefineries: An opportunity to produce fungal biopigments. Crit Rev Biotechnol 44:1–20. https://doi.org/10.1080/07388551.2024.2349581
7. Avhad MR, MarchettiJM (2019) Uses of enzymes for biodiesel production. In: Advanced bioprocessing for alternative fuels, biobased chemicals, and bioproducts. Elsevier, pp 135–152. https://doi.org/10.1016/B978-0-12-817941-3.00007-3
8. Balachandar G, Varanasi JL, Singh V, Singh H, Das D (2020) Biological hydrogen production via dark fermentation: a holistic approach from lab-scale to pilot-scale. Int J Hydrogen Energy 45(8):5202–5215. https://doi.org/10.1016/j.ijhydene.2019.09.006
9. Balbino TR, Sánchez-Muñoz S, Díaz-Ruíz E, Moura Rocha T, Mier-Alba E, Inácio SC, Castro-Alonso MJ, Santos-Ebinuma VC, Pereira JFB, Santos JC, da Silva SS (2023) Lignocellulosic biorefineries as a platform for the production of high-value yeast derived pigments—a review. Biores Technol 386:129549. https://doi.org/10.1016/j.biortech.2023.129549
10. Bashir T, Majeed H, Iftikhar T (2022) Isolation and screening of thermophilic bacteria and its subsequent evaluation for lipases production. Pak J Botany 56(2). https://doi.org/10.30848/PJB2024-2(3)
11. Bhatia SK (2023) Microbial biopolymers: trends in synthesis, modification, and applications. Polymers 15(6):1364. https://doi.org/10.1016/j.bej.2024.109539
12. Bokelmann C, Bromley J, Takors R (2025) Pros and cons of airlift and bubble column bioreactors: how internals improve performance. Biochem Eng J 213:109539. https://doi.org/10.1016/j.bej.2024.109539
13. Boodhoo KVK, Flickinger MC, Woodley JM, Emanuelsson EAC (2022) Bioprocess intensification: a route to efficient and sustainable biocatalytic transformations for the future. Chem Eng Process—Process Intens 172:108793. https://doi.org/10.1016/j.cep.2022.108793

14. Cardona CA, Sánchez ÓJ (2007) Fuel ethanol production: process design trends and integration opportunities. Biores Technol 98(12):2415–2457. https://doi.org/10.1016/j.biortech.2007.01.002
15. Carrillo-Reyes J, Buitrón G, Arcila JS, López-Gómez MO (2021) Thermophilic biogas production from microalgae-bacteria aggregates: biogas yield, community variation and energy balance. Chemosphere 275:129898. https://doi.org/10.1016/j.chemosphere.2021.129898
16. Chandel AK, da Silva SS, Carvalho W, Singh OV (2011) Sugarcane bagasse and leaves: foreseeable biomass of biofuel and bio-products. J Chem Technol Biotechnol 87(1):11–20. https://doi.org/10.1002/jctb.2742
17. Cheng F, Brewer CE (2021) Conversion of protein-rich lignocellulosic wastes to bio-energy: review and recommendations for hydrolysis+ fermentation and anaerobic digestion. Renew Sustain Energy Rev 146:111167. https://doi.org/10.1016/j.rser.2021.111167
18. Cinar A, Parulekar SJ, Undey C, Birol G (2003) Batch fermentation: modeling: monitoring, and control. CRC Press. https://doi.org/10.1016/j.jclepro.2017.03.123
19. Clark JH, Luque R, Matharu AS (2012) Green chemistry, biofuels, and biorefinery. Annu Rev Chem Biomol Eng 3(1):183–207. https://doi.org/10.1146/annurev-chembioeng-062011-081014
20. Crater, J. S., Lievense, J. C. (2018). Scale-up of industrial microbial processes. FEMS Microbiol Lett 365(13):fny138. https://doi.org/10.1093/femsle/fny138
21. Cruz-Santos MM, Antunes FAF, Arruda GL, Shibukawa VP, Prado CA, Ortiz-Silos N, Castro-Alonso MJ, Marcelino PRF, Santos JC (2023) Production and applications of pullulan from lignocellulosic biomass: challenges and perspectives. Biores Technol 385:129460. https://doi.org/10.1016/j.biortech.2023.129460
22. Delavar MA, Wang J (2022) Bioreactor concepts, types, and modeling. In: Advanced methods and mathematical modeling of biofilms: vol VI. https://doi.org/10.1016/b978-0-323-85690-4.00004-x
23. de Mello AFM, Vandenberghe LPS, Herrmann LW, Letti LAJ, Burgos WJM, Scapini T, Manzoki MC, de Oliveira PZ, Soccol CR (2024) Strategies and engineering aspects on the scale-up of bioreactors for different bioprocesses. Syst Microbiol Biomanuf 4(2):365–385. https://doi.org/10.1007/s43393-023-00205-z
24. Dhagat S, Jujjavarapu SE (2024) Economics of fermentation processes. In: Recent advances in bioprocess engineering and bioreactor design. Springer, Singapore, pp 287–298
25. Drobnjakovic M, Hart R, Kulvatunyou BS, Ivezic N, Srinivasan V (2023) Current challenges and recent advances on the path towards continuous biomanufacturing. Biotechnol Prog 39(6):e3378. https://doi.org/10.1002/btpr.3378
26. Fortuin J, Hoffmeester LJ, Minnaar LS, Haan R (2025) Advancing cellulose utilization and engineering consolidated bioprocessing yeasts: current state and perspectives. Appl Microbiol Biotechnol 109:43. https://doi.org/10.1007/s00253-025-13426-0
27. Garg S, Behera S, Ruiz HA, Kumar S (2023) A review on opportunities and limitations of membrane bioreactor configuration in biofuel production. Appl Biochem Biotechnol 195:5497–5540. https://doi.org/10.1007/s12010-022-03955-z
28. Guajardo N, Schrebler RA (2024) Upstream and downstream bioprocessing in enzyme technology. Pharmaceutics 16(1):38. https://doi.org/10.3390/pharmaceutics16010038
29. Guo J, Zhang Y, Fang J, Ma Z, Li C, Yan M, Qiao N, Liu Y, Bian M (2024) Reduction and reuse of forestry and agricultural bio-waste through innovative green utilization approaches: a review. Forests 15(8):1372. https://doi.org/10.3390/f15081372
30. Hoey M, Clark R, Webb E, Lynd L (2025) Scale sensitivity of ethanol production via consolidated bioprocessing with consideration of feedstock cost. Biofuels, Bioprod Biorefin 19(1):9–17. https://doi.org/10.1002/bbb.2691
31. Holyavka MG, Evstigneev MP, Artyukhov VG, Savin VV (2016) Development of heterogeneous preparation with inulinase for tubular reactor systems. J Mol Catal B Enzym 129:1–5. https://doi.org/10.3109/07388551.2013.798254
32. Huang J, Li X, Zhao J, Qu Y (2024) Fed-batch strategy achieves the production of high concentration fermentable sugar solution and cellulosic ethanol from pretreated corn stover and corn cob. Int J Mol Sci 25(22):12306. https://doi.org/10.3390/ijms252212306

33. Igbokwe VC, Ezugworie FN, Onwosi CO, Aliyu GO, Obi CJ (2022) Biochemical biorefinery: a low-cost and non-waste concept for promoting sustainable circular bioeconomy. J Environ Manage 305:114333. https://doi.org/10.1016/j.jenvman.2021.114333
34. Igwebuike CM, Awad S, Andrès Y (2024) Renewable energy potential: second-generation biomass as feedstock for bioethanol production. Molecules 29(7):1619. https://doi.org/10.3390/molecules29071619
35. Isoko K, Cordiner JL, Kis Z, Moghadam PZ (2024) Bioprocessing 4.0: a pragmatic review and future perspectives. Digital Discovery. 3:1662–1681. https://doi.org/10.1039/D4DD00127C
36. Jaibiba P, Vignesh SN, Hariharan S (2020) Working principle of typical bioreactors. In: Bioreactors. Sustainable design and industrial applications in mitigation of GHG emissions. Elsevier, pp 145–173. https://doi.org/10.1016/B978-0-12-821264-6.00010-3
37. Jin M, Balan V, Gunawan C, Dale BE (2011) Consolidated bioprocessing (CBP) performance of *Clostridium phytofermentans* on AFEX-treated corn stover for ethanol production. Biotechnol Bioeng 108(6):1290–1297. https://doi.org/10.1002/bit.23059
38. Jin Q, Yang L, Poe N, Huang H (2018) Integrated processing of plant-derived waste to produce value-added products based on the biorefinery concept. Trends Food Sci Technol 74:119–131. https://doi.org/10.1016/j.tifs.2018.02.014
39. Jofre FM, Prado CA, Shibukawa VP, Rodrigues BG, Sarangi PK, Chandel AK (2025) Critical analysis of process parameters towards smart bioreactors development in biorefinery for biorenewables production. Int J Biol Macromol 305(P1):140957. https://doi.org/10.1016/j.ijbiomac.2025.140957
40. Joshi A, Kanthaliya B, Meena S, Khan F, Arora J (2021) Process consolidation approaches for cellulosic ethanol production. In: Sustainable biofuels. Opportunities and challenges. Academic Press,, pp 43–72. https://doi.org/10.1016/B978-0-12-820297-5.00013-X
41. Joy SP, Krishnan C (2022) Modified organosolv pretreatment for improved cellulosic ethanol production from sorghum biomass. Ind Crops Prod 177:114409. https://doi.org/10.1016/j.indcrop.2021.114409
42. Kang L, Wang W, Lee YY (2010) Bioconversion of kraft paper mill sludges to ethanol by SSF and SSCF. Appl Biochem Biotechnol 161:53–66. https://doi.org/10.1007/s12010-009-8893-4
43. Khonngam T, Salakkam A (2019) Bioconversion of sugarcane bagasse and dry spent yeast to ethanol through a sequential process consisting of solid-state fermentation, hydrolysis, and submerged fermentation. Biochem Eng J 150:107284. https://doi.org/10.1016/j.bej.2019.107284
44. Koller M (2018) A Review on established and emerging fermentation schemes for microbial production of polyhydroxyalkanoate (PHA) biopolyesters. Fermentation 4(2):30. https://doi.org/10.3390/fermentation4020030
45. Koppram R, Olsson L (2014) Combined substrate, enzyme and yeast feed in simultaneous saccharification and fermentation allow bioethanol production from pretreated spruce biomass at high solids loadings. Biotechnol Biofuels Bioprod 7:54. https://doi.org/10.1186/1754-6834-7-54
46. Kosamia NM, Samavi M, Piok K, Rakshit SK (2022) Perspectives for scale up of biorefineries using biochemical conversion pathways: technology status, techno-economic, and sustainable approaches. Fuel 324(A):124532. https://doi.org/10.1016/j.fuel.2022.124532
47. Koutinas AA, Vlysidis A, Pleissner D, Kopsahelis N, Garcia IL, Kookos IK, Papanikolaou S, Kwan TH, Lin CSK (2014) Valorization of industrial waste and by-product streams via fermentation for the production of chemicals and biopolymers. Chem Soc Rev 43(8):2587–2627. https://doi.org/10.1039/c3cs60293a
48. Kukreti N, Kag S, Kumar P, Kataria R (2023) Potential of waste stream in conversion into sustainable metabolites: an overview and update. Bioresour Technol Rep 22:101502. https://doi.org/10.1016/j.biteb.2023.101502
49. Ishola MM, Jahandideh A, Haidarian B, Brandberg T, Taherzadeh ML (2013) Simultaneous saccharification, filtration and fermentation (SSFF): a novel method for bioethanol production from lignocellulosic biomass. Biores Technol 133:68–73

50. Le Y, Zhang M, Wu P, Wang H, Ni J (2024) Biofuel production from lignocellulose via thermophile-based consolidated bioprocessing. Eng Microbiol 4(4):100174. https://doi.org/10.1016/j.engmic.2024.100174
51. Lynd LR, Guss AM, Himmel ME, Beri D, Herring C, Holwerda EK, Murphy SJ, Olson DG, Paye J, Rydzak T, Shao X, Tian L, Worthen R (2017) Advances in consolidated bioprocessing using *Clostridium thermocellum* and *Thermoanaerobacter saccharolyticum*. In: Industrial biotechnology: microorganisms, vol 2, issue 10. Wiley, pp 365–394. https://doi.org/10.1002/9783527807796.ch10
52. Matano YHT, Kondo A (2013) Cell recycle batch fermentation of high-solid lignocellulose using a recombinant cellulase-displaying yeast strain for high yield ethanol production in consolidated bioprocessing. Bioresource Technol 135:403–409. https://doi.org/10.1016/j.biortech.2012.07.025.
53. Mears L, Stocks SM, Sin G, Gernaey KV (2017) A review of control strategies for manipulating the feed rate in fed-batch fermentation processes. J Biotechnol 245:34–46. https://doi.org/10.1016/j.jbiotec.2017.01.008
54. Mgeni ST, Mero HR, Gervas JB, Mtashobya LA, Emmanuel JK (2025) Bioethanol production from dilute fruit wastes using sorghum as additional fermentable sugar. Biofuels 16(1):1–10. https://doi.org/10.1080/17597269.2024.2371211
55. Mishra A, Kumar M, Bolan NS, Kapley A, Kumar R, Singh L (2021) Multidimensional approaches of biogas production and up-gradation: Opportunities and challenges. Biores Technol 338:125514. https://doi.org/10.1016/j.biortech.2021.125514
56. Mitra S, Murthy GS (2022) Bioreactor control systems in the biopharmaceutical industry: a critical perspective. Syst Microbiol Biomanuf 2:1–22. https://doi.org/10.1007/s43393-021-00048-6
57. Mutungwazi A, Awosusi A, Matambo TS (2023) Comparative functional microbiome profiling of various animal manures during their anaerobic digestion in biogas production processes. Biomass Bioenerg 170:106728. https://doi.org/10.1016/j.biombioe.2023.106728
58. Nikita S, Mishra S, Gupta K, Runkana V, Gomes J, Rathore AS (2023) Advances in bioreactor control for production of biotherapeutic products. Biotechnol Bioeng 120(5):1189–1214. https://doi.org/10.1002/bit.28346
59. Ning P, Yang G, Hu L, Sun J, Shi L, Zhou Y, Wang Z, Yang J (2021) Recent advances in the valorization of plant biomass. Biotechnol Biofuels Bioprod 14:102. https://doi.org/10.1186/s13068-021-01949-3
60. Olguin-Maciel E, Singh A, Chable-Villacis R, Tapia-Tussell R, Ruiz HA (2020) Consolidated bioprocessing, an innovative strategy towards sustainability for biofuels production from crop residues: an overview. Agronomy 10(11):1834. https://doi.org/10.3390/agronomy10111834
61. Olofsson K, Bertilsson M, Lidén G (2008) A short review on SSF – an interesting process option for ethanol production from lignocellulosic feedstocks. Biotechnol Biofuels Bioprod 1:7. https://doi.org/10.1186/1754-6834-1-7
62. Ou MS, Awasthi D, Nieves I, Wang L, Erickson J, Vermerris W, Ingram LO, Shanmugam KT (2016) Sweet sorghum juice and bagasse as feedstocks for the production of optically pure lactic acid by native and engineered *Bacillus coagulans* strains. BioEnergy Res 9:123–131. https://doi.org/10.1007/s12155-015-9670-6
63. Palladino F, Marcelino PRF, Schlogl AE, José ÁHM, Rodrigues RDCLB., Fabrino DL, Santos IJB, Rosa CA (2024) Bioreactors: applications and Innovations for a sustainable and healthy future—A critical review. Appl Sci 14(20):9346. https://doi.org/10.3390/app14209346
64. Pandey K, Pandey M, Kumar V, Aggarwal U, Singhal B (2024) Bioprocessing 4.0 in biomanufacturing: paving the way for sustainable bioeconomy. Syst Microbiol Biomanuf 4:407–424. https://doi.org/10.1007/s43393-023-00206-y
65. Pérez-Rangel M, Barboza-Corona JE, Valdez-Vazquez I (2023) Effect of the organic loading rate and temperature on hydrogen production via consolidated bioprocessing of raw lignocellulosic substrate. Int J Hydrogen Energy 48(92):35907–35918. https://doi.org/10.1016/j.ijhydene.2023.05.329

66. Poornima S, Manikandan S, Prakash R, Deena SR, Subbaiya R, Karmegam N, Kim W, Govarthanan M (2024) Biofuel and biochemical production through biomass transformation using advanced thermochemical and biochemical processes—A review. Fuel 372:132204. https://doi.org/10.1016/j.fuel.2024.132204
67. Prado CA, Cunha MLS, Arruda GL, Cruz-Santos MM, Antunes FAF, Shibukawa VP, Terán-Hilares R, da Silva SS, Santos JC (2024) Hydrodynamic cavitation-assisted acid pretreatment and fed-batch simultaneous saccharification and co-fermentation for ethanol production from sugarcane bagasse using immobilized cells of Scheffersomyces parashehatae. Biores Technol 394:130234
68. Prakash P, Pandey S, Jha SK, Singh HR (2022) Development of bioreactors: current scenario and future challenges. In: Techno-economics and life cycle assessment of bioreactors. Elsevier, pp 37–54. https://doi.org/10.1016/B978-0-323-89848-5.00012-3
69. Rial RC (2024) Biofuels versus climate change: Exploring potentials and challenges in the energy transition. Renew Sustain Energy Rev 196:114369. https://doi.org/10.1016/j.rser.2024.114369
70. Rodionova MV, Bozieva AM, Zharmukhamedov SK, Leong YK, Lan JCW, Veziroglu A, Veziroglu TN, Tomo T, Chang JS, Allakhverdiev SI (2022) A comprehensive review on lignocellulosic biomass biorefinery for sustainable biofuel production. Int J Hydrogen Energy 47(3):1481–1498. https://doi.org/10.1016/j.ijhydene.2021.10.122
71. Saratale RG, Cho SK, Saratale GD, Kumar M, Bharagava RN, Varjani S, Kadam AA, Ghodake GS, Palem RR, Mulla SI, Kim DS, Shin HS (2021) An overview of recent advancements in microbial polyhydroxyalkanoates (PHA) production from dark fermentation acidogenic effluents: a path to an integrated bio-refinery. Polymers 13(24):4297. https://doi.org/10.3390/polym13244297
72. Sasaki C, Kushiki Y, Asada C, Nakamura Y (2014) Acetone–butanol–ethanol production by separate hydrolysis and fermentation (SHF) and simultaneous saccharification and fermentation (SSF) methods using acorns and wood chips of *Quercus acutissima* as a carbon source. Ind Crops Prod 62:286–292. https://doi.org/10.1016/j.indcrop.2014.08.049
73. Schmidell W (2018) Agitação e Aeração em Biorreatores. In: Schmidell W, Lima UA, Aquarone E, Borzani W (eds) Biotecnologia industrial, vol 2, 8th edn., Cap. 14. Blucher, pp 241–310
74. Shahab RL, Luterbacher JS, Brethauer S, Studer MH (2018) Consolidated bioprocessing of lignocellulosic biomass to lactic acid by a synthetic fungal-bacterial consortium. Biotechnol Bioeng 115(5):1208–1215. https://doi.org/10.1002/bit.26541
75. Shanmugam MK, Mandari V, Devarai SK, Gummadi SN (2022) Types of bioreactors and important design considerations. In: Current developments in biotechnology and bioengineering. Advances in bioprocess engineering. Elsevier, pp 3–30. https://doi.org/10.1016/B978-0-323-91167-2.00008-3
76. Sharma S, Swain MR, Mishra A, Mathur AS, Gupta RP, Puri SK, Ramakumar SSV, Sharma AK (2021) High solid loading and multiple-fed simultaneous saccharification and co-fermentation (mf-SSCF) of rice straw for high titer ethanol production at low cost. Renewable Energy 179:1915–1924
77. Sharma MD, Sharma S., Mishra P, Kulshrestha S (2022) Economic aspects of bioreactors: current trends and future perspective. In: Techno-economics and life cycle assessment of bioreactors. Elsevier, pp 55–68 https://doi.org/10.1016/B978-0-323-89848-5.00003-2
78. Sharma M, Tellili N, Kacem I, Rouissi T (2024) Microbial biopolymers: from production to environmental applications—A review. Appl Sci 14(12):5081. https://doi.org/10.3390/app14125081
79. Sharma S, Nair A, Sarma SJ (2021) Biorefinery concept of simultaneous saccharification and co-fermentation: challenges and improvements. Chem Eng Process-Process Intens 169:108634. https://doi.org/10.1016/j.cep.2021.108634
80. Singh R, Langyan S, Rohtagi B, Darjee S, Khandelwal A, Shrivastava M, Kothari R, Mohan H, Raina S, Kaur J, Singh A (2022) Production of biofuels options by contribution of effective and suitable enzymes: technological developments and challenges. Mat Sci Energy Technol 5:294–310. https://doi.org/10.1016/j.mset.2022.05.001

81. Singhania RR, Ruiz HA, Awasthi MK, Dong CD, Chen CW, Patel AK (2021) Challenges in cellulase bioprocess for biofuel applications. Renew Sustain Energy Rev 151:111622. https://doi.org/10.1016/j.rser.2021.111622
82. Singhania RR, Patel AK, Singh A, Haldar D, Soam S, Chen CW, Tsai ML, Dong CD (2022) Consolidated bioprocessing of lignocellulosic biomass: technological advances and challenges. Biores Technol 354:127153. https://doi.org/10.1016/j.biortech.2022.127153
83. Sinharoy A, Kumar M, Pakshirajan K (2020) An overview of bioreactor configurations and operational strategies for dark fermentative biohydrogen production. In: Bioreactors. Sustainable design and industrial applications in mitigation of GHG emissions. Elsevier, pp 249–288. https://doi.org/10.1016/B978-0-12-821264-6.00014-0
84. Sun H, Wang Y, He Y, Liu B, Mou H, Chen F, Yang S (2023) Microalgae-derived pigments for the food industry. Mar Drugs 21(2):82. https://doi.org/10.3390/md21020082
85. Tao Q, Huang H, Yang M, Zou Y, Harder MK, Yan Q, Liang B, Ntaikou I, Antonopoulou G, Lyberatos G, Zhang Y (2024) Comparison of batch, fed-batch and continuous operation modes for scalable polyhydroxyalkanoate (PHA) production and carbon sequestration from phenol. J Water Process Eng 60:105147. https://doi.org/10.1016/j.jwpe.2024.105147
86. Terán-Hilares R, Ienny JV, Marcelino PF, Ahmed MA, Antunes FAF, da Silva SS, dos Santos JC (2017) Ethanol production in a simultaneous saccharification and fermentation process with interconnected reactors employing hydrodynamic cavitation-pretreated sugarcane bagasse as raw material. Biores Technol 243:652–659. https://doi.org/10.1016/j.biortech.2017.06.159
87. Tyagi, A., Banerjee, B., Tyagi, S. (2025). A holistic approach toward the maximum conversion of lignocellulose into biofuels. Biofuels and Sustainability. Life Cycle Assessments, System Biology, Policies, and Emerging Technologies Woodhead Series in Bioenergy. 81–103. Woodhead Publishing. https://doi.org/10.1016/B978-0-443-21433-2.00014-1
88. Varriale L, Ulber R (2023) Fungal-based biorefinery: from renewable resources to organic acids. ChemBioEng Rev 10(3):272–292. https://doi.org/10.1002/cben.202200059
89. Verma ML, Kumar S, Jeslin J, Dubey NK (2020) Microbial production of biopolymers with potential biotechnological applications. In: Biopolymer-based formulations. Biomedical and food application. Elsevier, pp 105–137. https://doi.org/10.1016/B978-0-12-816897-4.00005-9
90. Vijayendra SVN, Shamala TR (2014) Film forming microbial biopolymers for commercial applications—A review. Crit Rev Biotechnol 34(4):338–357. https://doi.org/10.3109/07388551.2013.798254
91. Yang O, Qadan M, Ierapetritou M (2020) Economic analysis of batch and continuous biopharmaceutical antibody production: a review. J Pharm Innov 15:182–200. https://doi.org/10.1007/s12247-018-09370-4
92. Zeng AP (2019) New bioproduction systems for chemicals and fuels: Needs and new development. Biotechnol Adv 37(4):508–518. https://doi.org/10.1016/j.biotechadv.2019.01.003
93. Zhang B, Li J, Liu X, Bao J (2022) Continuous simultaneous saccharification and co-fermentation (SSCF) for cellulosic L-lactic acid production. Ind Crops Prod 187:115527. https://doi.org/10.1016/j.indcrop.2022.115527

Customized Bioreactors Development for Enhanced Fuels Production from Microalgal Feedstock

Matheus Henrique Alcântara de Lima Cardozo, Cristiano Eduardo Rodrigues Reis, Gisella Lamas-Samanamud, and Mateus de Souza Amaral

Abstract Microalgae are considered promising raw materials for the sustainable production of biofuels, as they offer advantages over traditional agricultural crops, such as higher productivity, use of wastewater or brackish water, high CO_2 capture capacity, and potential for the generation of various biofuels, such as biodiesel, bioethanol, biogas, biohydrogen, etc. However, microalgae productivity depends on many operational parameters of the bioreactors, including light intensity, temperature, pH, nutrients, gas transfer, choice of algal strain, etc., indicating the need for customized bioreactor designs. Thus, this chapter presents a summary of the main parameters that impact microalgae productivity and biofuel production in bioreactors.

Keywords Microalgae · Bioreactors · Biofuels

M. H. A. de Lima Cardozo · M. de Souza Amaral (✉)
Environmental and Chemical Technology Group, Department of Chemistry, Federal University of Ouro Preto, Ouro Preto, Brazil
e-mail: mateus.amaral@ufop.edu.br

M. H. A. de Lima Cardozo
e-mail: matheus.cardozo@aluno.ufop.edu.br

C. E. R. Reis
Department of Chemical Engineering, Howard University, Washington, DC, USA
e-mail: cristiano.reis@howard.edu

G. Lamas-Samanamud
Department in Chemical Engineering, University of Kentucky, Paducah, KY, USA
e-mail: gisella.lamas@uky.edu

M. de Souza Amaral
Research Group on Waste Treatment and Management Processes, Department in Chemical Engineering, Federal University of Sao João Del Rei, Ouro Branco, Brazil

© The Author(s), under exclusive license to Springer Nature Singapore Pte Ltd. 2025
A. K. Chandel and S. Sundaramurthy (eds.), *Biorefineries in Circular Bioeconomy*, Green Energy and Technology, https://doi.org/10.1007/978-981-95-3913-0_9

1 Introduction

Biofuel production has emerged as a crucial component of global strategies to reduce greenhouse gas emissions and dependence on fossil fuels. Liquid biofuels are particularly attractive, because they can directly replace petroleum fuels in transportation and are easier to store and distribute than intermittent energy sources like solar or wind [3, 13]. In this context, microalgae have gained considerable attention as a sustainable feedstock for biofuel production. These photosynthetic microorganisms utilize sunlight and CO_2 to accumulate energy-rich compounds (such as lipids and starch) that can be converted into various biofuels [17, 29]. Unlike first-generation biofuel crops (e.g., corn or sugarcane) that compete with food agriculture, microalgae are considered as the third-generation biofuel resources, avoiding the food-versus-fuel dilemma [13, 23]. They can be cultivated on nonarable land and in brackish or wastewater [17]. Microalgae also exhibit far higher biomass productivities than terrestrial plants, with some strains showing photosynthetic efficiencies up to three times greater [3]. These characteristics position microalgae as a promising feedstock for sustainable fuel production.

Despite their potential, significant challenges remain in scaling up microalgal biofuels. The dilute nature of algal cultures, coupled with limitations in light penetration and nutrient delivery, makes it difficult to achieve the high biomass yields needed for economical fuel production [13, 17]. Moreover, the cultivation systems (bioreactors) for microalgae are complex and expensive, leading to high production costs that currently render algal biofuels less competitive than fossil fuels [14, 29]. Key challenges include the need to optimize light distribution, gas exchange, and mixing in large-scale cultures, as well as preventing contamination by invasive organisms. Recent techno-economic assessments highlight that improvements in bioreactor design and operation will be essential to close the gap toward commercial viability [13, 23]. This chapter provides an overview of microalgae as a biofuel feedstock and examines how customized bioreactor designs can enhance fuel production.

2 Microalgae as a Feedstock for Biofuel Production

Microalgae offer several key advantages over traditional land-based biomass as a source of biofuels. First, they have extremely high growth rates and productivity. Many microalgae can double their biomass in a matter of hours under optimal conditions, enabling multiple harvests in a single year [13]. In terms of oil yield, microalgae far outperform oil crops—they can produce oil yields per unit area at least 7 times greater than crops like soybean or palm oil [29]. For example, algae are reported to accumulate lipids up to 50–70% of their dry weight under stress conditions, vastly exceeding the oil content of most crops [14]. This superior productivity means that fuel production from microalgae requires much less land. Indeed, microalgae can be grown on nonarable land (e.g., desert areas or saline flats) and do not compete with food agriculture [23]. This alleviates the land-use pressure and food security concerns associated with first-generation biofuels.

Additionally, many microalgae strains can be cultivated using brackish water or wastewater, dramatically reducing freshwater demands [17]. They can uptake nutrients (nitrogen and phosphorus) from wastewater, simultaneously treating waste streams while producing biomass [14, 29]. Microalgae also actively capture CO_2—including direct fixation of industrial flue gases—which offers the dual benefit of carbon sequestration and enhanced growth [3]. This carbon capture potential means algal biofuels could achieve a smaller carbon footprint compared to fossil fuels or even other biofuels.

The biochemical composition of microalgal biomass is particularly suited for producing a variety of biofuels. Microalgae cells are rich in lipids, carbohydrates, and proteins, with the relative proportions depending on the species and cultivation conditions [17, 29]. Many algae naturally synthesize high levels of triacylglycerols (TAGs)—neutral lipids stored as energy reserves—which can be directly converted into biodiesel via transesterification [3, 14]. Under nutrient limitation (such as nitrogen deprivation), certain microalgae divert metabolic energy from protein production to lipid or starch synthesis, resulting in markedly higher storage lipid contents [13]. *Chlorella* and *Scenedesmus* strains under nitrogen stress can accumulate well over 20% of their dry weight as lipids [29]. Some species (e.g., *Botryococcus braunii*) are known to reach lipid fractions above 50% or more of biomass dry weight in favorable conditions [14]. These lipids are excellent precursors for biodiesel, yielding a fuel with properties comparable to conventional diesel [29]. Microalgae also produce substantial carbohydrates (such as starches and polysaccharides), which can be fermented into ethanol or butanol, or thermochemically converted into bio-oil. In fact, certain algal strains (e.g., *Chlamydomonas, Chlorella*) can accumulate > 25–50% of dry weight as carbohydrates when subjected to appropriate stresses [14]. These carbohydrates can be enzymatically saccharified and fermented to produce ethanol and other biofuels [23]. The remaining protein fraction of microalgae, often constituting 30–50% of dry biomass in many species, can serve as a by-product for animal feed or be anaerobically digested to generate biogas [29]. Some high-protein algae (such as *Galdieria sulphuraria*) have over 60% protein content, although maximizing protein is usually not the goal for fuel production [3]. Nevertheless, protein-rich residual biomass after lipid extraction can be utilized in a biorefinery approach to improve overall process economics.

The diversity in biochemical composition means microalgae can be a feedstock for multiple fuel pathways: biodiesel from lipids, bioethanol from carbohydrates, biogas from residual biomass, and even bio-jet fuel or biocrude oil via thermochemical processes like pyrolysis and hydrothermal liquefaction [13]. This versatility underscores the value of microalgal biomass in an integrated biofuel production system, where nearly all components of the biomass are converted into energy products or value-added coproducts [29]. Table 1 presents different bioreactor configurations for biofuel production from microalgae.

Table 1 Bioreactors for biofuel production

Bioreactor	Operation mode	pH	Microorganism	Agitation	References
Two-stage immobilized bioreactor	Continuous (in series)	7.0–10.0	*Chlorella vulgaris*	Magnetic stirring	Javed et al. [12]
Photobioreactor-Biofilter + Hermetic Bioreactor	Batch (2 stages)	6.0–9.0	*Chlorella kessleri* + H_2 bacteria	Aeration: 2×2.5 L/min	Velmozhina et al. [31]
Three-stage photobioreactor	Batch	6.5	*Spirulina platensis*	Manual mixing	Nazarpour et al. [22]
Open reactor + fermentation	Batch	4–5.5 (fermentation)	*Chlorella vulgaris* + *Saccharomyces cerevisiae*	Aeration: 4 L/min	Silva et al. [25]
Anaerobic membrane bioreactor (AnMBR) + anaerobic reactor (AnR)	Continuous (in series)		Microalgae (unspecified) + primary sludge (62%-38% VS)	Biogas mixing	Serna-Garcia et al. [27]
Glass photobioreactor	Continuous / batch	7.12 to 7.93	*Chlorella vulgaris* FSP-E;	Magnetic stirring at 100 rpm	Condor et al. [4]
Concentric tubes PBR	Continuous	–	*Chlorella* sp.	Air supply (1 NL/min)	Vargas-Rodríguez [31]
Top-lit gas-lift bioreactor	Continuous	–	*Scenedesmus dimorphus*	–	Hosseini et al. [28]

(continued)

Table 1 (continued)

Bioreactor	Operation mode	pH	Microorganism	Agitation	References
Bubble column bioreactor	Photoautotrophic, heterotrophic, mixotrophic	6.3–6.5	*Chlorella vulgaris*	Shaking incubator (150 rpm preculture), continuous air in reactor (0.12 L/min)	Javed et al. [11]

Bioreactor	Flow (L/min)	Biofuel	Yield	Other parameters	References
Two-stage immobilized bioreactor	0.00035	$BioH_2$	222.7 mL/L	Z-medium + 10 g/L glucose	Javed et al. [12]
Photobioreactor-Biofilter + Hermetic Bioreactor	–	$BioH_2$	44.24 mL/L	Wastewater + 30% CO_2	Velmozhina et al. [31]
Three-stage photobioreactor	–	$BioH_2$	1.8 mL/g-VS	Sulfur addition (0.5–1%)	Nazarpour et al. [22]
Open reactor + fermentation	–	Bioethanol	68 °GL (136.6 g/L)	1.5% H_2SO_4 pretreatment; acid hydrolysis	Silva et al. [25]
Anaerobic membrane bioreactor (AnMBR) + anaerobic reactor (AnR)	–	Biomethane	130 mL $CH_4 \cdot d^{-1} \cdot L^{-1}$	73% biodegradability; total SRT: 200 d (100 d AnMBR + 100 d AnR); mesophilic (35 °C)	Serna-Garcia et al. [27]

(continued)

Table 1 (continued)

Bioreactor	Flow (L/min)	Biofuel	Yield	Other parameters	References
Glass photobioreactor	Air + CO_2 at 0.2 vvm	Bioethanol	0.22 ± 0.01 g ethanol per g biomass	300 μmol/m^2/s continuous light; BG-11 medium with nutrient variations; SWW supplement; biomass with 60% carbohydrates	Condor et al. [4]
Concentric tubes PBR	0.0000417 (60 mL/day)	Biomethane	Methane: 14 mL/day	LED light (1500 lux, 38.5 μmol m^{-2} s^{-1}), temp: 26°C, volume: 2 L, semisynthetic wastewater (4 mL urine/L)	Vargas-Rodríguez [31]
Top-lit gas-lift bioreactor	4–14 (optimized at 6.4 L/min)	Lipids for Biodiesel	Biomass: 32.1 gdw m^{-2} d^{-1}; Lipids: 198.4 gLipid m^{-2}	Depth 1.17 m; feed CO_2: 6.4%; top light (1000 μmol m^{-2} s^{-1}); 12h light/dark; half-strength BBM; optimized for lipid productivity	Hosseini et al. [28]
Bubble column bioreactor	. 0.12	Biodiesel (FAME)	Mixotrophic: 61.65 mg FAME/g biomass	Real RTWW (25–100%); best yield at 50% dilution; 1 g/L glucose added; LED 3W light; Temp: 25 ± 2 °C	Javed et al. [11]

2.1 Overview of Microalgae Cultivation Systems

Microalgae can be cultivated through open or closed systems, each characterized by distinct design considerations, operational requirements, and productivity potentials [29]. The most traditional approach involves the use of open pond systems, which can include natural water bodies such as lakes and lagoons or artificial raceway and circular ponds. Because these configurations are exposed to the surrounding environment, factors like climate and potential contamination have a direct influence on culture performance [29]. Raceway ponds typically feature shallow oval channels, approximately 0.2–0.3 meters deep, in which a paddlewheel continuously circulates the culture to facilitate mixing [14, 29]. These open ponds generally incur lower capital and operational costs, mainly requiring gentle mixing, and can be expanded efficiently by increasing their surface area [29]. However, open pond systems lack fine-tuned environmental controls. Variables such as sunlight, ambient temperature, and water evaporation fluctuate based on prevailing weather conditions, while undesirable organisms—including competing algae, protozoa, or pathogenic bacteria—can easily invade and crash the culture [23, 29]. Consequently, species that tolerate extreme salinity or pH levels are more suitable for open pond environments in order to minimize contamination risks [13]. Although open ponds are cost-effective and simple to operate, they typically yield lower cell densities. This low density necessitates extensive land area, leading to biomass harvesting being energy-intensive [13].

In contrast, closed photobioreactors (PBRs) provide a sealed environment with easier control over cultivation parameters. Common configurations of PBRs include tubular systems made of interconnected transparent pipes, flat-plate designs, and vertical columns [3, 29]. Because the culture remains isolated from external air, contamination is minimized, enabling stable monocultures [13, 23]. Additionally, these reactors allow precise regulation of light intensity, temperature, nutrient delivery, and pH, creating an optimal and consistent year-round production environment. Tubular PBRs have been employed in temperate regions for high-value algal production, such as *Haematococcus* for astaxanthin in Germany and *Chlorella* in Israel, often with higher productivities than those typically achieved in open pond systems [29]. Flat-plate PBRs likewise exhibit efficient light distribution, with some of the highest recorded photosynthetic efficiencies in the literature [13, 23]. Despite these advantages, closed PBRs carry substantial capital and operational expenses, in part due to specialized materials, precision instrumentation, and additional energy requirements for pumping and cooling [3, 29]. Scaling these systems for high-volume production presents further challenges such as fouling, elevated oxygen concentrations, and difficulties in heat dissipation, especially in large reactors [13, 23].

Hybrid cultivation approaches seek to balance the benefits of open and closed systems. One common strategy is to grow dense, contamination-free cultures in PBRs initially, then transfer them to open ponds for the bulk growth phase, thus reducing contamination risk while leveraging the cost advantages of open environments [17].

Another cultivation mode involves heterotrophic growth, where microalgae are fed organic carbon sources in conventional fermentation tanks. This process allows certain strains to thrive without light but increases the total expense of feedstock and negates some of the carbon sequestration benefits characteristic of photosynthetic cultures [13]. Ultimately, selecting an optimal cultivation system for a given application depends on considerations such as resource availability, operational scale, and the specific fuel targets.

3 Principles of Bioreactor Design

Designing an efficient bioreactor for microalgae involves understanding both the biological requirements of the algae and the engineering aspects needed to meet those requirements at scale. A bioreactor for microalgal cultivation must provide a suitable environment for photosynthesis and cell growth, which includes adequate light, nutrients, gas exchange, and proper hydrodynamics.

3.1 Essential Components of Microalgae Bioreactors

An effective microalgal bioreactor typically comprises several fundamental components and subsystems:

a. *Light Supply*: Because most biofuel-oriented cultivations use photosynthetic algae, provision of light is a central design consideration. In open ponds, sunlight is the primary source, whereas closed photobioreactors use either natural sunlight (often enhanced by orientation or reflective surfaces) or artificial illumination (LEDs, fiber-optic light delivery) [23]. The reactor configuration (depth of ponds, diameter of tubes, etc.) is designed to maximize light exposure while minimizing self-shading. Some advanced PBR designs incorporate internal light guides or thin optical fibers to distribute light evenly through dense cultures [13]. Effective light management ensures that algae receive sufficient irradiance for photosynthesis without causing photo-inhibition from excess light.
b. *Mixing System*: Adequate mixing is required to circulate algal cells so that they have equal access to light and nutrients. Mixing also prevents cells from settling and maintains uniform environmental conditions (temperature, pH) throughout the reactor. In open raceway ponds, a paddlewheel continuously circulates the water around the loop [29]. In closed PBRs, mixing can be achieved by mechanical pumps or airlift systems that bubble gas through the culture [23, 29]. For instance, vertical column PBRs often use air sparging at the base to generate bubbles that both aerate and mix the culture, creating a rising circulation pattern [17]. The mixing system must be optimized to avoid dead zones (poorly mixed regions) while not imparting excessive shear stress that could damage delicate

algal cells [29]. Properly designed mixing ensures that each cell experiences a light/dark cycle as it moves through illuminated and dark regions, which can actually enhance photosynthetic efficiency compared to static illumination [3, 13].

c. *Gas Exchange Mechanism*: Microalgae need a supply of carbon dioxide for photosynthesis and must expel the oxygen they produce. Therefore, bioreactors are equipped with gas exchange interfaces. In open systems, the broad water surface facilitates passive CO_2 absorption and O_2 release to the atmosphere, though often supplemental CO_2 is bubbled in to saturate the culture with carbon [29]. In closed PBRs, dedicated aeration is provided—typically CO_2-enriched air (or pure CO_2) is injected through spargers or gas diffusers [23]. Tubular PBRs may have CO_2 introduced at various points along the tube length and separate degassing units to vent oxygen [29]. Efficient gas exchange is critical: insufficient CO_2 leads to carbon limitation, while oxygen accumulation can inhibit algal growth (and even become dangerous in high concentrations in a closed system). Modern designs often include pH control linked to CO_2 addition—CO_2 injection lowers pH, while algal uptake raises it, so maintaining pH within an optimal range (typically ~7–8) ensures both adequate carbon supply and healthy physiological conditions [3, 13].
d. *Temperature Control*: Microalgae grow best within certain temperature ranges (often around 20–30 °C for many mesophilic species). Bioreactors require mechanisms to dissipate excess heat (from solar heating or metabolic activity) or to retain heat in cold conditions. Open ponds rely on their large thermal mass and sometimes use cooling towers or spray systems in hot climates [29]. Photobioreactors may be equipped with heat exchangers—for example, water jackets around reactor vessels or coils within flat-panel reactors to circulate cooling water [29]. Temperature control is especially important in closed systems, since they can heat up quickly under strong sunlight.
e. *Materials and Configuration*: The construction materials (transparent plastics or glass for PBRs, lined ponds for open systems) and geometry of the bioreactor are also key design aspects. Materials must be selected to allow good light transmission, resist UV degradation, and not leach any toxic substances into the culture. The configuration—whether a series of flat panels, a network of tubes, or a shallow pond—is chosen based on the trade-offs between surface area for light capture, ease of scale-up, mixing efficiency, and cost. For instance, flat-plate PBRs provide a high surface-to-volume ratio (excellent light distribution) but can be more difficult to scale and clean [23, 29]. Tubular PBRs are easier to scale linearly by adding more tube length, but they can suffer from gradients along the flow path (e.g., CO_2 depletion or pH drift from inlet to outlet) [13, 17]. Each design must incorporate ports for nutrient addition, harvesting, and cleaning, as well as monitoring sensors (for pH, dissolved oxygen, optical density, etc.) to enable feedback control.

3.2 Factors Affecting Microalgal Productivity in Bioreactors

The productivity of a microalgae culture (usually measured as biomass produced per volume per day) is influenced by a complex interplay of physical, chemical, and biological factors.

a. *Light Intensity and Photoperiod*: Both the intensity and the duration/distribution of light exposure (photoperiod) critically impact growth. If light is too limiting (in deep cultures or at high cell densities), photosynthesis is underpowered and growth slows. Conversely, if light is too intense at the surface, it can cause photoinhibition while cells deeper in the culture receive little light due to shading [23]. Bioreactor designs aim to maximize average light availability per cell, for example by mixing (so cells alternate between light and dark zones) or using diffuse light. In practice, many reactors operate at sub-saturating light intensities for the algae to avoid wasteful excess. The photoperiod (light/dark cycle) can also be adjusted – some systems use continuous light to maximize productivity, while others find that giving algae a short dark period can relieve photostress and allow for cellular respiration and repair. Providing the appropriate spectrum of light (e.g. red and blue wavelengths that algae absorb efficiently) is another consideration, especially for LED-lit systems [14]. Overall, effective light management (intensity, duration, spectrum) is one of the most significant challenges in scaling up algal bioreactors [14, 23].
b. *Temperature and pH*: Each microalgal strain has an optimal temperature range for growth. Deviations can slow metabolic rates or alter the biochemical composition of the cells. Maintaining culture temperature is easier in controlled PBRs (with active cooling/heating) and more difficult in open ponds subject to diurnal and seasonal swings. Temperature not only affects growth rates but also influences lipid accumulation—for instance, some algae produce more saturated fatty acids at lower temperatures (beneficial for biodiesel quality) [29]. pH affects nutrient availability and enzyme function, most algae prefer neutral to slightly alkaline conditions. As algae consume CO_2, the pH can rise, and as CO_2 is added or as respiration produces CO_2 at night, pH can drop. Bioreactors often include pH control systems that modulate CO_2 injection to stabilize pH, thereby optimizing carbon uptake and preventing extreme pH shifts that could harm the culture [3, 13].
c. *Nutrient Concentration*: Sufficient supplies of macronutrients (nitrogen, phosphorus, silicon for diatoms) and micronutrients (iron, trace metals, vitamins) must be maintained. In continuous or semicontinuous cultivation, nutrients are replenished to avoid depletion. Interestingly, nutrient levels can be deliberately manipulated to trigger accumulation of desired storage compounds—a well-known strategy is nitrogen starvation near the end of a cultivation cycle to induce lipid accumulation in algae, which can significantly increase biodiesel yield from the biomass [13]. However, this comes at the expense of growth (since nitrogen limitation stops protein synthesis and cell division), so a balance must be struck

or a two-stage process used (growth phase with full nutrients, followed by a stress phase) [13].

d. *Gas Transfer and Mixing*: The availability of CO_2 and the removal of O_2 were mentioned as design components; their effectiveness greatly influences productivity. In dense cultures, algae can quickly deplete dissolved CO_2 during daylight, leading to carbon limitation if CO_2 is not replenished fast enough [23]. Similarly, oxygen produced via photosynthesis can accumulate and inhibit further photosynthesis (the O_2 can competitively interfere with the CO_2-fixing enzyme Rubisco and also cause oxidative stress). Thus, vigorous gas exchange (high aeration rates or large surface area) is needed to keep the CO_2/O_2 levels optimal [23, 29]. Effective mixing ensures that all cells have access to nutrients and CO_2 and that local hotspots of O_2 or waste metabolites do not occur. Poor mixing can lead to gradients—for example, in a long tubular reactor, cells near the gas injection point might get ample CO_2, while those farther along become starved [29]. Proper hydrodynamic design (e.g., bubble size in airlift mixers, impeller design in stirred tanks) can enhance mass transfer of gases and also keep cells in suspension without damaging them [13].
e. *Biological Factors* (*Strain and Culture Density*): The choice of microalgal strain inherently determines the baseline productivity and the biochemical profile of the biomass. Different species have different maximum growth rates, light and nutrient preferences, and stress tolerances. Selecting high-yield strains (or genetically improving them) for biofuel traits is an active area of research [3, 14]. Additionally, the inoculum quality and age can affect performance—healthy, exponentially growing inocula lead to better outcomes than older, stationary-phase cultures. Culture density (biomass concentration) is another factor: while higher densities mean more biomass is produced in the reactor, too high a density causes self-shading and nutrient competition, ultimately plateauing the growth. Most systems have an optimal cell density that balances light utilization with growth rate. Harvesting (biomass removal) can be timed to keep the culture in its most productive phase (e.g., continuous reactors harvest a portion of culture regularly to maintain a near-steady-state density) [3, 13]. Finally, in nonaxenic outdoor systems, invasive species or grazers (like rotifers) can bloom and decimate the target algal culture [23, 29]. This risk is mitigated by closed reactors or by cultivating in extreme conditions, but it remains a constant management concern affecting reliability and productivity. Table 2 presents different bioreactor configurations for microalgae cultivation.

3.3 Design of Experiment as a Useful Tool To Customized Bioreactors

The ability to understand and apply advanced statistical tools has a prominent role in the decision-making of various industries. Real world processes contain a large volume of data, and the use of univariate statistics is virtually impossible [9]. The

Table 2 Microalgae cultivation in photobioreactors

PBR	Operation mode	Microorganism species	Light intensity	Carbon source	Aeration	Other parameters	Productivity	References
Airlift (closed)	Continuous	*Scenedesmus dimorphus*	12h light / 12h dark	Atmospheric CO_2 (autotrophic)	Air bubbling for mixing	Electric energy for mixing: 50 W/m^3	0.084 g/L/day (biomass)	Chandra et al. [2]
Raceway pond (open)	Continuous	*Scenedesmus dimorphus*	Not directly specified (natural local conditions)	Atmospheric CO_2 (autotrophic)	Natural/ mechanical aeration not detailed	Area: 5.54 m^2, depth 0.15 m, electric energy 15 W/m^3	0.5 kg/m^3 concentration / 0.006 kg/m^2/day (biomass)	
Bubble column PBR 5 L	Continuous (aerated mixing)	*Tisochrysis lutea*	100 μmol photons m^{-2} s^{-1}	F/2 medium + starch 3900 g/m^3	13 L/min (3.25 vvm)	Temp: 24 ± 1 °C; online pH control; 16:8 light:dark cycle	Fucoxanthin: 24.96 mg/L/day	Mohamadnia et al. [19]
Bubble column PBR (BCPR)	Continuous, with CO2 supply (0.03%, 5%, 10%)	*Chlorella* sp. S-01, *Chlorella* sp. S-02, Co-CC (consortium)	2000 lx, 12h light / 12h dark	BBM, with CO2 at 0.03%, 5%, 10%	CO2 from pressurized cylinder	Temp 25 °C, 5 L volume, chlorophyll-a and lipid measurement	1.41–1.87 g/L DCW (max on day 20 with 10% CO2)	Mathivanan et al. [18]
Sequential-flow bubble column PBR (SFB-PBR)	Sequential batch in series	*Chlorella vulgaris*	60–70 μmol m^{-2} s^{-1} (cool-white fluorescent)	Organic nutrient (Tani compost), 8% v/v	9 L/min (air)	pH 3–3.5, 5 bioreactors in series, temp 25 ± 5 °C, inoculum 8% v/v	Biomass: 0.83–0.86 g/L/day; Lipid: 50.4% (in recycled medium)	Dasan et al. [5]

(continued)

Table 2 (continued)

PBR	Operation mode	Microorganism species	Light intensity	Carbon source	Aeration	Other parameters	Productivity	References
Flat plate (low-cost)	Batch	*Synechocystis* sp. CACIAM 05	100 μmol photons m^{-2} s^{-1}	BG-11 medium with 2 g/L $NaNO_3$	Compressed air, 4 L/min	Optimized system with internal static modules (PBR-OII); working volume 4.2 L; 20-day cultivation; 13:11 h light:dark	250 mg/L/day (dry biomass)	Rutiléia et al. (2024)
Flat plate PBR (IRF) 140 L	Continuous operation, air injected (variable vvm)	*Arthrospira platensis*	Up to 3122 μmol photons m^{-2} s^{-1} (PAR)	Zarrouk medium, non-sterile with potable water	Injected air: 0.17–0.27 vvm	Infrared reflective film for passive thermal control, culture depth 0.10 m, angle 32°	0.126 $g{\cdot}L^{-1}{\cdot}d^{-1}$ (dry biomass) at optimal aeration (0.21 vvm)	Amanna et al. [1]
Membrane PBR (MPBR) + airlift cylindrical (batch)	Batch and continuous	*Chlorella vulgaris*	135–140 μmol photons m^{-2} s^{-1}	Mixed CO2 (4%) + nutrients in simulated water (N: 40 mg/L NH4+-N, P: 5 mg/L DIP) + Cr(VI) 0–5 mg/L	0.5–1.0 L/min CO2 mixed with air	SRT and HRT adjusted (SRT = 10, 20, 40 d; HRT = 2, 3 d); ultrafiltration membrane (0.01 μm)	Max biomass: 0.21 mg/L/day (total Cr removal); growth stimulated at 0.5–1 mg/L Cr(VI), inhibited above 2 mg/L	Lu et al. [15]

(continued)

Table 2 (continued)

PBR	Operation mode	Microorganism species	Light intensity	Carbon source	Aeration	Other parameters	Productivity	References
Membrane PBR (MPBR) + Sequencing PBR (SPBR)	Continuous, SRT 20 days, HRT 12–36 h	Mixed microalgae: *Chlorella* sp. and *Spirulina* sp.	~80 μmol/(m^2·s)	AnMBR permeate (treated municipal sewage)	No aeration	MPBR with 0.1 μm ceramic membrane, 4.5 L volume, cyclic permeation (8 min suction / 2 min rest)	Microalgal biomass: 1.58–1.98 g/L; CO2 fixation: 22.9–43.4 g(C)/(m^3·d); N removal: 24.9–49.3 g(N)/(m^3·d); P removal: 3.4–8.1 g(P)/(m^3·d); methane: 89–93%	Ding et al. [6]
Modified raceway pond	Continuous	*Spirulina platensis*	1500 μE/m^2/s	$NaHCO_3$ (16.8 g/L), $NaNO_3$ (2.5 g/L)	2 m^3/h (7 spargers with 63 holes of 3 mm)	Initial pH: 9.2; maintained pH: 10.5; temp: 30 °C; linear velocity: 0.25 m/s; depth: 1 m; impeller speed: 125 rpm	21.22 g/m^2/day dry biomass	Sawant et al. [26]
Outdoor raceway	Semicontinuous	*Nannochloropsis gaditana*	Not specified (sunlight)	Flue gas (10.9% CO_2) and concentrate (20%-50%)	Air: 0.3 v/v/min, Flue gas: 0.05 v/v/min	pH 8.0, depth 0.11 m, D: 0.2 or 0.3 d^{-1}	32.42 g/m^2·day biomass	Romero-Villegas et al. [24]
Greenhouse raceway	Semicontinuous	*Scenedesmus* sp. (CCAP 276/24)	Seasonally variable; average irradiance calculated	CO_2 injected on demand	Controlled with online monitoring and CO_2 injection	pH controlled, volume 11.8 m^3, depth 0.135 m; temp and DO monitored	25.1 g·m^{-2}·day^{-1} in summer (dilution 0.2 d^{-1})	Morillas-España et al. [21]

(continued)

Table 2 (continued)

PBR	Operation mode	Microorganism species	Light intensity	Carbon source	Aeration	Other parameters	Productivity	References
Air-lift PBR (ALPBR)	Continuous	*Chlorella vulgaris, C. protothecoides*	50 µmol m^{-2} s^{-1}	5% CO_2	0.1–0.4 vvm	Temp: 28°C; medium: BG11; CFD simulation for hydrodynamics	*C. vulgaris*: Max DCW at 0.2 vvm; lipids ~15% DCW; *C. protothecoides*: max DCW at 0.3 vvm	Ding et al. [7]
Horizontal Tubular (air-lift)	Continuous / batch	*Spirulina* sp. LEB 18	40.5 µmol m^2/s (40 W fluorescent)	10% CO_2 (v/v), 6.5 L/min for 5 min every 40 min	Air: same flow as CO_2	Temp: 30 °C, 12h light/12h dark, 50% reduced nitrogen	Biomass: 0.02 $g{\cdot}L^{-1}{\cdot}day^{-1}$; CO_2 fixation: 160 $mg{\cdot}L^{-1}{\cdot}day^{-1}$	Morais et al. [20]

Design of Experiment (DOE) methodology, developed by R.A. Fisher in the late twentieth century, is structured and involves a series of steps, including planning, executing experiments, and analyzing collected data using statistical methods and, it is useful to customized bioreactors. The DOE method is classified into comparative (significant factor), screening (selecting main effects), and optimal fitting of regression models (factor interactions and quadratic effects). It can be widely used for product development, process improvement, and scientific research. The selection of the optimal DOE becomes a challenge when there are so many options. The most employed designs for characterization in science and industry are definitive screening design (DSD), Placket-Burmann design (PBD), Taguchi design (TD), Response Surface Methodology (RSM), and regression analysis [10].

DOEs are essential in bioprocess engineering to maximize experimental information content while minimizing costs and time. It helps refining and improving interdependent parameters within a unit operation or between units in a bioprocess sequence, which involves the transformation of nutrients and other medium components into biomass and/or molecular products. Model-based DOE uses a model to predict the information gain of a design, with initial model parameterization obtained from literature or regression of existing data. In pharmaceutical bioprocessing, DOE has improved the speed of development and helped to define manufacturing processes for high-quality products. The application of DOE in biotechnological processes includes antibiotic fermentation, expanded bed optimization, virus vector transfection of insect cell cultivation, feed profile adaptation, embryonic stem cell expansion protocols, mammalian cell harvesting, and biofuel productions [8, 16].

4 Conclusion

Advances in bioreactor design and operation, such as improved light management efficiency, precise environmental control, and nutritional stress strategies, are essential to increase productivity and reduce production costs. In addition, the integration of approaches such as microalgae cultivation in wastewater and full biomass utilization in biorefineries can contribute to commercial viability. The use of DOE is also suggested as a tool to achieve optimization of the many bioreactor operating parameters.

References

1. Amanna B et al (2025) Enhancing Arthrospira platensis productivity by optimizing mixing rates in a self-cooling flat plate photobioreactor. Algal Res 88:104035
2. Chandra TS et al (2018) Comparative life cycle assessment of microalgae-mediated CO_2 capture in open raceway pond and airlift photobioreactor system. Clean Technol Environ Policy 20:2357–2364

3. Chanquia SN, Vernet G, Kara S (2022) Photobioreactors for cultivation and synthesis: specifications, challenges, and perspectives. Eng Life Sci 22(7):712–724
4. Condor BE et al (2024) Effects of carbon dioxide concentration and swine wastewater on the cultivation of *Chlorella vulgaris* FSP-E and bioethanol production from microalgae biomass. Appl Energy 369:123617, 1 set
5. Dasan YK et al (2020) Cultivation of *Chlorella vulgaris* using sequential-flow bubble column photobioreactor: a stress-inducing strategy for lipid accumulation and carbon dioxide fixation. J CO_2 Utilization 41:101226, 1 out
6. Ding M et al (2022) Enhanced nutrient removal and bioenergy production in microalgal photobioreactor following anaerobic membrane bioreactor for decarbonized wastewater treatment. Bioresource Technol 364:128120
7. Ding N et al (2020) Biotechnology & biotechnological equipment evaluation of an enclosed air-lift photobioreactor (ALPBR) for biomass and lipid biosynthesis of microalgal cells grown under fluid-induced shear stress
8. Ellert A, Vikström C (2014) Design of experiments with small-scale bioreactor systems. In: Innovations in cell culture, vol 10
9. Freitas L, De Freitas APBR (eds) (2013) Multivariate analysis in management, engineering and the sciences. BoD–Books on Demand
10. Jankovic A, Chaudhary G, Goia F (2021) Designing the design of experiments (DOE)—an investigation on the influence of different factorial designs on the characterization of complex systems. Energy Build 250:111298
11. Javed F et al (2022) Real textile industrial wastewater treatment and biodiesel production using microalgae. Biomass Bioenergy 165:106559, 1 out
12. Javed MA et al (2024) A novel two-stage immobilized bioreactor for biohydrogen production using a partial microalgal-bacterial (*Chlorella vulgaris* and wastewater activated sludge) co-culture. Sustain Energy Technol Assess 62:103624, 1 fev 2024
13. Khan S, Das P, Quadir MA, Thaher MI, Mahata C, Sayadi S, Al-Jabri H (2023) Microalgal feedstock for biofuel production: recent advances, challenges, and future perspective. Fermentation 9(3):281
14. Khoo KS, Ahmad I, Chew KW, Iwamoto K, Bhatnagar A, Show PL (2023) Enhanced microalgal lipid production for biofuel using different strategies including genetic modification of microalgae: a review. Progr Energy Combustion Sci 96:101071
15. Lu MM et al (2021) Response of microalgae *Chlorella vulgaris* to Cr stress and continuous Cr removal in a membrane photobioreactor. Chemosphere 262:128422, 1 Jan 2021
16. Luna MF, Martínez EC (2018) Optimal design of dynamic experiments in the development of cybernetic models for bioreactors. Chem Eng Res Des 136:334–346
17. López-Pacheco IY, Ayala-Moreno VG, Mejia-Melara CA, Rodríguez-Rodríguez J, Cuellar-Bermudez SP, González-González RB, Coronado-Apodaca KG, Farfán-Cabrera LI, González-Meza GM, Iqbal HMN, Parra-Saldívar R (2023) Growth behavior, biomass composition and fatty acid methyl esters (FAMEs) production potential of *Chlamydomonas reinhardtii* and *Chlorella vulgaris* cultures. Marine Drugs 21(8):450
18. Mathivanan K et al (2023) Bubble column photobioreactor (BCPR) for cultivating microalgae and microalgal consortium (Co-CC) with additional CO_2 supply for enhancing biomass, lipid, and preferable fatty acids production. Environ Res 238:117284, 1 dez 2023
19. Mohamadnia S, Tavakoli O, Faramarzi MA (2022) Production of fucoxanthin from the microalga *Tisochrysis lutea* in the bubble column photobioreactor applying mass transfer coefficient. J Biotechnol 348:47–54, 20 Mar 2022
20. Morais E et al (2018) Evaluation of CO_2 biofixation and biodiesel production by spirulina (Arthospira) cultivated in air-lift photobioreactor. Braz Arch Biol Technol 61:18161339
21. Morillas-España A et al (2021) Year-long evaluation of microalgae production in wastewater using pilot-scale raceway photobioreactors: assessment of biomass productivity and nutrient recovery capacity. Algal Res 60:102500, 1 dez 2021
22. Nazarpour M et al (2022) Optimization of biohydrogen production from microalgae by response surface methodology (RSM). Energy 253:124059, 15 ago 2022

23. Penloglou G, Pavlou A, Kiparissides C (2024) Recent advancements in photo-bioreactors for microalgae cultivation: a brief overview. Processes 12(6):1104
24. Romero-Villegas GI et al (2018) Utilization of centrate for the outdoor production of marine microalgae at the pilot-scale in raceway photobioreactors. J Environ Manage 228:506–516, 15 dez 2018
25. SILVA G et al (2023) Cultivation of microalgae *Chlorella vulgaris* in open reactor for bioethanol production. Phycology 3(2):325–336, 12 Jun 2023
26. Sawant SS et al (2018) Computational and experimental studies of high depth algal raceway pond photo-bioreactor. Renew Energy 118:152–159, 1 abr 2018
27. Serna-García R et al (2020) Co-digestion of harvested microalgae and primary sludge in a mesophilic anaerobic membrane bioreactor (AnMBR): methane potential and microbial diversity. Bioresource Technol 298:122521, 1 fev 2020
28. Seyed Hosseini N, Shang H, Scott JA (2018) Optimization of microalgae-sourced lipids production for biodiesel in a top-lit gas-lift bioreactor using response surface methodology. Energy 146:47–56, 1 Mar 2018
29. Tan JS, Lee SY, Chew KW, Lam MK, Lim JW, Ho S-H, Chen W-H, Show PL (2020) A review on microalgae cultivation and harvesting, and their biomass extraction processing using ionic liquids. Bioengineered 11(1):116–129
30. Vargas-Rodríguez JE et al (2024) Simultaneous methane production and atmospheric carbon fixation during nutrient recycling from yellow wastewater in a continuously fed PBR-UASB system. Environ Technol Innov 33:103458, 1 fev 2024
31. Velmozhina K et al (2023) Production of biohydrogen from microalgae biomass after wastewater treatment and air purification from CO_2. Processes 11(10):2978, 14 out 2023

Detoxification Strategies to Remove Inhibitors from Lignocellulose Hydrolysates: Challenges and New Prospects

Sarha Lucia Murillo-Franco, Carla V. dos Santos de Vechi, Lara Maria Biancheti, Thais S. Milessi, Priscila V. Arruda, Erica Daniely Santos, Debora D. Virginio da Silva, and Kelly Johana Dussán

Abstract The growing global concern for environmental preservation and the demand for sustainable technologies have stimulated the development of biotechnological processes that enable the production of energy and chemicals in an environmentally responsible manner. Among these, fermentation from lignocellulosic biomass has gained prominence as a biological alternative to achieve these goals. However, the conversion of lignocellulosic feedstocks presents significant challenges due to the generation of microbial inhibitors (phenolic compounds, furans, and organic acids) during the hydrolysis pretreatment. These compounds compromise

S. L. Murillo-Franco · K. J. Dussán (✉)
Institute of Chemical, Department of Chemical Engineering, São Paulo State University (Unesp), Araraquara, Brazil
e-mail: kelly.medina@unesp.br

Bioenergy Research Institute (IPBEN), São Paulo State University (Unesp), Araraquara, Brazil

C. V. dos Santos de Vechi · D. D. V. da Silva
Institute of Chemical, Department of Biochemistry and Organic Chemistry, São Paulo State University (Unesp), Araraquara, Brazil

L. M. Biancheti
Institute of Biosciences, Humanities and Exact Sciences, University of São Paulo (UNESP), São José Do Rio Preto, São Paulo, Brazil

T. S. Milessi
Department of Chemical Engineering, Federal University of São Carlos, São Carlos, Brazil

T. S. Milessi · E. D. Santos
Graduate Program of Chemical Engineering, Federal University of São Carlos (PPGEQ-UFSCar), São Carlos, SP, Brazil

P. V. Arruda
Department of Bioprocess Engineering and Biotechnology - COEBB/TD, Federal University of Technology of Paraná, Toledo, PR, Brazil

K. J. Dussán
Center for Monitoring and Research of the Quality of Fuels, Biofuels, Crude Oil, and Derivatives (CEMPEQC), Institute of Chemistry, São Paulo State University (Unesp), Araraquara, Brazil

© The Author(s), under exclusive license to Springer Nature Singapore Pte Ltd. 2025
A. K. Chandel and S. Sundaramurthy (eds.), *Biorefineries in Circular Bioeconomy*, Green Energy and Technology, https://doi.org/10.1007/978-981-95-3913-0_10

microbial viability and fermentation efficiency, making detoxification a necessary step. This chapter addresses this issue by first reviewing the physical, chemical, and enzymatic pretreatment methods that lead to the formation of inhibitors, characterizing their behavior in the fermentation systems. Then it presents a comprehensive overview of available detoxification strategies, including chemical, biological, and physical methods, following for the potential reactor configurations. Furthermore, the chapter explores sustainability aspects by examining economic and environmental assessments associated with detoxification processes, including life-cycle analysis (LCA) studies that quantify their impact on resources. Finalizing with the technical challenges and new prospects. By integrating these perspectives, the chapter highlights the importance of selecting appropriate detoxification strategies based not only on technical performance but also on their economic and environmental feasibility. It concludes with a discussion of the remaining overall challenges and future perspectives for improving detoxification technologies and their role in advancing sustainable biorefinery processes.

Keywords Detoxification · Inhibitors · Fermentation · Sustainable bioprocesses · Bioeconomy

1 Introduction

The sustainable use of resources is a crucial topic in the modern world due to concerns about resource scarcity and environmental degradation [1]. In this context, the bioeconomy has emerged as an economic strategy with the potential to reduce greenhouse gas emissions and promote economic growth through the utilization of biological resources [2]. To facilitate this economic model, biorefineries have been identified as the pathway to achieving it, as they contribute to resource efficiency by extracting maximum value from biomass while minimizing environmental impact [3]. The concept of biorefinery was introduced to describe an efficient approach to producing high-value-added products from various biomass sources, such as lignocellulosic biomass, algae, food waste, microbially treated residues, and manure, by implementing relevant conversion technologies [4].

These technologies are classified into three categories: chemical, thermochemical, and biochemical processes. The biochemical route is further divided into digestion and fermentation. This last topic enables the production of a wide range of bioproducts derived from biomass by employing microorganisms. Through this process, various valuable products, including biofuels and organic acids, can be obtained [5]. However, despite the potential to obtain these high-value products, issues remain in the fermentation process. Prior to fermentation, the biomass must be converted into a sugar-rich solution, which requires a pretreatment step carried out by a hydrolysis process. This stage is often challenging due to the formation of certain compounds that act as inhibitors of the fermentation process, preventing microbial growth and the production of the target products [6].

This problem has been studied extensively over the years, and detoxification has been the strategy for overcoming it [7]. Nevertheless, technological advances continue to drive new detoxification approaches, presenting ongoing challenges and variations that will be addressed in this chapter.

Accordingly, this chapter begins with an explanation of hydrolysis techniques, detailing the various inhibitors encountered in the process. It then discusses existing detoxification alternatives, potential reactors and designs to facilitate detoxification, and an analysis of these strategies in economic and environmental terms as a contribution to sustainability. The chapter ends with an overview of the remaining technical challenges and new perspectives, followed by final conclusions.

2 Hydrolysis: Methods and Inhibitors

Hydrolysis is an efficient way to convert complex organic matter into simpler compounds to produce biofuels and bioproducts. The process involves breaking chemical bonds using acid, base, or enzymatic agents. The main objectives are to eliminate lignin, improve sugar formation, and increase the porosity of the material [8]. In this sense, it has been widely used in the conversion of lignocellulosic biomass, composed of cellulose, hemicellulose, and lignin, into fermentable sugars used in bioethanol production [9]. Furthermore, the controlled degradation of polysaccharides enables the production of cellulose microfibers and nanofibers, which have promising applications in the packaging, cosmetics, and composite materials industries, promoting the development of sustainable alternatives [10].

Hydrolysis efficiency depends directly on the type of substrate used, the applied reaction conditions, and the desired final product. Factors such as the choice of agent, temperature, reaction time, and presence of inhibitors can significantly influence the yield of the process. Thus, optimizing hydrolysis conditions is crucial to ensure its technical and economic viability [11]. Therefore, selecting the most suitable hydrolysis method, whether chemical, physical, or enzymatic—requires a comprehensive understanding of these parameters to maximize efficiency and minimize inhibitory effects. The following sections describe these hydrolysis techniques in detail, highlighting their mechanisms, advantages, and limitations.

2.1 Hydrolysis Techniques

Hydrolysis is a chemical reaction in which a water molecule breaks bonds in another substance. However, complete hydrolysis of many substances is difficult to achieve in the presence of water alone, as the reaction may be slow or inefficient. To increase its speed and efficiency, it is common to use controlled conditions, such as increased temperature and pressure, along with the addition of acids or bases, which provide H^+ or OH^- ions to facilitate the breaking of chemical bonds. In addition, specific enzymes can be employed as biological catalysts, accelerating the hydrolysis process and increasing its selectivity [12]. Furthermore, physical hydrolysis, which involves mechanical or thermal methods, can aid in the disaggregation of substances without necessarily altering their chemical composition, offering a valuable approach in certain applications [13].

2.1.1 Chemical Hydrolysis

When reagents are added to the reaction, hydrolysis is referred to as chemical hydrolysis. The use of acids and bases aids in breaking the glycosidic bonds of polysaccharides present in biomass.

Acid Hydrolysis

In acid hydrolysis, the molecule is decomposed by the action of water in an acidic medium, which provides cations for the reaction, such as hydronium (H_3O^+). This is one of the most traditional approaches for converting lignocellulosic biomass into fermentable sugars [9]. This method uses strong acids, such as hydrochloric and sulfuric acids, and can be performed in concentrating and diluted conditions.

- **Concentrated Acid Hydrolysis**: Use of strong acids in high concentrations and occurs at moderate temperatures (30–50 °C). It yields a high amount of sugar, but acid consumption is also extremely high. Additionally, this technique is effective in solubilizing cellulose and hemicellulose but requires subsequent dilution and neutralization steps, posing environmental risks and equipment corrosion [14].
- **Diluted Acid Hydrolysis**: Uses acids in lower concentrations and occurs at elevated temperatures (120–210 °C) and elevated pressure. Although less corrosive and safer than concentrated acid hydrolysis, this approach can lead to the formation of inhibitory by-products, such as furfural and hydroxymethylfurfural (HMF), which hinder subsequent fermentation processes. It generates a high yield of hemicellulose but little cellulose, and consequently, low sugar yields in material with higher cellulose content [14].

Alkaline Hydrolysis

In alkaline hydrolysis, the molecule is decomposed by the action of water in an alkaline medium, which provides hydroxide anions (OH^-) for the reaction. This process is widely used for the breakdown of esters and amides, and in pretreatment to enhance cellulose digestion in enzymatic hydrolysis. Common reagents include strong bases, such as sodium hydroxide (NaOH) and potassium hydroxide (KOH), occurring under different operational conditions depending on the substrate and the reaction's purpose [15].

Alkaline hydrolysis can also occur at both high and low concentrations. When performed with dilute concentration, the reaction is slower and more controlled, preserving the structure of the original material to a greater extent, being preferentially used for partial hydrolysis and saponification. On the other hand, when the base

Table 1 Hydrolytic enzymes, substrates, and products [18]

Enzyme	Substrate	Product
α-Amylase	Starch, glycogen	Maltose, dextrin, oligosaccharides
β-Amylase	Starch, glycogen	Maltose
Glucoamylase	Starch, maltose, dextrin	Glucose
Cellulase	Cellulose	Glucose
Hemicellulase	Hemicellulose	Pentose (xylose, arabinose)
Lactase	Lactose	Glucose + lactate
Maltase	Maltose	Glucose
Invertase	Sacarose	Glucose + fructose

is concentrated, the reaction is more aggressive and can break additional bonds. It is used for protein digestion and for the complete hydrolysis of polysaccharides [16].

2.1.2 Enzymatic Hydrolysis

In enzymatic hydrolysis, enzymes are used to reduce the activation energy required to initiate a reaction, acting as biological catalysts. The use of enzymes allows for high selectivity and efficiency while maintaining mild conditions of temperature, pH, and pressure, making it an efficient option. Furthermore, by choosing the appropriate enzyme, the reaction can be controlled, preventing the formation of undesirable by-products [17]. Also, considering that the enzymes are selective catalysts, it is important to know the material that is interested in hydrolase. In this sense, Table 1 presents different hydrolytic enzymes, and the substrates or target material that they act upon, and the products resulting from these reactions.

The efficiency of enzymatic hydrolysis is linked to the accessibility of the substrates and potential interferences in the process, such as the presence of residual lignin and product inhibition. Lignin, for example, can interact with enzymes, reducing their activity. Additionally, molecules resulting from hydrolysis, such as glucose and xylose, can function as inhibitors, hindering the continuation of the reaction. To overcome these challenges, some approaches include the development of modified enzymes with lower affinity for lignin and the optimization of process conditions [17, 19].

2.1.3 Physical Hydrolysis

Physical hydrolysis is less commonly used than chemical and enzymatic hydrolysis due to its lower selectivity, high-energy consumption, and reduced efficiency in certain applications. It is typically employed as a complementary method to other techniques. This technique involves breaking bonds through elevated temperatures, pressure, and ultrasound in an aqueous medium. Thermal hydrolysis relies on high temperature and pressure to weaken chemical bonds, with water penetrating the material and reacting with specific bonds to fragment the molecule. Pressure can further enhance the reactivity of water, enabling bond breakage at lower temperatures. Sono-chemical hydrolysis, on the other hand, uses high-frequency ultrasonic waves to induce acoustic cavitation. These waves generate bubbles that expand and collapse, increasing temperature and pressure, which in turn boosts the reactivity of water and facilitates bond breaking [13].

2.1.4 Comparison of Hydrolysis Methods

Hydrolysis is an important process responsible for the breaking of bonds and the decomposition of molecules in an aqueous medium. The variations of hydrolysis are classified according to the applied conditions. The choice of hydrolysis type depends on the objective of the experiment, the available equipment, and the molecules to be broken down. In this sense, Table 2 presents a comparison between the methods, with the respective advantages and disadvantages, considering the information mentioned previously, as well as the perceived sustainability of each methodology.

2.2 *Inhibitors*

During the hydrolysis process the breakdown of bonds in the biomass generates smaller molecules, which can often interfere with the reaction itself and can generate the degradation of the catalysts. The hydrolysis methods play a significant role in

Table 2 Comparison between hydrolysis methods (chemical, physical, and enzymatic)

Type	Reagents	Conditions	Advantages	Disadvantages	Sustainability
Chemical	Strong acids or bases	pH, pressure, and temperature vary with dilution	High efficiency and speed vary with dilution	Generating toxic by-products, depending on the dilution, may corrode equipment	Low sustainability due to the use of strong reagents
Physical	Temperature, pressure, and ultrasonic waves	Elevated temperatures and pressure	Fast, no chemical reagents	High-energy consumption	Low sustainability due to high-energy consumption
Enzymatic	Enzymes	Moderate temperature, pressure, and pH	Specific, less aggressive	Excessive cost, products themselves may inhibit enzyme activity	Highly sustainable, uses biological catalysts

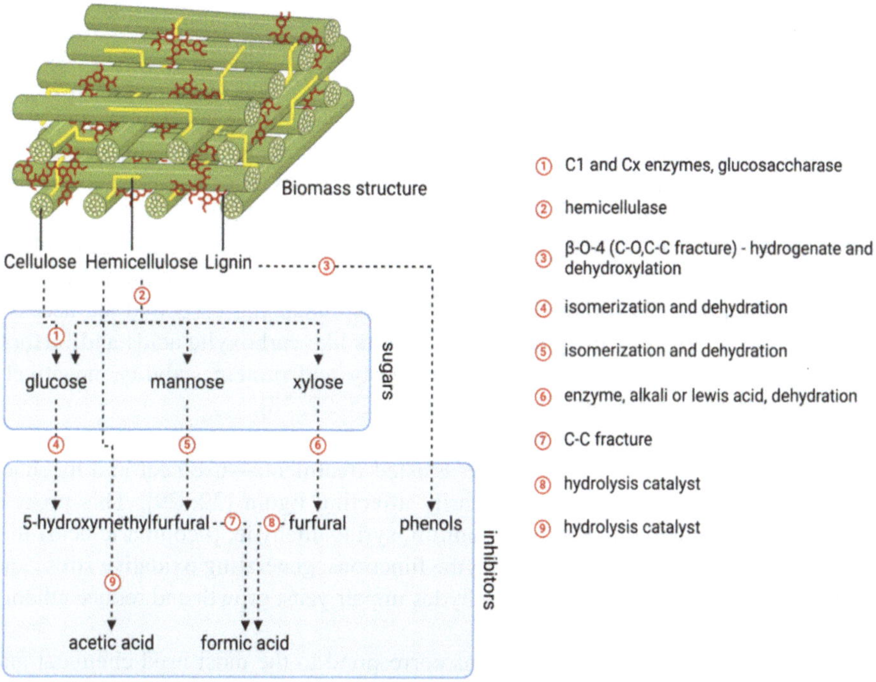

Fig. 1 Formation of inhibitors from pretreatment hydrolysate. Adapted from Guo et al. [20]

breaking down complex biomass structures. However, each method presents specific challenges, particularly in inhibitory compounds' formation that affect downstream microbial fermentation processes. Figure 1 presents the possible inhibitors that can appear in a hydrolysis process for a typical biomass structure.

2.2.1 Inhibitors from Chemical and Physical Hydrolysis

Chemical hydrolysis leads to the degradation of sugars and the formation of sugar-derived inhibitors such as furfural and 5-hydroxymethylfurfural (HMF) [20]. In contrast, physical hydrolysis, including mechanical and thermal methods, results in the release of lignin-derived phenolic compounds, which also pose toxicity issues for microbial systems [21, 22].

Chemical hydrolysis primarily employs acid catalysts like sulfuric or hydrochloric acid to break down cellulose and hemicellulose into fermentable sugars. However, the harsh conditions often degrade these sugars [23]. During acid hydrolysis, pentose sugars (e.g., xylose, arabinose) dehydrate into furfural, while hexose sugars (e.g., glucose, mannose) convert into HMF (Fülöp 2024). These compounds inhibit microbial metabolism by disrupting glycolytic enzymes and damaging cellular membranes, significantly hindering bioethanol production by reducing cell viability, prolonging lag phases, and lowering ethanol yields [20, 24]. Furfural and their derivates impacts microbial cells through multiple mechanisms, including oxidative stress via reactive oxygen species (ROS), which damage proteins, lipids, and DNA [20, 25]. It also disrupts energy metabolism by inhibiting key enzymes such as alcohol and aldehyde dehydrogenases, leading to impaired NADH regeneration and redox imbalance. Additionally, furfural integrates into cellular membranes, altering permeability and nutrient transport, further reducing microbial growth rates [26].

In addition to acid hydrolysis, alkaline treatments also play a significant role in biomass processing. Alkaline pretreatment employs bases such as sodium hydroxide or ammonia, primarily targets lignin disruption, enhancing carbohydrate accessibility. However, this process generates inhibitors like carboxylic acids and furfural derivatives, which can alter membrane integrity and protein stability, negatively affecting microbial fermentation performance [25, 27, 28].

On the other hand, physical hydrolysis methods—such as mechanical comminution, steam explosion, and microwave-assisted treatments—use heat and mechanical forces to degrade biomass, primarily affecting lignin [22, 29]. This process releases phenolic compounds (e.g., vanillin, syringaldehyde, p-coumaric acid) that disrupt microbial membranes and enzyme functions, generating oxidative stress and inhibiting fermentation. Phenolic aldehydes impair yeast growth and reduce ethanol productivity [30, 31].

As indicated above, the descriptions correspond to the most used chemical and physical hydrolysis processes, but other variations also have a significant impact on inhibitor generation. Table 3 provides a comprehensive overview of the various types of physicochemical pretreatments and the resulting inhibitors, along with their sources and mechanisms of action. In summary, chemical methods, such as acid hydrolysis, oxidation, and alkaline pretreatment, primarily generate sugar and lignin degradation products, including furans and organic acids, which affect cell viability and fermentation efficiency. Similarly, physical treatments yield lignin-derived phenolic compounds, which hinder enzyme function and disrupt the integrity of microbial membranes. Furthermore, hybrid physicochemical approaches, such as steam explosion and solvents, combine these effects, resulting in a broader spectrum of inhibitory compounds.

2.2.2 Inhibitors from Enzymatic Hydrolysis

Enzymatic hydrolysis of lignocellulosic biomass is essential for converting plant materials into fermentable sugars for biofuel and biochemical production. Cellulases and hemicellulases break down cellulose and hemicellulose into monomeric sugars but incomplete hydrolysis leads to the accumulation of oligomeric inhibitors, particularly xylooligomers and cellooligomers, which interfere with enzymatic efficiency and microbial fermentation. These inhibitors bind to enzyme active sites, reducing hydrolysis rates and overall sugar yield [41, 42].

Oligomeric inhibitors, consisting of short-chain sugar polymers or lignin-derived compounds, persist in hydrolysates after pretreatment and disrupt microbial growth,

Table 3 Types of pretreatments, the following inhibitors produced, their sources and action mechanism

Type	Pretreatment	Inhibitor	Source	Inhibitor's action mechanism	Reference
Chemical	Hydrothermal pretreatment	Phenols Monosaccharides Furans Carboxylic acids	Lignin Hemicellulose Cellulose	Highly toxic to fermenting microorganisms	[25, 32]
	Acid hydrolysis	Acetic acid Furfural HMF Pseudo-lignin	Cellulose Hemicellulose	Inhibition of cellulose activity	[28]
	Alkali pretreatment	Furan Acetic acid Formic acid	Hemicellulose	Disrupt DNA, proteins, and membrane structure of yeast cells	[28, 33]
	Oxidation and ozonation	Acetaldehyde Phenols	Hemicellulose	Inhibition of yeast cells	[11]
	Treatment with solvents	Weak acid furan	Hemicellulose	Inhibits enzyme and microbial activity	[34, 35]
	Ammonia fiber blasting	Phenols Furans	Hemicellulose Lignin	Inhibits enzyme and microbial activity	[25, 27]
Physical	Mechanical grinding	Phenols Furfural	Hemicellulose Lignin	Inhibition of enzyme activity	[36, 37]
	Acoustic degradation method (ultrasonic)	HMF	Hemicellulose	Inhibition of enzyme activity	[38, 39]
Physicochemical	Stem blasting	Organic acid Phenolic furan	Hemicellulose	Inhibition of cellulose and enzyme activity	[27, 40]

metabolism, and biofuel production [43, 44]. Unlike smaller inhibitors like furfural and HMF, oligomers primarily affect microbial cells by increasing hydrolysate viscosity and osmotic pressure, creating an unfavorable environment that limits nutrient availability and diffusion [45]. High osmotic pressure forces microbial cells to expend energy maintaining homeostasis, diverting resources from growth and fermentation [46]. Increased viscosity also hampers efficient mixing and mass transfer.

Oligomers, particularly lignin-derived ones, can integrate into microbial membranes, disrupting fluidity, permeability, and transport protein function [47]. This instability affects nutrient uptake and ion exchange while triggering stress responses that lead to reactive oxygen species (ROS) production, causing oxidative damage to proteins, lipids, and nucleic acids [48, 49]. Furthermore, sugar oligomers act as competitive inhibitors for carbohydrate-processing enzymes, reducing sugar metabolism efficiency, slowing fermentation, and causing metabolic imbalances that further inhibit microbial growth [50, 51]. In this sense, mitigating the accumulation of oligomeric inhibitors is crucial for improving enzymatic hydrolysis and fermentation performance.

2.2.3 Metal Ions as Inhibitors in the Hydrolysis and Fermentation Processes

The presence of heavy metal ions such as Fe, Cr, Ni, and Cu can originate from the corrosion of hydrolysis equipment or contaminants from milling equipment [7]. Chemical hydrolysis is significantly affected by metal ions and heavy metals in biomass processing [52]. In acid hydrolysis, metal ions, such as Fe^{3+} and Al^{3+}, can act as catalysts, stabilizing reactive intermediates and enhancing the breakdown of cellulose and hemicellulose into fermentable sugars. However, heavy metals like Pb^{2+} and Hg^{2+} may form insoluble precipitates with the acid catalysts, reducing efficiency and necessitating additional detoxification steps [28, 53]. In alkaline hydrolysis, metal ions such as Ca^{2+} and Mg^{2+} facilitate lignin removal by forming complexes with phenolic compounds, thereby improving cellulose accessibility [54, 55]. However, excessive heavy metals in the system can interfere with this process by disrupting solubilization and leading to undesired side reactions [21, 56].

Physical hydrolysis is also indirectly influenced by metal ions and heavy metals through their effects on reaction conditions and equipment interactions. In steam explosion, high temperatures and pressures can lead to the oxidation of metal ions, such as Fe^{2+} to Fe^{3+}, which may enhance biomass breakdown while promoting the formation of inhibitory compounds like furfural and HMF, already commented above [57, 58]. These heavy metals may also accelerate oxidative stress in microbial fermentation if not properly removed before downstream processing. In ultrasonication-assisted hydrolysis, metal ions such as Ti^{4+} can enhance cavitation effects, increasing biomass disintegration efficiency. However, heavy metals can also absorb ultrasonic energy, reducing the effectiveness of the process and leading to unexpected degradation reactions [59, 60].

Enzymes are particularly sensitive to metal ions and heavy metals; these compounds can significantly influence their stability and activity [52]. Some metal ions, including Mg^{2+}, Mn^{2+}, and Ca^{2+}, are essential cofactors that help maintain enzyme structure and facilitate substrate binding. For example, Mg^{2+} stabilizes ATP-dependent cellulases, while Mn^{2+} is critical for the functionality of certain lignin-degrading enzymes. Conversely, heavy metals such as Hg^{2+}, Pb^{2+}, and Cd^{2+} are potent inhibitors, as they bind to sulfhydryl groups in enzymes, disrupting catalytic function. Additionally, Cu^{2+} and Fe^{3+} contribute to oxidative stress, leading to enzyme degradation and reduced hydrolysis efficiency [61, 62].

3 Detoxification Strategies

Considering the wide range of inhibitors generated after biomass pretreatment and hydrolysis, a detoxification step must be performed to remove the toxic inhibitors and increase the performance of microorganisms and enzymes during biomass valorization [20]. In general, the ideal detoxification method must efficiently remove the inhibitor compounds without loss of fermentable sugars, be environmentally friendly and economically feasible [63]. The detoxification methods can be classified as physical, chemical and biological, which will be detailed in the following sections.

3.1 Physical Methods

Physical methods of detoxification involve the removal of inhibitory or toxic compounds without chemically modifying the remaining components, using techniques such as extraction and evaporation to purify the medium [64].

3.1.1 Evaporation

This methodology is also known as vacuum evaporation: it permits the reduction oof volatile compounds such as acetic acid, furfural, and vanillin, contributing to the improvement of fermentation conditions. For example, Kubisch and Ochsenreither [65] evaluated the impact of pressure on rotary evaporation for the lignocellulose hydrolysate detoxification. The efficiency of removal of inhibitory compounds was directly influenced by the applied pressure, with reductions of up to 79.4% in furfural at 400 mbar and complete removal of guaiacol at 300 mbar. On the other hand, Llano et al. [66] applied this technique to detoxify residual sulfite liquor from a pulp mill, achieving an 8.6% reduction in total inhibitors. However, this removal rate is relatively low, likely due to the persistence of nonvolatile toxic substances. In this sense, this approach has limitations, as it is not effective in removing some phenolic compounds and may increase the concentration of nonvolatile toxic substances, such as extractives, which negatively impact the fermentation process as they remain in the medium, requiring additional treatment steps for complete detoxification [67, 68].

3.1.2 Steam Stripping

Gas stripping is an attractive method for large-scale detoxification due to its relative simplicity and ability to remove volatile compounds from hydrolysates in clean forms (not contaminated with involatile material), and the possibility of using a low-cost stripping gas (CO_2) [69]. Zhu et al. [70] analyzed the detoxification of acid hydrolysates from corn bagasse by comparing vacuum evaporation with steam stripping. The results obtained indicated that the removal of inhibitors by steam stripping was superior, obtaining a removal of 80.83% of acetic acid, 58.79% of formic acid and a complete removal of furfural, in addition to a smaller number of degraded sugars. In the same way, Artifon et al. [71] performed the detoxification of sugarcane bagasse hydrolysate by air stripping and achieved 60% removal of acetic acid.

3.1.3 Extraction

Extraction is based on the differences in solubility between two immiscible phases, allowing the selective removal of compounds from the medium. The ideal solvent for extraction must have an adequate partition coefficient, be immiscible with the original solution and have an appropriate boiling point to facilitate evaporation, present selectivity and be low cost and safe [72]. Liquid–liquid extraction, although efficient in removing inhibitory compounds, presents challenges such as complexity in phase separation, emulsification problems and scalability restrictions, making its application in large-scale industrial processes difficult [68]. Roque et al. [64] investigated the detoxification of hemicellulosic hydrolysates from lignocellulosic biomass to optimize pentose fermentation in ethanol production. The study performed a liquid–liquid extraction with different solvents (1-butanol, isobutyl acetate, and methyl isobutyl ketone [MIBK]). 1-butanol demonstrated the greatest efficiency in the removal of inhibitory compounds, standing out in the extraction of acetic acid, with a removal 37.7% higher than that obtained with MIBK. Isobutyl acetate, in turn, showed low efficiency in this process due to its hydrolysis in an acidic medium. MIBK presented an intermediate removal, without surpassing the performance of 1-butanol. In the extraction of phenolic compounds, 1-butanol removed 82.1%, while MIBK and isobutyl acetate reached 53.3% and 45.1%, respectively.

3.1.4 Ionic Liquid

Ionic liquid detoxification methods use solvents composed of cations and anions, such as [EMIM]Ac, to recover inhibitors from biomass hydrolysates. Xavier et al. [73] applied a two-phase system based on cholinium ionic liquid and Tween 80 for the extraction of ferulic acid and p-coumaric acid from rice husk alkaline hydrolysate and achieved yields greater than 90% for both acids. However, challenges such as high cost, high viscosity and difficulties in recycling solvents limit the application of ionic liquids for detoxification at large-scale [74].

3.1.5 Plasma Treatment

Plasma, considered the fourth state of matter, is made up of high-energy particles, including ions, electrons, molecules, and free radicals. Gases such as air, oxygen, argon, and nitrogen can be converted into plasma when subjected to high-energy charges. Atmospheric Cold Plasma has applications in several sectors, such as semiconductors, food processing, medical sterilization, and waste degradation. In addition, it demonstrates great potential to optimize the use of substrates and increase efficiency in biomass production, without the need for the addition of chemicals or enzymes [75, 76]. An example is the study made by Santoso et al. [77], which investigated the detoxification of pineapple peel hydrolysate using atmospheric cold plasma, by assessing the reduction of toxic compounds, the preservation of fermentable sugars, and the production of bacterial cellulose. Treatment with an argon plasma reactor (R1) at 200 W for 15 min achieved a 74.06% reduction in furfural, 51.38% in HMF, and 21.81% in formic acid. In contrast, air plasma (R2) at 600 W removed 45.05%, 32.59%, and 60.41% of these compounds, respectively. Although reactor R1 was more effective in eliminating inhibitors, it also resulted in a greater loss of fermentable sugars compared to reactor R2. Additionally, plasma treatment presents the drawback of potentially generating undesirable gaseous compounds, such as CO, which may negatively impact fermentation processes. Also, the operational costs of this method can be higher compared to simpler alternatives, requiring an economic evaluation to make its industrial application viable [78].

3.1.6 Superficial Fluid Extraction (SFE)

Supercritical fluid extraction (SFE) corresponds to an efficient alternative to the use of organic solvents in biomass processing. Its main advantages include the reduction or elimination of organic solvents and the simplification of purification processes, since the solvent can be easily removed by expansion to atmospheric conditions. In addition, SFE allows adjustments to obtain highly concentrated extracts, ensuring superior selectivity compared to traditional extraction methods. In this sense, Persson et al. [79] applied Supercritical fluid extraction with carbon dioxide was applied for the detoxification of spruce hydrolysate, resulting in increased yeast fermentability and reduced concentrations of phenolics and furan derivatives. Despite having low impacts on the sugars in the environment, the method has a high operational cost, given the requirement for specific equipment for the severe temperature and pressure conditions, as well as the complexity of process optimization due to the variety of variables involved [80].

3.1.7 Molecular Sieve

Molecular sieves are porous materials made of hydrated aluminosilicates, with high surface area and regular pores that allow the selective adsorption of molecules of specific sizes and are widely used in the purification of gases and liquids [81]. Conventional molecular sieves are composed of silicon, aluminum, and oxygen, forming a three-dimensional network of tetrahedrons that determines their performance [82]. For example, Wang et al. [83] used a carbon molecular sieve to separate levulinic and formic acid by continuous fractional adsorption. Although they can confine active species and serve as good catalytic supports, challenges such as sintering, carbon deposition, and sulfur poisoning compromise their stability under severe industrial conditions, restricting their large-scale application [82].

3.2 Chemical Methods

The primary chemical detoxification methods used in the hydrolysate detoxification process involve the addition of reducing agents and pH adjustment, as well as alkali treatment, adsorption, nanomaterials, membranes, ion exchange, and activated charcoal, as described in the treatments below.

3.2.1 Alkali Treatment

Alkali treatment is widely recognized as an effective approach for the pretreatment and detoxification of lignocellulosic biomass, primarily aimed at enhancing biofuel production efficiency [84]. Persson et al. [85] evaluated several alkali-based detoxification methods, including treatments with calcium hydroxide, sodium hydroxide, and ammonia. According to these authors, calcium hydroxide and ammonia were particularly effective, reducing furaldehyde concentrations by up to 40%. This reduction was significantly greater than that achieved with sodium hydroxide treatment. Importantly, the study demonstrated that these alkaline treatments could substantially improve the fermentability of hydrolysates when using *Saccharomyces cerevisiae*, resulting in a five-fold increase in ethanol productivity compared to nondetoxified controls. The underlying mechanism of alkaline detoxification involves raising the pH to 9–10, followed by reacidification, which induces several chemical transformations. This pH swing effectively targets aldehyde and ketone inhibitors, including furfural and HMF, through chemical conversion or precipitation. The process not only removes these inhibitors but also enhances the overall fermentation performance of microbial strains [86].

Further evidence supporting the efficacy of alkaline detoxification comes from studies with *Mucor indicus*. Karimi et al. [87] investigated the impact of calcium hydroxide treatment on rice straw hydrolysates non-detoxified and detoxified and observed an increase in ethanol yield from 0.35 $g \cdot g^{-1}$ to 0.38 $g \cdot g^{-1}$, respectively. While this improvement might appear modest, it was achieved without compromising fermentation productivity, which remained stable at 0.04 $g/dm^3/h^1$. More recently, Kordala et al. [88] evaluated different detoxification methods for Miscanthus giganteus hydrolysates. Their study compared water rinsing, activated carbon, calcium hydroxide (overliming), and ammonium hydroxide treatments. The results demonstrated that the optimal detoxification strategy was microorganism-dependent: overliming yielded the highest ethanol production with *Saccharomyces cerevisiae*, while activated carbon proved most effective for *Mucor rouxii*. Notably, the activated carbon treatment eliminated the need for water rinsing, thereby minimizing substrate loss. While alkaline hydrolysis offers significant benefits, it can also lead to the degradation of valuable sugars, particularly xylose [86, 89]. This effect underscores the importance of carefully optimizing process parameters, including pH adjustment, treatment duration, and temperature. The challenge lies in achieving a delicate balance between effective inhibitor removal and preservation of fermentable sugars.

3.2.2 Adsorption Process Using Activated Charcoal

Activated charcoal has shown promise in detoxifying lignocellulosic hydrolysates for bioproduct production. Studies have demonstrated its effectiveness in removing inhibitory compounds, particularly phenolics, from various biomass sources including corn cobs, sugarcane trash, spruce, and olive tree pruning residues [90, 91].

Notably, activated charcoal can be produced from agricultural wastes like brewer's spent grain lignin, offering a sustainable approach to hydrolysate detoxification. These carbons effectively remove metallic ions, phenolic compounds, and color from hydrolysates, with surface areas up to 692 m^2/g and pore volumes up to 0.453 cm^3/g [92]. Other studies have demonstrated its effectiveness in reducing inhibitors in lignocellulosic hydrolysates. For instance, Seo et al. [93] treated acid hydrolysate from corn hull with activated charcoal, achieving a 92.3% reduction in phenolic compounds. This detoxification led to a significant improvement in bioethanol production using *Pachysolen tannophilus*, with enhanced sugar consumption rates and increased ethanol yields.

Similarly, Artifon et al. [71] investigated the combined use of activated charcoal adsorption and dried air stripping for detoxifying hydrolyzed lignocellulosic substrates and the results indicated that this approach effectively removed volatile and non-volatile inhibitors, resulting in improved bioethanol production metrics.

The effectiveness of activated charcoal in detoxifying lignocellulosic hydrolysates depends on several factors such as pH, which influences the adsorption of inhibitors [7] dosage and contact time [94].

According to Jin et al. [95], the elimination of inhibitors increased along with the adsorption effectiveness when the adsorption duration was extended. Nevertheless, there was a negative correlation between the treatment temperature and the adsorption effectiveness. Reducing adsorption efficiency was the outcome of higher temperatures. Although activated carbon shows superior adsorption at low temperatures, the reaction's release of heat and subsequent temperature increase were most likely the cause. At 30 °C for 50 min, furfural and 5-HMF removal rates reached 88.45% and 88.92%, respectively. However, as the temperature increased, their elimination efficiency declined, with removal rates dropping to 85.47% and 86.46% at 50 °C, being lower than those achieved at 30 °C.

Additionally, the regeneration and reuse of activated charcoal for multiple detoxification cycles can significantly reduce operational costs, making it a cost-effective method for industrial applications [96]. These findings highlight the potential of activated charcoal detoxification in improving the efficiency of bioproduct production from lignocellulosic biomass.

3.2.3 Adsorption Process Using Ion-Exchange Polymeric Resins

Ion-exchange resins offer another effective strategy for detoxifying lignocellulosic hydrolysates by removing charged inhibitory compounds. Studies have shown that combining cation- and anion-exchange resins can remove up to 100% of furfurals and over 80% of other inhibitors like acetic acid, phenolics, and lignosulfonates [66].

In the same line, Nilvebrant et al. [97] evaluated the performance of different ion-exchange resins, including anion and cation exchangers, in treating spruce hydrolysate. The study demonstrated that anion-exchange resins were particularly effective, especially at higher pH levels, in removing both charged and uncharged inhibitors through mechanisms such as hydrophobic interactions. This treatment resulted in enhanced ethanol production during fermentation with *Saccharomyces cerevisiae.* Han et al. [98] demonstrated that anion-exchange resin effectively removed up to 80% of acetic acid while minimizing sugar loss to less than 10%. Furthermore, the integration of resin recycling for detoxification and cell recycling for catalysis significantly enhanced the cost-effectiveness and sustainability of xylonic acid production from lignocellulosic biomass.

Furthermore, in-situ application of anion-exchange resins during bioconversion has shown potential in creating self-balancing biosystems through reversible competitive adsorption between inhibitors, sugars, and products [99]. This approach facilitates the direct utilization of crude lignocellulosic hydrolysates for bioproduct synthesis.

3.2.4 Use of Nanomaterials

Nanomaterials appear as adsorbents for detoxifying lignocellulosic hydrolysates. For example, Fe(0) nanoparticles deposited on activated carbon demonstrated effective removal of acid-soluble lignin and furfural through integrated adsorption-oxidation [100]. In the same line, Adebule et al. [101] achieved high removal efficiencies for various inhibitors without sugar loss, primarily through surface adsorption and complexation mechanisms with Fe_3O_4 nanoparticles. Other nanomaterials, such as hybrid chitosan-chitin nanofibers modified with polyethylenimine and silanized with methyl triethoxysilane, enable efficient detoxification of lignocellulosic hydrolysates, achieving 85.1% HMF and 99.0% furfural removal, eliminating most phenolic inhibitors while retaining 100% of glucose and xylose [102]. Another example includes lignin nanoparticles produced from enzymatic hydrolysis residue via hydrotrope fractionation, which show superior detoxification ability compared to the original residue, significantly improving fermentability [103]. These nanomaterial-based detoxification methods not only enhance bioethanol production but also offer potential for recovering valuable chemicals from the inhibitors [104].

3.2.5 Membranes

Membrane-based processes have shown potential for detoxifying lignocellulosic hydrolysates, addressing challenges in bioethanol production [105–107]. Vacuum membrane distillation, coupled with adsorption, effectively concentrates sugars while removing inhibitors like furans and acetic acid, resulting in significantly improved ethanol yields [108]. In another study, Greer et al. [106] reported successful pervaporation using block copolymer membranes, effectively removing furfural (99%) and acetic acid (27%) and enabling the fermentation of previously unsuitable hydrolysates.

Membrane extraction with microporous polypropylene hollow fibers and organic extractants demonstrates rapid removal of sulfuric acid, furfural, and 5-hydroxymethyl furfural, leading to increased ethanol yields compared to alkaline detoxification methods [105].

These membrane-based detoxification methods offer advantages such as selective removal of inhibitors, preservation of fermentable sugars, and improved fermentation efficiency. Their efficiency depends on factors such as membrane material, transport mechanism, and operating conditions such as temperature, pH, pressure, and feeding concentration. However, challenges remain, including optimizing removal efficiencies for various inhibitors and ensuring the economic viability of these processes for large-scale bioethanol production.

3.3 Biological Methods

Biological detoxification is an environmentally friendly strategy to remove and neutralize inhibitory compounds using microorganisms and enzymes. This strategy has several advantages such as their simplified operation, the requirement of milder temperature and pH conditions, the less stringent processing conditions that minimize the need for chemical additives, the reduced incidence of undesirable side reactions, and greater energy efficiency [109]. In addition, generally biological detoxification is performed in similar conditions to sequential fermentation, facilitating the next steps of biomass processing. However, slower reaction kinetics, limited selectivity for certain inhibitors and the loss of fermentable sugars, are challenges that still limit their viability for applications in biorefineries [110, 111].

3.3.1 Use of Enzymes

Biological detoxification using enzymes consists of oxidation and subsequent polymerization of low-molecular-weight phenolic compounds, aiming at their conversion into less reactive and less toxic forms. This transformation is catalyzed by oxidative enzymes, such as laccases and peroxidases, which act by promoting the removal of electrons and the formation of more complex and less soluble polymeric structures [110]. Laccases are oxidative enzymes that act on phenolic and nonphenolic compounds, facilitating their degradation and conversion into less reactive forms, while peroxidases promote oxidation reactions using peroxides as electron donors, playing a fundamental role in the detoxification of organic and inorganic compounds present in the hydrolysate [111]. The oxidation of phenolic compounds results in the formation of insoluble oligomers or polymers, which can subsequently be removed by physical processes, such as sedimentation and filtration [68].

Schneider et al. [112] studied the production of laccase secreted by the fungus *Marasmiellus palmivorus*, for the degradation and detoxification of eucalyptus residues. The results obtained indicated a 31% reduction in the lignin content of eucalyptus wood, in addition to a 70% removal of phenolic inhibitors. De La Torre et al. [113] also use Laccase but from *Trametes villose* for the detoxification of wheat straw hydrolysate and achieved a removal of 71% phenolic compounds. Other successful enzymatic detoxifications of biomass hydrolysates are summarized in Table 4, where it is possible to observe that the remotion of phenolic compounds can achieve between 60 and 100% of phenolic compounds.

One challenge on enzymatic detoxification can be the high price of laccase and peroxidase (about 36 USD/kg and 12 USD/kg, respectively) [20]. For this reason, enzyme immobilization is a strategy that can overcome this issue, as it allows for the

Table 4 Enzymatic detoxification strategy for lignocellulosic hydrolysates

Hydrolysate	Biological agent	Results	References
Sugarcane bagasse	Enzyme cocktail with peroxidase and laccase	64% of phenolic compounds removal	[34]
Maple wood chip	Laccase	79% of phenolic compounds removal	[114]
Wheat straw	Lacasse from *Trametes vilosa*	71% of phenolic compounds removal	[113]
Wheat straw	Laccase from *Pycnoporous cinnabarinus*	Complete removal of formic acid	[115]
Sugarcane bagasse	Laccase from *Cyathusc stercoreus*	Removal of phenolic compounds	[116]
Sugarcane bagasse	Laccase from *T. versicolorm*	80% removal of phenolic compounds	[117]
Corn stover	Laccase from *Ganoderma lucidum*	84% removal of phenolic compounds	[118]

easy recovery and reuse of the immobilized enzyme in successive batches and can increase enzyme stability by providing structural rigidity to the enzyme molecules [119]. In this sense, Li et al. [120] performed the immobilization of laccase on magnetic particles, aiming at the degradation of four phenolic inhibitors commonly found in lignocellulosic hydrolysates. The immobilized laccase demonstrated superior performance in the removal of phenolic inhibitors, reaching rates of 9.8% for vanillin, 67.3% for syringaldehyde, 100% for ferulic acid, and 38.5% for p-coumaric acid, while the free form of the enzyme showed significantly lower removals of 3.9%, 15.2%, 50.2%, and 7.3%, respectively.

3.3.2 Microbial Detoxification

The use of microorganisms for biological detoxification has other advantages over the use of enzymes, because it eliminates the costs associated with the production and purification of enzymes, making the process potentially more economically feasible. In this strategy, microorganisms with high tolerance to the inhibitors can grow, assimilate, and convert inhibitory compounds into less toxic ones [109]. Microbial detoxification can also occur in situ, simultaneously with the fermentation process, and has been suggested as a strategy to simplify production, increase yields, and reduce overall process costs [121]. Furthermore, during microbial growth, unwanted consumption of sugars from the hydrolysate may happen, which is a critical aspect that must be considered in process optimization [122]. For example, in the case of xylose-rich hydrolysates, the yeast *S. cerevisiae* is an interesting choice, as it can assimilate inhibitors such as furfural, HFM, and aromatic aldehydes without consuming the xylose, preserving the fermentable sugars [123].

In the same line, a wide range of other different microorganisms have been applied on hydrolysate detoxification, such as *Coniochaeta ligniaria, Trichoderma reesei, Escherichia coli, Pseudomonas putida,* and *Issatchenkia occidentalis,* as is presented in Table 5.

3.3.3 Encapsulation of Microorganisms

Encapsulation of Microorganisms is a technique that can play an essential role in the detoxification of hydrolysates as the support can protect the cells from inhibitory compounds, increase the durability of microbial cultures, and facilitate the separation of cells from the medium [132]. An example of that is the case of the use Ca-alginate to encapsulate *S. cerevisiae* in the simultaneous detoxification and ethanol production from spruce forest residue hydrolysate wherein a 71% of conversion of HFM and complete conversion of furfural was observed [133]. However, there are limitations associated with this strategy, such as difficulties in nutrient transfer, possible formation of biofilms that compromise efficiency, high production costs, and challenges for large-scale implementation due to operational complexity [134]. An interesting alternative is promoting cell retention, thus allowing a high cell density,

Table 5 Microbial detoxification strategy for lignocellulosic hydrolysates

Hydrolysate	Biological agent	Results	References
Olive stones	*Saccharomyces ceverisiae*	Almost completely removed the inhibitory compounds (acetic acid, HMF, and furfural)	[124]
Miscanthus giganteus acid hydrolysate	*Cupriavidus basilensis* ATCC BAA-699	Complete depletion of all inhibitors, except ferulic acid and cinnamaldehyde after 12 h	[125]
Sugarcane bagasse	*Issatchenkia occidentalis* CCTCC M 206087	Reduction of 67% on syringaldehyde, 7% on ferulic acid, 62% on furfural, and 85% on HMF, with minimal sugar consumption	[126]
Sugarcane bagasse	*Lactobacillus plantarum* CCT 3751	85% and 99% decrease in HMF and furfural, respectively	[127]
Corncob	*Rhodococcus aetherivorans* N1	Complete removal of syringaldehyde and 80.5% furfural, 50.7% HMF, and 71.5% phenolic compounds	[128]
Corn stover	*Rhodopseudomonas palustris* CGA009	Complete removal of inhibitors	[129]
Corn stover	*Coniochaeta ligniaria* C8	Removed inhibitors in furans, phenolic, and aromatic and aliphatic acids and improved xylose utilization	[130]
Corn stover	*Pseudomonas putida* KT2440	Removal of acetate, levulinic acid, and aldehydes	[131]

which may improve the ability of microorganisms to convert inhibitory compounds into less inhibitory groups during the fermentation of lignocellulosic hydrolysate [135].

3.3.4 Microbial Adaptation

An alternative to mitigate the effects of inhibitory compounds on microorganism's metabolism involves the adaptive evolution of selected microorganisms in media containing inhibitory compounds, such as hydrolysates resulting from pretreatment. This approach has allowed the development of strains with high tolerance to aliphatic acids, furanic aldehydes, and other small organic molecules, contributing to the improvement of productivity and yield in fermentation processes [136]. Adaptive evolution is a strategy that induces the selection of microorganisms that are more tolerant to inhibitors through progressive exposure to these compounds, allowing the accumulation of beneficial mutations over generations. This process facilitates the development of improved strains, which present greater growth, yield, and productivity, even under adverse conditions [137].

For example, Chen et al. [138] conducted the adaptive evolution of the fungus *Paecilomyces variotii* to improve its efficiency in the biodetoxification of inhibitors present in pretreated lignocellulosic substrates. This process resulted in the evolved strains AC70 and ZW70, with ZW70 demonstrating a greater capacity to degrade inhibitory compounds, such as acetic acid, HMF, and furfural, compared to the parental strain. The detoxification of the hydrolysates by strain ZW70 led to the production of L-lactic acid at concentrations close to 110 g L^{-1}, representing an increase of more than 20% compared to the original strain. Almeida et al. [139] used adaptive evolution to select three strains of *Papiliotrema laurentii* with increased tolerance to acetic acid. The evolved strains exhibited a significant reduction in the lag phase and higher specific growth rates relative to the parental strain, especially in environments with high concentrations of acetic acid.

3.3.5 Genetic Modification

Genetic engineering enables the creation of recombinant microorganisms that mitigate inhibitors in lignocellulosic hydrolysates. Recombinant DNA technology improves lignocellulolytic enzymes, enabling synthetic biology systems for targeted applications. Gene fusion from different organisms produces chimeric enzymes optimized for detoxification, affecting growth rate, lag phase, cell viability, and productivity [110, 122]. However, for it to be carried out in a promising way, this method requires precise control of gene expression, detailed knowledge of tolerance and inhibitory mechanisms, and the complex interaction of the inhibitor with biochemical or genetic networks [137].

Several studies demonstrate that modified strains present greater tolerance to inhibitory compounds. Chang et al. [140] through multi-omics techniques overexpressed the stress-responsive gene ycfR in *E. coli* and achieved increased multi-inhibitory tolerance levels. Borchert et al. [141] screened tolerant mutants of *Pseudomonas putida* based on random barcoded transposon insertion sequencing (RB-TnSeq) and achieved greater tolerance to lignin-derived inhibitors.

Other genetic modifications to increase inhibitors' tolerance correspond to the expression of laccase into microorganisms. In this sense, Ji et al. [142] expressed the laccase gene (*lacA*) from *Trametes sp* AH28-2 in *S. cerevisiae* CEN.PK strains and the recombinant yeast showed improved resistance to vanillin, with a shorter lag phase and a vanillin conversion rate of 0.069 $g/g_{biomass}/h$.

3.4 Combined Methods

By integrating complementary techniques, the detoxification of hydrolysates can be improved by the synergistic effects between different methods. For example, Milessi et al. [143] performed the detoxification of sugarcane bagasse acid hydrolysate by a combination of boiling and overliming to pH 10.0 and observed high furfural and

HMF removal (94% and 83%, respectively); however, a small reduction in lignin-derived compounds (14%). Theiri et al. [144] combined with chemical flocculation with a bacterial co-culture of *Cupriavidus taiwanensis* and *Ureibacillus thermosphaericus* led to detoxification of hardwood hemicelluloses hydrolysate, enabling acetone-butanol-ethanol fermentation. The authors observed the combined flocculation and biodetoxification produced resulted in a final concentration of 1.4 g/L of phenolic compounds, which is nearly equal to the flocculated hydrolysate's minimum inhibitory threshold of 1.1 g/L. Furthermore, Mussatto et al. [7] explored the impact of pH adjustment and activated charcoal treatment on rice straw hemicellulosic hydrolysate. The study found that increasing the pH to 2.0, followed by activated charcoal treatment and subsequent pH adjustment to 6.5, selectively removed lignin degradation products. This detoxification process led to a xylitol yield of 0.72 g/g xylose, highlighting the potential for improved fermentation outcomes.

These studies highlight the potential of combining various detoxification methods to enhance the efficiency of lignocellulosic biomass conversion processes. However, combined/integrated detoxification techniques also present some challenges, such as ensuring compatibility among different methods, optimizing operational parameters of each step for high efficiency, and managing increased costs and complexity. Designing modular detoxification units that can be easily integrated or scaled up is of major importance to reach detoxification strategy feasibility [145].

4 Reactor/Columns Configuration for Detoxification

Process configuration will directly influence hydrolysates detoxification efficiency. The detoxification can be performed in a batch or continuously. For this, a wide range of reactors, columns, and process configurations are reported in literature, which will depend on the detoxification methodology applied and hydrolysate characteristics.

Generally, classical reactor configurations are reported in literature, especially for biological and chemical strategies. Among the most studied reactors, stirred tank reactors, packed-bed reactors, and fluidized-bed reactors are extensively studied. The STR reactors operate in batch mode, where a fixed amount of hydrolysate is treated at a time. This type of bioreactor has the advantage of avoiding dead zones during mixer, is easy to operate, and allows precise control of reaction conditions. However, they may not be ideal for large-scale processes due to long cleaning, emptying, and loading times associated to this configuration.

In the packed bed configuration, also known as fixed bed reactors, resins or heterogeneous catalysts are placed in a fixed bed through which the hydrolysate passes, and the inhibitors are adsorbed or react during the residence time. This configuration is simple and effective, but can have mass transfer limitations, and regeneration of the adsorbent material and stability of the catalysts can be challenging [146]. Packed-bed reactors are particularly interesting when using immobilized enzymes or microorganisms. Niladevi and Prema [147] removed 70% of phenolic compounds using immobilized laccase in a packed bed reactor and successfully recycled the enzyme for eight successive runs. Maroldi et al. [148] applied residual *S. cerevisiae* cells from the brewing process in the adsorption of phenolic compounds and observed an adsorption capacity of 50 mg/g at pH 2,0.

Membrane reactors have been also explored for hydrolysate detoxification, in which semipermeable membranes are applied to separate the toxic compounds from the hydrolysate. Mahboubi et al. [107] demonstrated the effectiveness of a reverse membrane bioreactor (rMBR) for detoxifying inhibitory compounds in lignocellulosic hydrolysates while simultaneously fermenting sugars into ethanol. The membrane-based system creates a concentration gradient that allows gradual exposure of cells to inhibitors, reducing toxicity and enhancing detoxification. This approach enables high cell density and localized detoxification, improving fermentation robustness even under extreme inhibitory conditions (up to 16 g/L furfural). This technology, besides presenting high efficiency on inhibitors removal, can be expensive and requires regular maintenance.

The use of columns for the detoxification of hydrolysates is a common approach, especially for detoxification by adsorption, and different column configurations are used depending on various factors such as the nature of the compounds to be removed, the scale of the process, operating costs, and the desired efficiency. The choice of adsorption material for filling the column depends on the nature of the inhibitors. Ion-exchange (anion or cation) resins are extensively studied for the removal of ions and acidic compounds from the hydrolysate. Resins with sulfonate groups, for example, have a good affinity for acidic compounds such as acetic acid [149]. In addition, considering the advantages of adsorption with activated charcoal already mentioned, there are many works exploring columns of activated charcoal for compounds removal and purification [150, 151]. The use of Layered Adsorption Columns, that involves the use of multiple layers of different adsorbent materials in the same column, each optimized to remove specific types of contaminants, can increase the overall efficiency of the process, but can also increase complexity and cost [152]. Alternatively, sequential columns can be set in series to remove different inhibitors in different steps. Villarreal et al. [153] performed the detoxification of eucalyptus chips hydrolysate in a system composed of four resin columns in series (cation-exchanger type Applexion in H+ form; anion-exchanger type A-860S in Cl– form; cation-exchanger type C-150 in H+ form; and anion-exchanger type Applexion in OH– form). With this system configuration, the authors achieved a removal of 99% of HMF, 95% of furfural, 100% of acetic acid, and 86% of phenolic compounds.

Besides the wide range of reactors and columns explored, new configurations can be designed to adapt and optimize a detoxification method based on the specific characteristics of the lignocellulosic hydrolysate and the objectives of the detoxification process. The choice of the ideal configuration will depend on multifactor such as cost, detoxification efficiency, scalability, and the nature of the compounds to be removed.

5 Economic and Environmental Assessment

Previously, various strategies applicable to the detoxification of hydrolysates for fermentation processes were presented. Although these strategies have been extensively studied in numerous technical cases, a deeper analysis is still needed [63].

Detoxification enables the removal of inhibitors; however, for better results, it is essential to consider the type of raw material, the microorganisms involved, and the target product. Current demands require a broader approach, considering feasibility in real applications and their impact on bioeconomy and sustainability. Therefore, recent analyses must incorporate economic and environmental assessments, which remain underexplored and are currently limited to specific detoxification strategies, as presented below.

From an economic perspective, some studies have evaluated the impact of detoxification on production costs and capital investment. Baral and Shah [154] demonstrated that butanol production from corn stover via ABE fermentation increases production costs by only 1%. Additionally, eliminating the detoxification step would reduce capital costs by 0.4%, considering that the process involved using ammonium to remove inhibitors and sulfuric acid for neutralization.

Similarly, Gubicza et al. [155] assessed ethanol production costs from sugarcane bagasse, showing that adding a detoxification step increases capital costs due to additional equipment (by $0.10 per liter) and chemical consumption (by approximately $0.05 per liter), resulting in a 20% rise in total production costs. However, the study did not specify the detoxification method used. Klein et al. [156] also reported that in succinic acid production from sugarcane bagasse, detoxification accounts for 19% of production costs, with equipment for this stage representing 5% of capital investment when using techniques like overliming and biochar. Other studies have integrated detoxification costs within material pretreatment. In this context, Kosamia et al. [157] analyzed succinic acid production from corn stover, indicating that these costs can represent up to 42% of direct production costs. Although this percentage is high, the primary factor influencing production cost increases is typically the price of the main raw material.

Considering the results mentioned in the last paragraph, detoxification processes can increase production costs up to 20% and capital inversion up to 7% in cases where the chemical strategy is the main one.

Regarding environmental analysis, the most commonly used strategy is life-cycle assessment (LCA). This methodology aims to reduce environmental burdens by substituting processes and techniques with lower environmental impact [158]. Although the application of LCA to detoxification techniques has not been widely explored, some studies have begun evaluating its impact within the production process, identifying the most affected environmental categories. Other research has compared different detoxification strategies to determine which ones present lower environmental impacts for specific fermentation products.

For instance, Bhavana et al. [159] conducted an LCA of xylitol production from wheat bran, comparing two pretreatment methods: diluted acid (including overliming detoxification) and wet air oxidation. The results showed that diluted acid pretreatment had higher eutrophication and freshwater aquatic ecotoxicity potentials, mainly due to chemical and electrical energy demand, as well as the use of $CaCO_3$ and activated carbon for neutralization and detoxification. In global terms, the detoxification and neutralization stage accounted for only 0.112% of the global warming potential and 11.47% of acidification potential, relatively low within the overall process. However, its contribution to eutrophication potential reached 65.6% of the total impact, while marine and freshwater toxicity impacts were 35.6% and 17.24%, respectively. These effects are linked to nutrient-rich wastewater, which can lead to harmful algal blooms and biotoxin production affecting water sources.

In another study, Gontia and Janssen [160] applied LCA to the production of sodium polyacrylate from secondary pulp production streams to identify environmental hotspots. In this case, detoxification involved using sodium dithionite in some streams and sodium hydroxide in others to reduce inhibitory conditions during fermentation. The highest environmental impacts in this stage were related to acidification and photochemical ozone creation potentials, contributing 8% and 19%, respectively, to the overall environmental impact of the process.

Similarly, Janssen et al. [161] conducted an LCA for ethanol production from pine chips, evaluating four detoxification strategies: (1) adaptation of yeast cells to the spruce slurry environment, (2) cell adaptation using additional nutrients, (3) detoxification of the slurry with sodium dithionite ($Na_2S_2O_4$), and (4) washing the slurry followed using the solid fraction in fermentation. The results showed that increased ethanol yield was associated with reduced environmental impacts in terms of global warming potential, eutrophication, acidification, and photochemical ozone formation. Among the evaluated strategies, slurry washing demonstrated the lowest environmental impact across these categories.

These findings highlight the growing interest in understanding the environmental impact of detoxification, though limitations in its analysis persist. To date, the most applied approaches are chemical detoxification techniques, which have shown low global warming potential but greater influence on impact categories related to water quality and acidification potential. Therefore, these aspects should be prioritized when assessing the environmental footprint of the detoxification stage.

In this regard, selecting the optimal detoxification technique requires consideration of multiple factors beyond technical efficiency, including economic feasibility and environmental impact. Thus, multicriteria decision-making (MCDM) strategies can support the selection of a suitable option. This is demonstrated by Llano et al. [63], who applied an MCDM approach to evaluate various detoxification strategies to produce polyhydroxyalkanoates, ethanol, and xylitol from spent sulfite liquor, considering techno-economic, social, and environmental parameters. Their findings show that the best strategy depends on both the target product and the evaluation criteria.

From a technical perspective, anion-exchange resins were the most effective alternative, while overliming emerged as the most viable option from a techno-economic standpoint. Socio-environmentally, overliming membranes and ultrafiltration were the most favorable strategies. When integrating all criteria, overliming stood out as the most balanced option. Moreover, depending on the product yield, optimal strategies varied: anion resins for polyhydroxyalkanoate production, activated carbon for ethanol, and overliming for xylitol.

These findings demonstrate that there is no single answer for determining the best detoxification strategy; instead, it is a multifactorial system that requires thorough analysis to reach an informed decision.

6 Technical Challenges and Their New Perspectives

In recent years, the biotechnology and biochemical industries have made significant advances in the use of biological detoxification processes to manage inhibitory compounds in various applications. To improve the viability of lignocellulosic biorefineries, more effective detoxification strategies must be developed to remove or neutralize these inhibitors while minimizing costs and maintaining the sustainability of the process. Therefore, this section examines the challenges and new perspectives associated with detoxification, strategies to reduce the cost of biological detoxification, approaches to minimize inhibitor accumulation, and the development of inhibitor-tolerant microorganisms and enzymes.

Detoxification of lignocellulosic hydrolysates is essential to improve microbial fermentation performance. However, several challenges hinder the efficiency of detoxification methods, such as the high cost of detoxification, as many chemical and physical detoxification techniques require expensive reagents or additional processing steps and operating conditions [96, 146, 162], loss of hydrolysate volume, and consequently loss of fermentable sugars [72, 95, 163]. Many detoxification methods are unable to effectively remove all inhibitory compounds, and residual inhibitors can still affect microbial growth and fermentation performance [90, 164, 165]. Another point is that some detoxification processes, especially physiochemical methods, can generate additional waste products that further complicate microbial fermentation [166]. In addition, many detoxification approaches that work at laboratory scale are difficult to implement at industrial scale due to process complexity and economic feasibility [145].

A promising environmentally friendly alternative to overcome these disadvantages of using physico-chemical methods for detoxification of lignocellulosic hydrolysates is the use of microorganisms or enzymes to degrade inhibitors: biological detoxification. However, biological detoxification also has certain drawbacks that prevent its widespread use in practical applications, as discussed below.

One of the main challenges of biological detoxification is the high operational costs associated with maintaining optimal conditions for microbial growth and enzyme activity. This includes factors such as nutrient supplementation, pH control, and temperature regulation, which contribute significantly to the overall economics of the process [34, 109]. Therefore, overcoming the inhibitory effects while maintaining cost effectiveness remains a critical challenge.

The complexity of inhibitory compounds in hydrolysates is another point to consider. Inhibitory compounds such as phenolics, furans, and organic acids are ubiquitous in biomass-derived streams and these compounds can interfere with microbial adaptation and stability by inhibiting microbial growth and enzyme activity, thereby reducing process efficiency and product yield [109]. For biological detoxification, microorganisms must naturally tolerate inhibitory compounds, or it is possible to develop microbial strains capable of tolerating high concentrations of inhibitory compounds (via adaptation or engineering), and ensuring the stability and long-term performance of these strains under industrial conditions is a significant technical challenge. However, both strategies share the challenge of longer incubation times, which can be prohibitive in industrial contexts. In addition, microorganisms that assimilate and transform inhibitory compounds into less toxic ones can be selective in the utilization of toxic components and can consume sugars present in the hydrolysate, which is important to evaluate [109, 128, 167].

This presents a further challenge in identifying the endpoint of biodetoxification to ensure timely removal of inhibitors while maintaining unaltered fermentable sugars to ensure successful fermentability in the subsequent bioconversion step [168].

Biological detoxification involves the use of specific enzymes (ligninolytic enzymes such as laccase and peroxidase) carried out with free or immobilized on inert materials or the cell surface, which may have limited tolerance to various inhibitory compounds [169]. Challenges for the industrial application of enzymatic detoxification include protein stability, recovery and recycling [34, 169]. Improving substrate specificity and enzymatic efficiency against a broad spectrum of inhibitors is essential for scaling up biological detoxification processes [166]. It is also important to highlight that regulatory compliance and safety concerns regarding the use of genetically modified organisms (GMOs) in industrial processes add another layer of complexity. Addressing these regulatory challenges, while advancing biotechnological solutions, is critical for the widespread adoption of inhibitor-tolerant microbial species. The cost-effectiveness of biological detoxification needs to be improved to increase its industrial applicability, and several strategies can be used.

Enzymes from industrial by-products or microbial sources can be used for cost-effective detoxification. The use of waste-derived enzymes such as peroxidases, laccases, superoxide dismutase from fungal or bacterial cultures can degrade phenolic inhibitors in hydrolysates [34, 165, 170]. Initially, the use of enzymes on a large scale may seem contradictory as their production cost is still high, but the cost could be reduced by reusing the enzymes bound to carrier supports in immobilization systems [166].

Optimization of fermentation conditions by adjusting fermentation parameters such as pH, temperature, and oxygen levels [168, 171, 172] can increase microbial tolerance to inhibitors and reduce the need for costly detoxification steps. In addition, adaptive laboratory evolutions can be designed to enhance biodetoxification [138].

Another strategy that has shown encouraging results is the use of microbial consortia with complementary metabolic capabilities that can enhance the degradation of multiple inhibitors simultaneously, reducing the need for external detoxification agents. According to Liu et al. [173], the use of a synthetic microbial consortium composed of aerobic soil bacteria and *Saccharomyces cerevisiae* demonstrated a high bioconversion rate of lignocellulosic hydrolysate into furan derivatives. Filamentous fungi such as *Trichoderma reesei* and *Penicillium janthinellum* have also been studied in coculture for biodetoxification due to their ability to metabolize inhibitory compounds such as formic acid, acetic acid, HMF, and furfural from rice straw hydrolysate, providing a one-step process for fungal cellulase production and biodetoxification of lignocellulosic hydrolysate [174]. Morozova and Semyonov [175] evaluated the detoxification of lignocellulose hydrolysate by a consortium of microorganisms formed by specially adapted activated sludge, which proved to be effective and inexpensive. On the other hand, there are some obstacles to the widespread implementation of microbial consortia for biodetoxification, as consortium stability is difficult to achieve and the growth of one organism can easily outcompete the other [121].

Recently, new synthetic biology tools have been developed to design more effective biodetoxification strains, that is, to construct microorganisms with enhanced tolerance to complex industrial stress conditions [167]. Engineering microbial strains to coconsume inhibitors with sugars can avoid or reduce the need for external detoxification, positively impacting process costs [176, 177]. Certain yeast strains, such as *Saccharomyces cerevisiae*, have been engineered to tolerate and metabolize furfural and weak acids [178, 179]. In addition, AI-based computational models have been developed to guide synthetic biology efforts, and these technologies will be important for future research, making biological detoxification an increasingly accessible approach rather than the complex physical and chemical processes [177, 180].

It is also important to think in terms of process integration with hydrolysate production: designing pretreatment and hydrolysis steps to minimize inhibitor formation in the first place can significantly reduce detoxification costs. Preventing the accumulation of these compounds during lignocellulosic biomass processing is a key strategy for improving fermentation efficiency, and several approaches can be used to achieve this.

Although challenges remain in reducing the cost of biological detoxification and developing inhibitor-tolerant microbial species or enzymes, recent advances and innovative strategies offer promising prospects for overcoming these barriers. Continued research investment, technological innovation, and collaborative efforts are essential to realize the full potential of biological detoxification in sustainable biotechnology applications.

7 Conclusions

The detoxification of lignocellulosic hydrolysates is crucial for improving the efficiency and sustainability of biotechnological processes, especially in the context of biorefineries to produce biofuels and value-added bioproducts. This chapter provides a comprehensive review of the various inhibitors formed during lignocellulosic hydrolysis, including phenolic compounds, furans, organic acids, oligomers, and heavy metals, and discusses their negative impact on microbial viability and fermentation yields.

Detoxification strategies are categorized into physical, chemical, and biological methods, each with unique advantages and limitations. Physical methods, such as evaporation, steam stripping, and extraction, are effective but may face scalability and cost challenges. Chemical methods, such as alkaline treatments, activated carbon adsorption, ion-exchange resins, and nanomaterials, are highly effective in removing inhibitors, but require careful optimization to prevent sugar losses and minimize environmental impact. Biological methods using microbial detoxification, enzyme treatments, microbial encapsulation, adaptive evolution, and genetic modification offer environmentally friendly and potentially cost-effective solutions, but face challenges related to reaction kinetics, enzyme stability, and regulatory concerns.

Economic and environmental assessments have highlighted the need for a balanced selection of detoxification strategies, considering their impact on production costs, scalability, and environmental footprint. Life-cycle analysis and techno-economic assessments are essential to identify sustainable and economically viable detoxification methods.

Future research must focus on overcoming the technical and economic challenges associated with these processes, developing integrated and hybrid detoxification strategies, and improving microbial and enzyme robustness to inhibitors. Ultimately, continued advances in detoxification technologies will play a key role in achieving efficient, economically viable and environmentally sustainable lignocellulosic biorefineries.

Acknowledgements The authors appreciate the infrastructure support by the São Paulo State University (Unesp), the Institute of Chemical of Araraquara and Bioenergy Research Institute (IPBEN)— Araraquara. And financial support by the São Paulo Research Foundation (FAPESP) and the National Council for Scientific and Technological Development (CNPq).

Funding This work was supported by the São Paulo Research Foundation (FAPESP), grants #2022/03000-0, #2024/01232-7 and #2024/ 06741-7. And by the National Council for Scientific and Technological Development (CNPq) grant #316230/2023-5.

References

1. Wagh MS, Nath PC, Chakraborty A, Amrit R, Mishra B, Mishra AK, Mohanta YK (2024) Valorisation of agro-industrial wastes: circular bioeconomy and biorefinery process—a sustainable symphony. Process Safety Environ Protection 183:708–725. https://doi.org/10.1016/j.psep.2024.01.055
2. Gawel E, Pannicke N, Hagemann N (2019) A path transition towards a bioeconomy—the crucial role of sustainability. Sustainability 11:3005. https://doi.org/10.3390/su11113005
3. Maina S, Kachrimanidou V, Koutinas A (2017) A roadmap towards a circular and sustainable bioeconomy through waste valorization. Curr Opinion Green Sustain Chem 8:18–23. https://doi.org/10.1016/j.cogsc.2017.07.007
4. Ubando AT, Felix CB, Chen W-H (2020) Biorefineries in circular bioeconomy: a comprehensive review. Biores Technol 299:122585. https://doi.org/10.1016/j.biortech.2019.122585
5. Stikane A, Dace E, Stalidzans E (2022) Closing the loop in bioproduction: spent microbial biomass as a resource within circular bioeconomy. New Biotechnol 70:109–115. https://doi.org/10.1016/j.nbt.2022.06.001
6. Huang Y, Wang Y, Shang N, Li P (2023) Microbial fermentation processes of lactic acid: challenges, solutions, and future prospects. Foods 12:2311. https://doi.org/10.3390/foods12122311
7. Mussatto S (2004) Alternatives for detoxification of diluted-acid lignocellulosic hydrolyzates for use in fermentative processes: a review. Biores Technol 93:1–10. https://doi.org/10.1016/j.biortech.2003.10.005
8. Sun Y, Cheng J (2002) Hydrolysis of lignocellulosic materials for ethanol production: a review. Biores Technol 83:1–11. https://doi.org/10.1016/S0960-8524(01)00212-7
9. Magro FG, Decesaro A, Berticelli R, Colla LM (2016) Produção de Bioetanol Utilizando Microalgas: Uma Revisão. Semina: Tech Ex 37:159. https://doi.org/10.5433/1679-0375.2016v37n1p159
10. Kalia S, Kaith BS, Kaur I (2011) Cellulose fibers: bio- and nano-polymer composites: green chemistry and technology. Springer, Heidelberg
11. Alvira P, Tomás-Pejó E, Ballesteros M, Negro MJ (2010) Pretreatment technologies for an efficient bioethanol production process based on enzymatic hydrolysis: a review. Biores Technol 101:4851–4861. https://doi.org/10.1016/j.biortech.2009.11.093
12. Galaverna R, Pastre JC (2017) Production of 5-(Hydroxymethyl)furfural from biomass: synthetic challenges and applications as building block in polymers and liquid fuel production. Rev Virtual Quim 9:248–273. https://doi.org/10.21577/1984-6835.20170017
13. Polachini TC (2015) Determinação das melhores condições de hidrólise física com ultrassom de potência em resíduos de amendoim
14. Taherzadeh MJ, Karimi K (2007) Acid-based hydrolysis processes for ethanol from lignocellulosic materials: a review. BioRes 2:472–499. https://doi.org/10.15376/biores.2.3.472-499
15. Kim JS, Lee YY, Kim TH (2016) A review on alkaline pretreatment technology for bioconversion of lignocellulosic biomass. Biores Technol 199:42–48. https://doi.org/10.1016/j.biortech.2015.08.085
16. Do L, Souza S, Pereira A, Antônio M, Farias M, Lopes R, Oliveira, Duvoisin S, Nazareno J, Quaresma N (2019) Valorization of Andiroba (Carapa guianensis Aubl.) residues through optimization of alkaline pretreatment to obtain fermentable sugars. Bioresources 15:894–909. https://doi.org/10.15376/biores.15.1.894-909
17. Giovannoni M, Gramegna G, Benedetti M, Mattei B (2020) Industrial use of cell wall degrading enzymes: the fine line between production strategy and economic feasibility. Front Bioeng Biotechnol 8. https://doi.org/10.3389/fbioe.2020.00356
18. Nelson DL, Cox MM (2017) Lehninger principles of biochemistry, 7th edn. W. H. Freeman, New York
19. Shimizu F (2018) Remoção de lignina e hemicelulose: influência na acessibilidade à celulose e sacarificação enzimática. UNESP

20. Guo H, Zhao Y, Chang J-S, Lee D-J (2022) Inhibitor formation and detoxification during lignocellulose biorefinery: a review. Biores Technol 361:127666. https://doi.org/10.1016/j.biortech.2022.127666
21. Tejirian A, Xu F (2011) Inhibition of enzymatic cellulolysis by phenolic compounds. Enzyme Microb Technol 48:239–247. https://doi.org/10.1016/j.enzmictec.2010.11.004
22. Kim D (2018) Physico-chemical conversion of lignocellulose: inhibitor effects and detoxification strategies: a mini review. Molecules 23:309. https://doi.org/10.3390/molecules23020309
23. Binder JB, Raines RT (2010) Fermentable sugars by chemical hydrolysis of biomass. Proc Natl Acad Sci USA 107:4516–4521. https://doi.org/10.1073/pnas.0912073107
24. Schultze-Jena A, Vroon RC, Macleod AKA, Hreggviðsson GÓ, Adalsteinsson BT, Engelen-Smit NPE, De Vrije T, Budde MAW, Van Der Wal H, López-Contreras AM, Boon MA (2022) Production of acetone, butanol, and ethanol by fermentation of Saccharina latissima: cultivation, enzymatic hydrolysis, inhibitor removal, and fermentation. Algal Res 62:102618. https://doi.org/10.1016/j.algal.2021.102618
25. Baruah J, Nath BK, Sharma R, Kumar S, Deka RC, Baruah DC, Kalita E (2018) Recent trends in the pretreatment of lignocellulosic biomass for value-added products. Front Energy Res 6. https://doi.org/10.3389/fenrg.2018.00141
26. Moreno AD, Ibarra D, Alvira P, Tomás-Pejó E, Ballesteros M (2015) A review of biological delignification and detoxification methods for lignocellulosic bioethanol production. Crit Rev Biotechnol 35:342–354. https://doi.org/10.3109/07388551.2013.878896
27. Jaffur N, Jeetah P, Kumar G (2021) A review on enzymes and pathways for manufacturing polyhydroxybutyrate from lignocellulosic materials. 3 Biotech 11:483. https://doi.org/10.1007/s13205-021-03009-x
28. Sun Y, Li X, Wu L, Li Y, Li F, Xiu Z, Tong Y (2021) The advanced performance of microbial consortium for simultaneous utilization of glucose and xylose to produce lactic acid directly from dilute sulfuric acid pretreated corn stover. Biotechnol Biofuels 14:233. https://doi.org/10.1186/s13068-021-02085-8
29. Kumar R, Mago G, Balan V, Wyman CE (2009) Physical and chemical characterizations of corn stover and poplar solids resulting from leading pretreatment technologies. Biores Technol 100:3948–3962. https://doi.org/10.1016/j.biortech.2009.01.075
30. Stamogiannou I, Van Camp J, Smagghe G, Van De Walle D, Dewettinck K, Raes K (2021) Impact of phenolic compound as activators or inhibitors on the enzymatic hydrolysis of cellulose. Int J Biol Macromol 186:174–180. https://doi.org/10.1016/j.ijbiomac.2021.07.052
31. Honarmandrad Z, Kucharska K, Kaykhaii M, Gębicki J (2024) Removal of phenolic inhibitor compounds from hydrolysates and post-fermentation broths by using a hydrophobic magnetic deep eutectic solvent. J Environ Chem Eng 12:112621. https://doi.org/10.1016/j.jece.2024.112621
32. Maitra S, Singh V (2021) Balancing sugar recovery and inhibitor generation during energycane processing: coupling cryogenic grinding with hydrothermal pretreatment at low temperatures. Biores Technol 321:124424. https://doi.org/10.1016/j.biortech.2020.124424
33. Lu X, Li F, Zhou X, Hu J, Liu P (2022) Biomass, lignocellulolytic enzyme production and lignocellulose degradation patterns by Auricularia auricula during solid state fermentation of corn stalk residues under different pretreatments. Food Chem 384:132622. https://doi.org/10.1016/j.foodchem.2022.132622
34. Tramontina R, Brenelli LB, Sousa A, Alves R, Zetty Arenas AM, Nascimento VM, Rabelo SC, Freitas S, Ruller R, Squina FM (2020) Designing a cocktail containing redox enzymes to improve hemicellulosic hydrolysate fermentability by microorganisms. Enzyme Microb Technol 135:109490. https://doi.org/10.1016/j.enzmictec.2019.109490
35. Chin DWK, Lim S, Pang YL, Lim CH, Shuit SH, Lee KM, Chong CT (2021) Effects of organic solvents on the Organosolv pretreatment of degraded empty fruit bunch for fractionation and lignin removal. Sustainability 13:6757. https://doi.org/10.3390/su13126757
36. Cheng Y, Quan W, Qu T, He Y, Wang Z, Zeng M, Qin F, Chen J, He Z (2021) Effects of 60Co-irradiation and superfine grinding wall disruption pretreatment on phenolic compounds

in pine (Pinus yunnanensis) pollen and its antioxidant and α-glucosidase-inhibiting activities. Food Chem 345:128808. https://doi.org/10.1016/j.foodchem.2020.128808

37. Agrawal R, Verma A, Singhania RR, Varjani S, Di Dong C, Kumar Patel A (2021) Current understanding of the inhibition factors and their mechanism of action for the lignocellulosic biomass hydrolysis. Biores Technol 332:125042. https://doi.org/10.1016/j.biortech.2021.125042
38. Bhatia MS, Dora M, Jakhar SK (2019) Appropriate location for remanufacturing plant towards sustainable supply chain. Ann Oper Res. https://doi.org/10.1007/s10479-019-03294-z
39. Fadimu GJ, Farahnaky A, Gill H, Truong T (2022) Influence of ultrasonic pretreatment on structural properties and biological activities of lupin protein hydrolysate. Int J of Food Sci Tech 57:1729–1738. https://doi.org/10.1111/ijfs.15549
40. Brandt BA, García-Aparicio MDP, Görgens JF, Van Zyl WH (2021) Rational engineering of Saccharomyces cerevisiae towards improved tolerance to multiple inhibitors in lignocellulose fermentations. Biotechnol Biofuels 14:173. https://doi.org/10.1186/s13068-021-02021-w
41. Zangiabadi M, Zhao Y (2022) Synergistic hydrolysis of cellulose by a blend of cellulase-mimicking polymeric nanoparticle catalysts. J Am Chem Soc 144:17110–17119. https://doi.org/10.1021/jacs.2c06848
42. East NJ, Clifton BE, Jackson CJ, Kaczmarski JA (2022) The role of oligomerization in the optimization of cyclohexadienyl dehydratase conformational dynamics and catalytic activity. Protein Sci 31:e4510. https://doi.org/10.1002/pro.4510
43. Sjulander N, Kikas T (2020) Origin, impact and control of lignocellulosic inhibitors in bioethanol production—a review. Energies 13:4751. https://doi.org/10.3390/en13184751
44. Lyu Q, Dar RA, Baganz F, Smoliński A, Rasmey A-HM, Liu R, Zhang L (2025) Effects of lignocellulosic biomass-derived hydrolysate inhibitors on cell growth and lipid production during microbial fermentation of oleaginous microorganisms—a review. Fermentation 11:121. https://doi.org/10.3390/fermentation11030121
45. Beri D, Herring CD, Blahova S, Poudel S, Giannone RJ, Hettich RL, Lynd LR (2021) Coculture with hemicellulose-fermenting microbes reverses inhibition of corn fiber solubilization by clostridium thermocellum at elevated solids loadings. Biotechnol Biofuels 14:24. https://doi.org/10.1186/s13068-020-01867-w
46. Wood JM (2015) Bacterial responses to osmotic challenges. J Gen Physiol 145:381–388. https://doi.org/10.1085/jgp.201411296
47. Vermaas JV, Crowley MF, Beckham GT (2022) Molecular simulation of lignin-related aromatic compound permeation through gram-negative bacterial outer membranes. J Biol Chem 298:102627. https://doi.org/10.1016/j.jbc.2022.102627
48. Su L-J, Zhang J-H, Gomez H, Murugan R, Hong X, Xu D, Jiang F, Peng Z-Y (2019) Reactive oxygen species-induced lipid peroxidation in apoptosis, autophagy, and ferroptosis. Oxid Med Cell Longev 2019:5080843. https://doi.org/10.1155/2019/5080843
49. Juan CA, Pérez de la Lastra JM, Plou FJ, Pérez-Lebeña E (2021) The Chemistry of reactive oxygen species (ROS) revisited: outlining their role in biological macromolecules (DNA, lipids and proteins) and induced pathologies. Int J Mol Sci 22:4642. https://doi.org/10.3390/ijms22094642
50. Xue S, Uppugundla N, Bowman MJ, Cavalier D, Da Costa SL, Dale Bruce E, Balan V (2015) Sugar loss and enzyme inhibition due to oligosaccharide accumulation during high solids-loading enzymatic hydrolysis. Biotechnol Biofuels 8:195. https://doi.org/10.1186/s13068-015-0378-9
51. Westerbeke FHM, Attaye I, Rios-Morales M, Nieuwdorp M (2025) Glycaemic sugar metabolism and the gut microbiota: past, present and future. FEBS J 292:1421–1436. https://doi.org/10.1111/febs.17293
52. Mukherjee P, Pal S, Sivaprakasam S (2020) Process parameter controls for efficient enzymatic hydrolysis of cellulosic biomass. In: Bisaria V (ed) Handbook of biorefinery research and technology. Springer, Dordrecht, pp 1–29
53. Lv Y, Zhang Y, Xu Y (2024) Understanding and technological approach of acid hydrolysis processing for lignocellulose biorefinery: panorama and perspectives. Biomass Bioenerg 183:107133. https://doi.org/10.1016/j.biombioe.2024.107133

54. Xu L, Zhang S-J, Zhong C, Li B-Z, Yuan Y-J (2020) Alkali-based pretreatment-facilitated lignin valorization: a review. Ind Eng Chem Res 59:16923–16938. https://doi.org/10.1021/acs.iecr.0c01456
55. Saadan R, Hachimi Alaoui C, Ihammi A, Chigr M, Fatimi A (2024) A brief overview of lignin extraction and isolation processes: from lignocellulosic biomass to added-value biomaterials. Environ Earth Sci Proc 31:3. https://doi.org/10.3390/eesp2024031003
56. Mudhoo A, Kumar S (2013) Effects of heavy metals as stress factors on anaerobic digestion processes and biogas production from biomass. Int J Environ Sci Technol 10:1383–1398. https://doi.org/10.1007/s13762-012-0167-y
57. Chen H, Qin L, Yu B (2015) Furfural production from steam explosion liquor of rice straw by solid acid catalysts (HZSM-5). Biomass Bioenerg 73:77–83. https://doi.org/10.1016/j.biombioe.2014.12.013
58. Hermundsgård DH, Ghoreishi S, Tanase-Opedal M, Moe ST, Brusletto R, Barth T (2025) Maximizing yields of furfural and 5-hydroxymethylfurfural in side streams from steam explosion of lignocellulosic residues. Biofuels Bioprod Bioref 19:319–336. https://doi.org/10.1002/bbb.2719
59. Papa S, Alvarez-Romero M (2023) Biological activities in artificially heavy-metal-contaminated growing substrates. Soil Syst 7:111. https://doi.org/10.3390/soilsystems7040111
60. Gavrila AI, Vartolomei A, Calinescu I, Vinatoru M, Parvulescu OC, Psenovschi G, Chipurici P, Trifan A (2024) Ultrasound-assisted alkaline pretreatment of biomass to enhance the extraction yield of valuable chemicals. Agronomy 14:903. https://doi.org/10.3390/agronomy14050903
61. Pereira JC, Giese EC, de SouzaMoretti MM, Gomes AC S, Perrone OM, Boscolo M, da Silva R, Gomes E, Martins DAB, Pereira JC, Giese EC, SouzaMoretti MM, Gomes AC S, Perrone OM, Boscolo M, da Silva R, Gomes E, Martins DAB (2017) Effect of metal ions, chemical agents and organic compounds on lignocellulolytic enzymes activities. In: Enzyme inhibitors and activators. IntechOpen
62. Janusz G, Pawlik A, Sulej J, Świderska-Burek U, Jarosz-Wilkołazka A, Paszczyński A (2017) Lignin degradation: microorganisms, enzymes involved, genomes analysis and evolution. FEMS Microbiol Rev 41:941–962. https://doi.org/10.1093/femsre/fux049
63. Llano T, Rueda C, Dosal E, Andrés A, Coz A (2021) Multi-criteria analysis of detoxification alternatives: techno-economic and socio-environmental assessment. Biomass Bioenerg 154:106274. https://doi.org/10.1016/j.biombioe.2021.106274
64. Roque LR, Morgado GP, Nascimento VM, Ienczak JL, Rabelo SC (2019) Liquid-liquid extraction: a promising alternative for inhibitors removing of pentoses fermentation. Fuel 242:775–787. https://doi.org/10.1016/j.fuel.2018.12.130
65. Kubisch C, Ochsenreither K (2022) Detoxification of a pyrolytic aqueous condensate from wheat straw for utilization as substrate in Aspergillus oryzae DSM 1863 cultivations. Biotechnol Biofuels 15:18. https://doi.org/10.1186/s13068-022-02115-z
66. Llano T, Quijorna N, Coz A (2017) Detoxification of a lignocellulosic waste from a pulp mill to enhance its fermentation prospects. Energies 10:348. https://doi.org/10.3390/en10030348
67. Canilha L, Chandel AK, Suzane Dos Santos Milessi T, Antunes FAF, Luiz Da Costa Freitas W, Das Graças Almeida Felipe M, Da Silva SS (2012) Bioconversion of sugarcane biomass into ethanol: an overview about composition, pretreatment methods, detoxification of hydrolysates, enzymatic saccharification, and ethanol fermentation. J Biomed Biotechnol 2012:1–15. https://doi.org/10.1155/2012/989572
68. Devi A, Singh A, Bajar S, Pant D, Din ZU (2021) Ethanol from lignocellulosic biomass: an in-depth analysis of pre-treatment methods, fermentation approaches and detoxification processes. J Environ Chem Eng 9:105798. https://doi.org/10.1016/j.jece.2021.105798
69. Sonego JLS, Lemos DA, Rodriguez GY, Cruz AJG, Badino AC (2014) Extractive batch fermentation with CO_2 stripping for ethanol production in a bubble column bioreactor: experimental and modeling. Energy Fuels 28:7552–7559. https://doi.org/10.1021/ef5018797

70. Zhu J, Yong Q, Xu Y, Yu S (2009) Comparative detoxification of vacuum evaporation/steam stripping combined with overliming on corn stover prehydrolyzate. In: 2009 international conference on energy and environment technology, pp 240–243
71. Artifon W, Bonatto C, Bordin ER, Bazoti SF, Dervanoski A, Alves SL, Treichel H (2018) Bioethanol production from hydrolyzed lignocellulosic after detoxification via adsorption with activated carbon and dried air stripping. Front Bioeng Biotechnol 6:107. https://doi.org/10.3389/fbioe.2018.00107
72. Zhang Q-W, Lin L-G, Ye W-C (2018) Techniques for extraction and isolation of natural products: a comprehensive review. Chin Med 13:20. https://doi.org/10.1186/s13020-018-0177-x
73. Xavier L, Rocha M, Pisani J, Zecchi B (2024) Aqueous two-phase systems based on cholinium ionic liquids for the recovery of ferulic and p-coumaric acids from rice husk hydrolysate. Appl Food Res 4:100381. https://doi.org/10.1016/j.afres.2023.100381
74. Azizan A, Darim RA, Azmi IS, Jalil R, Manaf SFA (2025) Understanding crystallinity indexes benchmarking concept from optimized parametric ionic liquid pretreatment using 1-ethyl-3-methylimidazolium Acetate [EMIM]Ac on oil palm biomass. J Ionic Liquids 5:100136. https://doi.org/10.1016/j.jil.2025.100136
75. Lin S-P, Kuo T-C, Wang H-T, Ting Y, Hsieh C-W, Chen Y-K, Hsu H-Y, Cheng K-C (2020) Enhanced bioethanol production using atmospheric cold plasma-assisted detoxification of sugarcane bagasse hydrolysate. Biores Technol 313:123704. https://doi.org/10.1016/j.biortech.2020.123704
76. Mahmoud YA-G, Elkaliny NE, Darwish OA, Ashraf Y, Ebrahim RA, Das SP, Yahya G (2025) Comprehensive review for aflatoxin detoxification with special attention to cold plasma treatment. Mycotoxin Res. https://doi.org/10.1007/s12550-025-00582-5
77. Santoso SP, Lin S-P, Wang T-Y, Ting Y, Hsieh C-W, Yu R-C, Angkawijaya AE, Soetaredjo FE, Hsu H-Y, Cheng K-C (2021) Atmospheric cold plasma-assisted pineapple peel waste hydrolysate detoxification for the production of bacterial cellulose. Int J Biol Macromol 175:526–534. https://doi.org/10.1016/j.ijbiomac.2021.01.169
78. Yoosefian SH, Ebrahimi R, Hosseinzadeh Samani B, Maleki A (2022) Modification of bioethanol production in an innovative pneumatic digester with non-thermal cold plasma detoxification. Biores Technol 350:126907. https://doi.org/10.1016/j.biortech.2022.126907
79. Persson P, Larsson S, Jönsson LJ, Nilvebrant N, Sivik B, Munteanu F, Thörneby L, Gorton L (2002) Supercritical fluid extraction of a lignocellulosic hydrolysate of spruce for detoxification and to facilitate analysis of inhibitors. Biotech Bioeng 79:694–700. https://doi.org/10.1002/bit.10324
80. Escobar ELN, Da Silva TA, Pirich CL, Corazza ML, Pereira Ramos L (2020) Supercritical fluids: a promising technique for biomass pretreatment and fractionation. Front Bioeng Biotechnol 8:252. https://doi.org/10.3389/fbioe.2020.00252
81. Jia M, Li H, Zhang X, Han Y, Cordeiro IMDC, Yeoh GH (2024) Experimental investigation on suppression effect of molecular sieve powders on asphalt fume. J Clean Prod 476:143769. https://doi.org/10.1016/j.jclepro.2024.143769
82. Pei Z, Feng Z, Yao Z, Luo Y, Lu J (2025) Emerging molecular sieve regulation strategy for the adsorption and catalysis: silicon hydroxyl engineering. Molecular Catalysis 570:114667. https://doi.org/10.1016/j.mcat.2024.114667
83. Wang B, Cong H, Li X, Li H, Gao X (2022) Selective and efficient separation of biomass hydrolysate levulinic acid and formic acid from aqueous solution. Chem Eng Res Des 181:266–277. https://doi.org/10.1016/j.cherd.2022.03.027
84. Kumar P, Barrett DM, Delwiche MJ, Stroeve P (2009) Methods for pretreatment of lignocellulosic biomass for efficient hydrolysis and biofuel production. Ind Eng Chem Res 48:3713–3729. https://doi.org/10.1021/ie801542g
85. Persson P, Andersson J, Gorton L, Larsson S, Nilvebrant N-O, Jönsson LJ (2002) Effect of different forms of alkali treatment on specific fermentation inhibitors and on the fermentability of lignocellulose hydrolysates for production of fuel ethanol. J Agric Food Chem 50:5318–5325. https://doi.org/10.1021/jf025565o

86. Coz A, Llano T, Cifrián E, Viguri J, Maican E, Sixta H (2016) Physico-chemical alternatives in lignocellulosic materials in relation to the kind of component for fermenting purposes. Materials 9:574. https://doi.org/10.3390/ma9070574
87. Karimi K, Emtiazi G, Taherzadeh MJ (2006) Production of ethanol and mycelial biomass from rice straw hemicellulose hydrolyzate by Mucor indicus. Process Biochem 41:653–658. https://doi.org/10.1016/j.procbio.2005.08.014
88. Kordala N, Lewandowska M, Bednarski W (2023) Effect of the method for the elimination of inhibitors present in Miscanthus giganteus hydrolysates on ethanol production effectiveness. Biomass Conv Bioref 13:2089–2097. https://doi.org/10.1007/s13399-020-01255-2
89. Nilvebrant N-O, Persson P, Reimann A, De Sousa F, Gorton L, Jönsson LJ (2003) Limits for alkaline detoxification of dilute-acid lignocellulose hydrolysates. ABAB 107:615–628. https://doi.org/10.1385/ABAB:107:1-3:615
90. Guo X, Cavka A, Jönsson LJ, Hong F (2013) Comparison of methods for detoxification of spruce hydrolysate for bacterial cellulose production. Microb Cell Fact 12:93. https://doi.org/10.1186/1475-2859-12-93
91. Thontowi A, Mayangsari W, Kholida LN, Kanti A, Wardani AK, Hermiati E (2020) Evaluation of addition the activated charcoals and pH adjustment in the treatment of lignocellulosic hydrolisates for xylitol production. IOP Conf Ser Earth Environ Sci 439:012023. https://doi.org/10.1088/1755-1315/439/1/012023
92. Mussatto SI, Fernandes M, Rocha GJM, Órfão JJM, Teixeira JA, Roberto IC (2010) Production, characterization and application of activated carbon from brewer's spent grain lignin. Biores Technol 101:2450–2457. https://doi.org/10.1016/j.biortech.2009.11.025
93. Seo H-B, Kim S-S, Lee H-Y, Jung K-H (2009) Improved bioethanol production using activated carbon-treated acid hydrolysate from corn hull in *Pachysolen tannophilus*. Mycobiology 37:133–140. https://doi.org/10.4489/MYCO.2009.37.2.133
94. Wahyuhadi ME, Kusumadewi RA, Hadisoebroto R (2023) Effect of contact time on the adsorption process of activated carbon from banana peel in reducing heavy metal cd and dyes using a stirring tub (pilot scale). IOP Conf Ser Earth Environ Sci 1203:012035. https://doi.org/10.1088/1755-1315/1203/1/012035
95. Jin T, Xing X, Xie Y, Sun Y, Bian S, Liu L, Chen G, Wang X, Yu X, Su Y (2022) Evaluation of preparation and detoxification of hemicellulose hydrolysate for improved xylitol production from quinoa straw. IJMS 24:516. https://doi.org/10.3390/ijms24010516
96. Ahuja V, Kshirsagar S, Ghosh P, Sarkar B, Sutar A, More S, Dasgupta D (2022) Process development for detoxification of corncob hydrolysate using activated charcoal for xylitol production. J Environ Chem Eng 10:107097. https://doi.org/10.1016/j.jece.2021.107097
97. Nilvebrant N-O, Reimann A, Larsson S, Jönsson LJ (2001) Detoxification of Lignocellulose hydrolysates with ion-exchange resins. ABAB 91–93:35–50. https://doi.org/10.1385/ABAB:91-93:1-9:35
98. Han J, Xu B, Wang H, Huang G, Zhang X, Xu Y (2021) Facilitate cost-practical bioproduction of xylonic acid from acidic lignocellulosic hydrolysate by combined recycling-resin detoxification and recycling-cell catalysis. SSRN Journal. https://doi.org/10.2139/ssrn.3995349
99. Lv Y, Zhou S, Zhang X, Xu Y (2022) A smart self-balancing biosystem with reversible competitive adsorption of in-situ anion exchange resin for whole-cell catalysis preparation of lignocellulosic xylonic acid. Biores Technol 363:127998. https://doi.org/10.1016/j.biortech.2022.127998
100. Sajab MS, Santanaraj J, Mohammad AW, Kaco H, Harun S (2019) Detoxification of lignocellulosic hydrolysates by in situ formation of Fe(0) nanoparticles on activated carbon. BioRes 14:8614–8626. https://doi.org/10.15376/biores.14.4.8614-8626
101. Adebule AP, Sanusi IA, Kana EBG (2024) Nano-based surface adsorption detoxification of process inhibitors for improved bioethanol productivity. Bioresource Technol Rep 25:101783. https://doi.org/10.1016/j.biteb.2024.101783

102. Liu W, Liu P, Liu L, Sun H, Fan Y, Ma C, Ouyang J, Zheng Z (2024) Promoting microbial fermentation in lignocellulosic hydrolysates by removal of inhibitors using MTES and PEI-modified chitosan-chitin nanofiber hybrid aerogel. Carbohyd Polym 328:121766. https://doi.org/10.1016/j.carbpol.2023.121766
103. Zhu J, Zhang H, Jiao N, Xu G, Xu Y (2021) Lignin nanoparticles produced from enzymatic hydrolysis residue of biorefinery via hydrotrope fractionation for inhibitors detoxification and improve ethanol fermentability
104. Covarrubias-García I, Arriaga S (2022) Adsorbents for the detoxification of lignocellulosic wastes hydrolysates to improve fermentative processes to bioenergy and biochemicals production. In: Sinharoy A, Lens PNL (eds) Renewable energy technologies for energy efficient sustainable development. Springer International Publishing, Cham, pp 63–83
105. Grzenia DL, Schell DJ, Ranil Wickramasinghe S (2012) Membrane extraction for detoxification of biomass hydrolysates. Biores Technol 111:248–254. https://doi.org/10.1016/j.biortech.2012.01.169
106. Greer DR, Basso TP, Ibanez AB, Bauer S, Skerker JM, Ozcam AE, Leon D, Shin C, Arkin AP, Balsara NP (2014) Fermentation of hydrolysate detoxified by pervaporation through block copolymer membranes. Green Chem 16:4206–4213. https://doi.org/10.1039/C4GC00756E
107. Mahboubi A, Elyasi S, Doyen W, De Wever H, Taherzadeh MJ (2020) Concentration-driven reverse membrane bioreactor for the fermentation of highly inhibitory lignocellulosic hydrolysate. Process Biochem 92:409–416. https://doi.org/10.1016/j.procbio.2020.01.031
108. Zhang Y, Li M, Wang Y, Ji X, Zhang L, Hou L (2015) Simultaneous concentration and detoxification of lignocellulosic hydrolyzates by vacuum membrane distillation coupled with adsorption. Biores Technol 197:276–283. https://doi.org/10.1016/j.biortech.2015.08.097
109. Sodré V, Vilela N, Tramontina R, Squina FM (2021) Microorganisms as bioabatement agents in biomass to bioproducts applications. Biomass Bioenerg 151:106161. https://doi.org/10.1016/j.biombioe.2021.106161
110. Chandel AK, Da Silva SS, Singh OV (2013) Detoxification of lignocellulose hydrolysates: biochemical and metabolic engineering toward white biotechnology. Bioenerg Res 6:388–401. https://doi.org/10.1007/s12155-012-9241-z
111. Ujor VC, Okonkwo CC (2022) Microbial detoxification of lignocellulosic biomass hydrolysates: biochemical and molecular aspects, challenges, exploits and future perspectives. Front Bioeng Biotechnol 10:1061667. https://doi.org/10.3389/fbioe.2022.1061667
112. Schneider WDH, Fontana RC, Baudel HM, De Siqueira FG, Rencoret J, Gutiérrez A, De Eugenio LI, Prieto A, Martínez MJ, Martínez ÁT, Dillon AJP, Camassola M (2020) Lignin degradation and detoxification of eucalyptus wastes by on-site manufacturing fungal enzymes to enhance second-generation ethanol yield. Appl Energy 262:114493. https://doi.org/10.1016/j.apenergy.2020.114493
113. De La Torre M, Martín-Sampedro R, Fillat Ú, Eugenio ME, Blánquez A, Hernández M, Arias ME, Ibarra D (2017) Comparison of the efficiency of bacterial and fungal laccases in delignification and detoxification of steam-pretreated lignocellulosic biomass for bioethanol production. J Ind Microbiol Biotechnol 44:1561–1573. https://doi.org/10.1007/s10295-017-1977-1
114. Allard-Massicotte R, Chadjaa H, Marinova M (2017) Phenols removal from hemicelluloses pre-hydrolysate by laccase to improve butanol production. Fermentation 3:31. https://doi.org/10.3390/fermentation3030031
115. Moreno AD, Ibarra D, Ballesteros I, González A, Ballesteros M (2013) Comparing cell viability and ethanol fermentation of the thermotolerant yeast Kluyveromyces marxianus and Saccharomyces cerevisiae on steam-exploded biomass treated with laccase. Biores Technol 135:239–245. https://doi.org/10.1016/j.biortech.2012.11.095
116. Chandel AK, Kapoor RK, Singh A, Kuhad RC (2007) Detoxification of sugarcane bagasse hydrolysate improves ethanol production by Candida shehatae NCIM 3501. Biores Technol 98:1947–1950. https://doi.org/10.1016/j.biortech.2006.07.047
117. Martín C, Galbe M, Wahlbom CF, Hahn-Hägerdal B, Jönsson LJ (2002) Ethanol production from enzymatic hydrolysates of sugarcane bagasse using recombinant xylose-utilising

Saccharomyces cerevisiae. Enzyme Microbial Technol 31:274–282. https://doi.org/10.1016/S0141-0229(02)00112-6

118. Fang Z, Liu X, Chen L, Shen Y, Zhang X, Fang W, Wang X, Bao X, Xiao Y (2015) Identification of a laccase Glac15 from Ganoderma lucidum 77002 and its application in bioethanol production. Biotechnol Biofuels 8:54. https://doi.org/10.1186/s13068-015-0235-x
119. Corradini FAS, Milessi TS, Gonçalves VM, Ruller R, Sargo CR, Lopes LA, Zangirolami TC, Tardioli PW, Giordano RC, Giordano RLC (2021) High stabilization and hyperactivation of a recombinant β-Xylosidase through immobilization strategies. Enzyme Microb Technol 145:109725. https://doi.org/10.1016/j.enzmictec.2020.109725
120. Li S, Qi B, Luo J, Wan Y (2022) Degradation of phenolic inhibitors by laccase immobilized on tannic acid/polyethylenimine modified magnetic nanoparticles. Results Eng 15:100585. https://doi.org/10.1016/j.rineng.2022.100585
121. Nogueira CDC, Padilha CEDA, Dantas JMDM, Medeiros FGMD, Guilherme ADA, Souza DFDS, Santos ESD (2021) In-situ detoxification strategies to boost bioalcohol production from lignocellulosic biomass. Renew Energy 180:914–936. https://doi.org/10.1016/j.renene.2021.09.012
122. Parawira W, Tekere M (2011) Biotechnological strategies to overcome inhibitors in lignocellulose hydrolysates for ethanol production: review. Crit Rev Biotechnol 31:20–31. https://doi.org/10.3109/07388551003757816
123. Thomsen ST, Kádár Z, Schmidt JE (2014) Compositional analysis and projected biofuel potentials from common West African agricultural residues. Biomass Bioenerg 63:210–217. https://doi.org/10.1016/j.biombioe.2014.01.045
124. Romero-García JM, Fehér C, Cara C, Ruiz-Ramos E, Castro E (2022) Improved xylitol production from olive stones hydrolysates by biological detoxification. J Clean Prod 336:130408. https://doi.org/10.1016/j.jclepro.2022.130408
125. Agu CV, Ujor V, Gopalan V, Ezeji TC (2016) Use of *Cupriavidus basilensis* -aided bioabatement to enhance fermentation of acid-pretreated biomass hydrolysates by *Clostridium beijerinckii.* J Ind Microbiol Biotechnol 43:1215–1226. https://doi.org/10.1007/s10295-016-1798-7
126. Fonseca BG, Moutta RDO, Ferraz FDO, Vieira ER, Nogueira AS, Baratella BF, Rodrigues LC, Hou-Rui Z, Da Silva SS (2011) Biological detoxification of different hemicellulosic hydrolysates using Issatchenkia occidentalis CCTCC M 206097 yeast. J Ind Microbiol Biotechnol 38:199–207. https://doi.org/10.1007/s10295-010-0845-z
127. Alves De Oliveira R, Schneider R, Vaz Rossell CE, Maciel Filho R, Venus J (2019) Polymer grade l-lactic acid production from sugarcane bagasse hemicellulosic hydrolysate using Bacillus coagulans. Bioresource Technol Rep 6:26–31. https://doi.org/10.1016/j.biteb.2019.02.003
128. Jiang W, Lei Z, Gao H, Jiang Y, Lin CSK, Zhang W, Xin F, Jiang M (2024) Biodetoxification of lignocellulose hydrolysate for direct use in succinic acid production. BioDesign Res 6:0044. https://doi.org/10.34133/bdr.0044
129. Austin S, Kontur WS, Ulbrich A, Oshlag JZ, Zhang W, Higbee A, Zhang Y, Coon JJ, Hodge DB, Donohue TJ, Noguera DR (2015) Metabolism of multiple aromatic compounds in corn stover hydrolysate by *Rhodopseudomonas palustris.* Environ Sci Technol 49:8914–8922. https://doi.org/10.1021/acs.est.5b02062
130. Nichols NN, Sharma LN, Mowery RA, Chambliss CK, Van Walsum GP, Dien BS, Iten LB (2008) Fungal metabolism of fermentation inhibitors present in corn stover dilute acid hydrolysate. Enzyme Microb Technol 42:624–630. https://doi.org/10.1016/j.enzmictec.2008.02.008
131. Zou L, Ouyang S, Hu Y, Zheng Z, Ouyang J (2021) Efficient lactic acid production from dilute acid-pretreated lignocellulosic biomass by a synthetic consortium of engineered Pseudomonas putida and Bacillus coagulans. Biotechnol Biofuels 14:227. https://doi.org/10.1186/s13068-021-02078-7
132. Soares RC, Zangirolami TC, Giordano RLC, Demeke MM, Thevelein JM, Milessi TS (2022) Cell immobilization using alginate-based beads as a protective technique against stressful

conditions of hydrolysates for 2G ethanol production. Polymers 14:2400. https://doi.org/10.3390/polym14122400
133. Talebnia F, Taherzadeh MJ (2006) In situ detoxification and continuous cultivation of dilute-acid hydrolyzate to ethanol by encapsulated S. cerevisiae. J Biotechnol 125:377–384. https://doi.org/10.1016/j.jbiotec.2006.03.013
134. Sadeghi A, Ebrahimi M, Shahryari S, Assadpour E, Jafari SM (2024) Potential applications of encapsulated yeasts especially within alginate and chitosan as smart bioreactors and intelligent micro-machines. Carbohydrate Polymer Technol Appl 7:100513. https://doi.org/10.1016/j.carpta.2024.100513
135. Ylitervo P, Franzén C, Taherzadeh M (2013) Impact of furfural on rapid ethanol production using a membrane bioreactor. Energies 6:1604–1617. https://doi.org/10.3390/en6031604
136. Milessi TS, Sandri JP, Arruda PV, Esteves TD, Pinheiro LP, Kumar V, Chandel AK (2025) Role of non-genetically modified or native pentose fermenting microorganisms in establishing viable lignocellulosic biorefineries in the Brazilian context. Crit Rev Biotechnol 1–19. https://doi.org/10.1080/07388551.2025.2452628
137. Hemansi H, Patel AK, Saini JK, Singhania RR (2022) Development of multiple inhibitor tolerant yeast via adaptive laboratory evolution for sustainable bioethanol production. Biores Technol 344:126247. https://doi.org/10.1016/j.biortech.2021.126247
138. Chen A, Zhang B, Bao J (2024) Adaptive evolution of Paecilomyces variotii enhanced the biodetoxification of high-titer inhibitors in pretreated lignocellulosic feedstock. Biores Technol 411:131351. https://doi.org/10.1016/j.biortech.2024.131351
139. Almeida ELM, Ventorim RZ, Ferreira MAM, Costa MD, Mantovani HC, Silveira WB (2023) New Papiliotrema laurentii UFV-1 strains with improved acetic acid tolerance selected by adaptive laboratory evolution. Fungal Genet Biol 164:103765. https://doi.org/10.1016/j.fgb.2022.103765
140. Chang D, Islam ZU, Zheng J, Zhao J, Cui X, Yu Z (2021) Inhibitor tolerance and bioethanol fermentability of levoglucosan-utilizing Escherichia coli were enhanced by overexpression of stress-responsive gene ycfR: the proteomics-guided metabolic engineering. Synthetic Syst Biotechnol 6:384–395. https://doi.org/10.1016/j.synbio.2021.11.003
141. Borchert AJ, Bleem A, Beckham GT (2023) RB-TnSeq identifies genetic targets for improved tolerance of Pseudomonas putida towards compounds relevant to lignin conversion. Metab Eng 77:208–218. https://doi.org/10.1016/j.ymben.2023.04.007
142. Ji L, Shen Y, Xu L, Peng B, Xiao Y, Bao X (2011) Enhanced resistance of Saccharomyces cerevisiae to vanillin by expression of lacA from trametes sp. AH28-2. Biores Technol 102:8105–8109. https://doi.org/10.1016/j.biortech.2011.06.057
143. Milessi TS, Silva CR, Moraes GS, Aquino PM, Giordano RC, Giordano RLC, Zangirolami TC (2020) Continuous 2G ethanol production from xylose in a fixed-bed reactor by native Saccharomyces cerevisiae strain through simultaneous isomerization and fermentation. Cellulose 27:4429–4442. https://doi.org/10.1007/s10570-020-03108-7
144. Theiri M, Chadjaa H, Marinova M, Jolicoeur M (2019) Combining chemical flocculation and bacterial co-culture of *Cupriavidus taiwanensis* and *Ureibacillus thermosphaericus* to detoxify a hardwood hemicelluloses hydrolysate and enable acetone–butanol–ethanol fermentation leading to butanol. Biotechnol Prog 35:e2753. https://doi.org/10.1002/btpr.2753
145. Gyan K, Afedzi AEK, Tanypramphan P, Parakulsuksatid P (2024) A review of the advances in detoxification strategies of lignocellulosic hydrolysate for bio-based succinic acid production. Biocatal Agric Biotechnol 60:103323. https://doi.org/10.1016/j.bcab.2024.103323
146. Wang D, Zhang H, Wang Y, Pinelo M, Mazzei R, Fan R, Wan Y, Luo J (2025) Optimizing enzymatic bioreactors: the role of mass transfer in enhancing catalytic efficiency and stability. Chem Eng J 508:160844. https://doi.org/10.1016/j.cej.2025.160844
147. Niladevi KN, Prema P (2008) Immobilization of laccase from Streptomyces psammoticus and its application in phenol removal using packed bed reactor. World J Microbiol Biotechnol 24:1215–1222. https://doi.org/10.1007/s11274-007-9598-x
148. Maroldi WV, Maciel GM, Rossetto R, Bortolini DG, Fernandes IDAA, Haminiuk CWI (2022) Biosorption of phenolic compounds from *Plinia cauliflora* seeds in residual yeast: kinetic,

equilibrium, and bioaccessibility studies. Food Process Preservation 46:e17156. https://doi.org/10.1111/jfpp.17156
149. Roncal T, Aguirre A, Belaustegui Y, Andrés E (2025) Recovery and purification of acetic acid from extremely diluted solutions using a mixed bed ion exchange resin—technical feasibility. RSC Adv 15:477–488. https://doi.org/10.1039/D4RA08341E
150. Hong T, Wei L, Cui K, Dong Y, Li R, Zhang T, Zhao Y, Luo L (2021) Adsorption performance of volatile organic compounds on activated carbon fibers in a fixed bed column. J Environ Chem Eng 9:106347. https://doi.org/10.1016/j.jece.2021.106347
151. Payne EJR, Griffiths M, Harrison STL, Fagan-Endres MA (2025) Exploring the activated charcoal adsorption column to improve cosmetic grade C-phycocyanin purification from Spirulina. J Appl Phycol. https://doi.org/10.1007/s10811-025-03458-4
152. Rokanuzzaman M, Veawab A, Aroonwilas A (2017) Design method for layered-bed adsorption column for separation of CO_2 and N_2 from natural gas. Energy Proc 114:2441–2449. https://doi.org/10.1016/j.egypro.2017.03.1391
153. Villarreal MLM, Prata AMR, Felipe MGA, Almeida E Silva JB (2006) Detoxification procedures of eucalyptus hemicellulose hydrolysate for xylitol production by Candida guilliermondii. Enzyme Microbial Technol 40:17–24. https://doi.org/10.1016/j.enzmictec.2005.10.032
154. Baral NR, Shah A (2016) Techno-Economic analysis of cellulosic butanol production from corn stover through acetone–butanol–ethanol fermentation. Energy Fuels 30:5779–5790. https://doi.org/10.1021/acs.energyfuels.6b00819
155. Gubicza K, Nieves IU, Sagues WJ, Barta Z, Shanmugam KT, Ingram LO (2016) Techno-economic analysis of ethanol production from sugarcane bagasse using a liquefaction plus simultaneous saccharification and co-fermentation process. Biores Technol 208:42–48. https://doi.org/10.1016/j.biortech.2016.01.093
156. Klein BC, Silva JFL, Junqueira TL, Rabelo SC, Arruda PV, Ienczak JL, Mantelatto PE, Pradella JGC, Junior SV, Bonomi A (2017) Process development and techno-economic analysis of bio-based succinic acid derived from pentoses integrated to a sugarcane biorefinery. Biofuels Bioprod Bioref 11:1051–1064. https://doi.org/10.1002/bbb.1813
157. Kosamia NM, Sanchez A, Rakshit SK (2024) Scenario-based techno-economics and heat integration feasibility assessment of integrated multiproduct biorefineries with biosuccinic acid as the main product and various byproduct options. Biomass Conv Bioref 14:8729–8743. https://doi.org/10.1007/s13399-022-02945-9
158. Hiloidhari M, Baruah DC, Singh A, Kataki S, Medhi K, Kumari S, Ramachandra TV, Jenkins BM, Thakur IS (2017) Emerging role of geographical information system (GIS), life cycle assessment (LCA) and spatial LCA (GIS-LCA) in sustainable bioenergy planning. Biores Technol 242:218–226. https://doi.org/10.1016/j.biortech.2017.03.079
159. Bhavana BK, Mudliar SN, Debnath S (2023) Life cycle assessment of fermentative xylitol production from wheat bran: a comparative evaluation of sulphuric acid and chemical-free wet air oxidation-based pretreatment. J Clean Prod 423:138666. https://doi.org/10.1016/j.jclepro.2023.138666
160. Gontia P, Janssen M (2016) Life cycle assessment of bio-based sodium polyacrylate production from pulp mill side streams: case study of thermo-mechanical and sulfite pulp mills. J Clean Prod 131:475–484. https://doi.org/10.1016/j.jclepro.2016.04.155
161. Janssen M, Xiros C, Tillman A-M (2016) Life cycle impacts of ethanol production from spruce wood chips under high-gravity conditions. Biotechnol Biofuels 9:53. https://doi.org/10.1186/s13068-016-0468-3
162. Oriez V, Peydecastaing J, Pontalier P-Y (2019) Lignocellulosic biomass fractionation by mineral acids and resulting extract purification processes: conditions, yields, and purities. Molecules 24:4273. https://doi.org/10.3390/molecules24234273
163. Silva DDV, Arruda PV, Dussan KJ, Felipe MGA (2014) Adaptation of scheffersomyces stipitis cells as a strategy to the improvement of ethanol production from sugarcane bagasse hemicellulosic hydrolysate. Chem Eng Trans 38:427–432. https://doi.org/10.3303/CET1438072

164. Tavares APM, Gonçalves MJA, Brás T, Pesce GR, Xavier AMRB, Fernandes MC (2022) Cardoon hydrolysate detoxification by activated carbon or membranes system for bioethanol production. Energies 15:1993. https://doi.org/10.3390/en15061993
165. Yin J, Wang C, Li Y, Lv B, Lv H, Xue Y, Wu J, Zhang J (2023) Biological detoxification of the inhibitors in corncob acid hydrolysate using aspergillus niger. Fermentation 9:854. https://doi.org/10.3390/fermentation9090854
166. Saravanakumar T, Park H-S, Mo A-Y, Choi M-S, Kim D-H, Park S-M (2016) Detoxification of furanic and phenolic lignocellulose derived inhibitors of yeast using laccase immobilized on bacterial cellulosic nanofibers. J Mol Catal B Enzym 134:196–205. https://doi.org/10.1016/j.molcatb.2016.11.006
167. Abbate E, Andrion J, Apel A, Biggs M, Chaves J, Cheung K, Ciesla A, Clark-ElSayed A, Clay M, Contridas R, Fox R, Hein G, Held D, Horwitz A, Jenkins S, Kalbarczyk K, Krishnamurthy N, Mirsiaghi M, Noon K, Rowe M, Shepherd T, Tarasava K, Tarasow TM, Thacker D, Villa G, Yerramsetty K (2023) Optimizing the strain engineering process for industrial-scale production of bio-based molecules. J Ind Microbiol Biotechnol 50:kuad025. https://doi.org/10.1093/jimb/kuad025
168. Han T, Zhang B, Yang H, Liu X, Bao J (2023) Changes in pH values allow for a visible detection of the end point in submerged liquid biodetoxification during biorefinery processing. ACS Sustain Chem Eng 11:16608–16617. https://doi.org/10.1021/acssuschemeng.3c04986
169. Fernández-Sandoval MT, García A, Teymennet-Ramírez KV, Arenas-Olivares DY, Martínez-Morales F, Trejo-Hernández MR (2024) Removal of phenolic inhibitors from lignocellulose hydrolysates using laccases for the production of fuels and chemicals. Biotechnol Prog 40:e3406. https://doi.org/10.1002/btpr.3406
170. Rostami A, Abdelrasoul A, Shokri Z, Shirvandi Z (2022) Applications and mechanisms of free and immobilized laccase in detoxification of phenolic compounds—a review. Korean J Chem Eng 39:821–832. https://doi.org/10.1007/s11814-021-0984-0
171. Becerra ML, Lizarazo LM, Rojas HA, Prieto GA, Martinez JJ (2022) Biotransformation of 5-hydroxymethylfurfural and furfural with bacteria of bacillus genus. Biocatal Agric Biotechnol 39:102281. https://doi.org/10.1016/j.bcab.2022.102281
172. Li Z, Zhang L, He N, Zhang B, Bao J (2024) Competition between biodetoxification fungus and lactic acid bacterium in the biorefinery processing chain for production of cellulosic L-lactic acid. Bioresour Bioprocess 11:54. https://doi.org/10.1186/s40643-024-00772-6
173. Liu C, Choi B, Efimova E, Nygård Y, Santala S (2024) Enhanced upgrading of lignocellulosic substrates by coculture of *Saccharomyces cerevisiae* and *Acinetobacter baylyi* ADP1. Biotechnol Biofuels 17:61. https://doi.org/10.1186/s13068-024-02510-8
174. Panda SK, Maiti SK (2024) Fungus-yeast tri-culture system for in situ cellulase production, biodetoxification, and bioethanol production using rice straw with cyclic shifting of temperature strategy. Bioenerg Res 18:3. https://doi.org/10.1007/s12155-024-10806-8
175. Morozova T, Semyonov S (2016) Biological detoxification of lignocellulosic hydrolysates for improved biobutanol production. Key Eng Mater 683:525–530. https://doi.org/10.4028/www.scientific.net/KEM.683.525
176. Kannisto MS, Mangayil RK, Shrivastava-Bhattacharya A, Pletschke BI, Karp MT, Santala VP (2015) Metabolic engineering of *Acinetobacter baylyi* ADP1 for removal of *Clostridium butyricum* growth inhibitors produced from lignocellulosic hydrolysates. Biotechnol Biofuels 8:198. https://doi.org/10.1186/s13068-015-0389-6
177. Tian L, Qi T, Zhang F, Tran VG, Yuan J, Wang Y, He N, Cao M (2025) Synthetic biology approaches to improve tolerance of inhibitors in lignocellulosic hydrolysates. Biotechnol Adv 78:108477. https://doi.org/10.1016/j.biotechadv.2024.108477
178. Wang Y, Zhang Y, Cui Q, Feng Y, Xuan J (2024) Composition of lignocellulose hydrolysate in different biorefinery strategies: nutrients and inhibitors. Molecules 29:2275. https://doi.org/10.3390/molecules29102275
179. Li B, Liu N, Zhao X (2022) Response mechanisms of *Saccharomyces cerevisiae* to the stress factors present in lignocellulose hydrolysate and strategies for constructing robust strains. Biotechnology Biofuels Bioproducts 15:28. https://doi.org/10.1186/s13068-022-02127-9

180. Guan Y, Chen J, Nepovimova E, Long M, Wu W, Kuca K (2021) Aflatoxin detoxification using microorganisms and enzymes. Toxins (Basel) 13:46. https://doi.org/10.3390/toxins13010046

Recent Developments in Ethanol Recovery After Fermentation of 1G, 1.5G, and 2G Sugars

Álvaro Henrique Mello José, Thiago Queiroz Jardim Rodrigues, Bruna Green Rodrigues, Heitor Gonçalves da Silva, Lucas Ramos, Willian de Souza Matias Reis, and Carina Aline Prado

Abstract This chapter provides a comprehensive overview of the most recent advancements in ethanol recovery following fermentation of sugars from first-generation (1G), intermediate-generation (1.5G), and second-generation (2G) biomass. As the world seeks alternatives to fossil fuels, ethanol stands out for its potential in reducing greenhouse gas emissions and promoting sustainable energy systems. The chapter explores conventional and emerging separation technologies such as azeotropic and extractive distillation, adsorption, pervaporation, hybrid systems, and membrane technologies. It highlights specific challenges and innovations associated with each feedstock type, analyzing economic and environmental implications across production generations. Future trends are also discussed, including the integration of ethanol recovery into biorefinery systems, the role of artificial intelligence, and the development of low-energy and eco-friendly technologies. These insights contribute to understanding how technological innovation can

Á. H. M. José (✉) · T. Q. J. Rodrigues · B. G. Rodrigues · H. G. da Silva · L. Ramos · W. de S. M. Reis · C. A. Prado
Department of Biotechnology, Lorena School of Engineering-EEL, University of São Paulo, Lorena, SP, Brazil
e-mail: alvaro_he@usp.br

T. Q. J. Rodrigues
e-mail: thiago_queiroz@usp.br

B. G. Rodrigues
e-mail: bgreen@usp.br

H. G. da Silva
e-mail: heitorgoncalves2010@usp.br

L. Ramos
e-mail: lucas.eng12@usp.br

W. de S. M. Reis
e-mail: willian.matias@usp.br

C. A. Prado
e-mail: carinadoutor@gmail.com

© The Author(s), under exclusive license to Springer Nature Singapore Pte Ltd. 2025
A. K. Chandel and S. Sundaramurthy (eds.), *Biorefineries in Circular Bioeconomy*, Green Energy and Technology, https://doi.org/10.1007/978-981-95-3913-0_11

optimize ethanol production and make it a more viable and sustainable solution in the global energy transition.

Keywords 1G, 1.5G, and 2G bioethanol · Bioethanol purification · Separation technologies · Sustainable biorefineries · Low-energy processes

1 Introduction

The global population was estimated at 8.2 billion in 2024 and is projected to reach approximately 10.3 billion by the mid-2080s (United Nations, 2024). As the population grows, the demand for food, water, energy, and essential chemicals will also increase. Global energy consumption has risen sharply due to rapid industrialization, urbanization, and economic growth, leading to an overreliance on conventional fossil fuels such as coal, oil, and natural gas [1]. These fuels account for nearly 86% of the world's energy use, significantly contributing to greenhouse gas (GHG) emissions and climate change [2].

To achieve carbon neutrality by 2050, there is an urgent need to transition toward renewable energy sources [3]. Among these, bioethanol has emerged as a promising alternative to fossil fuels, offering lower toxicity, biodegradability, and the potential to reduce GHG emissions [4]. The global bioethanol market, valued at USD 83.4 billion in 2023, is projected to reach USD 114.7 billion by 2028, growing at a compound annual growth rate (CAGR) of 6.6% (OECD/FAO, 2024). Ethanol production is expected to reach 155 billion liters by 2033, with the United States and Brazil remaining the leading producers. In 2023, the main feedstocks for ethanol production were corn (59%), sugarcane (24%), molasses (6%), wheat (2%), and a mix of other raw materials (9%) [5].

Bioethanol is classified into different generations based on feedstock and production methods. First-generation (1G) biofuels are derived from food-based raw materials such as sugarcane, sugar beet, corn, and other starch-rich crops. Second-generation (2G) biofuels utilize lignocellulosic biomass, including agricultural residues. Third-generation (3G) biofuels are derived from algae and nonedible microorganisms, while fourth-generation (4G) biofuels involve genetically modified microorganisms [6]. Among these, 1G ethanol remains the most commercially viable, benefiting from established infrastructure, government policies, and technological advancements.

This chapter explores recent developments in ethanol recovery across different bioethanol production pathways. The discussion covers ethanol recovery techniques for 1G, 1.5G, and 2G sugars, evaluating their economic and environmental implications while highlighting emerging trends and innovations.

2 Ethanol Recovery Techniques: An Overview

The microbial conversion of sugars into bioethanol results in a highly diluted product, with over 80% of the fermentation broth consisting of water and ethanol concentrations ranging from 5 to 10%. This limitation arises from microbial inhibition at higher ethanol levels [7]. To meet fuel-grade ethanol specifications (95.6% or > 99%), water removal is required using recovery (preconcentration) and dehydration (purification) techniques [8]. Serra et al. (1987) categorize these into conventional distillation-based methods and nonconventional alternatives [9].

2.1 *Conventional Distillation Techniques*

Distillation is the most widely used method in biorefineries, initially concentrating ethanol to an azeotropic composition (~92.4% by mass or 95.6% by volume) in beer or preconcentration columns [8, 10]. Advanced distillation processes, such as azeotropic and extractive distillation, enable further dehydration by altering component volatilities using entrainers. The choice of entrainer depends on process feasibility, energy requirements, and environmental impact. Ionic liquids and inorganic salts are emerging as sustainable alternatives to traditional organic solvents [11].

Azeotropic distillation involves multiple stages, using a large solvent volume to achieve additional purification. Extractive distillation, performed under vacuum to prevent solvent degradation, requires less solvent and avoids forming new azeotropes [12, 13]. Despite their high ethanol recovery (>99%), these techniques face challenges such as high energy consumption and process complexity, leading to interest in nonconventional alternatives. As shown in Table 1, extractive distillation presents lower solvent consumption but may require vacuum, increasing process complexity.

2.2 *Nonconventional Separation Techniques*

To address the limitations of distillation, alternative methods such as adsorption, pervaporation, and hybrid approaches have been explored [14].

2.2.1 Adsorption

Adsorption selectively removes water using solid materials that can be regenerated via heating, solvent extraction, or vapor displacement [15]. The choice of adsorbent is crucial, with zeolites and ion-exchange resins being widely studied, alongside lignocellulosic materials (activated carbon, corn cobs, straw, bagasse) and starch-based adsorbents [8, 16]. Molecular sieves (3–4 Å zeolites) are the most common,

Table 1 Bioethanol separation technologies from raw fermentation broths: principle, advantages, disadvantages, and energy consumption

Technique	Principle of the technique	Advantage	Disadvantage	Energy consumption	References
Azeotropic distillation	Addition of an entrainer for azeotrope formation	Ease of operation, high recovery rate, well-established technology	High temperatures, azeotrope formation, culture death	High	[16]
Extractive distillation	Separation based on the difference in boiling points (ethanol vs. water) through the use of an entrainer	Possibility of using various entrainers (liquid, ionic liquid, salt, polymers). Lower solvent consumption. Well-established technology	High entrainer cost, increased process complexity. May require vacuum	Moderate to high	[12]
Vacuum fermentation	Change in the liquid–vapor equilibrium	Does not harm the microorganism, allows simultaneous and continuous fermentation/purification operation, easy handling	High initial investment (more complex equipment), strict process control, low selectivity	Moderate to high	[25]
Adsorption process	Use of selective adsorbents for ethanol removal	Operates at lower temperatures, does not harm the culture, easy handling	Requires adsorbent regeneration steps, limited capacity, high cost (depending on the adsorbent), and low selectivity	Moderate to low	[8, 15]
Pervaporation	Separation by selective membranes, where ethanol vaporizes due to the pressure difference	High selectivity, low damage to the culture	Possibility of membrane fouling, high acquisition and maintenance costs	Moderate to low	[26]

(continued)

Table 1 (continued)

Technique	Principle of the technique	Advantage	Disadvantage	Energy consumption	References
Liquid–liquid extraction	Use of a solvent immiscible with water for separation into two aqueous phases	Allows use in highly diluted solutions, enables solvent regeneration	Need for additional steps for solvent removal. Potential contamination with solvent residues	Moderate	[10]
Reverse osmosis	Use of semipermeable membranes based on pressure difference	Operates at room temperature, high selectivity, does not cause damage to the culture	Requires high pressure, may suffer from fouling, requires maintenance, high investment cost	Moderate to high	[27]

selectively retaining water while allowing ethanol to pass [15]. Table 1 highlights that adsorption operates at lower temperatures and does not harm microorganisms, although it has capacity limitations and requires adsorbent regeneration.

2.2.2 Membrane Separation (Pervaporation)

Pervaporation employs selective membranes to separate ethanol–water mixtures based on partial vapor pressure differences. The success of this method depends on membrane composition, thickness, and selectivity [17]. Membranes can be hydrophilic (e.g., poly(vinyl alcohol), cellulose, chitosan) or hydrophobic (e.g., PEBA, PTMSP), with mixed-matrix membranes (zeolite, silica, COF, MOF) showing promise in surpassing polymeric membranes' selectivity [18, 19]. While pervaporation remains a complementary process, recent advancements in membrane materials aim to enhance efficiency and cost-effectiveness [20, 21]. According to Table 1, one of the disadvantages of pervaporation is the membrane incrustation potential and the high acquisition and maintenance costs.

2.2.3 Hybrid Processes

To improve process efficiency, hybrid approaches integrate conventional distillation with nonconventional techniques, such as distillation-adsorption and distillation-membrane separation. More advanced strategies involve supercritical fluids and ionic

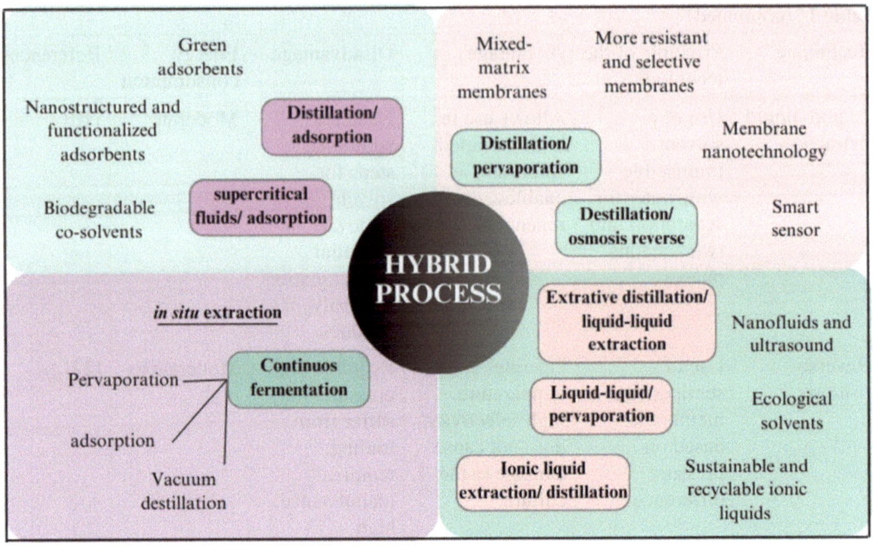

Fig. 1 Integration of advanced separation techniques in hybrid processes for ethanol recovery

liquids [22, 23]. Emerging trends focus on sustainable solvents, nanotechnology for enhanced selectivity, and in situ extraction to reduce energy consumption [24]. The concept of hybrid processes is illustrated in Fig. 1, highlighting key techniques such as distillation-adsorption, distillation-pervaporation, and liquid–liquid pervaporation, among others. These approaches incorporate green adsorbents, mixed-matrix membranes, and biodegradable cosolvents to improve efficiency and sustainability.

As shown in Fig. 1, hybrid processes leverage multiple separation techniques to optimize ethanol recovery while minimizing energy consumption. The integration of membrane nanotechnology, ionic liquid extraction, and continuous fermentation allows for innovative solutions that enhance ethanol purification. Additionally, sustainable strategies such as the use of biodegradable cosolvents and recyclable ionic liquids are gaining prominence, aligning with environmental sustainability goals.

Each ethanol recovery method has advantages and limitations, with economic feasibility remaining a key challenge for large-scale implementation. Table 1 summarizes the principles, benefits, drawbacks, and energy requirements of various separation techniques, including adsorption, pervaporation, vacuum fermentation, and liquid–liquid extraction. Future advancements in adsorbents, membranes, and hybrid systems will play a crucial role in optimizing bioethanol recovery for industrial applications [25]. The integration of novel materials and process intensification strategies will be essential for achieving a more efficient and sustainable ethanol production process.

3 Ethanol Recovery from 1G Sugars

First-generation (1G) ethanol is primarily produced from sugar- and starch-based feedstocks such as sugarcane, sugar beet, corn, and cassava. These crops provide readily fermentable sugars that undergo microbial fermentation to yield bioethanol. The efficiency of ethanol recovery is influenced by factors such as feedstock composition, fermentation conditions, and downstream processing technologies.

3.1 Sugarcane-Based Ethanol Production and Recovery

Sugarcane is one of the most widely used raw materials for bioethanol production, particularly in Brazil, which is expected to account for 81% of global sugarcane-based ethanol by 2033 [5]. With annual harvests and replanting cycles every five years, sugarcane can yield over 8000 L of ethanol per hectare per year. The production process begins with sugarcane milling to extract juice, which undergoes clarification and sterilization before fermentation. The resulting broth contains approximately 10.5% ethanol [28].

A key advantage of sugarcane-based ethanol is its cost competitiveness and environmental benefits. It has a theoretical ethanol yield of 617 L per metric ton of sucrose, with operational yields ranging between 510 and 530 L per ton. Furthermore, it can reduce GHG emissions by up to 80% compared to gasoline and is classified as an "advanced biofuel" by the U.S. Environmental Protection Agency (EPA) [29].

3.1.1 Sugar Beet as a Feedstock for Ethanol Recovery

Sugar beet is another major raw material for ethanol production, particularly in Europe and the United States. With a sucrose content of 16–20%, it can yield up to 120 L of ethanol per ton of beet, translating to approximately 5000 L/ha [30]. Under optimal conditions, ethanol yields may exceed 11,000 L/ha. However, sugar beet processing requires chemical or enzymatic hydrolysis, which can be energy-intensive and costly [31].

3.1.2 Starch-Based Ethanol Production and Recovery

Corn is the primary feedstock for bioethanol production in the United States, where 34% of total corn production was allocated to ethanol in 2023 [5]. Corn kernels contain approximately 60% starch, which must be hydrolyzed into glucose before fermentation. This is typically achieved through enzymatic processes involving α-amylase and glucoamylase.

Two main processing methods exist for corn-based ethanol recovery:

- **Dry milling**: The most common method, where corn is ground, mixed with water, and enzymes are added to convert starch into dextrose. After fermentation, ethanol is distilled and dehydrated to approximately 99.3%. The remaining solids (stillage) are processed into animal feed.
- **Wet milling**: Involves soaking corn in water and sulfuric acid for 24–48 h, allowing the separation of components such as starch, fiber, gluten, and germ. The extracted starch is converted into ethanol, while by-products are used in the food and feed industries [28].

Starch-based ethanol production is widely adopted due to its high yield and well-established infrastructure. However, enzymatic hydrolysis and fermentation processes involve high energy costs, particularly due to steam and cooling requirements [32]. Novel approaches integrating liquefaction, saccharification, and fermentation into a single step have shown potential to reduce processing costs [33].

3.2 Challenges in Ethanol Recovery

Regardless of the feedstock, ethanol recovery involves three main steps: fermentation, distillation, and dehydration. One of the primary challenges is the formation of an ethanol–water azeotrope at 95.6% ethanol by weight, which prevents complete separation through conventional distillation alone. To achieve anhydrous ethanol, additional separation techniques such as extractive distillation, azeotropic distillation, pervaporation, adsorption, or hybrid methods are required [34].

Optimizing ethanol recovery from 1G feedstocks involves improving fermentation efficiency, reducing energy consumption in distillation, and integrating advanced dehydration technologies. These improvements can enhance the overall sustainability and economic viability of bioethanol production.

4 Ethanol Recovery from 1.5G Sugars

4.1 Definitions of 1.5G Ethanol

A comprehensive literature review presents two conceptualizations of "1.5G Ethanol." The first defines it as an intermediate stage in the transition from 1 to 2G ethanol, where traditional 1G ethanol facilities integrate technologies to process lignocellulosic biomass residues, enhancing biofuel yield and economic viability [35–37]. Sugarcane and corn are the main industrial crops, with bagasse, straw, and corn stover being key residues [35–38].

Another, less established definition associates 1.5G ethanol with bioethanol production from nonfood crops or industrial residues [39]. Feedstocks include sweet sorghum, cassava, fiber feedstocks, and unconventional materials like durum clear flour, brewer's spent grain, olive stones, and rice hulls [38, 40–42]. Due to the predominance of the first concept in literature, it will be the focus of this chapter.

4.2 Overview of 1.5G Ethanol Production Characteristics

Sugarcane and corn are the most studied cases of 1.5G ethanol production, integrating 1G and 2G processes. Converting lignocellulosic residues into biofuels requires overcoming the natural recalcitrance of plant cell walls due to hemicellulose, cellulose, and lignin interactions [37, 43, 44]. This necessitates pretreatment to break linkages, facilitating hydrolysis. Commercial methods include autohydrolysis with high-temperature steam and dilute acid pretreatment, each producing distinct sugar and inhibitor profiles [43, 44].

Enzymatic hydrolysis further breaks down cellulose (C-6) and hemicellulose (C-5 and C-6) into fermentable sugars, often using separate or combined enzyme complexes at higher temperatures (45 °C–50 °C) than fermentation (30–35 °C) [37]. Besides that, biochemical routes from employed microorganisms must assure simultaneous C-6 and C-5 sugars fermentation into ethanol, a scenery in which economic gains are projected to be the greatest, from the perspective of technical–economic analysis [45–47]. Recent research efforts have been aiming toward obtaining yeast and bacteria recombinants that are able to produce ethanol from cellulose, such as *E. coli*, *Z. mobilis*, and *S. cerevisiae* strains [36, 48].

1.5G ethanol production shares infrastructure with 1G ethanol, including concentration, fermentation, and distillation units [35, 37]. Hydrolyzed liquor is mixed with sugarcane juice, integrating fermentation, distillation, and dehydration operations [35, 37]. Fermentation is typically batch-mode, using multiple fermenters in parallel. Challenges arise from solids in fermenters, requiring agitation or pump systems to mitigate mixing issues and inhibitor formation from phenolics [37, 43, 49].

4.3 Challenges of 1.5G Ethanol Recovery

Recovering 1.5G ethanol faces challenges like 1G and 2G processes, requiring separation and purification to isolate anhydrous ethanol from water, CO_2, and solids. Standard methods include distillation and dehydration [36, 49, 50].

Distillation separates ethanol from a binary ethanol–water azeotrope, an energy-intensive step. In the stripping column, distillate (~50 wt.% ethanol) is purged of residual CO_2 before rectification [51]. The rectification column purifies ethanol to hydrous ethanol, which is then dehydrated to fuel-grade (>99.5% purity) ethanol [49]. Traditional distillation raises ethanol concentration to 92.4–94%, with vinasse by-products directed to evaporation [42, 49]. Established ethanol recovery techniques include pressure-modulated distillation, solvent-assisted distillation (using solvents, salts, or moon polymers), azeotropic distillation, and hybrid approaches [42, 52].

However, conventional methods are energy-intensive, impacting economic feasibility and ethanol prices [42, 53].

4.4 Recent Developments in 1.5G Ethanol Recovery

Recent developments acknowledge the intensive energy costs due to heating operations carried out on ethanol recovery, therefore aiming to reduce energy expenditures with alternative distillation processes or substituting heating process through distinct techniques related to exploiting chemical affinities between Ethanol and other chemical compounds [42, 54, 55].

In distillation improvements, cyclic distillation offers lower energy consumption and enhanced separation by alternating vapor flow and liquid flow periods, which leads advantages, such as reduced energy requirements and increased separation performance [42, 54]. Pass-through distillation (PFD) introduces an absorption–desorption loop between evaporation and condensation, decoupling pressure effects and reducing separation costs with smaller equipment [55].

Membrane technologies also gain attention due to their low energy requirements and suitability for temperature-sensitive materials. Pervaporation is particularly promising for its simplicity, reduced energy use, and absence of additional chemicals. It utilizes impermeable membranes, such as polydimethylsiloxane (PDMS), block amides, and polypropylene, which are categorized as polymeric, inorganic, or mixed matrix membranes [42, 53, 55].

However, many membranes originate from fossil fuels, raising sustainability concerns. To address this, research explores alternative materials like recycled PET membranes, which offer competitive separation performance (factor of 10.45) compared to conventional fossil-based membranes [53].

5 Ethanol Recovery from 2G Sugars

5.1 Lignocellulosic Feedstocks and Their Role in 2G Ethanol Recovery

The feedstocks used to produce the 2G ethanol are the lignocellulosic biomass, including agricultural residues and forestry waste. The diversity of lignocellulosic residues is vast, especially in Brazil, an agro-industrial country with a large territorial extension. In addition to sugarcane, there are a wide range of agricultural crops spread across different regions of the country. While sugarcane is fully utilized, other crops can also be used to obtain fermentable sugars [56, 57].

Although the chemical composition of each biomass varies, pretreatment processes for delignification are well-established and can be applied to a wide range of biomass types. The type of biomass available is specific to each region, and its utilization should avoid competing with the food supply chain. Thus, leaves, branches, wood chips, corn cobs, grass, and other residues can be effectively utilized [56–58].

Brazil also has competitive advantages over other countries in second-generation ethanol, such as lower costs and the easy availability of large quantities of raw materials, particularly due to the possibility of using bagasse and sugarcane straw in the processing. This technology can be integrated into the existing infrastructure of production units, reducing installation and operating costs [56, 59].

Furthermore, the gradual adoption of mechanized harvesting, the prohibition of cane straw burning, and the increasing use of high-performance boilers will lead to surplus bagasse and straw, which can be converted into second-generation ethanol. In addition to bagasse and straw, other residues can also be utilized depending on seasonal availability [56].

5.2 Pretreatment Impact on Recovery: Effect of Inhibitors like Furfural and HMF

Currently, there are a variety of pretreatments for the physicochemical modification of lignocellulosic biomass. Pretreatments can be chemical, physical, biological, or even hybrid process. Each pretreatment modifies the biomass in a different way. For industrial second-generation ethanol production, steam explosion pretreatment is commonly used. The choice of pretreatment will depend on the process's objective, whether the interest is in the cellulose pulp or in the lignin [59]. Most of the traditional pretreatment methods (acids and alkali) have disadvantages including the release of the degradation products, lack of selectivity, low sugar yields, lower process efficiency, and higher processing costs. There are always open opportunities to develop a novel and green pretreatment method with low cost, lower energy consumption, lower waste, and new product production [59].

The ideal pretreatment should not result in sugar loss or degradation. It should produce minimal toxic compounds that would require additional steps for their removal. Among the inhibitory compounds formed during pretreatment are furan aldehydes (furfural, 5-hydroxy methyl furfural (HMF)), phenolics (vanillin), weak acids (acetic acid and formic acid), along with fermentable sugars, which inhibit microbial growth by causing intracellular acidification, energy drainage, and accumulation of reactive oxygen species [60, 61].

Biomass hydrolysate requires detoxification before it can be used in microbial fermentation to avoid negative effects during the fermentation process. Various physical (evaporation, membrane filtration), chemical (resin exchange, alkali detoxification, active carbon adsorption), and biological (microbial and enzymatic) detoxification methods have been reported to remove the inhibitor. The choice of method, or combination of methods, will depend on the concentration of inhibitors and the characteristics of the obtained hydrolysate [4]. Table 2 shows some methods of hydrolysate detoxification.

Ideally, the aim is to avoid adding high costs at this stage, pretreatment is optimized to minimize reaction severity, enhance delignification, reduce inhibitor formation, limit the degradation of hemicellulose and sugars, and keep costs low [62].

Table 2 Different purification techniques for removing inhibitors from lignocellulosic plant biomass hydrolysates, along with their benefits and limitations

Technique	Procedure	Advantages	Disadvantages	References
Liquid–liquid extraction	Ethyl acetate	Ease of scalability, solvent recovery, and selectivity	Toxicity of organic solvents	[63]
	Supercritical fluid extraction	Selectivity and efficient removal	High cost and high energy consumption	[64]
Liquid–solid extraction	Activated carbon	Low cost, simple process, and high adsorption capacity	Adsorption of desirable compounds and the need for charcoal regeneration	[62, 65]
	Ion exchange	Resins can be regenerated and reused. Successfully applied for removal of lignin-derived inhibitors	Pore diffusion is slow and requires high processing time and pressure. Scaling up is difficult	[66]
Chemical additives	Alkali	Easy application and control, low cost, and scalability	Risk of unwanted reactions and possible contamination	[67, 68]
	Reducing agents	Low cost and easy removal, high specificity, and efficiency	Possibility of degradation of beneficial compounds	[68]
Heating and vaporization	Evaporation	Able to remove volatile compounds (acetic acid, furfural, and vanillin)	Increasing the concentration of nonvolatile compounds	[69]
	Heating treatment	Fast process, ease scalability, and low cost	High energy cost and possibility of degradation of beneficial compounds	[70]
Biodetoxification	*Rhodococcus aetherivorans*	Ability to degrade organics complex compounds	Limited scalability and cultivation and maintenance costs	[4]
	Laccase	Sustainable process	High cost and difficulty in scaling up	[71]

5.3 *Technology Adaptations: Low-Energy Distillation, Advanced Membrane Systems, and Solvent Extraction*

Advances in 2G ethanol production, coupled with the need to reduce greenhouse gas emissions, drive a continuous search for innovative technologies with lower environmental impact. In the ethanol recovery stage through distillation, a low-energy distillation (LED) is employed to reduce energy consumption. This technology stands out as a reconfiguration of the ethanol recovery process that does not rely on a membrane-based approach and saves up to 50% of the steam generated per gallon of ethanol [72, 73]. The goal of LED is to arrange the hottest temperature and highest pressure on the cleanest fluids while reusing energy in multiple effects. Unnecessary phase changes are historically one of the largest users of energy in distillation, but LED removes them. This adjustment, while significant in the amount of energy conserved, is simple and operates similarly to existing distillation systems [73].

In addition to this technology, the use of the membrane system for dehydration via pervaporation is an alternative that reduces the unit's steam consumption. However, this technology is still costly, despite being highly efficient and reducing energy consumption. In the conventional distillation process, cyclohexane is used as a ternary agent, making it an energetically unfavorable process. However, this process is optimized through a multistage operation that uses three pressure levels, which increases distillation efficiency and reduces energy consumption. Among dehydration technologies, pervaporation is the most energy-efficient [57, 72, 74].

6 Economic and Environmental Implications of Ethanol Recovery

6.1 *Energy Efficiency: Comparing Energy Requirements Across Generations*

Energy efficiency is a key factor in reducing greenhouse gas emissions and achieving sustainable energy policies. Ethanol, which accounts for 65% of global biofuel production, plays a crucial role in energy security. The United States and Brazil lead global production in 2024, yielding 61.4 and 36.83 billion liters, respectively, primarily from corn and sugarcane [75, 76]. Despite its benefits, ethanol production faces challenges related to energy consumption and waste management, driving research into more efficient processes.

6.2 Environmental Benefits

Ethanol production impacts water use, emissions, and waste generation. The 1.5G process produces less vinasse and increases biomass utilization, improving sustainability. Corn-based ethanol has the highest water consumption, while sugarcane-based ethanol is more efficient due to its higher yield. 2G ethanol further reduces water use and produces valuable by-products like lignin, which can be used in bioproducts [77]. Regarding emissions, corn-based ethanol reduces GHG emissions by 50%, while sugarcane ethanol achieves up to 80% reduction compared to gasoline [78] (Table 3).

6.3 Economic Feasibility

Ethanol production costs vary across generations. 1G ethanol is dependent on feedstock prices and government subsidies. 1.5G ethanol has slightly higher costs due to additional processing but benefits from waste utilization. 2G ethanol faces the highest capital and operational expenses, primarily due to costly enzymes and pretreatment requirements. However, advanced recovery systems could improve long-term economic viability [83].

6.4 Advanced Recovery Systems: Costs and Benefits

Technologies like membrane separation and enzymatic hydrolysis enhance ethanol production efficiency. Membrane filtration reduces water effluents and operational costs by 20% compared to conventional distillation [84]. Additionally, lignin valorization in 2G ethanol production could offset enzyme costs if effectively commercialized. Biorefineries using bagasse for power generation further minimize waste disposal expenses [20].

6.5 Challenges and Future Perspectives

1.5G ethanol improves resource efficiency but requires integration with existing infrastructure. 2G ethanol, despite its environmental advantages, faces significant scalability barriers due to high enzyme costs and infrastructure demands. To make 2G ethanol competitive, cost reductions in enzyme production and advancements in recovery technologies are crucial [83].

While 1G ethanol remains the most cost-effective option, its long-term viability depends on feedstock availability and policy support. The 1.5G approach offers a

Table 3 Comparative different scenarios with 1 G, 1.5 G, and 2 G ethanol production

Generation	Feedstock type	Energy input (MJ/L)	Water use (L/L Ethanol)	Efficiency	Key challenges	Main waste products	Management strategies	References
1 G	Sugar/starch crops	10–15	1,000–3,000 (corn, wheat) 500–1000 (sugarcane)	Moderate	Food vs. fuel, high land use	Vinasse (sugarcane) Distillers' grains (corn)	Vinasse used as fertilizer, grains for animal feed	[79–81]
1.5 G	Residues + 1G crops	8–12	300–800	Higher	Process integration, limited scalability	Bagasse, lignin, vinasse	Bagasse used for energy, lower vinasse per liter	[82]
2 G	Lignocellulosic biomass	5–10	50–300	Potentially highest	High processing costs, technology development	Lignin, enzyme residues	Lignin used for bioenergy, enzymatic recycling under development	[77]

balanced alternative, increasing efficiency while leveraging existing infrastructure. Future advancements in recovery systems and process integration will be key to making 2G ethanol a viable large-scale solution [85–87].

7 Future Trends and Innovations

The production of biofuels, such as bioethanol, holds significant potential as a sustainable alternative to fossil fuels [88, 89]. However, the future of this sector depends on continuous research and technological advancements that improve the efficiency and economic viability of second-generation biofuels [90]. Future scenarios for bioethanol focus on integrating cost-effective and sustainable technological solutions. Biorefineries play a crucial role in the circular bioeconomy, offering environmental and economic benefits, but increasing process efficiency is essential to fully realize these advantages [91, 92]. Future research should prioritize the economic feasibility of biorefineries, emphasizing biofuel and biochemical integration, bioprocess optimization, scale-up strategies, and the use of computational tools and algorithms for bioprocess analysis [89, 93].

7.1 Integration with Biorefineries: Ethanol Recovery as Part of Holistic Biomass Processing

Biorefineries convert lignocellulosic biomass into a wide range of products, including biofuels, biochemicals, and energy [88, 90, 94]. The integration of biofuels from waste materials with innovative production processes can significantly reduce greenhouse gas emissions, enhance waste management, and promote sustainable resource utilization [92]. Expanding the range of biomass-derived products while improving their value and purity is a priority, alongside reducing costs associated with upstream and downstream unit operations and minimizing environmental impacts [95].

Upstream operations, such as microbial strain development, biomass collection, storage, and pretreatment, along with downstream processes, including product purification and waste recovery, are energy-intensive and contribute significantly to operational expenses. These economic challenges have been addressed through process integration (PI), which combines or redesigns operational units to reduce production costs, energy consumption, residence time, and waste generation, leading to more cost-effective and safer bioprocesses [88, 90].

For example, Vargas et al. (2023) demonstrated the integrated production of ethanol and xylitol, achieving productivities of 0.42 g of ethanol per gram of glucose and 0.61 g of xylitol per gram of xylose using elephant grass as biomass. The process involved alkaline pretreatment, enzymatic hydrolysis, and fermentation [96].

Another promising advancement is consolidated bioprocessing (CBP), which integrates enzyme production, saccharification, and fermentation within a single bioreactor using multiple microbial species. CBP is considered one of the most promising strategies for reducing costs, improving product quality and yield, minimizing environmental impact, and shortening production times [88, 90]. Gupte et al. (2024) utilized starch-rich municipal waste to produce bioethanol at nearly 100% of the theoretical yield by employing a novel yeast strain cosecreting glucoamylase and alpha-amylase enzymes in a CBP system [97].

7.1.1 Future Trends in Ethanol Purification

The purification of ethanol remains a critical step in bioethanol production, influencing both economic feasibility and environmental sustainability. Traditional purification methods, such as azeotropic distillation and molecular sieves, are energy-intensive, prompting the development of low-energy and innovative separation technologies [89, 98].

Emerging trends in ethanol purification include the following:

- **Membrane Separation Technologies**: Processes like pervaporation and membrane distillation are gaining attention due to their lower energy requirements and ability to selectively remove water from ethanol, reducing the need for high-temperature distillation [90].
- **Hybrid Separation Systems**: Combining conventional distillation with adsorption or membrane technologies improves efficiency by reducing energy consumption and increasing ethanol yield [91].
- **Integration of Purification with Biorefineries**: Novel biorefinery designs incorporate ethanol recovery within continuous fermentation and separation systems, minimizing losses and improving process economics [88].
- **AI and Process Modeling in Purification**: Computational tools and artificial intelligence-driven simulations optimize separation conditions, leading to more cost-effective ethanol dehydration processes [93].

Recent studies highlight the potential of these technologies in reducing purification costs and emissions, making ethanol production more sustainable. For example, Rodrigues et al. (2024) demonstrated that deep eutectic solvents (DES) could enhance ethanol–water separation while reducing energy inputs compared to conventional distillation. These advancements will be crucial for improving the economic and environmental aspects of large-scale ethanol production [98].

7.2 *Sustainable Technologies: Focus on Low-Energy and Eco-Friendly Recovery Systems*

Enhancing the economic viability of biorefineries requires optimizing operational processes. Key strategies include the use of efficient and cost-effective biocatalysts, improved pretreatment methods, reduced energy consumption, and environmentally friendly biomass logistics [99].

The efficiency of many chemical processes depends on industrial catalysts, and similarly, the development of potent biocatalysts can improve the efficiency of biochemical reactions in biorefineries. Although enzyme consortia and their use in CBP present exciting possibilities, further optimization is needed before laboratory-scale processes achieve commercial efficiency [91, 95].

Another critical factor in biorefinery optimization is pretreatment efficiency. Recent studies highlight the potential of novel pretreatments, such as Deep Eutectic Solvents (DES), in enhancing carbohydrate conversion into fermentable sugars while facilitating lignin utilization in biomaterial applications [98, 99]. For instance, Rodrigues et al. (2024) optimized a pretreatment using choline chloride/lactic acid (1:5) and response surface methodology, achieving 90.08% lignin extraction and a glucose yield of 68.77% from corncobs [98].

7.3 *Use of AI and Modeling: Optimizing Recovery Processes Through Simulations*

Artificial intelligence (AI) and computational modeling are increasingly vital in optimizing bioethanol production. By analyzing large datasets from various sources, these technologies can identify innovative solutions to sustainability challenges and improve process efficiency [89, 93]. AI-driven approaches, such as artificial neural networks (ANNs), can model complex processes and optimize pretreatment, fermentation, and conversion stages in bioethanol production systems [93].

For instance, Gegić et al. (2024) used simulation models to evaluate scenarios for the integrated production of bioethanol and biogas from municipal organic waste, aiming to determine optimal process conditions [100]. Their study estimated an annual bioethanol production capacity of 5 million liters, with all other outputs converted into by-products to generate additional revenue. Results indicated that simultaneous hydrolysis was more advantageous than sequential hydrolysis, while direct routing of organic waste to biogas was inefficient. Additionally, the simulation estimated the lowest plausible ethanol selling price, demonstrating its value as a tool for assessing economic feasibility.

8 Conclusions and Recommendation

The advancements in ethanol recovery and purification discussed in this chapter highlight the crucial role of technological innovations in enhancing the economic and environmental feasibility of bioethanol production. The analysis of first-, second-, and intermediate-generation ethanol recovery processes demonstrated that process optimization is essential to improve efficiency and reduce costs. The integration of ethanol recovery into biorefinery systems offers a holistic approach to biomass utilization, minimizing waste and maximizing the value of coproducts. Moreover, emerging purification technologies, such as membrane separation, hybrid systems, and deep eutectic solvents, have shown promise in reducing energy consumption and operational costs compared to conventional distillation methods.

The economic and environmental assessment of ethanol recovery underscores the need for sustainable processing solutions that balance profitability with ecological responsibility. The implementation of artificial intelligence and computational modeling in process optimization represents a transformative step, allowing for better decision-making and resource allocation. However, several challenges remain, particularly in scaling up novel purification methods and further integrating ethanol recovery with biorefinery processes. Future research should focus on refining separation technologies, improving biocatalyst efficiency, and developing predictive models for large-scale operations. Additionally, policy incentives and investments in sustainable biofuel technologies will be crucial in driving the transition to a circular bioeconomy. By addressing these key areas, ethanol production can continue evolving as a competitive and sustainable alternative to fossil fuels, supporting global efforts to reduce greenhouse gas emissions and promote renewable energy solutions.

References

1. Shahbeik H, Peng W, Kazemi Shariat Panahi H, Dehhaghi M, Guillemin GJ, Fallahi A, Amiri H, Rehan M, Raikwar D, Latine H, Pandalone B, Khoshnevisan B, Sonne C, Vaccaro L, Nizami AS, Gupta VK, Lam SS, Pan J, Luque R, Sels B, Tabatabaei M, Aghbashlo M (2022) Synthesis of liquid biofuels from biomass by hydrothermal gasification: a critical review. Renew Sustain Energy Rev 167. https://doi.org/10.1016/j.rser.2022.112833
2. Shafizadeh A, Rastegari H, Shahbeik H, Mobli H, Pan J, Peng W, Li G, Tabatabaei M, Aghbashlo M (2023) A critical review of the use of nanomaterials in the biomass pyrolysis process. J Clean Prod 400. https://doi.org/10.1016/j.jclepro.2023.136705
3. Allen M, Babiker M, Chen Y, de Coninck HC, de Kleijne K (2018) Summary for policymakers. In: Global warming of 1.5°C. Technology, innovation & society. Cambridge University Press, pp 1–24. https://doi.org/10.1017/9781009157940.001
4. Jiang W, Lei Z, Gao H, Jiang Y, Lin CSK, Zhang W, Xin F, Jiang M (2024) Biodetoxification of lignocellulose hydrolysate for direct use in succinic acid production. BioDesign Research 6. https://doi.org/10.34133/bdr.0044
5. OECD/FAO (2024) OECD-FAO agricultural outlook 2024–2033, OECD. https://doi.org/10.1787/4c5d2cfb-en

6. Velvizhi G, Jacqueline PJ, Shetti NP, Mohanakrishna G, Aminabhavi TM (2023) Emerging trends and advances in valorization of lignocellulosic biomass to biofuels. J Environ Manage 345. https://doi.org/10.1016/j.jenvman.2023.118527
7. Michaels W, Zhang H, Luyben WL, Baltrusaitis J (2018) Design of a separation section in an ethanol-to-butanol process. Biomass Bioenergy 109:231–238. https://doi.org/10.1016/j.biombioe.2017.12.031
8. Huang HJ, Ramaswamy S, Tschirner UW, Ramarao BV (2008) A review of separation technologies in current and future biorefineries. Sep Purif Technol 62:1–21. https://doi.org/10.1016/j.seppur.2007.12.011
9. Serra A, Poch M, Solà C (1987) A survey of separation systems for fermentation ethanol recovery. Process Biochem 22:154–158
10. Sanap PP, Mahajan YS (2023) Review on technologies to separate and purify ethyl alcohol from dilute aqueous solutions. Rev Chem Eng 39:297–328. https://doi.org/10.1515/revce-2020-0114
11. Kossack S, Kraemer K, Gani R, Marquardt W (2008) A systematic synthesis framework for extractive distillation processes. In: Chemical engineering research and design. Institution of Chemical Engineers, pp 781–792. https://doi.org/10.1016/j.cherd.2008.01.008
12. Sun S, Lü L, Yang A, Wei S, Shen W (2019) Extractive distillation: advances in conceptual design, solvent selection, and separation strategies. Chin J Chem Eng 27:1247–1256. https://doi.org/10.1016/j.cjche.2018.08.018
13. Lei Z, Li C, Chen B (2003) Extractive distillation: a review. Sep Purif Rev 32:121–213. https://doi.org/10.1081/SPM-120026627
14. Singh A, Rangaiah GP (2017) Review of technological advances in bioethanol recovery and dehydration. Ind Eng Chem Res 56:5147–5163. https://doi.org/10.1021/acs.iecr.7b00273
15. Mekala M, Neerudi B, Are PR, Surakasi R, Manikandan G, Kakara VR, Dhumal AA (2022) Water removal from an ethanol-water mixture at azeotropic condition by adsorption technique. In: Adsorption science and technology. https://doi.org/10.1155/2022/8374471
16. Saini S, Chandel AK, Sharma KK (2020) Past practices and current trends in the recovery and purification of first generation ethanol: a learning curve for lignocellulosic ethanol. J Clean Prod 268. https://doi.org/10.1016/j.jclepro.2020.122357
17. Wasewar KL (2023) Pervaporation as a promising approach for recovery of bioethanol. In: Lignocellulosic biomass refining for second generation biofuel production, 1st edn, pp 1–28
18. Imad M, Castro-Muñoz R (2023) Ongoing progress on pervaporation membranes for ethanol separation. Membranes (Basel) 13. https://doi.org/10.3390/membranes13100848
19. Mao G, Gao Y, Zhou H, Jin W (2022) Tuning of solvent evaporation to prepare PEBA membrane with high separation performance for the pervaporation of phenol aqueous solution. J Memb Sci 656. https://doi.org/10.1016/j.memsci.2022.120638
20. Zhang YM, Xie WW, Hu XX, Chen XR, Mao H, Xu LH, Shi XY, Wang T, Feng YN, Zhao ZP (2025) A large-area MOF-based membrane for challenging CO_2 purification and ethanol/CO_2/N_2 ternary mixture separation. Sep Purif Technol 356. https://doi.org/10.1016/j.seppur.2024.129818
21. Chai Z, Wang Y (2025) PDMS membranes with reaction-diffusion-driven turing structure for advanced alcohol recovery. J Memb Sci 722. https://doi.org/10.1016/j.memsci.2025.123854
22. Martínez AA, Saucedo-Luna J, Segovia-Hernandez JG, Hernandez S, Gomez Castro FI, Castro-Montoya AJ (2012) Dehydration of bioethanol by hybrid process liquid-liquid extraction/extractive distillation. Ind Eng Chem Res 5847–5855. https://doi.org/10.1021/ie200932g
23. Granjo JFO, Nunes DS, Duarte BPM, Oliveira NMC (2020) A comparison of process alternatives for energy-efficient bioethanol downstream processing. Sep Purif Technol 238. https://doi.org/10.1016/j.seppur.2019.116414
24. Chauhan PR, Baiju V, Asif Sha A, Tyagi SK (2024) Adsorption of ethanol onto novel and indigenous green adsorbents: synthesis, characterization, and applications. J Clean Prod 442. https://doi.org/10.1016/j.jclepro.2024.140978

25. Zentou H, Abidin ZZ, Yunus R, Biak DRA, Korelskiy D (2019) Overview of alternative ethanol removal techniques for enhancing bioethanol recovery from fermentation broth. Processes 7. https://doi.org/10.3390/pr7070458
26. Van der Bruggen B, Luis P (2015) Pervaporation. In: Progress in filtration and separation. Elsevier Ltd, pp 101–154. https://doi.org/10.1016/B978-0-12-384746-1.00004-5
27. Khalid A, Aslam M, Qyyum MA, Faisal A, Khan AL, Ahmed F, Lee M, Kim J, Jang N, Chang IS, Bazmi AA, Yasin M (2019) Membrane separation processes for dehydration of bioethanol from fermentation broths: recent developments, challenges, and prospects. Renew Sustain Energy Rev 105:427–443. https://doi.org/10.1016/j.rser.2019.02.002
28. Oliveira CM, Cruz AJG, Costa CBB (2014) Comparison among proposals for energy integration of processes for 1G/2G ethanol and bioelectricity production
29. Nogueira LAH, Souza GM, Cortez LAB, de Brito Cruz CH (2020) Biofuels for transport. In: Future energy: improved, sustainable and clean options for our planet. Elsevier, pp 173–197. https://doi.org/10.1016/B978-0-08-102886-5.00009-8
30. Alexiades A, Kendall A, Winans KS, Kaffka SR (2018) Sugar beet ethanol (*Beta vulgaris* L.): a promising low-carbon pathway for ethanol production in California. J Clean Prod 172:3907–3917. https://doi.org/10.1016/j.jclepro.2017.05.059
31. Marzo C, Díaz AB, Caro I, Blandino A (2018) Status and perspectives in bioethanol production from sugar beet. In: Bioethanol production from food crops: sustainable sources, interventions, and challenges. Elsevier, pp 61–79. https://doi.org/10.1016/B978-0-12-813766-6.00004-7
32. Šokarda Slavić M, Margetić A, Dojnov B, Vujčić M, Mišić M, Božić N, Vujčić Z (2023) Modified simultaneous saccharification and fermentation for the production of bioethanol from highly concentrated raw corn starch. Fuel 338. https://doi.org/10.1016/j.fuel.2022.127363
33. Dai K, Miao Y, Qu C, Wang C, Fu H, Wang J (2024) Adaptive laboratory evolution of Thermoanaerobacterium aotearoense for enhanced ethanol production from raw cassava starch. Renew Energy 233. https://doi.org/10.1016/j.renene.2024.121179
34. Karimi S, Karri RR, Tavakkoli Yaraki M, Koduru JR (2021) Processes and separation technologies for the production of fuel-grade bioethanol: a review. Environ Chem Lett 19:2873–2890. https://doi.org/10.1007/s10311-021-01208-9
35. Dias MOS, Junqueira TL, Cavalett O, Cunha MP, Jesus CDF, Rossell CEV, Maciel Filho R, Bonomi A (2012) Integrated versus stand-alone second generation ethanol production from sugarcane bagasse and trash. Bioresour Technol 103:152–161. https://doi.org/10.1016/j.biortech.2011.09.120
36. Kordala N, Walter M, Brzozowski B, Lewandowska M (2024) 2G-biofuel ethanol: an overview of crucial operations, advances and limitations. Biomass Convers Biorefin 14:2983–3006. https://doi.org/10.1007/s13399-022-02861-y
37. Bermejo PM, Raghavendran V, Gombert AK (2020) Neither 1G nor 2G fuel ethanol: setting the ground for a sugarcane-based biorefinery using an iSUCCELL yeast platform. FEMS Yeast Res 20. https://doi.org/10.1093/FEMSYR/FOAA027
38. Tao L, Foust TD (2021) Biorefinery ethanol upgrading: Opportunities and challenges. Joule 5:524–526. https://doi.org/10.1016/j.joule.2021.02.014
39. Huang J, Khan MT, Perecin D, Coelho ST, Zhang M (2020) Sugarcane for bioethanol production: potential of bagasse in Chinese perspective. Renew Sustain Energy Rev 133. https://doi.org/10.1016/j.rser.2020.110296
40. Sayaslan A, Koyuncu M, Türker S, Irklı Y, Serin A, Orhan FG (2018) Use of durum wheat clear flour in vital gluten and bioethanol production. J Cereal Sci 80:50–56. https://doi.org/10.1016/j.jcs.2018.01.014
41. Liguori R, Soccol CR, de Souza Vandenberghe LP, Woiciechowski AL, Faraco V (205) Second generation ethanol production from brewers' spent grain. Energies (Basel) 8:2575–2586. https://doi.org/10.3390/en8042575
42. Abdul Kareem Joyia M, Ahmad M, Chen YF, Mustaqeem M, Ali A, Abbas A, Ashraf Gondal M (2024) Trends and advances in sustainable bioethanol production technologies from first to fourth generation: a critical review. Energy Convers Manag 321. https://doi.org/10.1016/j.enconman.2024.119037

43. Saville BA, Griffin WM, MacLean HL (2016) Ethanol production technologies in the US: status and future developments. In: Global bioethanol: evolution, risks, and uncertainties. Elsevier Inc, pp 163–180. https://doi.org/10.1016/B978-0-12-803141-4.00007-1
44. Iram A, Cekmecelioglu D, Demirci A (2019) Optimization of dilute sulfuric acid, aqueous ammonia, and steam explosion as the pretreatments steps for distillers' dried grains with solubles as a potential fermentation feedstock. Bioresour Technol 282:475–481. https://doi.org/10.1016/j.biortech.2019.03.009
45. Moonsamy TA, Mandegari M, Farzad S, Görgens JF (2022) A new insight into integrated first and second-generation bioethanol production from sugarcane. Ind Crops Prod 188:115675. https://doi.org/10.1016/j.indcrop.2022.115675
46. Dias MOS, Junqueira TL, Cavalett O, Cunha MP, Jesus CDF, Rossell CEV, Maciel Filho R, Bonomi A (2012) Integrated versus stand-alone second generation ethanol production from sugarcane bagasse and trash. Bioresour Technol 103:152–161. https://doi.org/10.1016/j.biortech.2011.09.120
47. Bermejo PM, Raghavendran V, Gombert AK (2020) Neither 1G nor 2G fuel ethanol: setting the ground for a sugarcane-based biorefinery using an iSUCCELL yeast platform. FEMS Yeast Res 20:foaa027. https://doi.org/10.1093/femsyr/foaa027
48. Susmozas A, Martín-Sampedro R, Ibarra D, Eugenio ME, Iglesias R, Manzanares P, Moreno AD (2020) Process strategies for the transition of 1G to advanced bioethanol production. Processes 8. https://doi.org/10.3390/pr8101310
49. Moonsamy TA, Mandegari M, Farzad S, Görgens JF (2022) A new insight into integrated first and second-generation bioethanol production from sugarcane. Ind Crops Prod 188. https://doi.org/10.1016/j.indcrop.2022.115675
50. Susmozas A, Martín-Sampedro R, Ibarra D, Eugenio ME, Iglesias R, Manzanares P, Moreno AD (2020) Process strategies for the transition of 1G to advanced bioethanol production. Processes 8:1–45. https://doi.org/10.3390/pr8101310
51. Bechara R, Gomez A, Saint-Antonin V, Schweitzer JM, Maréchal F (2016) Methodology for the design and comparison of optimal production configurations of first and first and second generation ethanol with power. Appl Energy 184:247–265. https://doi.org/10.1016/j.apenergy.2016.09.100
52. Singh A, da Cunha S, Rangaiah GP (2019) Heat-pump assisted distillation versus double-effect distillation for bioethanol recovery followed by pressure swing adsorption for bioethanol dehydration. Sep Purif Technol 210:574–586. https://doi.org/10.1016/j.seppur.2018.08.043
53. Saeed B, Khan P, Laeeq Khan A, AlMohamadi H, Naeem A, Aslam M, Zaman M, Niazi MBK, Amjad Gilani M, Yasin M (2025) Experimental and theoretical insights into bioethanol recovery: valorizing waste PET bottles for sustainable pervaporation membranes. Sep Purif Technol 357. https://doi.org/10.1016/j.seppur.2024.130104
54. Bedryk O, Shevchenko A, Mishchenko OS, Maleta VN, Kiss AA (2023) Industrial experience in using cyclic distillation columns for food grade alcohol purification. Chem Eng Res Des 192:102–109. https://doi.org/10.1016/j.cherd.2023.02.026
55. Janković T, Straathof AJJ, McGregor IR, Kiss AA (2024) Bioethanol separation by a new pass-through distillation process. Sep Purif Technol 336. https://doi.org/10.1016/j.seppur.2024.126292
56. Embrapa (2024) Plano Diretor da Embrapa 2024–2030
57. Novacana (2025) Inpasa vai além do etanol de milho e considera produção de biodiesel e SAF
58. Raizen (2023) Etanol de segunda geração: potencial e oportunidades
59. Shibukawa VP, Ramos L, Cruz-Santos MM, Prado CA, Jofre FM, de Arruda GL, da Silva SS, Mussatto SI, dos Santos JC (2023) Impact of product diversification on the economic sustainability of second-generation ethanol biorefineries: a critical review. Energies (Basel) 16. https://doi.org/10.3390/en16176384
60. Jönsson LJ, Martín C (2016) Pretreatment of lignocellulose: formation of inhibitory by-products and strategies for minimizing their effects. Bioresour Technol 199:103–112. https://doi.org/10.1016/j.biortech.2015.10.009

61. Moreno AD, Carbone A, Pavone R, Olsson L, Geijer C (2019) Evolutionary engineered Candida intermedia exhibits improved xylose utilization and robustness to lignocellulose-derived inhibitors and ethanol. Appl Microbiol Biotechnol 103:1405–1416. https://doi.org/10.1007/s00253-018-9528-x
62. Yaverino-Gutierrez MA, Ramos L, Ascencio JJ, Chandel AK (2024) Enhanced production of clean fermentable sugars by acid pretreatment and enzymatic saccharification of sugarcane bagasse. Processes 12. https://doi.org/10.3390/pr12050978
63. Chen Q, Jin Y, Huang H, Su Z, Kim SK, Kim KM (2018) Enantioselective liquid-liquid extraction of underivatized amino acids with simple chiral aminophenyl-aldehyde. Bull Korean Chem Soc 39:960–964. https://doi.org/10.1002/bkcs.11533
64. Roque LR, Morgado GP, Nascimento VM, Ienczak JL, Rabelo SC (2019) Liquid-liquid extraction: a promising alternative for inhibitors removing of pentoses fermentation. Fuel 242:775–787. https://doi.org/10.1016/j.fuel.2018.12.130
65. Candido JP, Claro EMT, de Paula CBC, Shimizu FL, de Oliveria Leite DAN, Brienzo M, de Angelis DF (2020) Detoxification of sugarcane bagasse hydrolysate with different adsorbents to improve the fermentative process. World J Microbiol Biotechnol 36. https://doi.org/10.1007/s11274-020-02820-7
66. Santana NB, Teixeira Dias JC, Rezende RP, Franco M, Silva Oliveira LK, Souza LO (2018) Production of xylitol and bio-detoxification of cocoa pod husk hemicellulose hydrolysate by Candida boidinii XM02G. PLoS One 13. https://doi.org/10.1371/journal.pone.0195206
67. Llano T, Quijorna N, Coz A (2017) Detoxification of a lignocellulosic waste from a pulp mill to enhance its fermentation prospects. Energies (Basel) 10. https://doi.org/10.3390/en10030348
68. Cavka A, Jönsson LJ (2013) Detoxification of lignocellulosic hydrolysates using sodium borohydride. Bioresour Technol 136:368–376. https://doi.org/10.1016/j.biortech.2013.03.014
69. Coz A, Llano T, Cifrián E, Viguri J, Maican E, Sixta H (2016) Physico-chemical alternatives in lignocellulosic materials in relation to the kind of component for fermenting purposes. Materials 9. https://doi.org/10.3390/MA9070574
70. Andary J, Maalouly J, Ouaini R, Chebib H, Rutledge DN, Ouaini N (2021) Stability study of furans, glucose and xylose under overliming conditions: effect of sugar degradation products. Bioresour Technol Rep 15. https://doi.org/10.1016/j.biteb.2021.100722
71. Fernández-Sandoval MT, García A, Teymennet-Ramírez KV, Arenas-Olivares DY, Martínez-Morales F, Trejo-Hernández MR (2024) Removal of phenolic inhibitors from lignocellulose hydrolysates using laccases for the production of fuels and chemicals. Biotechnol Prog 40. https://doi.org/10.1002/btpr.3406
72. Botshekan M, Moheb A, Vatankhah F, Karimi K, Shafiei M (2022) Energy saving alternatives for renewable ethanol production with the focus on separation/purification units: a techno-economic analysis. Energy 239. https://doi.org/10.1016/j.energy.2021.122363
73. Embrapa (2019) Embrapa identifica gene da cana que pode aumentar produção de etanol 2G, APLA
74. Jyothi MS, Reddy KR, Soontarapa K, Naveen S, Raghu AV, Kulkarni RV, Suhas DP, Shetti NP, Nadagouda MN, Aminabhavi TM (2019) Membranes for dehydration of alcohols via pervaporation. J Environ Manage 242:415–429. https://doi.org/10.1016/j.jenvman.2019.04.043
75. U. da I. de C.-A. UNICA (2024) A landmark year for ethanol and sugar production. https://www.Sugarcane.Org/News/2024-a-Landmark-Year-for-Ethanol-and-Sugar-Production/
76. Renewable Fuels Association (2024) U.S. ethanol production and exports hit records in 2024, new data show. Ethanol Producer Magazine. https://Ethanolproducer.Com/Articles/Rfa-Us-Ethanol-Production-and-Exports-Hit-Records-in-2024-New-Data-Show
77. Zhao L, Ou X, Chang S (2016) Life-cycle greenhouse gas emission and energy use of bioethanol produced from corn stover in China: current perspectives and future prospectives. Energy 115:303–313. https://doi.org/10.1016/j.energy.2016.08.046
78. Kothare CB, Kongre S, Malwe P, Sharma K, Qasem NAA, Ağbulut Ü, Eldin SM, Panchal H (2023) Performance improvement and CO and HC emission reduction of variable compression

ratio spark-ignition engine using n-pentanol as a fuel additive. Alex Eng J 74:107–119. https://doi.org/10.1016/j.aej.2023.05.024
79. Guo L, Ma X, Xia Y, Xiang X, Wu X (2015) A novel method of production of ethanol by carbon dioxide with steam. Fuel 158:843–847. https://doi.org/10.1016/j.fuel.2015.06.020
80. Darda S, Papalas T, Zabaniotou A (2019) Biofuels journey in Europe: currently the way to low carbon economy sustainability is still a challenge. J Clean Prod 208:575–588. https://doi.org/10.1016/j.jclepro.2018.10.147
81. Mohanty SK, Swain MR (2019) Bioethanol production from corn and wheat: food, fuel, and future. In: Bioethanol production from food crops. Elsevier, pp 45–59. https://doi.org/10.1016/b978-0-12-813766-6.00003-5
82. Vennila T, Karuna MS, Srivastava BK, Venugopal J, Surakasi R (2023) New strategies in treatment and enzymatic processes. In: IGI global, pp 219–240. https://doi.org/10.4018/978-1-6684-4118-3.ch011
83. Sydney EB, Letti LAJ, Karp SG, Sydney ACN, Vandenberghe LPS, de Carvalho JC, Woiciechowski AL, Medeiros ABP, Soccol VT, Soccol CR (2019) Current analysis and future perspective of reduction in worldwide greenhouse gases emissions by using first and second generation bioethanol in the transportation sector. Bioresour Technol Rep 7. https://doi.org/10.1016/j.biteb.2019.100234
84. Petersen AM, Aneke MC, Görgens JF (2014) Techno-economic comparison of ethanol and electricity coproduction schemes from sugarcane residues at existing sugar mills in Southern Africa. Biotechnol Biofuels 7. https://doi.org/10.1186/1754-6834-7-105
85. Carpio RR, Miyoshi SC, Elias AM, Furlan FF, Giordano RC, Secchi AR (2021) Multi-objective optimization of a 1G-2G biorefinery: a tool towards economic and environmental viability. J Clean Prod 284:125431. https://doi.org/10.1016/j.jclepro.2020.125431
86. Junqueira TL, Chagas MF, Gouveia VLR, Rezende MCAF, Watanabe MDB, Jesus CDF, Cavalett O, Milanez AY, Bonomi A (2017) Techno-economic analysis and climate change impacts of sugarcane biorefineries considering different time horizons. Biotechnol Biofuels 10. https://doi.org/10.1186/s13068-017-0722-3
87. MacRelli S, Mogensen J, Zacchi G (2012) Techno-economic evaluation of 2 nd generation bioethanol production from sugar cane bagasse and leaves integrated with the sugar-based ethanol process. Biotechnol Biofuels 5. https://doi.org/10.1186/1754-6834-5-22
88. Igbokwe VC, Ezugworie FN, Onwosi CO, Aliyu GO, Obi CJ (2022) Biochemical biorefinery: a low-cost and non-waste concept for promoting sustainable circular bioeconomy. J Environ Manage 305. https://doi.org/10.1016/j.jenvman.2021.114333
89. Sahoo PK, Datta R, Rahman MM, Sarkar D (2024) Sustainable environmental technologies: recent development. Opportunities Key Challenges Appl Sci 14:10956. https://doi.org/10.3390/app142310956
90. Jofre FM, Prado CA, Shibukawa VP, Rodrigues BG, Sarangi PK, Chandel AK (2025) Critical analysis of process parameters towards smart bioreactors development in biorefinery for biorenewables production. Int J Biol Macromol 305:140957. https://doi.org/10.1016/j.ijbiomac.2025.140957
91. Afedzi AEK, Afrakomah GS, Gyan K, Khan J, Seidu R, Baidoo T, Sultan IN, Tareen AK, Parakulsuksatid P (2025) Enhancing economic and environmental sustainability in lignocellulosic bioethanol production: key factors, innovative technologies, policy frameworks, and social considerations. Sustainability (Switzerland) 17. https://doi.org/10.3390/su17020499
92. Barragán-Ocaña A, Merritt H, Sánchez-Estrada OE, Méndez-Becerril JL, del Pilar Longar-Blanco M (2023) Biorefinery and sustainability for the production of biofuels and value-added products: a trends analysis based on network and patent analysis. PLoS One 18. https://doi.org/10.1371/journal.pone.0279659
93. Owusu WA, Marfo SA (2023) Artificial intelligence application in bioethanol production. Int J Energy Res. https://doi.org/10.1155/2023/7844835
94. Ezeorba TPC, Okeke ES, Mayel MH, Nwuche CO, Ezike TC (2024) Recent advances in biotechnological valorization of agro-food wastes (AFW): optimizing integrated approaches for sustainable biorefinery and circular bioeconomy. Bioresour Technol Rep 26. https://doi.org/10.1016/j.biteb.2024.101823

95. Althuri A, Venkata Mohan S (2022) Emerging innovations for sustainable production of bioethanol and other mercantile products from circular economy perspective. Bioresour Technol 363. https://doi.org/10.1016/j.biortech.2022.128013
96. Vargas ACG, Dresch AP, Schmidt AR, Tadioto V, Giehl A, Fogolari O, Mibielli GM, Alves SL, Bender JP (2023) Batch fermentation of lignocellulosic elephant grass biomass for 2G ethanol and xylitol production. Bioenergy Res 16:2219–2228. https://doi.org/10.1007/s12155-022-10559-2
97. Gupte AP, Di Vita N, Myburgh MW, Cripwell RA, Basaglia M, van Zyl WH, Viljoen-Bloom M, Casella S, Favaro L (2024) Consolidated bioprocessing of the organic fraction of municipal solid waste into bioethanol. Energy Convers Manag 302. https://doi.org/10.1016/j.enconman.2024.118105
98. Rodrigues BG, José ÁHM, Prado CA, Rodrigues D, Rodrigues RCLB (2024) Optimizing corncob pretreatment with eco-friendly deep eutectic solvents to enhance lignin extraction and cellulose-to-glucose conversion. Int J Biol Macromol 283. https://doi.org/10.1016/j.ijbiomac.2024.137432
99. Wo´zniak AW, Kuligowski K, Lesław´swierczek LL, Cenian A (2025) Review of lignocellulosic biomass pretreatment using physical, thermal and chemical methods for higher yields in bioethanol production. https://doi.org/10.3390/su17010287/Copyright
100. Gegić B, Vučurović D, Dodić S, Bajić B (2024) Process modelling of integrated bioethanol and biogas production from organic municipal waste. Energies (Basel) 17. https://doi.org/10.3390/en17174286

Life Cycle Assessment in Biorefineries: Foundations, Methodologies, and Future Directions

Rafael Rodrigues Philippini, Razieh Shafiei-Alavijeh, Eveline Peeters, and Paulo Ricardo Franco Marcelino

Abstract Life Cycle Assessment (LCA) is a standardized methodology employed to quantify the environmental impacts associated with products and processes throughout their entire life cycle. In the context of biorefineries, facilities that convert biomass into biofuels, bio-based chemicals, and materials, LCA serves as an essential tool to evaluate sustainability trade-offs and guide environmentally sound decision-making. This chapter provides a comprehensive overview of the historical development and methodological foundations of LCA, with particular emphasis on its application in the biorefinery sector. Key methodological elements, including system boundaries, functional units, impact assessment models, and allocation strategies, are discussed in detail. Additionally, the chapter explores contemporary challenges such as data quality, multifunctionality, and regional differentiation, as well as the integration of emerging technologies including artificial intelligence and blockchain to enhance LCA accuracy and transparency. By critically assessing the role of LCA in biorefinery development, this work underscores its relevance for advancing sustainable innovation and policy formulation within the bioeconomy.

Keywords Life Cycle Assessment (LCA) · Biorefineries · Circular economy · Data quality · Sustainability

R. R. Philippini (✉) · R. Shafiei-Alavijeh · E. Peeters
Research Group of Microbiology, Department of Bioengineering Sciences, Vrije Universiteit Brussel, Brussels, Belgium
e-mail: rafael.philippini@vub.be

R. Shafiei-Alavijeh
e-mail: razieh.shafiei-alavijeh@vub.be

E. Peeters
e-mail: eveline.peeters@vub.be

P. R. F. Marcelino (✉)
Center for Natural and Human Sciences (CCNH), Federal University of ABC (UFABC), Santo André, São Paulo State, Brazil
e-mail: paulo.franco@ufabc.edu.br

© The Author(s), under exclusive license to Springer Nature Singapore Pte Ltd. 2025

A. K. Chandel and S. Sundaramurthy (eds.), *Biorefineries in Circular Bioeconomy*, Green Energy and Technology, https://doi.org/10.1007/978-981-95-3913-0_12

1 Introduction

Biorefineries are refineries that convert biomass into a diverse array of valuable products, including biofuels, chemicals, and materials. Biorefinery development leads to a reduced dependence on fossil resources while also promoting sustainability through the efficient repurposing of organic waste and side streams. Moreover, biorefineries play a pivotal role in the circular economy, supporting the transition to a more self-sufficient economic model [1], forming closed-loop systems in resource utilization, in which each output serves as an input for subsequent processes [2, 3]. As biorefineries evolve toward more sophisticated and integrated systems, addressing challenges related to coproduct distribution and its influence on sustainability assessments becomes increasingly important [4].

Life Cycle Assessment (LCA) is a systematic approach to evaluate the environmental impacts of products, processes, or services throughout their entire life span. In the context of biorefineries, LCA provides insights into resource use, emissions, and sustainability performance, thereby supporting a transition to more efficient and environmentally responsible production systems [5]. The application of LCA not only illuminates areas of inefficiency in a biorefinery but also facilitates well-informed decisions on feedstock selection and technology deployment. For instance, LCA can be used for a comparative analysis of different biomass inputs, for example, agricultural residues or municipal solid waste, thereby aiding in discerning those alternatives with minimal environmental impact while ensuring elevated process efficiency [6, 7]. An incorporation of more sustainable energy sources into these systems, for example, bioelectricity derived from waste, can aid in further diminishing greenhouse gas emissions connected to production [8]. This holistic approach to environmental evaluation and resource management catalyzes innovation within the sector, thereby facilitating the transition toward more sustainable practices that stimulate a global movement away from fossil fuels and toward climate-resilient solutions [9]. Due to the importance of LCA in the development of bioprocesses in biorefineries, in the last decade, there has been a significant increase in academic publications, as can be seen in Fig. 1.

In this chapter, we aim to elucidate the historical development of LCA in the context of biorefineries, outlining its purpose, core methodology, and the evolution of key approaches used in practice. Commonly applied LCA methodologies will be explored, as well as available software tools and databases that support LCA implementation, followed by a discussion of current challenges and future perspectives for advancing LCA as a decision-making tool in the transition toward sustainable biorefinery systems.

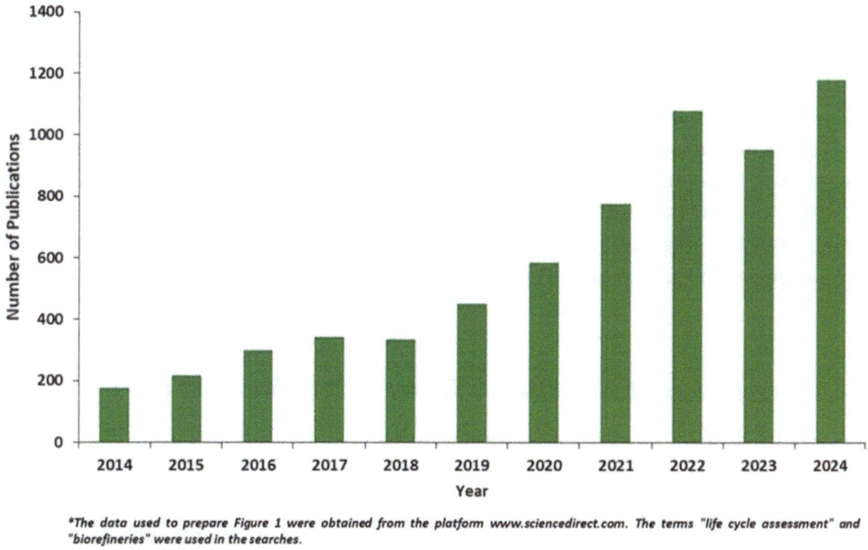

Fig. 1 Evolution of the number of publications on life cycle assessment and biorefineries in the last decade (2014–2024)

2 Historical Perspectives on LCA Development

Biorefineries are facilities designed to convert biomass into energy, fuels, and valuable bulk or fine chemicals. As biorefinery systems grow in complexity while being expected to meet ambitious environmental goals, robust methods are required for assessing their sustainability. LCA stands out as a comprehensive tool for measuring environmental impacts of biorefinery systems throughout their entire life cycle [9]. In this part, the historical development of LCA will be reviewed from its early foundations, through its expansion, to advancements of the 2010s until how it has evolved into the sophisticated methodology it is today.

2.1 Early Foundations of LCA

The conceptual groundwork for LCA was laid in the late 1960s and early 1970s in the United States, driven by the energy crises of the time [10]. Initial efforts focused on energy and resource flow analyses to evaluate packaging materials and industrial processes. However, it was not until the 1990s that LCA processes matured into a coherent analytical framework, supported by organizations such as the Society of Environmental Toxicology and Chemistry (SETAC), which helped establish structural and terminological standards. In the late 1990s, the ISO 14040 series further formalized the methodology, thereby outlining principles and guidelines that remain relevant today [11].

However, in this period, application of LCA remained comparatively restricted in the context of biorefineries. The methodology was predominantly employed in the evaluation of early biofuel systems, including first-generation ethanol and biodiesel production, focusing on energy balances [12] and global warming potential (GWP). Nevertheless, despite their limited use, these initial LCAs were pivotal in elucidating both the potential benefits and limitations of bio-based systems, particularly in relation to land usage and agricultural inputs [13, 14].

2.2 *Expansion of Biorefinery LCAs in the Twenty-First Century*

The early 2000s saw a surge in interest in biomass-derived products, driven by growing environmental awareness and global climate commitments. The term "biorefinery" began to gain prominence, representing an integrated transformation of biomass into diverse products and energy streams [15]. In this period and context, a more structured LCA method was employed to examine the ecological performance of biofuels including corn ethanol, sugarcane ethanol, and soybean biodiesel [16]. These assessments illuminated critical issues associated with indirect land use change (ILUC), nutrient runoff, and the trade-offs inherent between energy production and food security [17].

In the 2010s, more complex biorefinery models were developed. Although these models were still predominantly centered on single-product systems, they were adept at generating fuels, chemicals, and materials derived from lignocellulosic and alternative nonfood biomass resources. Concomitantly, LCA strategies experienced a progression in their sophistication, with the attention being broadened from climate change to tackle multiple impact sectors, including eutrophication, acidification, ecotoxicity, and water scarcity [18, 19]. Notably, this decade also marked the amalgamation of LCA with various other decision-support instruments, such as techno-economic analysis (TEA), social LCA, and risk assessment. Prospective LCAs and consequential LCAs emerged as prominent methodologies, facilitating the evaluation of innovative technologies and the systemic ramifications of policy or market transformations. Moreover, biorefineries started to evolve toward more integrated systems, challenges were now also related to coproduct distribution, and its influence on sustainability assessments becomes increasingly important [4].

Entering the 2020s, the significance of the use of LCA in the advancement of biorefinery initiatives has escalated to unprecedented levels. The use of LCA within biorefineries began to investigate novel dimensions such as nutrient cycling, carbon capture, and utilization (CCU) and circularity [20]. With ambitious policy frameworks, exemplified by the European Green Deal and the U.S. Bioeconomy

Blueprint, LCA is increasingly employed as a tool to inform investment decisions, foster innovation, and shape regulations within the bio-based sector [21, 22].

Highlighted that in the last decade, the concepts of LCA in synergy with ESG (Environmental, Social and Governance) have become a tool of fundamental importance in the industrial field in the development of new processes and products, going beyond the biorefinery models. This integration between LCA and ESG encourages research, development, and innovation in the various departments of industries, especially, since the development of processes and products that make efficient use of natural resources, the reduction of emissions and waste, and the acquisition of clean technologies are fundamental to their environmental goals.

2.3 Current Scenarios for Biorefinery LCAs

Contemporary biorefinery LCAs increasingly focus on the evaluation of comprehensive value chains, thereby integrating the cascading use of biomass, maximal valorization of residues, and implementation of closed-loop systems. The move toward a circular bioeconomy is pushing LCA practitioners to go beyond traditional cradle-to-grave evaluations and to explore circularity metrics alongside dynamic modeling approaches [23]. Advances in digitalization, machine learning, and big data are also transforming LCA methodologies, making them more responsive and forward-looking. Additionally, LCA is becoming integral to facilitate carbon accounting in green certification frameworks, such as the International Sustainability and Carbon Certification (ISCC) and regulatory instruments like the Renewable Energy Directive II (RED II) [24–26].

3 The Role of LCAs in Biorefineries and Circular Bioeconomy

The integration of LCA findings in biorefinery development assists in the development of more sustainable solutions in which resource utilization can be refined according to the principles of a circular economy [27, 28] (Fig. 2). The integration of such comprehensive assessments into strategic decision-making processes not only assists in cost reduction but also alleviates environmental repercussions, thereby aligning with regulatory mandates and consumer anticipations. By fostering collaboration among researchers, policymakers, and industry leaders, there exists considerable potential not only to refine biorefinery operations but also to establish robust frameworks: for example, by converting waste into valuable commodities [9].

With the rising intricacies of multifunctional biorefinery systems, it has become more important to understand how various allocation methodologies can affect LCA outcomes [8, 29] and to develop standardized methodologies, and the development of

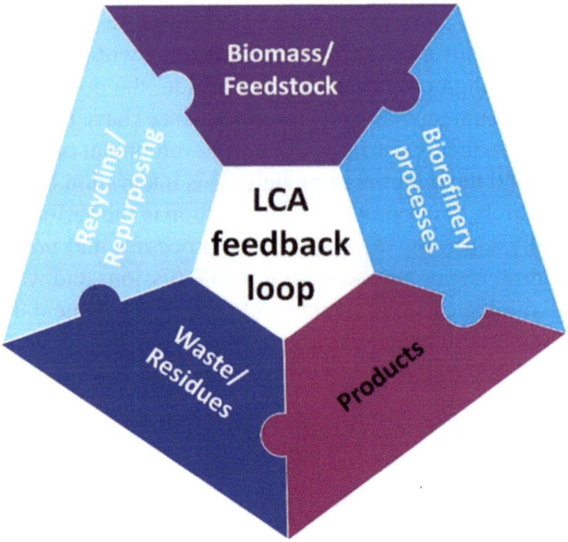

Fig. 2 LCA feedback loop biorefinery design

consensus among researchers. Several allocation methods, including system expansion, partitioning based on mass or economic value, and hybrid approaches, are carefully examined to evaluate their influence on environmental performance results [8, 30]. The ISO standards offer a fundamental basis for LCA practice, supporting comprehensive and comparable assessments across different studies [8]. In addition, continuous discussions within the LCA community seek to resolve methodological issues, such as the clear definition of functional units, precise system boundaries, and reliable inventory data collection procedures [28]. This standardization will facilitate the comparison of performance metrics across various biorefinery models, while ensuring that advancements are both ecologically sustainable and economically feasible [28].

3.1 LCA for Scale-Up

The LCA scale-up in biorefineries represents a pivotal progression in enhancing the sustainability and ecological efficacy of these systems. It provides a thorough basis for assessing the ecological effects tied to biorefineries, incorporating everything from the choice of feedstock to the production of goods, and is essential for spotting chances for growth and refinement in the scaling journey. LCA helps optimize biorefinery designs by identifying sustainable biomass sources and technologies that reduce emissions and energy use [7, 20], supporting decision-making by integrating environmental, economic, and social aspects [31]. LCA also highlights the benefits of renewable energy and multiproduct outputs, promoting more efficient and sustainable systems (Sreekumar et al. 2020).

The methodological choices made in LCA, including the delineation of system boundaries and allocation methodologies, are imperative for precise impact evaluations, as evidenced by research conducted across diverse feedstock categories [27]. The combination of numerous products in biorefineries, especially apparent in lignocellulosic waste biorefineries, can significantly boost resource efficiency and sustainability; however, it calls for detailed life cycle assessment to investigate the environmental trade-offs involved [32]. The application of anticipatory LCA methodologies can also inform the design phase of biorefineries, thereby ensuring that environmental impacts are mitigated from the initial stages [33].

3.2 Prospective LCA and Technology Forecasting

LCA provides a critical framework for evaluating the environmental sustainability of biorefineries across different generations: from first-generation systems using food-based crops to more advanced second- and third-generation approaches that rely on lignocellulosic residues and algae. Advanced biorefineries, particularly those that utilize organic waste and remnants or those that facilitate both environmental performance and economic viability through the coproduction of bio-based fuels and chemicals [33], have been gaining attention not only because they facilitate both environmental performance and economic viability but also for their promise to mitigate the negative ecological consequences connected to first-generation biorefineries.

For example, the establishment of waste biorefineries is critical for fostering a circular economy, with LCA offering a structured framework to assess environmental impacts and discern opportunities for enhancement [34]. Notwithstanding the potential of biorefineries in waste reduction and sustainability promotion, challenges such as scalability and cost-effectiveness remain, necessitating continuous technological innovations and the optimization of resource recovery systems [35].

Predictive LCA models, which utilize decision trees and artificial neural networks, provide a systemic approach to embed LCA considerations early in the design trajectory, thereby conserving time and resources while pinpointing sustainable design alternatives [36]. These models necessitate meticulous consideration of functional units, system boundaries, and allocation methodologies to effectively manage multi-functionality [28]. In this context, it is also important to integrate prospective LCA and technology forecasting within biorefineries.

3.3 Case Studies and Applications

First-generation biorefineries, which employ consumable biomass such as corn and sugarcane, encounter scrutiny for their utilization of arable land and their role in the food versus fuel discourse. Such systems face environmental obstacles resulting from fertilizer usage, leading to detrimental effects on indicators including eutrophication and acidification [37].

Biorefineries of the second generation that draw upon lignocellulosic biomass and agricultural byproducts provide a more sustainable solution by taking advantage of nonfood biomass, hence minimizing the competition for food resources [32]. Lignocellulosic feedstocks, including sugarcane bagasse and various agricultural residues, are both plentiful and renewable; however, they necessitate rigorous pretreatment processes for effective conversion, potentially demanding substantial chemical and energy inputs [32, 38]. The LCA of these systems accentuates the significance of refining bioconversion processes and incorporating sustainable energy sources to mitigate environmental impacts [7].

Algae-based biorefineries, representing third-generation systems, present considerable promise owing to their elevated productivity and capacity to generate bioactive compounds in conjunction with biofuels [38]. Nonetheless, the environmental efficacy of these systems remains under scrutiny, facing obstacles related to large-scale implementation and energy consumption [39]. In summary, the LCA of biorefineries highlights the necessity for methodological uniformity and comprehensive data to accurately evaluate their sustainability and inform future advancements within the bioeconomy [28, 40]. Table 1 summarizes some comparative analysis of biorefinery systems, with feedstock examples, key products, and environmental performance.

Table 1 Comparative analysis of LCA in different biorefinery systems

Biorefinery type	Feedstock	Key products	Environmental performance	References
First generation	Corn, sugarcane	Ethanol, sugar, lactic acid	Moderate GHG emissions, high resource use (land, water, fertilizers), but improved with circular economy practices	[41–43]
Second generation	Lignocellulosic (Agro-industrial residues & forest residues)	Ethanol, biogas, biochemicals	Lower GHG emissions, resource efficient, challenges with feedstock variability, and high pretreatment energy demand	Arruda et al. (2024) Börjesson et al. (2013) [44, 45]
Algae based	Microalgae	Biodiesel, proteins, polysaccharides	High potential for circularity and CO_2 capture; energy-intensive processes (e.g., harvesting, drying) limit current scalability	Bussa et al. (2021) [46, 47]

4 LCA Multidimensional Perspective

LCA is traditionally focused on environmental impacts; its concept can be expanded to incorporate broader sustainability dimensions such as social, economic, and policy perspectives. This multidimensional approach is particularly relevant in the context of biorefineries, where Life Cycle Sustainability Assessment (LCSA) enables a more comprehensive evaluation of sustainability performance throughout the entire production chain [48]. By integrating tools such as Social Life Cycle Assessment (S-LCA) and Life Cycle Costing (LCC), LCA becomes more capable of capturing the complex interactions among environmental, social, and economic factors [49].

4.1 Environmental Perspective

An LCA is executed to determine the environmental implications that come with every phase of a product's lifecycle, from obtaining raw materials through production, usage, and eventual waste disposal (Kalkanis et al. 2024). These impacts could include cases like eutrophication (nutrient pollution in water bodies), acidification (soil and water acid damage), land use (occupation and transformation of land), water scarcity (consumption in stressed regions), fossil resource depletion (use of nonrenewable resources), and global warming (greenhouse gas emissions).

Recent studies highlight that applying an LCA framework to biorefinery models reveals substantial differences in environmental impacts depending on the choice of feedstock and process technologies, underscoring the need for tailored strategies that consider specific contexts and objectives [7, 27]. Such an approach enhances the precision and relevance of environmental assessments while enabling more informed decisions by a wide range of stakeholders, including researchers, industry professionals, and policymakers [50]. Adapting LCA methodologies in this manner can help advance biorefineries as genuine sustainable alternatives to fossil-based systems, promoting innovations that balance environmental impact with economic feasibility and social considerations.

4.2 Economic Perspective

In conjunction with environmental and policy considerations, the economic implications of technological innovation are increasingly central to enhancing LCA practices within biorefineries. Life Cycle Costing (LCC) is an economic assessment tool integrated with LCA to provide a more comprehensive sustainability evaluation [51]. By including economic impacts throughout a product's life cycle, LCC helps ensure decisions are both environmentally sound and economically viable; however, its implementation can be challenging due to the need for detailed data and assumptions, which may introduce uncertainties [49].

The differences in system boundaries might be traced back to the unpredictable nature of technological advancements, all of which can affect the accuracy and reliability of cost estimations. The execution of well-crafted policies can stimulate the integration of renewable energy technologies and advocate for investigative efforts into novel feedstock applications, thus augmenting both ecological performance and economic feasibility [7].

The integration of advanced methodologies such as machine learning and real-time data analytics not only strengthens environmental assessments but also provides valuable economic insights by identifying cost-saving opportunities, improving process efficiency, and optimizing resource allocation [52]. These innovations support the development of more resilient and economically viable production systems, enabling stakeholders to better respond to market fluctuations and resource constraints. As such, the convergence of digital tools and economic analysis is vital for scaling up sustainable and profitable biorefinery models [53].

Furthermore, socially inclusive strategies that ensure stakeholder engagement and transparent benefit-sharing can enhance the economic viability of biorefineries by fostering local acceptance and contributing to stable, long-term operations [54]. The integration of economic and environmental, and policy factors through LCA, alongside socially inclusive strategies, are essential to ensure a sustainable, economically viable, and widely accepted biorefinery operations.

4.3 Social Policies Perspective

The use of social policy instruments to influence LCA methodologies within biorefineries is deep, as they provide essential guidance for stakeholders navigating the complexities of sustainability evaluations. The Social Life Cycle Assessment (S-LCA) complements traditional LCA and LCC, by incorporating social dimensions such as labor conditions, community impact, and human rights, offering a more holistic sustainability evaluation [49].

While its implementation presents challenges due to the qualitative and context-specific nature of social data, integrating S-LCA into biorefinery assessments can guide research and development toward more socially responsible and informed decisions [55]. Involving a diverse array of parties, including local communities, policymakers, and industry specialists, can yield critical insights that encapsulate social dimensions, frequently neglected in conventional assessments, as, for example, the impact on workers and local communities, including health and safety issues, fair wages, and local employment opportunities [50], catalyzing enhancements across the sector.

This alignment between social policy efforts and LCA practices is vital for propeling the transition toward a sustainable bioeconomy that not only satisfies present requirements but also safeguards ecological integrity for forthcoming generations [56].

The combination of socially inclusive strategies, the alignment of policy frameworks with LCA methodologies strengthens the foundation for sustainable, economically viable, and widely accepted biorefinery operations [49, 55]. This integration supports inclusive, well-informed decisions that balance environmental, economic, and social priorities, ultimately fostering a resilient bioeconomy that meets today's needs while safeguarding long-term sustainability goals.

5 LCA Methodologies in Biorefinery

Standardization ensures consistency, credibility, and comparability across studies and sectors, enabling stakeholders to make informed decisions based on robust and transparent data. This section explores the international standards and norms that underpin LCA practices, beginning with an overview of the general guidelines, followed by a deeper dive into ISO standards and the fundamental principles and structure they promote.

5.1 Standards and Norms Guiding LCA

The international standardization of LCA methodologies is essential for accurately evaluating the environmental sustainability of biorefineries and ensuring comparability with traditional fossil fuel systems. Standardized LCA approaches facilitate consistent assessments of environmental impacts, highlighting key contributors like electricity use, transportation and manufacturing to energy consumption and greenhouse gas emissions and are guided by ISO frameworks [8, 40, 57].

Moreover, fostering collaborations between academia and industry can lead to the development of more refined LCA methodologies, which are tailored to specific feedstocks and operational contexts, thereby improving transparency and accuracy in sustainability assessments [27]. This collaborative approach not only addresses the complexities associated with coproduct allocation but also encourages a culture of continuous improvement within the sector, ultimately driving innovations that align with circular economy principles and promote long-term ecological resilience. The main approached phases are illustrated in Fig. 3.

Fig. 3 Phases of LCA methodologies in biorefinery context

5.2 *ISO 14040 and ISO 14044 Standards*

When discussing key ISO standards and norms, it is essential to delve into the specific standardized approach and principles that these standards embody. Two pivotal standards in this realm are ISO 14040 and ISO 14044, both of which are integral to the LCA process [58].

ISO 14040 serves as the foundational standard for LCA, outlining the principles and framework necessary for conducting a comprehensive assessment of a product's environmental impact throughout its entire life cycle. This standard emphasizes the importance of a systematic approach to evaluating the environmental aspects of a product, from raw material extraction and manufacturing to distribution, use and eventual disposal or recycling. By establishing a clear structure, ISO 14040 helps organizations identify opportunities for reducing environmental impacts and improving sustainability practices [59].

ISO 14044, on the other hand, builds upon the principles set forth in ISO 14040 by providing detailed requirements and guidelines for conducting an LCA. This standard elaborates on the four phases of an LCA: goal and scope definition, inventory analysis, impact assessment and interpretation. It emphasizes the need for transparency and reproducibility in the assessment process, ensuring that results are credible and can be used for informed decision-making [60]. Additionally, ISO 14044 addresses the importance of the engagement authorities and the need to consider various environmental impact categories, such as greenhouse gas emissions, resource depletion and ecological toxicity [61]. Together, ISO 14040 and ISO 14044 not only provide a robust standardized approach for assessing environmental impacts but also promote a culture of continuous improvement in sustainability practices across various industries [59, 61]. By adhering to these standards, organizations can enhance their environmental performance, foster innovation and demonstrate their commitment to sustainable development.

As the integration of innovative technologies continues to reshape the biorefinery landscape, it is crucial to also consider the implications of policy normative models that support these advancements. For instance, government incentives for adopting renewable energy sources can significantly enhance the economic viability of biorefineries while simultaneously reducing greenhouse gas emissions by up to 83% compared to traditional fossil fuels [7].

5.3 Goal and Scope Definition

The goal definition serves as the foundational pillar of LCA, guiding methodological accurate and ensuring alignment with intended applications. The scope defines system boundaries, excluding irrelevant processes (e.g., social impacts of raw material harvesting) to maintain focus on environmentally significant factors. According to ISO 14040 and 14044 standards, this phase mandates clear expression of the study's purpose, scope, audience, and compliance requirements for public comparative assertions [62]. It establishes the functional unit, system boundaries, and impact categories while determining whether attributional (ALCA) or consequential (CLCA) modeling is appropriate.

On one hand, in ALCA, the inputs and outputs are methodically ascribed to the functional unit of a product system through the association and/or division of the unit processes of the system, adhering to a normative protocol. On the other hand, CLCA represents a systematic methodology wherein activities are incorporated into the product system(s) under evaluation solely to the degree that they are anticipated to be altered as a result of fluctuations in demand for the functional unit [56].

This clarity is essential for the correct interpretation of results and methodological choices throughout the LCA process. The goal clarifies why the study is conducted (e.g. product optimization, policy support or public claims) and for whom (e.g. internal stakeholders, policymakers or consumers) [63]. For public comparative assertions, ISO 14044 requires rigorous critical review to prevent greenwashing and ensure transparency. Ambiguous goals may lead to misaligned system boundaries or irrelevant data collection. For instance, applying CLCA to product comparisons introduces unnecessary complexity, while using ALCA for policy analysis overlooks broader market effects [62]. Best practices involve iterative refinement of goals during interpretation phases and adherence to ISO guidelines for critical reviews, particularly in public disclosures.

5.4 Functional Unit and Reference Flow LCA

The functional unit (FU) and reference flow are critical components in LCA, serving as the foundation for comparative analyses of environmental impacts across different products and systems. It has been shown that varying functional units can significantly influence LCA outcomes, particularly when durability metrics are included (Panesar et al. 2017). Indeed, the choice of functional unit and reference flow can alter the perceived environmental benefits of bioenergy pathways, highlighting the importance of clear definitions in LCA methodologies (Choudhary et al. 2014). Thus, establishing robust criteria for these elements is essential for accurate and comparable LCA results.

The FU quantifies the function of a product or service, serving as a consistent basis for measuring and comparing environmental impacts in LCA [64]. The reference flow corresponds to the amount of product required to fulfill the FU. Commonly, the FU is defined as the production of 1 kg of a product or the provision of a specific service over a given time period [27]. Its selection must reflect the primary function of the system under study and account for contextual factors such as geographical location, user behavior, and market conditions to ensure relevance to real-world applications. For example, in dairy systems, the FU may be defined as 1 kg of energy-corrected milk, though inconsistency across studies can introduce bias if standardization is not applied (Schüler et al. 2017).

In packaging assessments, the FU might be defined as the delivery of an equivalent volume of product, facilitating equitable comparisons across alternatives [2]. Ensuring consistency in the FU across scenarios supports informed decision-making and comparability. Furthermore, the FU can be extended to encompass attributes like product lifespan, service efficiency, or output quality, enriching the analysis and promoting a more comprehensive understanding of environmental impacts [56]. This nuanced approach enhances the reliability of LCA results and informs improvements in product design, service delivery, and production processes [62].

5.5 System Boundary

The system boundary serves as a critical delineation that defines the limits between the product or system being analyzed and the external systems that influence or

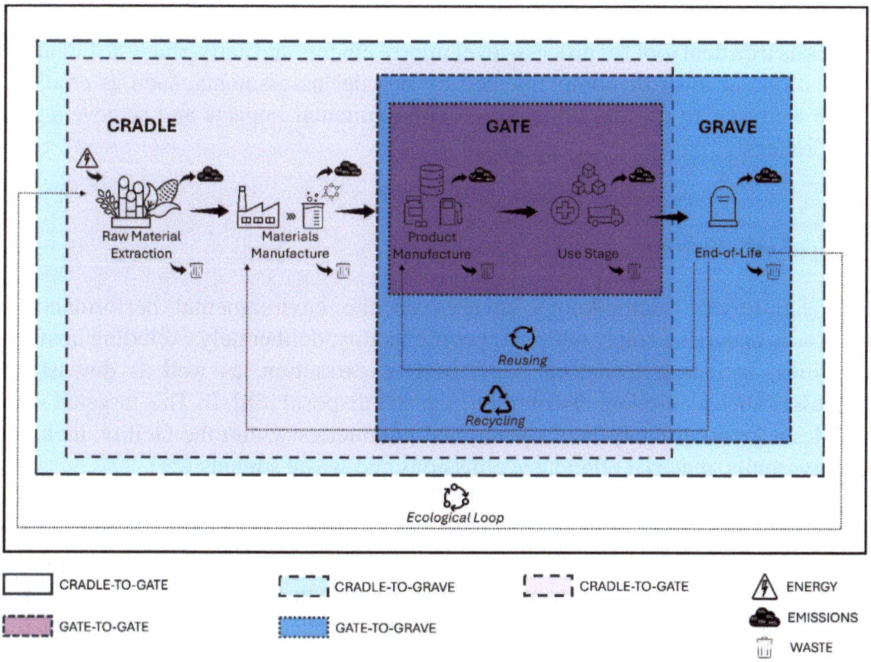

Fig. 4 Integration of system boundaries in biorefinery in LCA methodologies

interact with it [62]. The establishment of these boundaries is paramount for ensuring that the assessment captures all relevant environmental impacts associated with the product throughout its life cycle [27]. The comprehensiveness of an LCA is largely influenced by the chosen system boundaries, which dictate what aspects are included or excluded from the evaluation [56]. Within the biorefinery systems examined in the current study, five principal system boundaries have been identified and shown on Fig. 4.

5.5.1 Cradle-to-Gate

The cradle-to-gate approach assesses the environmental impacts of a product from the extraction of raw materials "cradle" up to the point at which the product exits the manufacturing facility "gate.“ This partial LCA excludes downstream stages such as distribution, use, and end-of-life disposal [65]. By focusing exclusively on production, this approach enables a detailed evaluation of resource use, energy consumption, emissions, and waste generation associated with manufacturing processes. Cradle-to-gate assessments are particularly valuable for generating Environmental Product Declarations (EPDs) and optimizing sustainability performance in industrial operations. However, the limited system boundaries inherent to this approach preclude a full understanding of a product's environmental burden [66]. As such, while it serves as a critical tool for process-level improvements and early-stage sustainability evaluations, it must be complemented by broader assessments, such as cradle-to-grave analyses, to capture downstream environmental impacts and achieve a more comprehensive sustainability appraisal [67].

5.5.2 Gate-to-Gate

The gate-to-gate methodology focuses on the environmental performance of processes occurring solely within a specific facility, deliberately excluding upstream activities, including feedstock cultivation or extraction, as well as downstream processes such as product distribution, use and disposal [68]. It. This targeted scope enables a granular analysis of operational parameters within the facility, including energy inputs, material efficiency, emissions and waste streams [27].

Gate-to-gate assessments are particularly useful for identifying process bottlenecks, enhancing operational efficiencies and supporting technology optimization within the boundaries of the industrial site [69]. This approach can inform process intensification and foster innovation aimed at increasing the economic and environmental viability of bio-based production systems. Nonetheless, its limited scope may mask the broader environmental and social impacts associated with the full product life cycle [58]. Therefore, gate-to-gate analyses are best utilized in conjunction with more expansive frameworks to ensure a holistic understanding of sustainability across the value chain.

5.5.3 Cradle-to-Cradle

The cradle-to-cradle model represents a paradigm shift in sustainability, promoting a regenerative, circular economy model that contrasts sharply with the conventional linear "take-make-dispose" production model [70]. Cradle-to-cradle emphasizes the design of products and systems in which all materials are intended either for biological cycles (biodegradation) or technical cycles (reuse or recycling without loss of quality) [69].

This philosophy underlines the importance of safe material use, design for disassembly and continuous cycling of materials, thereby minimizing waste generation and environmental degradation. In practice, cradle-to-cradle encourages the use of nontoxic, biodegradable inputs and closed-loop manufacturing systems that retain the value of resources. By embedding circularity into product design from the outset, the approach not only reduces environmental impact but also drives innovation and long-term economic resilience [70, 71]. Cradle-to-cradle thus provides an approach for rethinking industrial systems in harmony with ecological principles, aiming to transform products into resources for future use rather than waste.

5.5.4 Gate-To-Grave

Gate-to-grave approach encompasses the downstream stages of a product's life cycle, starting from the point it leaves the production facility and extending through its use phase and eventual disposal [27]. In the context of biorefineries, this assessment methodology captures the environmental impacts associated with product utilization and end-of-life treatment, such as incineration, landfilling or recycling. This methodology is essential for understanding how products perform environmentally after leaving the production site and provides critical insights into their contribution to emissions, energy consumption, and waste during the post-manufacturing phases.

By accounting for these downstream effects, gate-to-grave assessments help participants, including manufacturers, users and policymakers, identify opportunities to reduce environmental burdens, promote circularity and improve overall system sustainability [2, 48]. While the gate-to-grave perspective offers valuable information about the fate of products, it must be integrated with upstream assessments (e.g. cradle-to-gate) to yield a complete evaluation of a product's life cycle impacts.

5.5.5 Cradle-to-Grave

Cradle-to-grave represents the most comprehensive LCA framework, encompassing all stages of a product's life, from raw material extraction and manufacturing to distribution, use, and final disposal [72]. This holistic assessment captures the cumulative environmental impacts associated with the entire life cycle, including energy use, emissions, water consumption, and waste generation. In the context of biorefineries, cradle-to-grave analyses provide a critical foundation for evaluating the true environmental performance of bio-based products, informing decision-making across research, policy, and industrial practice. By revealing the trade-offs and hotspots across the value chain, this approach supports the development of strategies aimed at minimizing life cycle impacts and enhancing sustainability outcomes. Furthermore, it fosters interdisciplinary collaboration among engineers, environmental scientists, industry stakeholders, and regulators, promoting shared responsibility in advancing sustainable innovation [30, 62]. Cradle-to-grave assessments are thus indispensable for achieving an integrated understanding of environmental performance and guiding the transition to a more sustainable and circular bioeconomy.

5.5.6 Alternative System Boundaries in LCA

In addition to the more prevalently utilized system boundaries, a variety of LCA methodologies exist that cater to particular analytical objectives, yet are infrequently implemented. For instance, the Well-to-Wheel approach, commonly employed within the transportation and energy domains, evaluates the environmental repercussions stemming from resource extraction (such as crude oil or biomass) through the stages of fuel production and ultimately to the operational phase of the vehicle, thereby providing insights into the efficiency and emissions associated with diverse fuel pathways [2]. Seed-to-Gate represents a variant that holds particular significance within agricultural evaluation system, initiating with seed cultivation and extending to the farm gate, thereby facilitating the evaluation of agricultural inputs and practices [73].

The Well-to-Tank and Tank-to-Wheel methodologies are frequently applied in a complementary manner to dissect the life cycles of fuels into upstream (fuel production) and downstream (vehicle utilization) components, respectively [74]. An additional niche concept is Grave-to-Cradle, which emphasizes the recovery and reintegration of materials post-product disposal, thereby bolstering strategies aligned with the circular economy [33]. Although these methodologies are less commonly applied within the realm of biorefineries, they proffer invaluable insights for particular sectors and can be incorporated into more comprehensive assessments to tackle distinct sustainability inquiries.

Finally, when defining goals and scope, it is essential to explicitly state all underlying assumptions, as they significantly influence the study's outcomes and interpretation. These assumptions may relate to system boundaries, data sources, product lifespan, or functional equivalence [75]. Clearly articulating them ensures transparency in how key variables are handled, thereby facilitating reproducibility and informed interpretation. Equally important is the acknowledgment of methodological and data-related limitations, which may affect the comprehensiveness and reliability of the results [18]. Transparent reporting of both assumptions and limitations enhances the credibility of the assessment and supports critical evaluation by entities, enabling more robust decision-making and communication of environmental implications.

5.6 *Life Cycle Inventory (LCI)*

The Life Cycle Inventory (LCI) analysis focuses on the systematic collection and validation of data regarding energy and material inputs, as well as environmental outputs throughout a product's life cycle. This process encompasses various stages, including raw material extraction, manufacturing, transportation, usage, and end-of-life disposal, thereby providing a comprehensive overview of resource use and emissions associated with each phase [76].

The LCI serves as a foundational repository that enables parties to identify environmental impacts and opportunities for optimization, facilitating informed decision-making in product development and sustainability strategies [13, 77]. The methodology involves several steps, such as constructing flow charts, gathering data, and conducting calculations, which culminate in a detailed inventory of elementary flows that informs subsequent impact assessments [70]. By integrating these elements, organizations can achieve a nuanced understanding of the interrelations among life cycle stages and their collective environmental repercussions [70, 76].

5.6.1 Data Collection

The data collection is essential in the LCI phase, as it underpins the accurate quantification of environmental exchanges throughout a product's life cycle. This process involves gathering input data, such as raw materials and energy, and output data, including emissions and waste, across various stages from extraction to disposal [76, 78]. The integration of diverse data sources, ranging from direct measurements to established LCI databases, enhances the precision and relevance of the collected data [79].

Engaging stakeholders, particularly suppliers, is crucial for obtaining context-specific operational data, which can significantly improve data quality [27]. Furthermore, addressing temporal and geographical variability is vital, as environmental impacts vary widely based on location and time frame (Krstić et al. 2014). A robust data collection strategy not only elevates the quality of the LCI but also bolsters the credibility of the overall LCA, facilitating informed environmental decision-making [27].

5.6.2 Mass and Energy Balance

The conduction of a mass and energy balance is mandatory to account for all material and energy inputs and outputs within a delineated system. This methodology, rooted in the principles of mass and energy conservation, guarantees that no materials are lost or left unrecorded, whereas each input must be traceable to an output, storage, or transformation [80].

The fundamental objective of these calculations is to quantify all streams, raw materials, energy, emissions, and wastes, within the designated process boundaries, thus facilitating a comprehensive understanding of system efficiency and environmental ramifications. To accomplish this, practitioners are required to delineate precise system boundaries and to compile extensive data pertinent to the functional unit [81]. Such information is crucial not only for the identification of inefficiencies or emissions that could potentially remain undetected but also for underpinning applications such as environmental permitting and energy optimization [56].

For instance, in a fermentation process, mass and energy balances may account for variables such as the heat of reaction, biomass composition, and electron flows [2]. Furthermore, when subsystems are equilibrated in this manner, LCI data can be seamlessly integrated into an expedited LCA, wherein input–output values are multiplied by eco-costs, thereby streamlining the evaluation process by circumventing intricate impact calculations [81].

5.6.3 System Modeling of Integrated Biorefinery Chains

Modeling integrated biorefinery systems requires a comprehensive and multidisciplinary approach that considers both economic viability and environmental performance. Strategies such as mixed integer linear programming (MILP) can optimize supply chains by evaluating multiple biomass feedstocks, storage options, transportation routes and conversion technologies, supported by robust database systems for informed decision-making [82]. The system modeling can also be used for improving the reliability of environmental assessments by incorporating uncertainty analysis, as seen in the BioSTEAM-LCA normative model, which allows for more accurate and flexible projections of biorefinery impacts [83].

Comparative assessments of sugarcane-based biorefineries suggest that scenarios maximizing ethanol output, particularly those integrating first- and second-generation ethanol production, can achieve the best environmental and economic outcomes [2]. The importance of process synthesis and systematic screening in early design stages is also highlighted as a means to enhance flexibility and optimize network configurations [84]. Finally, broader sustainability evaluations that go beyond traditional LCA, incorporating tools such as multicriteria decision analysis, offer a more holistic understanding of trade-offs and synergies in biorefinery development (Tuazon and Gnansounou 2017). These combined efforts underscore the value of integrating diverse modeling techniques and assessment tools to support more sustainable biorefinery systems.

5.6.4 Allocation Strategies for Multioutput Systems

The strategies allocation in LCI for multioutput biorefinery systems are inherently complex, demanding a nuanced understanding of process interactions to ensure accurate and consistent environmental assessments. Traditional allocation methods often fall short, as they may not fully account for all coproducts, introducing variability and inconsistency in LCI results. One recommended approach is to model these processes in their unallocated form, which allows for the development of a consistent set of allocation factors that can be applied uniformly across all outputs [33].

In lignocellulosic biorefineries, matrix-based allocation methods have been explored to better trace environmental burdens, particularly emphasizing the impact of internal energy use and upstream greenhouse gas emissions in lignin-derived products [85]. For lignin extraction, the choice of allocation method can significantly alter the outcomes of climate impact assessments, with variables such as market values and system-driving factors playing a decisive role [86]. To address the limitations of conventional allocation strategies, a hybrid mass-energy (HMEN) method has been introduced, offering a more balanced distribution of impacts across products by integrating both mass and energy considerations [87].

In parallel, machine learning techniques such as deep neural networks have been proposed to create adaptive allocation systems that respond to changes in feedstock composition and market demands, thereby optimizing both cost and environmental performance [52]. These varied approaches highlight the necessity of context-specific allocation strategies tailored to the structure, outputs, and sustainability objectives of each biorefinery.

5.6.5 Data Quality Assessment

A data quality assessment is important as it directly impacts the reliability and validity of the results. Key criteria for evaluating data quality include completeness, representativeness, consistency, and accuracy. Completeness ensures that all necessary data points are collected, preventing analytical gaps that could underestimate environmental impacts [76]. Representativeness reflects actual conditions, accounting for variability across geographic, temporal, and technological dimensions [88]. Consistency ensures uniform methodologies in data collection to avoid bias and maintain comparability, while accuracy verifies alignment with real-world values, often requiring validation through empirical measurements or authoritative sources [89]. Recent approaches emphasize a broader perspective on data quality, integrating input data characteristics and overall model quality, which enhances the robustness of LCA outcomes [78].

5.6.6 Life Cycle Impact Assessment (LCIA)

The impact assessment phase of LCA is fundamental to translating inventory data into potential environmental consequences. This stage begins with the classification of elementary flows, compiled during the LCI into specific impact categories such as climate change, acidification, eutrophication, and human toxicity [81]. Each category reflects a distinct environmental mechanism linked to emissions and resource consumption occurring across the life cycle of a product or service.

Characterization Models

LCIA characterization enables the expression of environmental impacts on a standardized scale across various categories. This process involves applying characterization factors to inventory data, which facilitates the aggregation of multiple environmental flows and enhances the comparability of different products and systems [90]. The development of normalization factors (NFs) is crucial for contextualizing LCA results, as they provide reference points against which environmental profiles can be compared [91, 92].

Overall, while characterization enhances the scientific rigor of LCA, ongoing efforts are necessary to address data gaps and improve the consistency of characterization factors across various impact categories [91, 92]. Moreover, the choice of impact assessment methods, such as ReCiPe, focused on how the choice of perspective and time frame affects result variability, or CML-IA, focused on how interpretation and methodological rigor ensure result reliability and usefulness, can lead to varying conclusions, emphasizing the need for careful selection based on the study's goals and data characteristics [93]. Ultimately, a thorough interpretation process, which includes sensitivity analysis, enhances decision-making by clarifying significant issues and guiding eco-innovation efforts [94, 95].

Normalization

A normalization serves as a critical step for contextualizing environmental impacts by comparing them to reference values, such as per capita emissions or resource use at various scales. This process enhances the understanding of the relative significance of different impact categories, as demonstrated by studies that highlight the importance of normalization references (NRs) in interpreting LCA results, particularly in sectors like waste management and renewable energy [96, 97].

However, the application of normalization is fraught with methodological challenges, including data gaps and uncertainties stemming from the selection of data sources and classification of emissions [91, 98]. Moreover, traditional normalization approaches may not adequately reflect sustainability limits, potentially downplaying significant impacts [99]. Therefore, while normalization aids in clearer communication and prioritization, it necessitates careful consideration of its methodological foundations to ensure robust and meaningful interpretations [96, 98].

Weighting

In LCIA, weighting is considered a critical, yet controversial, phase that integrates various environmental impact categories into a single score, facilitating decision-making amidst trade-offs between diverse environmental issues [100–102]. This process can draw on expert judgment, participant input, and socioeconomic indicators, enhancing the relevance and acceptability of LCA results when conducted transparently [100, 103].

However, the subjective nature of weighting raises concerns, particularly regarding its compatibility with international standards, as it may reflect the values of specific stakeholders or regions, complicating public comparisons [102, 104]. Despite these challenges, effective weighting can simplify complex assessments, making it particularly useful in eco-design and eco-efficiency calculations, where a unified environmental score aids in material selection and cost–benefit analyses [102, 103].

5.6.7 Midpoint Versus Endpoint Impact Models in Biorefinery LCAs

In biorefinery LCAs, selecting between midpoint and endpoint impact models is a critical decision that shapes the depth and scope of the environmental analysis.

The midpoint models focus on specific environmental mechanisms, such as climate change, eutrophication, acidification, or water scarcity. These mechanisms provide detailed, problem-oriented information that supports targeted mitigation strategies [105, 106].

A midpoint level of granularity is particularly valuable in identifying environmental trade-offs and hotspots within complex biorefinery systems. In contrast, endpoint models aggregate these intermediate impacts into broader areas of concern like human health, ecosystem quality, and resource availability. While less detailed, this approach offers a more synthesized and communicable perspective, often facilitating interpretation in policy or managerial contexts [105, 106]. The ReCiPe 2016 framework illustrates the integration of both approaches, providing a consistent set of midpoint and endpoint characterization factors that enable assessments at varying levels of abstraction [105].

In addition, endpoint models can incorporate economic metrics, such as the valuation of dissipated resources based on market prices, which is useful for evaluating economic sustainability in resource-intensive systems like biorefineries [107]. Nevertheless, endpoint assessments, while offering simplicity and comprehensibility, may obscure critical process-level insights available through midpoint indicators [108]. This way, the choice between models should be aligned with the goals of the LCA, and in many cases, a hybrid approach that combines the strengths of both may provide the most balanced and actionable results.

Relevance of Regionalized Impact Assessment

The regionalized impact assessment is essential for the precise evaluation of environmental repercussions, as it incorporates geographical disparities in environmental conditions and population densities that can substantially affect the results of LCA analyses. The necessity for regional differentiation is emphasized by the heterogeneity of environmental impacts, such as water scarcity and land utilization, which exhibit considerable variation across distinct regions and necessitate regionally specific life cycle inventory (LCI) data for accurate evaluation [109].

The advancement of regionalized life cycle impact assessment (LCIA) methodologies, exemplified by IMPACT World + , underscores the significance of integrating spatial variability into impact categories such as climate change, particulate matter formation, and water consumption, all of which considerably contribute to global human health and ecosystem quality concerns [110]. Despite progress, challenges persist in harmonizing regionalized LCIA methodologies, particularly in reconciling spatial resolutions and addressing uncertainties related to input parameters and model systems [111]. The IMPACT North America model illustrates the advantages of geographically differentiated evaluations, revealing that particulate matter has a profound impact on human health, thereby necessitating comprehensive regional modeling [112].

Additionally, regionalized assessments have demonstrated a capacity to yield more precise outcomes for categories such as acidification and eutrophication in comparison to site-generic assessments, thereby highlighting the imperative for enhanced geographic information within LCA databases [113]. In summary, regionalized impact assessment amplifies the accuracy and significance of LCA by capturing the spatial heterogeneity of environmental impacts, thus facilitating more informed decision-making in the realms of environmental management and policy formulation.

5.7 *Interpretation of LCA Results*

5.7.1 Identifying Critical Hotspots in Biorefinery Life Cycles

The interpretation phase of LCA is essential for translating technical findings into actionable insights that inform environmental decision-making. This phase involves a systematic evaluation of results from inventory analysis and impact assessment, focusing on the completeness, consistency, and relevance of findings to identify significant environmental issues and hotspots, such as stages that contribute disproportionately to greenhouse gas emissions or resource depletion [94, 114].

The process is iterative, requiring practitioners to revisit earlier phases to ensure alignment with the study's goals and to validate the robustness of conclusions drawn [115]. Moreover, effective interpretation is often hindered by inadequate sensitivity analyses and visualization techniques, which can obscure critical insights for the interested parties [95]. Ultimately, a thorough interpretation phase enhances the reliability of LCA results and supports targeted sustainability actions, assessing completeness, consistency, and relevance, while addressing limitations like weak sensitivity analyses. Practitioners can identify key environmental hotspots and ensure robust, actionable insights.

5.7.2 Sensitivity and Uncertainty

Sensitivity and uncertainty analyses play a crucial role in LCA by evaluating how uncertainties, such as those found in life cycle inventory (LCI) data and modeling, can influence study outcomes. Sensitivity analysis enhances the robustness of results by examining the effects of varying parameters and assumptions while uncertainty analysis focuses on quantifying the influence of these uncertainties on final results, often through statistical methods like Monte Carlo simulations. Although many LCA studies incorporate sensitivity analysis, fewer include uncertainty analysis, despite its importance in future-proofing conclusions, particularly in the context of biorefinery systems that are often at early stages of technology readiness (TRL) [27].

5.7.3 Reporting

Finally, reporting presents as a critical phase that integrates findings from various analytical stages, including goal and scope definition, inventory analysis, and impact assessment, to provide actionable insights for decision-makers [116]. This phase is inherently iterative, requiring practitioners to revisit earlier stages to ensure alignment with the study's objectives and the reliability of the data [115]. Furthermore, effective interpretation is essential for accurate communication of results, which can significantly influence decision-making in both industrial and policy contexts [95, 114]. Despite its importance, systematic approaches to interpretation, including sensitivity and uncertainty analyses, are often lacking, suggesting a need for improved methodologies and visualization techniques to enhance the understanding and usability of LCA findings [95].

5.7.4 Importance of Peer Review for Public LCA Disclosures

Peer review plays a vital role in ensuring the credibility, accuracy, and scientific integrity of public LCA disclosures. Peer review plays a crucial role in validating methodologies, identifying errors, and upholding the standards necessary to influence policy and public perception; its importance is underscored by John Bohannon's hoax, where a fake paper was widely accepted, exposing the risks of inadequate scrutiny and the need for rigorous evaluation [117]. In complex LCA applications, such as those involving electricity networks or construction materials, peer review ensures methodological soundness and data reliability, while fostering methodological advancement and continuous improvement through constructive feedback and the identification of research gaps [118, 119]. Overall, peer review strengthens the scientific foundation of LCA, supporting its role in sustainable decision-making.

6 Software and Databases

Software and databases are integral components in LCA, as they furnish the requisite tools and data essential for executing thorough evaluations. LCA software applications, including SimaPro, GaBi, and OpenLCA, provide capabilities for life cycle inventory (LCI), life cycle impact assessment (LCIA), and result interpretation. These technological instruments enhance the depiction of biorefinery functions, the review of material and energy exchanges, and the appraisal of environmental effects, including greenhouse gas emissions, resource scarcity, and modifications in land application [120–122].

On the other hand, databases serve as the cornerstone of data necessary for LCA investigations. They encompass critical information regarding the environmental repercussions associated with diverse processes, materials, and energy sources, which are vital for the development of life cycle inventories. Notable examples of frequently

Table 2 Key features and limitations of LCA tools and database

Tool/database	Key features	Limitations	References
SimaPro	User-friendly interface, supports both consequential and attributional LCA, integrates with various databases	Limited customization options for advanced users	[120, 122]
GaBi	Advanced features for life cycle engineering, robust database integration, sensitivity, and scenario analyses	Steeper learning curve due to complex functionalities	Don et al. (2022) [122]
OpenLCA	Open-source, flexible and customizable, supports various databases and models	Limited user support compared to commercial tools	Rajić et al. (2024) [122]
Aspen Plus	Process simulation, energy and mass balance calculations, integration with LCA tools	Primarily focused on process simulation rather than LCA	[123, 124]
ecoinvent Centre	Comprehensive coverage of processes and materials, regularly updated	Requires subscription for full access	[121, 122]
USLCI Database	Publicly available, transparent, ease of use	Limited to North American systems	[121, 122]
Agri-Footprint	Specializes in agricultural and food systems, detailed environmental impact data	Limited coverage of nonagricultural processes	[121, 122]

utilized LCA databases include the ecoinvent Centre, the United States Life Cycle Inventory (USLCI) database and the Agri-Footprint database, which are particularly focused on agricultural and bio-based systems [121, 122]. There are several LCA databases and tools available, presenting distinct features, strengths, and limitations. Table 2 compares these tools and databases showing the essential key features to understand their applicability in biorefinery LCA studies.

7 Future Perspectives in LCA

7.1 *Emerging Technologies and Innovations on LCA in Biorefineries*

Several emerging technologies and innovations are poised to significantly enhance LCA methodologies. As the biorefinery sector continues its trajectory of evolution, the incorporation of sophisticated data analytics and machine learning methodologies represents a promising frontier for the enhancement of LCA methodologies. These technological advancements provide real-time insights into operational efficiencies and environmental ramifications, enabling flexible feedstock management and process refinement [51]. For example, by utilizing predictive modeling, facilities could more effectively evaluate the ramifications of diverse coproduct allocation strategies on sustainability metrics, thereby addressing the intricacies highlighted in contemporary studies regarding multifunctionality and system boundaries [8, 28].

Advanced data analytics and artificial intelligence can streamline the processing of vast amounts of qualitative and quantitative data, enabling more accurate predictions of environmental impacts throughout the production process [125]. As biorefineries continue to increasingly embrace and implement cutting-edge advanced technologies in their operations and processes, the immense potential for significantly enhancing sustainability through the development and improvement of various waste management practices becomes ever more pronounced and evident. For instance, the integration of anaerobic digestion alongside gasification techniques not only effectively addresses the numerous challenges posed by municipal solid waste (MSW) but also facilitates the remarkable transformation of these by-products into highly valuable resources, thereby aligning perfectly with the overarching principles of a circular economy [57, 126].

Furthermore, the exploration of decentralized biorefinery models has the potential to democratize and broaden access to sustainable energy production while simultaneously fostering the growth and development of local economies, thus playing a crucial role in mitigating certain environmental burdens that are often associated with centralized systems of production [127]. This significant shift toward localized solutions underscores the critical importance of adopting and implementing adaptive management strategies that can respond effectively to the specific needs of individual communities as well as the availability of local resources, ultimately promoting resilience across both economic and ecological dimensions within the comprehensive bioeconomy model [128]. By strategically leveraging such groundbreaking innovations, various entities can create a far more integrated and cohesive approach to resource utilization that not only enhances operational efficiency but also promotes greater social equity within the community.

7.1.1 Machine Learning and AI in LCA Biorefineries

AI is playing an increasingly pivotal role in enhancing LCA models for biorefineries by improving predictive analytics, resource optimization, and data accuracy. Through the use of machine learning (ML) algorithms, AI enables dynamic LCA models that process large, complex datasets across the entire lifecycle, from raw material extraction to processing and end-of-life scenarios, offering real-time insights into environmental impacts [129]. In terms of operational efficiency, AI can optimize resource utilization by identifying the most effective biomass valorization pathways, such as determining optimal conditions for volatile fatty acid (VFA) recovery, a key step in producing biofuels and biochemicals [130]. Furthermore, AI contributes to sustainable waste management by simulating and evaluating various treatment scenarios, thereby supporting circular economy strategies [131]. Another critical advantage of AI lies in addressing data quality challenges: it can automate the integration of diverse data sources and utilize historical records to detect patterns, fill data gaps, and improve the overall reliability and accuracy of LCA outcomes [129, 131].

7.1.2 Blockchain in LCA

Blockchain technology presents a transformative potential for enhancing LCA in biorefineries by improving transparency, traceability, and data integrity across the environmental assessment process. Its decentralized and immutable ledger enables tamper-proof recording of environmental data and transactions, fostering trust and accountability among stakeholders throughout the supply chain [132, 133]. In addition to ensuring data transparency, blockchain-enabled smart contracts can automate compliance checks with environmental regulations—such as verifying carbon credits or monitoring sustainable biomass sourcing—thus streamlining regulatory adherence and reducing administrative burdens [132, 134]. This technology also enhances supply chain management through real-time tracking of materials and products, supporting operational efficiency and circular economy practices [135, 136]. While the integration of blockchain with LCA remains in its early stages, several studies highlight significant technical and organizational challenges that must be addressed for broader adoption [133]. A blockchain-oriented LCA framework tailored for building material selection has been proposed, emphasizing multicriteria evaluation grounded in ISO standards [137]. A combined framework using blockchain, IoT, and advanced data analytics has also been introduced, aiming to improve supply chain sustainability through certified data gathering and analysis [138]. The inadequate evolution of blockchain technology and supply chain organization has been noted as a critical barrier to implementation [139]. Furthermore, a broader digital integration—including blockchain, AI, IoT, and big data—has been proposed to synergize these technologies across LCA phases and enhance overall methodological effectiveness [140]. Collectively, these insights underscore blockchain's promise in advancing LCA for biorefineries, while also indicating the need for further empirical research to fully realize its capabilities.

7.1.3 Comparative Analysis of AI and Blockchain in LCA Biorefineries

Traditional LCA models in biorefineries often rely on static datasets and manual data collection, which can result in inefficiencies, reduced accuracy, and limited capacity to handle dynamic systems or real-time data. These limitations make traditional models less suitable for the complexity of modern biorefinery operations and hinder transparency and traceability in data origin and accountability [140, 141].

In contrast, AI-enhanced LCA models offer dynamic, adaptive assessments capable of processing large datasets, predicting environmental impacts, and optimizing resource use, key advantages for biorefinery applications. Nonetheless, their effectiveness hinges on the quality and availability of training data, which can present a bottleneck [129]. Blockchain-enhanced LCA models complement these advancements by providing robust transparency, traceability, and data security, which are crucial for ensuring the integrity of environmental reporting and regulatory compliance. However, high resource requirements and a lack of real-world implementation still challenge their scalability [132, 133].

A synergistic integration of AI and Blockchain technologies presents a promising pathway forward, AI contributes predictive power and data-handling efficiency, while Blockchain guarantees the trustworthiness and immutability of environmental data. Together, these technologies can overcome individual limitations and establish a comprehensive, resilient framework for sustainable and accountable biorefinery operations [132, 134].

7.1.4 Challenges and Limitations

Despite the promising potential of AI and blockchain technologies in enhancing LCA for biorefineries, several challenges and limitations must be addressed to ensure their effective implementation. One of the most pressing issues is data quality and availability, AI models require extensive and high-quality datasets for accurate training, while blockchain systems depend on reliable and precise data inputs to maintain integrity and functionality [133, 141].

Scalability and interoperability also pose significant concerns; blockchain technology, still in its developmental phase, often struggles to support large-scale operations, and its integration with existing infrastructures can be technically complex and resource-intensive [132, 136]. Furthermore, regulatory and ethical considerations add another layer of complexity. The deployment of these technologies in LCA raises concerns about data privacy, algorithmic bias, and even the environmental footprint of digital systems themselves, all of which need to be carefully managed to ensure responsible innovation and sustainable outcomes [136, 142].

The integration of AI and blockchain into LCA for biorefineries offers a promising path to improve sustainability, data transparency, and system efficiency. While challenges remain, their combined use can overcome key limitations of traditional LCA approaches. A summary of their comparative strengths and limitations is presented in Table 3.

8 Conclusions

The implementation of LCA within biorefinery systems is fundamental for quantifying environmental burdens, identifying trade-offs, and supporting the optimization of biomass valorization pathways. As biorefineries increasingly adopt integrated, multifunctional configurations, the complexity of environmental assessments escalates, necessitating robust methodological frameworks capable of addressing coproduct allocation, regional specificity, and temporal variability. Standardized

Table 3 Comparative analysis of traditional, AI, and blockchain-enhanced LCA models

Aspect	Traditional LCA	AI-enhanced LCA	Blockchain-enhanced LCA	References
Data handling	Relies on static datasets and manual data collection	Can process large datasets and predict missing data points	Ensures data integrity and traceability	[133, 141]
Transparency	Lack transparency and accountability	Limited transparency, dependent on data sources	Provides transparency and accountability through decentralized ledger systems	[129, 132, 140]
Predictive capabilities	Limited ability to handle dynamic systems and real-time data	Offers dynamic and adaptive assessments with predictive analytics	Limited predictive capabilities, primarily focused on data security and traceability	[132, 134]
Scalability	Suitable for small-scale operations	Can handle large-scale operations with advanced computational capabilities	Scalability issues due to technological limitations	[132, 136]
Compliance	Manual compliance checks and limited automation	Automated compliance checks through predictive analytics	Automated compliance checks through smart contracts	[136, 142]

ISO protocols such as ISO 14040 and ISO 14044 provide essential guidance; however, further harmonization is required to enhance cross-study comparability and methodological transparency.

Emerging digital tools, including machine learning and blockchain technologies, offer significant opportunities to improve data acquisition, impact modeling, and decision-support capabilities. These technologies can facilitate dynamic and predictive LCA applications, particularly valuable during early-stage design and scale-up of novel biorefinery concepts. Nevertheless, challenges such as data scarcity, methodological uncertainty, and integration of social and economic dimensions remain critical areas for advancement.

Overall, LCA constitutes a pivotal instrument in steering biorefinery systems toward sustainability, yet it must overcome methodological challenges related to space, time, and overall sustainability. These challenges can be addressed by advancing prospective and consequential LCA methodologies, improving regionalized impact assessments through systems modeling integration, and incorporating social and economic aspects via LCSA approaches. By fostering interdisciplinary collaboration and embedding LCA at the core of biorefinery research, development, and policy, we can support informed, evidence-based transitions toward a circular and low-carbon bioeconomy.

References

1. Wang S (2021) Holistic approach for sustainable product development and service. No further bibliographic information available.
2. Mazuchi M (2018) Life cycle assessment of sugarcane biorefinery: energy use, environmental assessment, and other sustainability indicators. In: Chandel AK, Silveira MHL (eds) Advances in sugarcane biorefinery technologies, commercialization, policy issues, and paradigm shift for bioethanol and by-products. Elsevier, pp 213–239. https://doi.org/10.1016/B978-0-12-804534-3.00011-2
3. Pandey N, Mishra A (2022) Biorefineries and the circular bioeconomy for the management of agro-industrial byproducts. In: Anal AK, Panesar PS (eds) Valorization of agro-industrial byproducts, 1st edn. Taylor & Francis Group, p 378. https://doi.org/10.1201/9781003125679-15
4. Mohan SV, Dahiya S, Amulya K, Katakojwala R, Vanitha TK (2019) Can circular bioeconomy be fueled by waste biorefineries? A closer look. In: Bioresource technology reports. https://doi.org/10.1016/J.BITEB.2019.100277
5. Roy S (2017) Applying life cycle assessment (LCA) in process industry—the chemours experience. https://doi.org/10.1016/B978-0-12-409548-9.10073-9
6. Hosseinzadeh-Bandbafha H, Tabatabaei M, Aghbashlo M, Rehan M, Nizami A-S (2020) Determining key issues in life-cycle assessment of waste biorefineries. In: Biorefinery for Bioenergy. Elsevier, pp 271–294. https://doi.org/10.1016/B978-0-12-818228-4.00019-8
7. Kumar B, Verma P (2021) Life cycle assessment: blazing a trail for bioresources management. Ecol Model 449:100063. https://doi.org/10.1016/J.ECMX.2020.100063
8. Karka P, Papadokonstantakis S, Hungerbühler K, Kokossis AC (2015) Life cycle assessment of biorefinery products based on different allocation approaches. In: Sustainable biorefineries. Elsevier, pp 347–365. https://doi.org/10.1016/B978-0-444-63576-1.50123-0
9. Lammert F (2022) Life-cycle assessment of biorefinery. In: Biorefinery design for sustainability. Elsevier, pp 99–115. https://doi.org/10.1016/B978-0-323-90034-8.00005-1
10. El Haggar SM (2005) Rural and developing country. Solutions. https://doi.org/10.1016/B978-012088441-4/50015-0
11. Finkbeiner M (2014) The international standards as the constitution of life cycle assessment: the ISO 14040 series and its offspring. https://doi.org/10.1007/978-94-017-8697-3_3
12. Rocha MH, Lora EES, Venturin OJ, Escobar JCP, Santos JJCS, Moura AG (2010) Use of the life cycle assessment (LCA) for comparison of the environmental performance of four alternatives for the treatment and disposal of bioethanol stillage. Int Sugar J 112(1343):611–616. https://doi.org/10.5555/20103351101
13. Levy M (2017) Life cycle analysis—strengths and limitations of LCA. In: Sustainability in industrial engineering. Elsevier, pp 223–239. https://doi.org/10.1016/B978-0-12-409548-9.10062-4
14. Mukamwi M, Somorin T, Soloha R, Dace E (2023). Databases for biomass and waste biorefinery: a mini-review and SWOT analysis. Bioengineered 14(1). https://doi.org/10.1080/21655979.2023.2286722
15. Cherubini F, Ulgiati S (2010) Crop residues as raw materials for biorefinery systems—a LCA case study. Appl Energy 87(4):1194–1202. https://doi.org/10.1016/J.APENERGY.2009.08.024
16. Pradhan A, Shrestha DS (2006) An update on life cycle study of soybean oil biodiesel production. https://doi.org/10.13031/2013.28030

17. Gawel E, Ludwig G (2011) The iLUC dilemma: how to deal with indirect land use changes when governing energy crops? Land Use Policy. https://doi.org/10.1016/J.LANDUSEPOL.2011.03.003
18. Ahlgren S, Bjorklund A, Ekman A, Karlsson H, Berlin J, Börjesson P, Ekvall T, Finnveden G, Janssen M, Strid I (2015) Review of methodological choices in LCA of biorefinery systems—key issues and recommendations. Biofuels Bioprod Bioref 9(5):606–619. https://doi.org/10.1002/bbb.1563
19. Azapagic A, Perdan S, Clift R (2005) Sustainable development in practice: case studies for engineers and scientists. Wiley, Chichester. https://doi.org/10.1002/9780470972847
20. Caetano NS, Corrêa PS, Morais Júnior WG de, Mata TM, Martins AA, Branco-Vieira M (2022) LCA: a tool to develop sustainable microalgal biorefineries. In: de Morais Jr WG, Ramos FA (eds) Biofuels and biorefinery technologies. CRC Press, pp 185–198. https://doi.org/10.1201/9781003195429-10
21. The White House (2012) National bioeconomy blueprint. https://obamawhitehouse.archives.gov/sites/default/files/microsites/ostp/national_bioeconomy_blueprint_april_2012.pdf
22. European Bioeconomy Alliance (EUBA) (2024) The bioeconomy blueprint: building a circular and resilient Europe. https://www.bioeconomyalliance.eu/
23. Sala S, Cerutti A, Pant R (2018) Development of a weighting approach for the environmental footprint. Publications Office of the European Union. https://doi.org/10.2760/945290
24. European Union (2018) Directive (EU) 2018/2001 of the European Parliament and of the Council of 11 December 2018 on the promotion of the use of energy from renewable sources (recast). https://eur-lex.europa.eu/legal-content/EN/TXT/?uri=uriserv:OJ.L_.2018.328.01.0082.01.ENG
25. ISCC (2024) ISCC carbon footprint certification (Version 1.0). https://www.iscc-system.org/wp-content/uploads/2024/03/ISCC-CFC_incl-CCS_v1.0.pdf
26. Kätelhön A, Meys R, Deutz S, Suh S, Bardow A (2019) Climate change mitigation potential of carbon capture and utilization in the chemical industry. Proc Natl Acad Sci (PNAS) 116(23):11187–11194. https://doi.org/10.1073/pnas.1821029116
27. Gaffey J, Collins MN, Styles D (2024) Review of methodological decisions in life cycle assessment (LCA) of biorefinery systems across feedstock categories. J Environ Manage. https://doi.org/10.1016/j.jenvman.2024.120813
28. Righi S (2019) Life cycle assessments of waste-based biorefineries—a critical review. In: Sustainable biorefineries. Springer, Heidelberg, pp 177–193. https://doi.org/10.1007/978-3-319-93740-3_9
29. Wellisch M, Jungmeier G, Karbowski A, Patel MK, Rogulska M (2010) Biorefinery systems—potential contributors to sustainable innovation. Biofuels Bioprod Biorefin 4(3):275–286. https://doi.org/10.1002/bbb.217
30. Cherubini F, Strømman AH, Ulgiati S (2011) Influence of allocation methods on the environmental performance of biorefinery products—a case study. Res Conserv Recycl 55(9):818–829. https://doi.org/10.1016/J.RESCONREC.2011.06.001
31. Bezergianni S, Chrysikou LP (2020) Application of life-cycle assessment in biorefineries. In: Kamm B, Kamm M (eds) Biorefinery: from biomass to chemicals and fuels. Elsevier, pp 377–388. https://doi.org/10.1016/B978-0-12-818228-4.00017-4
32. Venkatramanan V, Shah S, Prasad R, Shah M (2021) Life cycle assessment of lignocellulosic waste biorefinery. In: Handbook of sustainable chemical processes. Springer, Heidelberg, pp 301–318. https://doi.org/10.1007/978-981-15-9696-4_15
33. Wang H, Man S, Wang H, Presser V, Yan Q, Zhang Y (2023) Grave-to-cradle upcycling of harmful algal biomass into atomically dispersed iron catalyst for efficient ammonia electrosynthesis from nitrate. Appl Catal B 332:122778. https://doi.org/10.1016/j.apcatb.2023.122778
34. Liu Y, Liu Y, Lyu Y, Tian J, Zhao J, Ye N, Zhang Y, Chen L (2021) Review of waste biorefinery development towards a circular economy: from the perspective of a life cycle assessment. Renew Sustain Energy Rev 135:110716. https://doi.org/10.1016/J.RSER.2021.110716

35. Lee Y, Masimukku S, Li I-C, Keharika K, Pothu R, Boddula R, Al-Qahtani N, Huang B-W, Chang-Chien G (2024) Enhancing sustainable energy through cutting-edge waste biorefinery technologies. Reactions 5(4):59. https://doi.org/10.3390/reactions5040059
36. Karka P, Papadokonstantakis S, Kokossis AC (2019) Predictive LCA—a systems approach to integrate LCA decisions ahead of design. In: Sustainable chemical engineering. Elsevier, pp 185–203. https://doi.org/10.1016/B978-0-12-818634-3.50017-5
37. Bello S, Salim I, Feijoo G, Moreira MT (2021) Inventory review and environmental evaluation of first- and second-generation sugars through life cycle assessment. Environ Sci Pollut Res. https://doi.org/10.1007/S11356-021-12405-Y
38. Naira VR, Mahesh R, Panda SK, Maiti SK (2020) Biorefinery approaches for the production of fuels and chemicals from lignocellulosic and algal feedstocks. https://doi.org/10.1007/978-981-15-1804-1_7
39. Garlapati VK, Tewari S, Ganguly R (2019) Life cycle assessment of first-, second-generation, and microalgae. Biofuels. https://doi.org/10.1016/B978-0-12-817937-6.00019-9
40. Farzad S, Mandegari M, Görgens JF (2020) Life cycle assessment of lignocellulosic biorefineries. Elsevier, pp 259–277. https://doi.org/10.1016/B978-0-12-818223-9.00010-2
41. Latiza RJP, Rubi RVC (2024) Circular economy integration in 1G+2G sugarcane bioethanol production: application of carbon capture, utilization and storage, closed-loop systems, and waste valorization for sustainability. Appl Sci Eng Progress 17(7):1005–1021. https://doi.org/10.14416/j.asep.2024.07.005
42. Pachón ER, Vaskan P, Görgens JF, Gnansounou E (2020) Process design, techno-economic, and life-cycle assessments of selected sugarcane-based biorefineries: a case study in the South African context. https://doi.org/10.1016/B978-0-12-818996-2.00026-0
43. Vaskan P, Pachón ER, Gnansounou E (2017) Life cycle assessment of sugar crops and starch-based integrated biorefineries. In: Sustainability metrics: life cycle assessment and green design in polymers. Elsevier, pp 145–192. https://doi.org/10.1016/B978-0-444-63585-3.00004-8
44. Götz M, Lask J, Lewandowski I, Kruse A (2023) Comparative LCA studies of simulated HMF biorefineries from maize and miscanthus as an example of first- and second-generation biomass as a tool for process development. In: GCB Bioenergy. https://doi.org/10.1111/gcbb.13079
45. Qian Q, Wei Q, Shi J, Li L (2024) Life cycle assessment and techno-economic analysis of wood-based biorefineries for cellulosic ethanol production. Biores Technol. https://doi.org/10.1016/j.biortech.2024.130595
46. Lopes T, Ortigueira J, Matos CT, Costa LA, Ribeiro C, Reis A, Gírio FM (2023) Conceptual design of an autotrophic multi-strain microalgae-based biorefinery: preliminary techno-economic and life cycle assessments. Fermentation 9(3):255. https://doi.org/10.3390/fermentation9030255
47. Solis CMA, Juan JLGS, Mayol AP, Sy CL, Ubando AT, Culaba AB (2021) A multi-objective life cycle optimization model of an integrated algal biorefinery toward a sustainable circular bioeconomy considering resource recirculation. Energies. https://doi.org/10.3390/EN14051416
48. Schebek L, Mrani O (2014) Environmental and sustainability assessment of biorefineries. https://doi.org/10.1533/9780857097385.1.67
49. Collotta M, Champagne P, Tomasoni G, Mabee W (2025) Life cycle approach for the sustainability assessment of intensified biorefineries. In: Sustainable biorefineries: Advances in biochemical engineering/biotechnology, vol 174. Elsevier, pp 235–268. https://doi.org/10.1016/B978-0-12-824117-2.00013-2
50. Pérez-López P, Rajaonison A, Zebian B, Bouallou C, González-Fernandez C, Greses S, Gómez-Serrano C, Acién FG (2025) Exploiting outcomes of life cycle costing to conduct coherent screening social life cycle assessments of emerging systems: a case study of microalgae biorefineries. Int J Life Cycle Assess. https://doi.org/10.1007/s11367-024-02427-2

51. Vance C, Sweeney JS, Murphy F (2022) Space, time, and sustainability: the status and future of life cycle assessment frameworks for novel biorefinery systems. Renew Sustain Energy Rev 168:112259. https://doi.org/10.1016/j.rser.2022.112259
52. Chan WM, Le DVK, Chen Z, Tan J, Chew IML (2021) Resource allocation in multiple energy-integrated biorefinery using neuroevolution and mathematical optimization. J Sustain Bioenergy Syst 1(2):122–137. https://doi.org/10.1007/S41660-020-00151-6
53. Roy B, Tu Q (2022) A review of system dynamics modeling for the sustainability assessment of biorefineries. J Ind Ecol. https://doi.org/10.1111/jiec.13291
54. Santibañez-Aguilar JE, González-Campos JB, Ponce-Ortega JM, Serna-González M, El-Halwagi MM, El-Halwagi MM (2014) Optimal planning and site selection for distributed multiproduct biorefineries involving economic, environmental and social objectives. J Clean Prod. https://doi.org/10.1016/J.JCLEPRO.2013.08.004
55. Mármol C, Martín-Mariscal A, Picardo A, Peralta E (2023) Social life cycle assessment for industrial product development: a comprehensive review and analysis. Heliyon 9(11):e22861. https://doi.org/10.1016/j.heliyon.2023.e22861
56. Bishop G, Styles D, Lens PNL (2021) Environmental performance comparison of bioplastics and petrochemical plastics: a review of life cycle assessment (LCA) methodological decisions. Res Conserv Recycl 168:105451. https://doi.org/10.1016/j.resconrec.2021.105451
57. Duan Y, Tarafdar A, Kumar V, Ganeshan P, Rajendran K, Giri BS, Gómez-García R, Li H, Zhang Z, Sindhu R, Binod P, Pandey AK, Taherzadeh MJ, Sarsaiya S, Jain A, Awasthi M (2022) Sustainable biorefinery approaches towards circular economy for conversion of biowaste to value added materials and future perspectives. Fuel. https://doi.org/10.1016/j.fuel.2022.124846
58. Karka P, Papadokonstantakis S, Hungerbühler K, Kokossis AC (2014) Environmental impact assessment of biorefinery products using life cycle analysis. In: Advances in biorefinery. Elsevier, pp 147–166. https://doi.org/10.1016/B978-0-444-63433-7.50075-4
59. Schaubroeck T (2022) Sustainability assessment of product systems in dire straits due to ISO 14040–14044 standards: five key issues and solutions. J Ind Ecol. https://doi.org/10.1111/jiec.13330
60. Koffler C, Amor B, Carbajales-Dale M, Cascio J, Conroy A, Fava J, Gaudreault C, Gloria TP, Hensler CD, Horvath A, Humbert S, Manzardo A, Margni M, Osset P, Sinistore JC, Vigon B, Wallace M, Wang M, Prox M (2020) On the reporting and review requirements of ISO 14044. Int J Life Cycle Assess 25(5):1033–1046. https://doi.org/10.1007/S11367-019-01706-7
61. Finkbeiner M, Inaba A, Tan RBH, Christiansen K, Klüppel H-J (2006) The new international standards for life cycle assessment: ISO 14040 and ISO 14044. Int J Life Cycle Assess. https://doi.org/10.1065/LCA2006.02.002
62. Schaubroeck T, Schrijvers D, Schaubroeck S, Moretti C, Zamagni A, Pelletier N, Huppes G, Brandão M (2022) Definition of product system and solving multifunctionality in ISO 14040–14044: inconsistencies and proposed amendments—toward a more open and general LCA framework. https://doi.org/10.3389/frsus.2022.778100
63. Baitz M (2017) Attributional life cycle assessment. In: Udo de Haes H, Reap T (eds) Life cycle sustainability analysis. Springer, Heidelberg, pp 45–61. https://doi.org/10.1007/978-94-024-0855-3_3
64. Arzoumanidis I, Raggi A, Petti L (2019) Life cycle assessment of honey: considering the pollination service. Admin Sci 9(1):27. https://doi.org/10.3390/admsci9010027
65. Jiménez-González C, Curzons AD, Constable DJC, Cunningham VL (2004) Cradle-to-gate life cycle inventory and assessment of pharmaceutical compounds. Int J Life Cycle Assess 9(6):348–354. https://doi.org/10.1007/BF02978570
66. Sharma RK, Sarkar P, Singh H (2020) Assessing the sustainability of a manufacturing process using life cycle assessment technique—a case of an Indian pharmaceutical company. Clean Technol Environ Policy. https://doi.org/10.1007/S10098-020-01865-4
67. Karka P, Papadokonstantakis S, Kokossis A (2017) Cradle-to-gate assessment of environmental impacts for a broad set of biomass-to-product process chains. Int J Life Cycle Assess 22:1418–1440. https://doi.org/10.1007/s11367-016-1252-0

68. Brehmer B, Sanders JPM (2009) Methodology description behind implementing an energetic and exergetic cradle-to-gate analysis for valorizing the optimal utilization of biomass. Int J Green Energy 6(3):321–335. https://doi.org/10.1080/15435070902880992
69. Jamekhorshid A, Azin R (2023) Application of life cycle assessment on enhanced oil recovery processes. In: Challenges and recent advances in sustainable oil and gas recovery and transportation. Elsevier, pp 227–242. https://doi.org/10.1016/B978-0-323-99304-3.00013-3
70. Bjørn A, Bjørn A, Hauschild MZ (2018) Cradle to cradle and LCA. In: Hauschild MZ, Rosenbaum RK, Olsen SI (eds) Life cycle assessment: theory and practice. Springer, Heidelberg, pp 309–324. https://doi.org/10.1007/978-3-319-56475-3_25
71. Pauw I, Karana E, Kandachar P (2013) Cradle to cradle in product development: a case study of closed-loop design. https://doi.org/10.1007/978-981-4451-48-2_8
72. Farrell AE, Plevin RJ, Turner BT, Jones AD, O'Hare M, Kammen DM (2006) Ethanol can contribute to energy and environmental goals. Science 311(5760):502–508. https://doi.org/10.1126/science.1121416
73. Adler PR, Del Grosso SJ, Inman D, Jenkins R, Spatari S, Zhang Y (2012) Mitigation opportunities for life-cycle greenhouse gas emissions during feedstock production across heterogeneous landscapes. In: Ullmann's encyclopedia of industrial chemistry. Wiley-VCH, pp 1–7. https://doi.org/10.1016/B978-0-12-386897-8.00012-7
74. Hjelkrem OA, Arnesen P, Ingebrigtsen Bø T, Snefuglli Sondell R (2020) Estimation of tank-to-wheel efficiency functions based on type approval data. Appl Energy 258:115463. https://doi.org/10.1016/J.APENERGY.2020.115463
75. Curran MA (2017) Goal and scope definition in life cycle assessment. https://doi.org/10.1007/978-94-024-0855-3
76. Rodrigues TO, Belizario-Silva F, Braga TEN, Matsuura MISF (2021) LCA—life cycle assessment of a bioproduct derived from sugarcane bagasse. Environ Sci Pollut Res 28(6):6823–6835. https://doi.org/10.1007/s11356-020-10979-6
77. Arvidsson R, Ciroth A (2021) Introduction to "Life Cycle Inventory Analysis." In: The future of life cycle assessment. Springer, Heidelberg, pp 1–15. https://doi.org/10.1007/978-3-030-62270-1_1
78. Ciroth A, Burhan S (2021) Life cycle inventory data and databases. In: Hauschild MZ, Rosenbaum RK, Olsen SI (eds) Life cycle assessment: theory and practice. Springer, Heidelberg, pp 121–134. https://doi.org/10.1007/978-3-030-62270-1_6
79. Huizenga NG, Brown DC, Ehrenberg M (2011) Automating life cycle inventory data collection. J Ind Ecol 15(4):558–569. https://doi.org/10.1111/j.1530-9290.2011.00366.x
80. Rout PK, Behera BK (2021) Mass balance concept in livestock farming. In: Sustainable livestock farming. Springer, Heidelberg, pp 9–27. https://doi.org/10.1007/978-981-33-4343-6_2
81. Carneiro MLNM, Pradelle F, Braga SL, Gomes MSP, Martins ARFA, Turkovics F, Pradelle RNC (2017) Potential of biofuels from algae: comparison with fossil fuels, ethanol, and biodiesel in Europe and Brazil through life cycle assessment (LCA). Renew Sustain Energy Rev 73:632–653. https://doi.org/10.1016/j.rser.2017.01.152
82. Paulo H, Barbosa-Póvoa AP, Relvas S (2013) Modeling integrated biorefinery supply chains. https://doi.org/10.1016/B978-0-444-63234-0.50014-2
83. Shi R, Guest JS (2020) BioSTEAM-LCA: an integrated modeling framework for agile life cycle assessment of biorefineries under uncertainty. ACS Sustain Chem Eng. https://doi.org/10.1021/ACSSUSCHEMENG.0C05998
84. Stuart P, El-Halwagi MM (2012) Integrated biorefineries: design, analysis, and optimization. CRC Press. https://doi.org/10.1201/B13048
85. Obydenkova SV, Kouris PD, Smeulders D, Boot M, van der Meer Y (2021) Modeling life-cycle inventory for multi-product biorefinery: tracking environmental burdens and evaluation of uncertainty caused by allocation procedure. Biofuels Bioprod Biorefin. https://doi.org/10.1002/BBB.2214
86. Hermansson F, Janssen M, Svanström M (2020) Allocation in life cycle assessment of lignin. Int J Life Cycle Assess 25(7):1159–1172. https://doi.org/10.1007/S11367-020-01770-4

87. Djomo SN, Knudsen MT, Parajuli R, Andersen MS, Ambye-Jensen M, Jungmeier G, Gabrielle B, Hermansen JE (2017) Solving the multifunctionality dilemma in biorefineries with a novel hybrid mass–energy allocation method. GCB Bioenergy. https://doi.org/10.1111/GCBB.12461
88. Henriksen T, Astrup TF, Damgaard A (2021) Data representativeness in LCA: a framework for the systematic assessment of data quality relative to technology characteristics. J Ind Ecol 25(1):51–66. https://doi.org/10.1111/jiec.13048
89. Ciroth A (2021) Data quality in life cycle inventories. In: Hauschild MZ, Rosenbaum RK, Olsen SI (eds) Life cycle assessment: theory and practice. Springer, Heidelberg, pp 101–119. https://doi.org/10.1007/978-3-030-62270-1_5
90. Sonnemann G, Tsang M, Castells F (2018) Life-cycle. Impact Assessment. https://doi.org/10.1201/9780429436949-3
91. Crenna E, Secchi M, Benini L, Sala S (2019) Global environmental impacts: data sources and methodological choices for calculating normalization factors for LCA. Int J Life Cycle Assess. https://doi.org/10.1007/S11367-019-01604-Y
92. Ryberg M, Vieira M, Zgola M, Bare JC, Rosenbaum RK (2014) Updated US and Canadian normalization factors for TRACI 2.1. In: Clean technologies and environmental policy. https://doi.org/10.1007/S10098-013-0629-Z
93. Rybaczewska-Błażejowska M, Jezierski D (2024) Comparison of ReCiPe 2016, ILCD 2011, CML-IA baseline and IMPACT 2002+ LCIA methods: a case study based on the electricity consumption mix in Europe. Int J Life Cycle Assess. https://doi.org/10.1007/s11367-024-02326-6
94. Hauschild MZ, Bonou A, Olsen SI (2018) Life cycle interpretation. In: Handbook of life cycle assessment. Springer, Heidelberg, pp 205–224. https://doi.org/10.1007/978-3-319-56475-3_12
95. Sala S, Andreasson J (2018) Improving interpretation, presentation and visualisation of LCA studies for decision making support. https://doi.org/10.1007/978-3-319-66981-6_37
96. Cucurachi S, Seager TP, Prado V (2017) Normalization in comparative life cycle assessment to support environmental decision making. J Ind Ecol. https://doi.org/10.1111/JIEC.12549
97. Kim J, Yang Y, Bae J, Suh S (2013) The importance of normalization references in interpreting life cycle assessment results. J Ind Ecol 17(6):871–879. https://doi.org/10.1111/J.1530-9290.2012.00535.X
98. Benini L, Sala S (2016) Uncertainty and sensitivity analysis of normalization factors to methodological assumptions. Int J Life Cycle Assess. https://doi.org/10.1007/S11367-015-1013-5
99. Matuštík J, Paulu A, Kočí V (2024) Is normalization in life cycle assessment sustainable? Alternative approach based on natural constraints. J Clean Prod 395:141234. https://doi.org/10.1016/j.jclepro.2024.141234
100. Bengtsson M, Steen B (2000) Weighting in LCA—approaches and applications. Environ Prog 19(2):208–215. https://doi.org/10.1002/EP.670190208
101. Gloria TP (2017) Weighting of different environmental impact categories. https://doi.org/10.1016/B978-0-12-409548-9.10070-3
102. de Haes HAU (1999) Weighting in life-cycle assessment: is there a coherent perspective? J Ind Ecol 3(1):17–28. https://doi.org/10.1162/108819899569629
103. Gjalt H, Lauran VO (2011) Background review of existing weighting approaches in Life Cycle Impact Assessment (LCIA). https://doi.org/10.2788/88828
104. Edmunds JA (1978) Weighting environmental effects: Analytic survey with operational evaluation methods and a meta-method. Int J Life Cycle Assess 13(5):374–383. https://doi.org/10.1007/s11367-012-0415-x
105. Huijbregts MAJ, Huijbregts MAJ, Steinmann ZJN, Elshout PMF, Stam G, Verones F, Vieira M, Zijp MC, Hollander A, van Zelm R (2017) ReCiPe2016: a harmonised life cycle impact assessment method at midpoint and endpoint level. Int J Life Cycle Assess 22(8):1381–1393. https://doi.org/10.1007/S11367-016-1246-Y

106. Kägi T, Dinkel F, Frischknecht R, Humbert S, Lindberg J, Mester SD, Ponsioen T, Sala S, Schenker U (2016) Session "Midpoint, endpoint or single score for decision-making?"—SETAC Europe 25th annual meeting, May 5th, 2015. Int J Life Cycle Assess 21(7):1090–1096. https://doi.org/10.1007/S11367-015-0998-0
107. Poncelet AC, Loubet P, Helbig C, Beylot A, Muller S, Villeneuve J, Laratte B, Thorenz A, Tuma A, Sonnemann G (2022) Midpoint and endpoint characterization factors for mineral resource dissipation: methods and application to 6000 data sets. Int J Life Cycle Assess. https://doi.org/10.1007/s11367-022-02093-2
108. Weidema BP (2015) Comparing three life cycle impact assessment methods from an endpoint perspective. J Ind Ecol 19(2):349–357. https://doi.org/10.1111/jiec.12162
109. Pfister S, Oberschelp C, Sonderegger T (2020) Regionalized LCA in practice: the need for a universal shapefile to match LCI and LCIA. Int J Life Cycle Assess. https://doi.org/10.1007/S11367-020-01816-7
110. Bulle C, Bulle C, Margni M, Patouillard L, Boulay A-M, Bourgault G, Bruille VD, Cao V, Hauschild MZ, Henderson AD, Humbert S, Kashef-Haghighi S, Kounina A, Laurent A, Levasseur A, Liard G, Jolliet O et al (2019) IMPACT World+: a globally regionalized life cycle impact assessment method. Int J Life Cycle Assess. https://doi.org/10.1007/S11367-019-01583-0
111. Mutel CL, Liao X, Patouillard L, Bare JC, Fantke P, Frischknecht R, Hauschild MZ, Jolliet O, de Souza DM, Laurent A, Pfister S, Verones F (2019) Overview and recommendations for regionalized life cycle impact assessment. Int J Life Cycle Assess. https://doi.org/10.1007/S11367-018-1539-4
112. Humbert S (2009) Geographically differentiated life-cycle impact assessment of human health (Doctoral dissertation, University of California, Berkeley). eScholarship. https://escholarship.org/uc/item/1xv927gv
113. Mutel CL, Hellweg S (2009) Regionalized life cycle assessment: computational methodology and application to inventory databases. Environ Sci Technol. https://doi.org/10.1021/ES803002J
114. Saade MRM, Gomes V, Silva MG (2021) LCA—interpretation of results. https://doi.org/10.1007/978-3-030-78044-9_6
115. Curran MA (2023) Interpretation, critical review, and reporting: scientific outline of interpretation. https://doi.org/10.1007/978-3-031-35727-5_2
116. Grahl B, Gensch C (2023) Why do we cover interpretation as well as critical review and reporting in one volume? In: Springer handbook of environmental and sustainability economics. Springer, Heidelberg, pp 1–8. https://doi.org/10.1007/978-3-031-35727-5_1
117. Wenning RJ, Burton GA, Ward H, Lynch JM (2014) The importance of scientific peer review at SETAC. Integr Environ Assess Manag 10(3):289–291. https://doi.org/10.1002/ieam.1501
118. Gargiulo A, Girardi P, Temporelli A (2017) LCA of electricity networks: a review. Int J Life Cycle Assess. https://doi.org/10.1007/S11367-017-1279-X
119. Martínez-Rocamora A, Solís-Guzmán J, Marrero M (2016) LCA databases focused on construction materials: a review. Renew Sustain Energy Rev 58:970–981. https://doi.org/10.1016/J.RSER.2015.12.243
120. Luo C, Yao XH, Zhang Y, Zhou H (2023) An empirical study on the impact of different structural systems on carbon emissions of prefabricated buildings based on SimaPro. World J Eng Technol 11(03):434–453. https://doi.org/10.4236/wjet.2023.113032
121. Ramakrishna S, Brindha R (2024) Life cycle assessment and tools. In: Handbook of materials circular economy. Springer, Heidelberg, pp 31–64. https://doi.org/10.1007/978-981-97-0589-4_2
122. Rice G, Clift R, Burns R (1997) LCA software review: comparison of currently available European LCA software. Int J Life Cycle Assess 2(1):53–59. https://doi.org/10.1007/BF02978725
123. Dhabliya D, Sasirekha R, Adnan MM, Sapaev JB, Gopalakrishnan S, Badhoutiya A, Swathi D (2023) Advancing bioenergy sustainability: integrating sustainability assessment tools and frameworks for bioenergy and bioproduct systems. https://doi.org/10.1109/icteasd57136.2023.10585054

124. García A, Alriols MG, Llano-Ponte R, Labidi J (2011) Energy and economic assessment of soda and Organosolv biorefinery processes. Biomass Bioenerg. https://doi.org/10.1016/J.BIOMBIOE.2010.10.002
125. Álvarez-Rodríguez C, Martín-Gamboa M, Iribarren D (2022) Enhancing life cycle management through the symbiotic use of data envelopment analysis: novel advances in LCA + DEA. In: Singh A, Yadav R (eds) Advances in green energy. Springer, Heidelberg, pp 361–378. https://doi.org/10.1007/978-3-030-77127-0_23
126. Hamby T (2022) Upgrading the value of anaerobic fermentation via renewable chemicals production: a sustainable integration for circular bioeconomy. Sci Total Environ 832:150312. https://doi.org/10.1016/j.scitotenv.2021.150312
127. Maina K (2022) Environmental and economic assessment of decentralized bioenergy and biorefinery networks treating urban biowaste. Resour Conserv Recycl 176:105898. https://doi.org/10.1016/j.resconrec.2021.105898
128. Sakapaji SC (2022) Integrating local and indigenous ecological knowledge (IEK) systems into climate adaptation policy for resilience building, and sustainability in agriculture. Int J Sustain Dev Res. https://doi.org/10.11648/j.ijsdr.20220801.12
129. Bassey KE, Juliet AR, Stephen AO (2024) AI-enhanced lifecycle assessment of renewable energy systems. Eng Sci Technol J. https://doi.org/10.51594/estj.v5i7.1254
130. Koubaa M (2024) Integrated biorefinery for a next-generation methanization process focusing on volatile fatty acid valorization: a critical review. Molecules 29(11):2477. https://doi.org/10.3390/molecules29112477
131. Mehdipour Z (2024) The application of AI for electronic LCA data and assessment toward the circular economy. Electronics Goes Green. https://doi.org/10.23919/egg62010.2024.10631236
132. Shareef AMA, Seçkiner S, Eid B, Abumeteir H (2024) Integration of blockchain with artificial intelligence technologies in the energy sector: a systematic review. Frontiers Energy Res. https://doi.org/10.3389/fenrg.2024.1377950
133. Zhang L, Fröhling M (2024) Integration of blockchain and life cycle assessment: a systematic literature review. Int J Life Cycle Assess. https://doi.org/10.1007/s11367-024-02371-1
134. Heston TF (2024) A blockchain AI solution to climate change. Int J Sci Res Archive 11(2):471–480. https://doi.org/10.30574/ijsra.2024.11.2.0471
135. Dar AA, Reegu F, Ahmed S, Hussain G (2024) Blockchain technology and artificial intelligence based integrated framework for sustainable supply chain management system. https://doi.org/10.23919/indiacom61295.2024.10498149
136. Hong Z, Xiao K (2024) Digital economy structuring for sustainable development: the role of blockchain and artificial intelligence in improving supply chain and reducing negative environmental impacts. Dental Sci Rep. https://doi.org/10.1038/s41598-024-53760-3
137. Gaaliche N, Elseoud MA (2024) A blockchain-based life cycle assessment (LCA) framework for building materials selection, pp 85–102. https://doi.org/10.1007/978-981-99-8533-3_5
138. Zhang A, Zhong RY, Farooque M, Kang K, Venkatesh VG (2020) Blockchain-based life cycle assessment: an implementation framework and system architecture. Resour Conserv Recycl 152:104512. https://doi.org/10.1016/j.resconrec.2019.104512
139. Carrières V, Lemieux A-A, Pellerin R (2021) Opportunities of blockchain traceability data for environmental impact assessment in a context of sustainable production. In: Sousa J, Alves C, Ferreira T (eds) Sustainable production and life cycle engineering. Springer, Heidelberg, pp 197–213. https://doi.org/10.1007/978-3-030-85874-2_13
140. Popowicz M, Katzer NJ, Kettele M, Schöggl J, Baumgartner RJ (2024) Digital technologies for life cycle assessment: a review and integrated combination framework. Int J Life Cycle Assess. https://doi.org/10.1007/s11367-024-02409-4
141. Hariyani D, Hariyani P, Mishra S, Sharma MK (2024) Leveraging digital technologies for advancing circular economy practices and enhancing life cycle analysis: a systematic literature review. Waste Manage Bull. https://doi.org/10.1016/j.wmb.2024.06.007
142. Chauhan M, Sahoo DR (2024) Towards a greener tomorrow: exploring the potential of AI, blockchain, and IoT in sustainable development. Nat Environ Pollut Technol 23(2):444–451. https://doi.org/10.46488/nept.2024.v23i02.044

Economic Analysis of Changes in Bioreactors Configuration Yielding Ethanol and Biochemicals

Arshia Akhtar and Haroon

Abstract Economic analysis of changes in bioreactor configurations for ethanol and biochemical production involves evaluating the financial impact of modifications on reactor design, operation, and efficiency. By altering bioreactor configurations, such as introducing new mixing strategies, optimizing nutrient delivery, or improving temperature and pH control, the overall productivity of ethanol and other biochemicals can be increased, potentially lowering the cost of production. However, these changes often require capital investment in new equipment, technology upgrades, and possibly additional labor costs. From an economic perspective, the key benefits include enhanced yields, reduced processing times, and energy savings, all of which can contribute to cost reductions. For instance, more efficient bioreactor systems can minimize resource consumption (e.g., energy and raw materials) while maximizing output. On the other hand, the potential for increased operational complexity or the need for specialized skills in managing new configurations may increase maintenance and training costs. Ultimately, the economic viability of these changes depends on a careful cost-benefit analysis, considering both short-term investment and long-term returns. The balance between the initial outlay and ongoing operational savings determines whether the change will lead to overall profitability in ethanol or biochemical production. Continuous monitoring and optimization of the bioreactor's performance are essential to maximize economic returns.

Keywords Bioreactor · Lignocellulosic biomass · Biochemicals · Ethanol

A. Akhtar (✉)
Department of Chemistry, Shri Ram Swaroop Memorial University, Dewa Road, Barabanki 225003, Uttar Pradesh, India
e-mail: arshia.chem@gmail.com

Haroon
Department of Physics, Integral University, Kursi Road, Lucknow 226026, Uttar Pradesh, India
e-mail: haroons@iul.ac.in

© The Author(s), under exclusive license to Springer Nature Singapore Pte Ltd. 2025
A. K. Chandel and S. Sundaramurthy (eds.), *Biorefineries in Circular Bioeconomy*, Green Energy and Technology, https://doi.org/10.1007/978-981-95-3913-0_13

1 Introduction

In recent years, renewable energy sources have gained significant importance, driven by rising concerns over climate change and the depletion of fossil fuel reserves. Among these alternatives, biofuels emerge as a promising solution, as they can be produced from diverse biomass resources. Interest in liquid biofuels derived from renewable biomass continues to grow as a strategy to reduce dependence on crude oil. The ongoing expansion of the global economy has intensified energy demand, alongside heightened awareness of greenhouse gas (GHG) emissions and their role in climate change. Consequently, many countries are investing in renewable energy technologies, including biofuel production. Biofuels are defined as energy carriers obtained from biomass, such as organic waste [1], and are regarded as having a reduced environmental impact compared to conventional fossil fuels [2]. Among them, bioethanol emerges as a leading example, with global production projected to surpass 130 billion liters per year [3], with the United States of America and Brazil accounting for the majority of this output [4]. Bioethanol is an alcohol produced by microbial fermentation of carbohydrates derived from plants or algae, such as wheat straw [5, 6], corn [7], sugarcane, and lignocellulosic materials. Ethanol obtained from lignocellulosic biomass (LCB) is considered a sustainable and environmentally friendly substitute for fossil fuels in transportation.

Lignocellulose can serve as a source for producing ethanol as well as other biofuels like butanol, bio-gasoline, and dimethylfuran through various reaction routes [8, 9]. Among these, research into bioethanol production is the most advanced, while many other biofuel technologies are still in their early development stages. Ethanol production from lignocellulosic materials generally follows two main process pathways:

(i) A biochemical method that includes chemical or enzymatic hydrolysis followed by microbial fermentation, and
(ii) A thermochemical route, where LCB is gasified and then further processed through either microbial fermentation or catalytic upgrading [10]. Worldwide, biomass ranks as the fourth most commonly used energy source, following traditional fuels like oil, coal, and natural gas [11].

The biorefinery (BRF) concept focuses on generating biofuels and high-value products from biomass using a combination of physical, chemical, biochemical, and thermochemical processes. This integrated strategy supports the attainment of four major Sustainable Development Goals (SDGs):

(i) Promoting decent work opportunities and economic growth.
(ii) Ensuring affordable and clean energy.
(iii) Advancing industry, innovation, and infrastructure.
(iv) Taking action against climate change by enabling the efficient and economical production of both energy and valuable chemicals, biorefineries open the door to innovation, sustainable practices, and economic progress. This is

achieved by maximizing resource use and reducing waste through advanced BRF technologies [12].

The main objective of BRFs is to establish efficient and sustainable methods for transforming biomass into fuels and high-value chemicals, adopting principles comparable to those applied in traditional petroleum refineries [13]. To make a significant impact, these methods must be economically viable compared to fossil-fuel-based technologies, while also addressing global concerns around climate change and energy reliability [14]. Biomass ranks as the fourth most widely used energy source globally, following oil, coal, and natural gas [11]. A key principle of BRF is the cascading use of biomass, which ensures maximum utilization of its components and minimizes waste [15]. Ultimately, the economic feasibility and cost analysis are crucial in evaluating the long-term viability of BRF systems [12, 16].

In the last twenty years, the bioeconomy (BE) concept has gained significant attention from researchers and industry experts because of its ability to reduce environmental impacts and open new business prospects. The expansion of BE has frequently been shaped by associated models, including the Circular Economy (CE) and the Circular Bioeconomy (CBE) [17]. CE focuses on replacing the traditional linear economic model by promoting waste reduction, resource efficiency, and environmental sustainability through strategies like reducing, reusing, and recycling [18]. Following this principle, the CBE approach involves channeling biomass through multiple stages, where valuable products are progressively extracted at each step.

1.1 BE, CE and CBE: A Comprehensive Overview

1.1.1 BE

The BE encompasses the sustainable generation and utilization of renewable biological resources, as well as their conversion—together with biological waste—into value-added products. Such products may include food, animal feed, bio-based materials, and sources of bioenergy [18].

Key aims of the BE include

- Offering nutritious, safe, and resource-efficient food and feed, along with innovative dietary supplements
- Generating novel chemicals, bio-based polymers, and functional materials with enhanced properties
- Producing bioenergy and alternative fuels that reduce reliance on fossil resources
- Advancing sustainable agricultural and marine practices, refining bioprocessing and BRF technologies, and supporting the development of industrial biotech solutions
- Providing environmentally friendly alternatives through green chemistry approaches

The BE also plays a crucial role in

- Lowering GHG emissions by replacing petroleum-based products with renewable alternatives across their life cycles
- Serving as the primary renewable carbon source, given that direct CO_2 utilization is still in early developmental stages
- Stimulating economic activity, attracting investment, and creating jobs in rural, coastal, and marine regions, thereby supporting local economies and small to medium enterprises

In essence, the BE is centered on generating renewable biological resources and converting them into high-value commodities that span sectors such as food, energy, and materials [19].

1.1.2 CE

As outlined in the European Commission's CE Action Plan, a CE seeks to maintain the value of resources, materials, and products for the longest possible duration while reducing waste generation to a minimum [20]. Rather than focusing solely on end-of-life disposal, this approach emphasizes the efficient use of both ecological and economic resources.

Within this framework, waste is not seen as a by-product, but as a resource with the potential to be reintegrated into the production cycle. Special attention is given to two key areas that intersect with the BE: reducing food waste across the supply chain and improving the efficient use of biomass.

The Ellen MacArthur Foundation additionally describes the CE as a regenerative and restorative system in which materials circulate continuously in loops "Nothing is wasted, and everything contributes to a new cycle" [21].

The CBE serves as a bridge between the CE and BE, as illustrated in Fig. 1, by integrating their sustainability goals [22]. CBE follows the principles of CE by using biomass to generate a range of biochemicals and bioenergy through BRF processes. However, transitioning from a fossil-fuel-based economy to a bio-based economy requires substantial scientific research and innovation. Tools like techno-economic assessments and life cycle analyses are essential in steering the advancement of CBE.

LCB is among the most plentiful and renewable bio-based resources found on Earth. Recognized as a carbon-neutral energy source, it has the potential to significantly lower CO_2 emissions and help mitigate environmental pollution that in turn offers great potential for producing biofuels and valuable chemicals. Many studies have investigated the technical methods used to transform LCB into a range of biofuels and high-value products through BRF techniques [23]. However, the path to successful commercial-scale BRF operations is fraught with challenges and risks. For example, Beta Renewables, formerly the world's largest ethanol production facility, was inaugurated in Crescentino, Italy, in 2013 with an annual capacity of 40 kilotonnes [24].

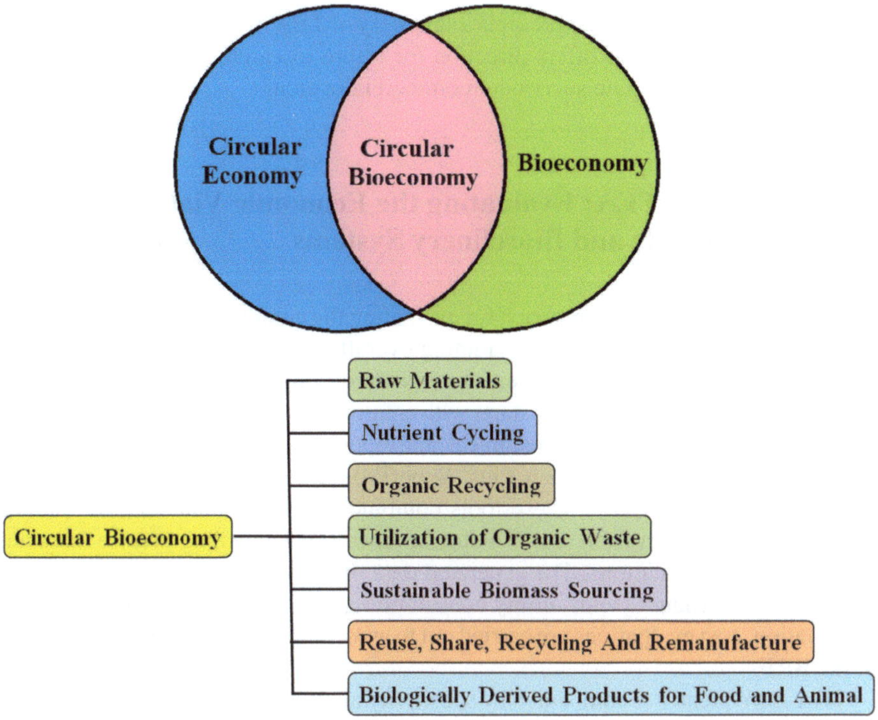

Fig. 1 The elements and framework of the CBE [19]

Techno-economic analysis (TEA) is essential for assessing the competitiveness of new technologies compared to existing options. TEA provides a comprehensive framework for evaluating both economic indicators and technical parameters throughout the entire production process [25]. It aids in estimating potential profitability by taking into account various factors, including plant scale, market pricing, raw material costs, product and by-product values, processing methods, and separation strategies [26]. Kamlesh et al. investigated the biochemical transformation of agricultural and algal biomass into biofuels, emphasizing TEA [27]. Their findings indicated that large-scale commercialization of biofuels from these sources is currently unfeasible, emphasizing the need for further research in areas such as the selection of fermentation methods, microbial strains, and suitable substrates. Similarly, Okolie et al. examined the potential of LCB to generate biofuels and high-value products, highlighting essential technical factors and the primary challenges involved in the process [28].

This chapter examines the potential of sustainable BRFs to produce biofuels and high-value products from LCB. It focuses particularly on TEA studies related to the production of bioethanol and valuable biochemicals. Along with a detailed review of TEA studies, the chapter discusses manufacturing processes, major production challenges, and possible solutions. It also considers the impact of policy frameworks on the TEA outlook. The chapter presents a summary of TEA in the context of integrated lignocellulosic BRFs and offers direction for future research, policy development, and innovation to facilitate successful commercialization.

2 Advances in TEA: Evaluating the Economic Viability of Bioproducts and Biorefinery Systems

TEA serves as an essential approach for evaluating the commercial viability of lignocellulosic BRFs. In addition to determining overall economic feasibility, it is instrumental in pinpointing system bottlenecks and identifying the primary factors driving costs [29]. The process generally starts with creating a comprehensive process model that accounts for all essential technical and operational parameters. Commercial simulation platforms like Aspen Plus, SuperPro, and CHEMCAD are frequently employed for this purpose. These tools combine data from mathematical simulations with information from actual industrial operations, enabling precise mass and energy balance estimations. The generated data offers valuable insights into design specifications, product yields, utility consumption, and waste generation. Using this information, cost evaluations are carried out to assess the overall economic feasibility of the BRF.

TEA is frequently accompanied by sensitivity analysis, which examines the effect of variations in key parameters on the overall feasibility of a BRF. Typical factors evaluated include feedstock cost, product price, market demand, and conversion efficiency. By exploring multiple scenarios reflecting changes in these variables, sensitivity analysis aids in comprehensive risk evaluation and supports well-informed decision-making.

A recent review by Clauser et al. outlines a framework for designing and assessing BRFs, stressing that integrating processes in terms of both mass and energy, along with process intensification, plays a crucial role in enhancing overall efficiency, economic feasibility, and environmental outcomes [30]. Additional insights into BRF process simulations are available in a separate review [31]. Key technical considerations relevant to conducting a TEA of BRF processes are summarized in Table 1.

In recent years, researchers have increasingly turned to open-source process simulation tools for performing TEAs, thereby avoiding reliance on commercial chemical engineering modeling software, as noted in the review by Scown et al. [48]. In TEA studies, various input parameters are considered, starting with the cost of raw materials, annual run time, and project site location, followed by the project duration and the time required for construction and commissioning; financial aspects such as the corporate tax rate, loan agreement terms, and potential devaluation are then factored in, along with a detailed costing breakdown, performance indicators like internal rate of return (IRR) (%) and Minimum Attractive Rate of Return (%), and finally the selection of a comparison year [48–53]. One notable example is

Table 1 Common technical parameters evaluated in TEA of BRF processes [32]

Input	Interpretation	References
Raw materials	Sources of renewable raw materials include agriculture, forestry, agro-industrial processes, and aquaculture. The distinction between crop-based and waste-based feedstocks is becoming increasingly significant. Often, a combination or integration of various raw materials spanning both first-generation and second-generation materials is also taken into account	[33]
Type of product	BRFs can be designed to focus on product output, such as platform chemicals or polymers, energy generation, such as biofuels, electricity, and heat, or a combination of both	[13, 34]
Source of energy	BRFs are frequently engineered to be energy self-sufficient by utilizing a portion of the raw materials or alternative renewable energy sources, such as biogas, to meet their energy requirements	[35]
Development of facilities	Greenfield facilities, constructed entirely from scratch, generally require a larger capital investment. On the other hand, brownfield facilities are integrated into existing production sites, for example, setting up a new bioethanol plant within an active sugarcane factory, enabling the sharing of existing utilities and infrastructure	[36, 34]
Manufacturing scale	This parameter defines the production capacity of the intended product, with reported values in the literature varying between 2.5 and 90 kilotonnes annually	[30]
Processes of conversion	These processes generally encompass thermochemical methods like combustion, gasification, and pyrolysis; biochemical approaches such as fermentation, anaerobic digestion, aerobic digestion, and enzymatic reactions; chemical techniques including oxidation, hydrogenation, hydrolysis, esterification, and steam reforming; along with physical or mechanical operations such as milling and separation, or, in many cases, a combination of these methods	[37]
Strategic approach to integration	An integrated BRF refers to a system in which renewable raw materials are transformed into one or more value-added products through multiple conversion pathways, thereby minimizing both raw material consumption and energy use. Adapting or enlarging existing industrial plants for bio-based chemical production can result in substantial cost savings	[38–40]
Downstream processing in BRFs can focus on producing either a single product or multiple products	The process of separating organic acids can be both costly and technically challenging	[41]

(continued)

Table 1 (continued)

Input	Interpretation	References
Centralization/ Decentralization	Centralized facilities, which are generally smaller, are often located closer to the biomass source. On the other hand, decentralized, larger facilities can source biomass from various locations, enabling them to capitalize on economies of scale	[42]
Capital and operational expenditures (CAPEX and OPEX)	TEA must evaluate capital investment costs (e.g., equipment, installation) and operational costs (e.g., raw materials, labor, maintenance, utilities). OPEX significantly influences project feasibility, especially for scale-up	[43]
Logistics and supply chain	The cost and reliability of biomass collection, storage, and transportation can impact profitability. Locational factors, seasonality, and moisture content of biomass affect supply chain design	[44]
Policy and regulatory framework	Government incentives, carbon credits, feed-in tariffs, and environmental regulations significantly affect project viability. Stable policy support encourages investment	[45]
Life cycle environmental impacts	TEA is often linked with Life Cycle Assessment (LCA) to quantify emissions, energy demand, and environmental benefits/risks across the supply chain	[46]
Market demand and price volatility	The success of a BRF depends on stable markets for bio-products. Price fluctuations in fossil-based alternatives or oversupply of bio-based products may affect economic returns	[47]

the BRF Simulation and Techno-Economic Analysis Modules (BioSTEAM), which enables the design, simulation, and economic evaluation of BRFs under uncertainty. BioSTEAM is reported to produce results comparable to those generated by proprietary software [54]. Numerous TEA studies utilizing BioSTEAM have since been published. Comparing results from various TEA studies in the literature can be challenging due to differences in technical methodologies, simulation models, and the underlying economic and technical assumptions. Even studies assessing the feasibility of producing the same bioproduct may sometimes present conflicting outcomes (Table 1). The complexity increases further when comparing TEA results across different products. To effectively identify which products should be prioritized for commercial deployment or further technological development, it is essential to adopt a standardized technical and economic framework for evaluating multiple bioproducts derived from the same feedstock [55–58]

3 Comparative Economics of 1G, 2G, and Integrated BRFs

As discussed above and in Table 1, a variety of technical parameters in TEA, including raw materials and product choices, production methods, operational duration, scale, geographical location, and integration strategies, significantly influence the TEAs of BRFs. This section presents and compares TEA findings from a wide range of studies to illustrate how these factors impact the economic viability of first-generation (1G), second-generation (2G), and integrated 1G2G BRFs.

TEA studies often evaluate various raw materials to determine the most cost-effective option for producing a specific bioproduct. For example, Koutinas et al. analyzed the economic feasibility of producing 2,3-butanediol (2,3-BDO) using different raw materials, namely sucrose, glycerol, and sugarcane molasses, under varying raw materials costs and production scales [59].

3.1 BRFs Utilizing LCB as Raw Materials

LCB is the world's most abundant renewable biomass resource, with an estimated global production of about 200 billion tons each year. In China, the annual production of agricultural straw reaches up to 900 million tons, with over 700 million tons being readily collectible. In recent decades, LCB has gained considerable interest as a sustainable resource for both energy production and energy storage applications. The rising interest is primarily fueled by the decline in fossil fuel reserves and the environmental consequences, particularly climate change, resulting from extensive fossil fuel combustion [60]. It is primarily generated through photosynthesis, whereby atmospheric carbon dioxide and water are converted into plant matter. LCB typically consists of three major components: cellulose (ranging from 9 to 80%), hemicellulose (10 to 50%), and lignin (5 to 35%) (Table 2).

Cellulose is a straight-chain polysaccharide composed of glucose units linked through β-(1–4) glycosidic bonds [66, 67]. Cellulose, lignin, and hemicellulose are interconnected both physically and through strong covalent cross-links as well as noncovalent interactions, creating a complex macroscopic network, as shown in Fig. 2. The hydroxyl groups within cellulose chains form extensive hydrogen bonding, giving rise to a highly ordered and crystalline arrangement. In contrast, hemicellulose is an irregular polymer consisting of diverse sugar monomers, including hexoses and pentoses [68]. These sugars are frequently substituted with functional groups such as methyl, acetyl, glucuronic acid, cinnamic acid, and galacturonic acid.

Lignin is an amorphous, water-insoluble aromatic polymer that provides structural reinforcement to plant cell walls and improves their resistance against microbial degradation. It is primarily synthesized from three kinds of phenylpropanoid monomers: sinapyl alcohol, coniferyl alcohol, and p-coumaryl alcohol.

Table 2 The proportions of cellulose, lignin, and hemicellulose present in primary LCB sources

Biomass	Cellulose (%)	Lignin (%)	Hemicellulose (%)	References
Corn cob	42–45	14–15	35–39	[61]
Sugarcane bagasse	42–48	20–42	19–25	[62]
Barley straw	31–45	14–19	27–38	[63]
Corn stover	38–40	7–19	24–26	
Wheat straw	33–38	17–19	26–32	
Sweet sorghum bagasse	34–45	14–21	34–45	
Rice straw	28–36	12–14	23–28	
Cotton	85–95	0	5–15	[64]
Grasses	25–40	10–30	35–50	
Hardwood stems	40–55	18–25	24–40	[65]
Newspaper	40–55	18–30	25–40	
Leaves	15–20	0	80–85	
Sorted refuse	60	20	20	
Nutshells	25–30	30–40	25–30	
Solid cattle manure	1.6–4.7	2.7–5.7	1.4–3.3	
Paper	85–99	0–15	0	
Softwood stems	45–50	25–35	25–35	

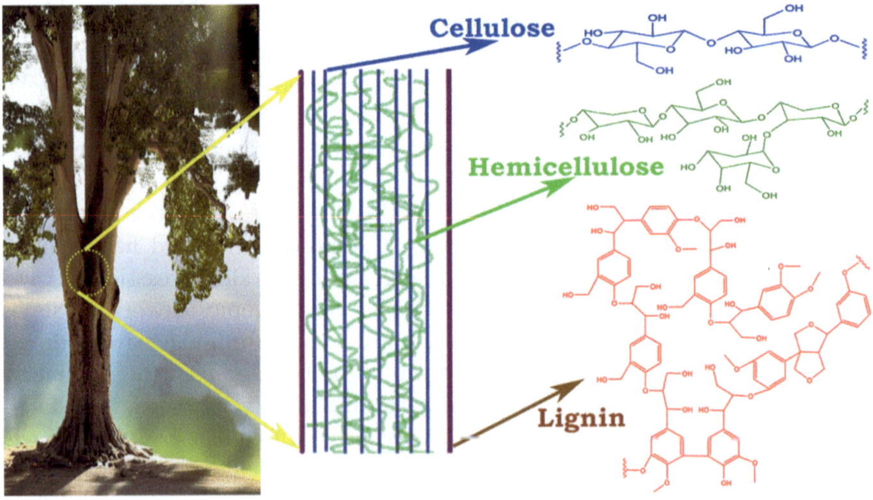

Fig. 2 Illustrates the plant cell wall together with the molecular structures of cellulose, lignin, and hemicellulose [70]

Depending on the source material, LCB is broadly categorized into several types, including dedicated energy crops (e.g., perennial grasses), Crop waste (such as bagasse, corn stover, and cereal straws), the biodegradable portion of municipal solid waste, along with residues from forestry operations (including hardwood, softwood, prunings, and bark residues) [69].

The bioconversion process of LCB comprises several key stages: biomass pretreatment and fractionation, generation of effective cellulolytic and hemicellulolytic enzymes, enzymatic hydrolysis of the pretreated material, fermentation of the obtained hydrolysate using suitable microorganisms, and, ultimately, downstream processing.

In this context, a TEA is presented for three major types of bio-based products: biofuels, bioethanol, and biochemicals. Each is accompanied by a brief overview of its production pathway and the key challenges associated with it.

3.2 *Biofuels*

Biofuels are renewable energy sources obtained from organic matter such as crops, agricultural residues, forestry by-products, and different forms of biomass. They serve as a sustainable substitute for fossil fuels, as their production can utilize diverse feedstocks. This adaptability reduces dependence on petroleum and aids in lowering GHG emissions.

The generation of biofuels from LCB requires the decomposition of its intricate structure into simpler sugars, which are then fermented into fuel products. However, the sustainability and feasibility of using LCB for biofuel production remain subjects of ongoing research and debate.

3.3 *Bioethanol*

Bioethanol is considered one of the most promising alternative fuels, because it can substitute gasoline and decrease dependence on conventional fossil fuels. It is typically mixed with gasoline in different proportions, commonly between 5% (E5) and 20% (E20). India successfully reached its E10 (10% ethanol blend) target in 2022 and plans to gradually implement E20 blending between 2023 and 2025.

In 2021, global ethanol production reached around 27 billion gallons, with the United States and Brazil accounting for the majority, producing approximately 55% and 27%, respectively. The global bioethanol market is currently valued at between USD80 and USD90 billion and is projected to grow at a compound annual growth rate (CAGR) of 6.6% through 2028.

Compared to gasoline, bioethanol offers several benefits, including a higher octane rating; greater heat of vaporization; wider flammability range; and reduced emissions of carbon monoxide (CO), nitrogen oxides (NO_x), and unburned hydrocarbons. The

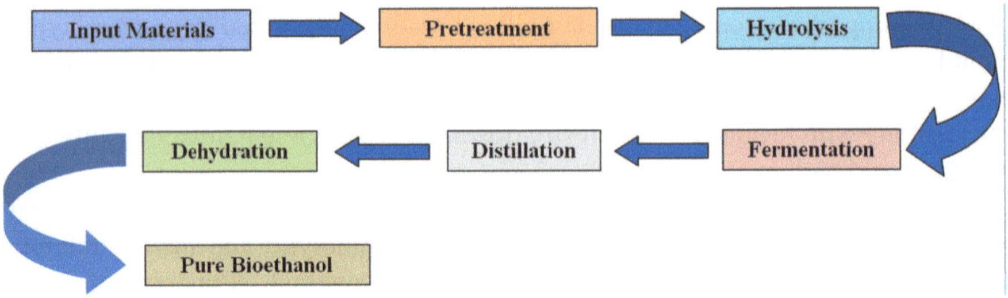

Fig. 3 Illustration of the phases in bioethanol fuel production

process of producing bioethanol from LCB generally consists of four primary stages: pretreatment, hydrolysis, fermentation, and distillation, as shown in Fig. 3.

After the fermentation process, ethanol is extracted from water using distillation. To achieve higher purity, a dehydration process is employed, resulting in ethanol with a concentration of approximately 99.5% by volume.

A summary of research conducted by various scholars on the production of bioethanol from LCB, particularly from the perspective of chemical and biochemical engineering (CBE), is presented in Table 3. These studies primarily explore the economic viability of producing bioethanol from LCB, focusing on the use of diverse biomass raw materials, various pretreatment techniques, and a range of solvents.

Bioethanol obtained from LCB is considered a highly promising renewable fuel. According to referenced studies, its Minimum Selling Price (MSP) is projected to range from USD 0.5 to USD 1.8 per liter, making it largely competitive with the prevailing market rate of approximately USD 1.0 per liter [71]. However, several analyses indicate that bioethanol is still not consistently cost-competitive with petroleum-derived ethanol. One of the main challenges to large-scale commercialization lies in the complex and expensive pretreatment processes, which often require the use of enzymes, acids, alkalis, or other chemicals [72, 73].

3.4 Operational and Seasonal Impacts on Biorefinery Economics

If the raw material is tied to seasonal agricultural harvests, BRF operations would be restricted to the harvesting period, typically around 4000–6450 h per year in the case of sugarcane mills. Alternatively, maintaining year-round operations (approximately 8000 h annually) would require storing the raw materials. However, prolonged storage may result in microbial degradation of the biomass, potentially causing significant and uncontrollable material losses. It may be concluded that increasing the annual operational duration improves the economic feasibility of a BRF scenario.

Based on the studies reviewed, 1G bio-based production methods are currently more economically favorable in most cases [86]. However, the economic viability

Table 3 An overview of the methods employed and the main outcomes reported in TEA studies on bioethanol production from LCB [74]

S. No.	Raw materials	Methodology	Detecting	Indicator of techno-economic performance	References
1	LCB	A comparative analysis was conducted between on-site enzyme setups and off-site enzyme production from an external facility	The use of ionic liquids in pretreatment was evaluated with a focus on cost reduction, lower loading, and improved recycling. Reducing ionic liquid use proved more effective than recycling, cutting capital costs, energy use, and working capital. Moreover, marketing lignin as a coproduct can help reduce the MSP of ethanol and enhance its economic feasibility. Nonetheless, handling ionic liquid waste and associated expenses continues to pose a major challenge	The MSP is USD 1/gal, based on lignin at USD 2/kg and ionic liquid at USD 2.50/kg	[43]
2	Spruce	A comparative analysis was conducted between organosolv and dilute acid pretreatment methods	The organosolv process uses more energy than dilute acid but yields a higher ethanol concentration (11.1% vs. 5.4%). While dilute acid has lower current costs, organosolv could reduce ethanol MSP by up to 42.8% with further optimization	The ethanol MSP is estimated at USD 0.72/kg, with annualized costs of USD 243.17 M for organosolv and USD 163.91 M for dilute acid pretreatment	[75]

(continued)

Table 3 (continued)

S. No.	Raw materials	Methodology	Detecting	Indicator of techno-economic performance	References
3	Spruce	An economic evaluation was performed through simulation and analysis (utilizing organosolv pretreatment) of six scenarios, focusing on the recovery of solvents and the integration of processes in bioethanol purification	The most cost-effective bioethanol purification uses distillation, neutralization, and evaporation, yielding the lowest MSP. Heating utilities dominated costs (~60%), this was followed by raw materials (approximately 19%), whereas process integration demonstrated significant potential for cost reduction	Total annual cost: USD 240.2 M; MSP: USD 1.27/ kg ethanol	[76]
4	Dry spruce chips	A comparison, conducted between two integrated enzyme fermentation setups (A and B) in terms of MSP and GHG emissions	Configuration A incorporated additional nutrients, whereas Configuration B utilized molasses. The integrated process reduced GHG emissions but requires more data for a robust comparison with off-site production	MSP ranges from €0.568 to €0.622 per liter	[77]

(continued)

Table 3 (continued)

S. No.	Raw materials	Methodology	Detecting	Indicator of techno-economic performance	References
5	Sugarcane bagasse, coffee cut stems, empty fruit bunches, and rice husks	A comparative study of four LCB raw materials for bioethanol production in Colombia, incorporating cost assessments both with and without cogeneration systems	Empty fruit bunches yielded the highest ethanol output (313.83 L/ton) at the lowest production cost (USD 0.45/L with cogeneration and USD 0.58/L without). This advantage was attributed to their lower raw material costs, higher holocellulose content, and reduced energy requirements, positioning them as a promising feedstock in Colombia. For all raw materials, production expenses were primarily influenced by raw material and utility costs	Production cost = USD 0.45/L	[78]
6	Corn stover	The methodology involved evaluating and comparing three pretreatment methods: dilute acid, liquid hot water, and ammonia fiber explosion	The MSP of ethanol amounted to set by optimizing ethanol concentration and costs; liquid hot water pretreatment yielded the lowest MSP, cutting costs by 30% versus dilute acid, with ethanol concentration being a key factor affecting MSP competitiveness	MSP = USD 1.78/L	[79]

(continued)

Table 3 (continued)

S. No.	Raw materials	Methodology	Detecting	Indicator of techno-economic performance	References
7	Corn stover	Two models were formulated: (i) the Bioethanol Plant Gate Price Assessment Model (BPAM) and (ii) the Feedstock Cost Estimation Model (FCEM)	A study in China showed bioethanol's MSP is not competitive with gasoline; key price factors include cellulose conversion, sugar yield, raw material costs, and CAPEX, prompting recommendations for tax breaks and subsidies to aid commercialization	MSP = USD 4.68-0.05/gal	[80]
8	Weeping love grass	A comparative evaluation was carried out between Separate Hydrolysis and Fermentation (SHF) and Simultaneous Saccharification and Fermentation (SSF) reactor configurations	Acid Mine Drainage (AMD) pretreatment was compared to sulfuric acid; only SHF with sulfuric acid showed potential profitability, but had a long 80.7-year payback due to low ethanol yield, indicating the need for better enzymes. AMD integration with SHF improved both economic and environmental impact	The payback period (PBP) is 80.7 years, with a CAPEX of USD 281 million and an OPEX of USD 45.82 million per year	[81]

(continued)

Table 3 (continued)

S. No.	Raw materials	Methodology	Detecting	Indicator of techno-economic performance	References
9	Spruce, salix, and corn stover	The methodology involved evaluating and comparing three LCBs based on bioethanol production costs and energy requirements	A process using SO_2-catalyzed pretreatment and SSF was evaluated, revealing CAPEX and OPEX, especially enzyme costs, as key drivers of ethanol costs. Using the pentose fraction and higher solids content improved economics, with process design impacting costs more than raw material choice	The production cost was calculated at SEK 4.15/L for ethanol from spruce and SEK 4.25/L for ethanol from corn stover	[82]
10	Red oak	The approach compared the efficiency of three solvents, namely, tetrahydrofuran (THF), acetone, and 1,4-dioxane, in extracting biomass sugars for bioethanol production	The THF-based BRF had the lowest ethanol MSP, highly influenced by the solvent-to-biomass ratio; reducing this ratio and boosting sugar yield can lower MSP further. Life cycle assessment results indicated a 25% reduction in GHG emissions for the THF process and a 45% reduction for the acetone process	MSP was USD 2.98/gal, with a CAPEX of USD 386.6 million and an OPEX of USD 103.3 million	[83]

(continued)

Table 3 (continued)

S. No.	Raw materials	Methodology	Detecting	Indicator of techno-economic performance	References
11	Sugarcane residue	The approach included performing an economic assessment of a BRF that generates electricity while producing bioethanol as a coproduct	Ethanol production employed both vacuum and conventional distillation methods, whereas electricity was produced using either high-pressure steam systems or a biomass-integrated gasification combined cycle (BIGCC). Vacuum distillation combined with BIGCC achieved the highest energy efficiency (32.91%), and lignin-based power was more cost-effective than electricity alone	IRR = 29.42%	[84]
12	Wet hardwood chips	The approach compared enzymatic hydrolysis coupled with fermentation and gasification combined with fermentation, evaluating them in terms of ethanol yield, power generation, and overall economic feasibility	Enzymatic hydrolysis combined with fermentation yielded ethanol at a lower MSP compared to gasification, with raw material expenses being the primary cost contributor (35.5% vs. 67%). However, even with reduced costs, its market price is still not competitive with petroleum-derived ethanol	MSP was €0.80/L, with a PBP of 5 years, an NPV of €9.4 million, an IRR of 11.0%, and a return on investment of 32.5%	[85]

of 2G BRFs can be improved through several strategies, such as increasing biomass crop yields, replacing diesel with biomethane, and introducing energy cane as a feedstock [32, 87]. Additionally, reducing enzyme costs and improving the efficiency of pretreatment processes can further enhance the competitiveness of 2G technologies. In cases involving low energy requirements, integrated 1G and 2G BRFs may offer a cost advantage over standalone 1G facilities due to economies of scale [88].

Bastos et al. examined how the production scale, ranging between 200 and 800 kilotonnes per year (kt/year), impacts the economic viability of manufacturing bio-based monoethylene glycol (bio-MEG) from sugarcane via bioethanol and bioethylene routes [89]. They developed simplified models to estimate CAPEX and MSP based on production capacity. Their findings showed that increasing production capacity led to a nearly linear rise in CAPEX from approximately 150 million USD at 200 kt/year to 480 million USD at 800 kt/year. In contrast, the MSP dropped from about 540 USD per tonne to 420 USD per tonne as capacity increased.

Although economies of scale favor larger BRFs, many facilities are constrained by the limited availability of raw materials, which restricts their production capacity. Sourcing biomass from distant regions can reduce the scale-related cost advantages due to higher logistics and supply chain expenses [30]. Furthermore, market size plays a crucial role in determining feasible production volumes, producing beyond market demand may drive down product prices and compromise economic sustainability [90]. Therefore, determining the optimal capacity of a BRF requires careful consideration of raw materials availability, logistics, and market demand.

Economic factors like labor costs, inflation, capital expenses, government bond rates, corporate taxes, carbon pricing, land value, and electricity costs vary widely between countries, making TEA results highly dependent on location [51]. Emphasized the influence of geographic location on economic outcomes by analyzing the feasibility of producing high-density liquid fuels in 12 different countries, including the Netherlands, China, Australia, Spain, India, Turkey, the United Arab Emirates, the United States, Brazil, Kenya, Mexico, and Tunisia. While the study specifically focused on third-generation (3G) feedstocks such as microalgae, it effectively demonstrates that the location of a biorefinery (BRF) can significantly affect the economic viability of its products [51].

The lower investment costs associated with the Portuguese BRF were primarily attributed to economies of scale resulting from higher raw materials availability. To isolate the impact of scale, a comparison was made between the net present values (NPVs) of the Portuguese and Chilean BRFs at a similar processing capacity (0–100 kilotons per year). Even at comparable scales, the Chilean BRF consistently showed higher NPVs than its Portuguese counterpart. The results indicate that the economic feasibility of a BRF is highly dependent on its geographical location, meaning that performing TEA for an identical facility in different countries can lead to differing economic outcomes.

3.5 Future Work

Looking ahead, achieving targeted cost reductions and reducing uncertainty in critical areas of the modeled processes will require addressing several key challenges and areas for further development, as outlined below. Many of these issues mirror those identified in NREL's 2013 biological design case and remain highly relevant today:

- Enhancing knowledge of advanced bioreactor designs: Both fuel production pathways evaluated, namely, the acids and BDO (2,3-butanediol) routes, may demand innovative bioreactor configurations and novel operating strategies. For example, the acids pathway could benefit from protractive fermentation systems, while the BDO pathway may require microaerophilic fermentation, particularly if the host organism cannot be engineered to eliminate its dependence on oxygen for redox balancing. Current designs and cost estimates for such advanced systems are less developed than those for conventional anaerobic or aerobic reactors previously analyzed by NREL, highlighting the need for further design optimization and validation.
- Improving sugar utilization and microbial metabolic efficiency: Additional work is needed in strain development for the *Zymomonas mobilis* BDO-producing strain, focusing on enhanced uptake of pentose sugars (xylose and arabinose), reduction of unwanted by-products, and elimination of oxygen requirements, potentially through alternative redox balancing mechanisms such as hydrogen or succinic acid coproduction. Likewise, *Clostridium* strains used for acid production may benefit from improved sugar conversion efficiencies and, more critically, from demonstrating economically viable fermentation processes that minimize pH control costs. A better understanding of optimal sugar concentrations and the impact of inhibitors present in hydrolysates is also needed for both pathways.

3.6 Enhance Catalyst Efficiency for Intermediate Conversion

The current process models incorporate multiple catalytic steps aimed at upgrading both fuel and coproduct intermediates. While research in this area is actively progressing, most catalytic upgrading processes remain in the early stages of development, particularly concerning reactor operating conditions, space velocities, yields, selectivity, and catalyst costs, areas largely supported by the ChemCatBio Consortium. As more data become available on these performance metrics and as optimization opportunities are identified, the models will be updated accordingly, with improvements potentially influencing future technology benchmarks. Additionally, the integration of these biochemically derived intermediates into existing refinery infrastructure has not been thoroughly explored. This presents a potential opportunity for cost savings, such as centralizing the final hydrotreating of fuel precursors within a refinery setting, rather than performing it on a smaller scale at distributed facilities.

4 Conclusion

This chapter highlights the techno-economic potential of advanced biorefinery (BRF) systems for producing renewable hydrocarbon fuels and high-value bioproducts from lignocellulosic biomass (LCB), with a specific focus on bioethanol and lignin-derived coproducts. The analysis integrates process innovations such as alkaline extraction, mechanical refining, and continuous enzymatic hydrolysis, aiming to maximize fermentable sugar recovery while minimizing losses. Additionally, counter-current alkaline extraction of DMR black liquor and catalytic upgrading of lignin to bio-based adipic acid demonstrate the capacity to enhance both product diversity and greenhouse gas (GHG) reduction potential.

Most biological and catalytic conversion steps evaluated already meet, or are on track to meet, 2030 technical targets, supported by ongoing research, federal funding, and private investment in crop genetics, enzyme efficiency, and microbial performance. However, lignin-to-coproduct pathways remain at an earlier development stage, requiring advancements in deconstruction, bioconversion yields, and integration with existing refinery infrastructure.

Techno-economic assessments (TEAs) underscore the importance of feedstock selection, process integration, scale optimization, and location-specific economic factors in determining commercial viability. While fuel yields and costs may be less favorable than cellulosic ethanol alone, the BRF model offers strategic advantages by producing infrastructure-compatible drop-in fuels alongside high-value chemicals. Moreover, potential enhancements such as waste gas utilization and alternative carbohydrate/lignin conversion routes could further improve system efficiency, profitability, and sustainability.

Ultimately, achieving widespread deployment will depend on continued innovation, process intensification, and supportive policy frameworks, ensuring these integrated pathways can deliver competitive, low-carbon fuels and bioproducts by 2030.

References

1. Xu J, Li M, Ni T (2015) Feedstock for bioethanol production from a technological paradigm perspective. BioResources 10(3):6285–6304. https://doi.org/10.15376/BIORES.10.3.XU
2. Stoeglehner G, Narodoslawsky M (2009) How sustainable are biofuels? answers and further questions arising from an ecological footprint perspective. Biores Technol 100(16):3825–3830. https://doi.org/10.1016/J.BIORTECH.2009.01.059
3. FAO, OECD (2015) OECD/food and agriculture organization of the United Nations (2015), OECD-FAO agricultural outlook. OECD Publishing, Paris
4. Annual ethanol production | renewable fuels association (nd)
5. Erdei B, Hancz D, Galbe M, Zacchi G (2013) SSF of steam-pretreated wheat straw with the addition of saccharified or fermented wheat meal in integrated bioethanol production. Biotechnol Biofuels 6(1):1–14. https://doi.org/10.1186/1754-6834-6-169/TABLES/4

6. Westman JO, Wang R, Novy V, Franzén CJ (2017) Sustaining fermentation in high-gravity ethanol production by feeding yeast to a temperature-profiled multifeed simultaneous saccharification and co-fermentation of wheat straw. Biotechnol Biofuels 10(1):1–16. https://doi.org/10.1186/S13068-017-0893-Y/FIGURES/8
7. Uppugundla N, Da Costa L, Sousa SPS, Chundawat XY, Simmons B, Singh S, Gao X et al (2014) A comparative study of ethanol production using dilute acid, ionic liquid and AFEX^{TM} pretreated corn stover. Biotechnol Biofuels 7(1):1–14. https://doi.org/10.1186/1754-6834-7-72/FIGURES/5
8. Ezeji TC, Qureshi N, Blaschek HP (2004) Acetone Butanol Ethanol (ABE) production from concentrated substrate: reduction in substrate inhibition by fed-batch technique and product inhibition by gas stripping. Appl Microbiol Biotechnol 63(6):653–658. https://doi.org/10.1007/S00253-003-1400-X
9. Huber GW, Chheda JN, Barrett CJ, Dumesic JA (2005) Production of liquid alkanes by aqueous-phase processing of biomass-derived carbohydrates. Science 308(5727):1446–1450. https://doi.org/10.1126/SCIENCE.1111166
10. Kamm B, Gruber PR, Kamm M (2008) Biorefineries-industrial processes and products: status quo and future directions. Biorefineries Ind Process Prod Status Quo Futur Dir 1–2:1–959. https://doi.org/10.1002/9783527619849
11. Islas J, Manzini F, Masera O, Vargas V (2018) Solid biomass to heat and power. In: The role of bioenergy in the emerging bioeconomy: resources, technologies, sustainability and policy, pp 145–177. https://doi.org/10.1016/B978-0-12-813056-8.00004-2
12. Kumar M, Rathour R, Gupta J, Pandey A, Gnansounou E, Thakur IS (2020) Bacterial production of fatty acid and biodiesel: opportunity and challenges. In: Refining biomass residues for sustainable energy and bioproducts: technology, advances, life cycle assessment, and economics, pp 21–49. https://doi.org/10.1016/B978-0-12-818996-2.00002-8
13. Chew KW, Yap JY, Show PL, Suan NH, Juan JC, Ling TC, Lee DJ, Chang JS (2017) Microalgae biorefinery: high value products perspectives. Biores Technol 229:53–62. https://doi.org/10.1016/J.BIORTECH.2017.01.006
14. Yamakawa CK, Qin F, Mussatto SI (2018) Advances and opportunities in biomass conversion technologies and biorefineries for the development of a bio-based economy. Biomass Bioenerg 119:54–60. https://doi.org/10.1016/J.BIOMBIOE.2018.09.007
15. Chandra R, Iqbal HMN, Vishal G, Lee HS, Nagra S (2019) Algal biorefinery: a sustainable approach to valorize algal-based biomass towards multiple product recovery. Biores Technol 278:346–359. https://doi.org/10.1016/J.BIORTECH.2019.01.104
16. Karagoun (2018) (PDF) Circular bioeconomy: do we really need another concept?
17. Ubando AT, Felix CB, Chen WH (2020) Biorefineries in circular bioeconomy: a comprehensive review. Bioresour Technol 299:122585. https://doi.org/10.1016/J.BIORTECH.2019.122585
18. Publications Office of the European Union (2012) Innovating for sustainable growth: a bioeconomy for Europe. https://doi.org/10.2777/6462
19. Carus M, Dammer L (2018) The circular bioeconomy—concepts, opportunities, and limitations. Ind Biotechnol 14(2):83–91. https://doi.org/10.1089/IND.2018.29121.MCA/ASSET/IND.2018.29121.MCA.FP.PNG_V03
20. European Commission (2015) Communication from the commission to the European Parliament, the council, the European economic and social committee and the committee of the regions
21. What is a circular economy? | Ellen MacArthur Foundation (nd)
22. Hadley Kershaw E, Hartley S, McLeod C, Polson P (2021) The sustainable path to a circular bioeconomy. Trends Biotechnol 39(6):542–545. https://doi.org/10.1016/J.TIBTECH.2020.10.015
23. Ashokkumar V, Venkatkarthick R, Jayashree S, Chuetor S, Dharmaraj S, Kumar G, Chen WH, Ngamcharussrivichai C (2022) Recent advances in lignocellulosic biomass for biofuels and value-added bioproducts—a critical review. Biores Technol 344(Pt B):126195. https://doi.org/10.1016/J.BIORTECH.2021.126195

24. Padella M, O'Connell A, Prussi M (2019) What is still limiting the deployment of cellulosic ethanol? Analysis of the current status of the sector. Appl Sci 9(21):4523. https://doi.org/10.3390/APP9214523
25. Davidson MG, Elgie S, Parsons S, Young TJ (2021) Production of HMF, FDCA and their derived products: a review of life cycle assessment (LCA) and techno-economic analysis (TEA) studies. Green Chem 23(9):3154–3171. https://doi.org/10.1039/D1GC00721A
26. Koutinas AA, Vlysidis A, Pleissner D, Kopsahelis N, Garcia IL, Kookos IK, Papanikolaou S, Kwan TH, Lin CSK (2014) Valorization of industrial waste and by-product streams via fermentation for the production of chemicals and biopolymers. Chem Soc Rev 43(8):2587–2627. https://doi.org/10.1039/C3CS60293A
27. Kamalesh R, Shaji A, Saravanan A, Vickram AS, Yaashikaa PR (2024) Advances in engineered microbes for sustainable biofuel production: current research and future outlook on lignocellulose utilization. Ind Crop Prod 222:119988. https://doi.org/10.1016/J.INDCROP.2024.119988
28. Okolie JA, Mukherjee A, Nanda S, Dalai AK, Kozinski JA (2021) Next-generation biofuels and platform biochemicals from lignocellulosic biomass. Int J Energy Res 45(10):14145–14169. https://doi.org/10.1002/ER.6697
29. Kumar D, Brown T, Bhatia SK, Kumar V (2022) Techno-economic analysis for evaluating biorefinery strategies. In: Biomass utilization: conversion strategies, pp 229–245. https://doi.org/10.1007/978-3-031-05835-6_9
30. Clauser NM, Felissia FE, Area MC, Vallejos ME (2021) A framework for the design and analysis of integrated multi-product biorefineries from agricultural and forestry wastes. Renew Sustain Energy Rev 139:110687. https://doi.org/10.1016/J.RSER.2020.110687
31. Chaturvedi T, Torres AI, Stephanopoulos G, Thomsen MH, Schmidt JE (2020) Developing process designs for biorefineries—definitions, categories, and unit operations. Energies 13(6):1493. https://doi.org/10.3390/EN13061493
32. Louw J, Dogbe ES, Yang B, Görgens JF (2023) Prioritisation of biomass-derived products for biorefineries based on economic feasibility: a review on the comparability of techno-economic assessment results. Renew Sustain Energy Rev 188:113840. https://doi.org/10.1016/J.RSER.2023.113840
33. Pandey A, Höfer R, Taherzadeh M, Nampoothiri KM, Larroche C (2015) Industrial biorefineries and white biotechnology. In: Industrial biorefineries and white biotechnology, pp 1–710. https://doi.org/10.1016/c2013-0-19082-4
34. Gunukula S, Klein SJW, Pendse HP, DeSisto WJ, Wheeler MC (2018) Techno-economic analysis of thermal deoxygenation based biorefineries for the coproduction of fuels and chemicals. Appl Energy 214:16–23. https://doi.org/10.1016/J.APENERGY.2018.01.065
35. Petersen AM, Van der Westhuizen WA, Mandegari MA, Görgens JF (2018) Economic analysis of bioethanol and electricity production from sugarcane in South Africa. Biofuels Bioprod Biorefin 12(2):224–238. https://doi.org/10.1002/BBB.1833
36. Farzad S, Mandegari MA, Görgens JF (2017) Integrated techno-economic and environmental analysis of butadiene production from biomass. Biores Technol 239:37–48. https://doi.org/10.1016/J.BIORTECH.2017.04.130
37. Aristizábal-Marulanda V, Cardona-Alzate CA (2019) Methods for designing and assessing biorefineries: review. Biofuels Bioprod Biorefining 13(3):789–808. https://doi.org/10.1002/BBB.1961
38. Crippa M, De Wilde B, Koopmans R, Leyssens J, Muncke J, Ritschkoff A-C, Van Doorsselaer K (2019) A circular economy for plastics | enhanced reader
39. Jong ED, Stichnothe H, Bell G, Jorgensen H (2020) Bio-based chemicals: A 2020 update. In: IEA bioenergy task 42 biorefinery, pp 1–79
40. Velvizhi G, Balakumar K, Shetti NP, Ahmad E, Pant KK, Aminabhavi TM (2022) Integrated biorefinery processes for conversion of lignocellulosic biomass to value added materials: paving a path towards circular economy. Biores Technol 343:126151. https://doi.org/10.1016/J.BIORTECH.2021.126151

41. Biddy MJ, Scarlata C, Kinchin C (2016) Chemicals from biomass: a market assessment of bioproducts with near-term potential
42. López-Molina A, Sengupta D, Shi C, Aldamigh E, Alandejani M, El-Halwagi MM (2020) An integrated approach to the design of centralized and decentralized biorefineries with environmental, safety, and economic objectives. Processes 8(12):1682. https://doi.org/10.3390/PR8121682
43. Klein-Marcuschamer D, Simmons BA, Blanch HW (2011) Techno-economic analysis of a lignocellulosic ethanol biorefinery with ionic liquid pre-treatment. Biofuels Bioprod Biorefin 5(5):562–569. https://doi.org/10.1002/BBB.303
44. Cambero C, Sowlati T (2014) Assessment and optimization of forest biomass supply chains from economic, social and environmental perspectives—a review of literature. Renew Sustain Energy Rev 36:62–73. https://doi.org/10.1016/J.RSER.2014.04.041
45. Cherubini F, Strømman A (2011) Life cycle assessment of bioenergy systems: state of the art and future challenges. Biores Technol 102(2):437–451. https://doi.org/10.1016/J.BIORTECH.2010.08.010
46. Ferdous J, Bensebaa F, Pelletier N (2023) Integration of LCA, TEA, process simulation and optimization: a systematic review of current practices and scope to propose a framework for pulse processing pathways. J Clean Prod 402:136804. https://doi.org/10.1016/J.JCLEPRO.2023.136804
47. Gnansounou E, Dauriat A (2010) Techno-economic analysis of lignocellulosic ethanol: a review. Biores Technol 101(13):4980–4991. https://doi.org/10.1016/J.BIORTECH.2010.02.009
48. Scown CD, Baral NR, Yang M, Vora N, Huntington T (2021) Technoeconomic analysis for biofuels and bioproducts. Curr Opin Biotechnol 67:58–64. https://doi.org/10.1016/J.COPBIO.2021.01.002
49. Humbird D, Davis R, Tao L, Kinchin C, Hsu D, Aden A, Schoen P et al (2002) Process design and economics for biochemical conversion of lignocellulosic biomass to ethanol: dilute-acid pretreatment and enzymatic hydrolysis of corn stover
50. Makepa DC, Chihobo CH, Ruziwa WR, Musademba D (2023) A systematic review of the techno-economic assessment and biomass supply chain uncertainties of biofuels production from fast pyrolysis of lignocellulosic biomass. Fuel Communications 14:100086. https://doi.org/10.1016/J.JFUECO.2023.100086
51. Roles J, Yarnold J, Hussey K, Hankamer B (2021) Techno-economic evaluation of microalgae high-density liquid fuel production at 12 international locations. Biotechnol Biofuels 14(1):1–19. https://doi.org/10.1186/S13068-021-01972-4/FIGURES/4
52. Ruegg RT, Marshall HE (1990) Internal rate-of-return (IRR). In: Building economics: theory and practice, pp 67–78. https://doi.org/10.1007/978-1-4757-4688-4_5
53. Wood C, Rosentrater KA, Muthukumarappan K (2014) Techno-economic modeling of a corn based ethanol plant in 2011/2012. Ind Crops Prod 56:145–155. https://doi.org/10.1016/J.INDCROP.2014.03.001
54. Cortes-Peña Y, Kumar D, Singh V, Guest JS (2020) BioSTEAM: a fast and flexible platform for the design, simulation, and techno-economic analysis of biorefineries under uncertainty. ACS Sustain Chem Eng 8(8):3302–3310. https://doi.org/10.1021/ACSSUSCHEMENG.9B07040/ASSET/IMAGES/LARGE/SC9B07040_0007.JPEG
55. Li Y, Bhagwat SS, Cortés-Peña YR, Ki D, Rao CV, Jin YS, Guest JS (2021) Sustainable lactic acid production from lignocellulosic biomass. ACS Sustain Chem Eng 9(3):1341–1351. https://doi.org/10.1021/ACSSUSCHEMENG.0C08055/ASSET/IMAGES/LARGE/SC0C08055_0006.JPEG
56. Sanchis-Sebastiá M, Gomis-Fons J, Galbe M, Wallberg O (2020) Techno-economic evaluation of biorefineries based on low-value feedstocks using the BioSTEAM software: a case study for animal bedding. Processes 8(8):904. https://doi.org/10.3390/PR8080904
57. Satam CC, Realff MJ (2020) Comparison of two routes for the bio-based production of economically important C4 streams. J Adv Manuf Process 2(3):e10054. https://doi.org/10.1002/AMP2.10054

58. Viswanathan MB, Raj Raman D, Rosentrater KA, Shanks BH (2020) A technoeconomic platform for early-stage process design and cost estimation of joint fermentative–catalytic bioprocessing. Processes 8(2):229. https://doi.org/10.3390/PR8020229
59. Koutinas AA, Yepez B, Kopsahelis N, Freire DMG, de Castro AM, Papanikolaou S, Kookos IK (2016) Techno-economic evaluation of a complete bioprocess for 2,3-butanediol production from renewable resources. Bioresour Technol 204:55–64. https://doi.org/10.1016/J.BIORTECH.2015.12.005
60. Wang F, Ouyang D, Zhou Z, Page SJ, Liu D, Zhao X (2020) Lignocellulosic biomass as sustainable feedstock and materials for power generation and energy storage. J Energy Chem 57:247–280. https://doi.org/10.1016/J.JECHEM.2020.08.060
61. Prasad S, Singh A, Joshi HC (2007) Ethanol as an alternative fuel from agricultural, industrial and urban residues. Resour Conserv Recycl 50(1):1–39. https://doi.org/10.1016/J.RESCONREC.2006.05.007
62. Kim M, Day DF (2011) Composition of sugar cane, energy cane, and sweet sorghum suitable for ethanol production at Louisiana sugar mills. J Ind Microbiol Biotechnol 38(7):803–807. https://doi.org/10.1007/S10295-010-0812-8
63. Saini JK, Saini R, Tewari L (2015) Lignocellulosic agriculture wastes as biomass feedstocks for second-generation bioethanol production: concepts and recent developments. 3 Biotech 5(4):337–353. https://doi.org/10.1007/S13205-014-0246-5/TABLES/8
64. Menon V, Rao M (2012) Trends in bioconversion of lignocellulose: biofuels, platform chemicals & biorefinery concept. Prog Energy Combust Sci 38(4):522–550. https://doi.org/10.1016/j.pecs.2012.02.002
65. Chandra R, Takeuchi H, Hasegawa T (2012) Methane production from lignocellulosic agricultural crop wastes: a review in context to second generation of biofuel production. Renew Sustain Energy Rev 16(3):1462–1476. https://doi.org/10.1016/J.RSER.2011.11.035
66. Motta IL, Miranda NT, Filho RM, Maciel MRW (2018) Biomass gasification in fluidized beds: a review of biomass moisture content and operating pressure effects. Renew Sustain Energy Rev 94:998–1023. https://doi.org/10.1016/J.RSER.2018.06.042
67. Yousuf A, Pirozzi D, Sannino F (2019) Fundamentals of lignocellulosic biomass. In: Lignocellulosic biomass to liquid biofuels. https://doi.org/10.1016/B978-0-12-815936-1.00001-0
68. Brandt A, Gräsvik J, Hallett JP, Welton T (2013) Deconstruction of lignocellulosic biomass with ionic liquids. Green Chem 15(3):550–583. https://doi.org/10.1039/C2GC36364J
69. Patel A, Shah AR (2021) Integrated lignocellulosic biorefinery: gateway for production of second generation ethanol and value added products. J Bioresour Bioprod 6(2):108–128. https://doi.org/10.1016/J.JOBAB.2021.02.001
70. Seah CC, Tan CH, Arifin NA, Hafriz RSRM, Salmiaton A, Nomanbhay S, Shamsuddin AH (2023) Co-pyrolysis of biomass and plastic: circularity of wastes and comprehensive review of synergistic mechanism. Results Eng 17:100989. https://doi.org/10.1016/J.RINENG.2023.100989
71. Ethanol price index—businessanalytiq (nd)
72. Nanda S, Mohammad J, Reddy SN, Kozinski JA, Dalai AK (2014) Pathways of lignocellulosic biomass conversion to renewable fuels. Biomass Conversion and Biorefinery 4(2):157–191. https://doi.org/10.1007/S13399-013-0097-Z
73. Saadabadi SA, Thattai AT, Fan L, Lindeboom REF, Spanjers H, Aravind PV (2019) Solid oxide fuel cells fuelled with biogas: potential and constraints. Renew Energy 134:194–214. https://doi.org/10.1016/J.RENENE.2018.11.028
74. Patel R, Rajaraman TS, Rana PH, Ambegaonkar NJ, Patel S (2025) A review on techno-economic analysis of lignocellulosic biorefinery producing biofuels and high-value products. Results in Chemistry 13:102052. https://doi.org/10.1016/J.RECHEM.2025.102052
75. Gurgel da Silva R (nd) Techno-economic analysis of organosolv pretreatment process from lignocellulosic biomass. https://doi.org/10.1007/s10098-017-1389-y
76. Silva D, Rodrigues A, Gurgel, Errico M, Rong, Guang B (nd) Aalborg universitet evaluation of organosolv pretreatment for bioethanol production from lignocellulosic biomass solvent recycle and process integration. https://doi.org/10.1007/s13399-017-0292-4

77. Olofsson J, Barta Z, Börjesson P, Wallberg O (2017) Integrating enzyme fermentation in lignocellulosic ethanol production: life-cycle assessment and techno-economic analysis. Biotechnol Biofuels 10(1):1–14. https://doi.org/10.1186/S13068-017-0733-0/TABLES/3
78. Quintero JA, Moncada J, Cardona CA (2013) Techno-economic analysis of bioethanol production from lignocellulosic residues in Colombia: a process simulation approach. Biores Technol 139:300–307. https://doi.org/10.1016/J.BIORTECH.2013.04.048
79. Silva ARGD, Ortega CET, Rong BG (2016) Techno-economic analysis of different pretreatment processes for lignocellulosic-based bioethanol production. Bioresour Technol 218:561–570. https://doi.org/10.1016/J.BIORTECH.2016.07.007
80. Zhao L, Zhang X, Xu J, Ou X, Chang S, Wu M (2015) Techno-economic analysis of bioethanol production from lignocellulosic biomass in China: dilute-acid pretreatment and enzymatic hydrolysis of corn stover. Energies 8(5):4096–4117. https://doi.org/10.3390/EN8054096
81. Burman NW, Sheridan C, Harding KG (nd) Feasibility assessment of the production of bioethanol from lignocellulosic biomass pretreated with acid mine drainage (AMD)
82. Sassner P, Galbe M, Zacchi G (2008) Techno-economic evaluation of bioethanol production from three different lignocellulosic materials. Biomass Bioenerg 32(5):422–430. https://doi.org/10.1016/J.BIOMBIOE.2007.10.014
83. Li W, Ghosh A, Bbosa D, Brown R, Wright MM (2018) Comparative techno-economic, uncertainty and life cycle analysis of lignocellulosic biomass solvent liquefaction and sugar fermentation to ethanol. ACS Sustain Chem Eng 6(12):16515–16524. https://doi.org/10.1021/ACSSUSCHEMENG.8B03622
84. Petersen AM, Aneke MC, Görgens JF (2014) Techno-economic comparison of ethanol and electricity coproduction schemes from sugarcane residues at existing sugar mills in Southern Africa. Biotechnol Biofuels 7(1):1–19. https://doi.org/10.1186/1754-6834-7-105/TABLES/8
85. Piccolo C, Bezzo F (2009) A techno-economic comparison between two technologies for bioethanol production from lignocellulose. Biomass Bioenerg 33(3):478–491. https://doi.org/10.1016/J.BIOMBIOE.2008.08.008
86. Junqueira TL, Chagas MF, Gouveia VLR, Rezende MCAF, Watanabe MDB, Jesus CDF, Cavalett O, Milanez AY, Bonomi A (2017) Techno-economic analysis and climate change impacts of sugarcane biorefineries considering different time horizons. Biotechnol Biofuels 10(1):1–12. https://doi.org/10.1186/S13068-017-0722-3/FIGURES/5
87. Usmani Z, Sharma M, Awasthi AK, Lukk T, Tuohy MG, Gong L, Nguyen-Tri P et al (2021) Lignocellulosic biorefineries: the current state of challenges and strategies for efficient commercialization. Renew Sustain Energy Rev 148:111258. https://doi.org/10.1016/J.RSER.2021.111258
88. Ratshoshi BK, Farzad S, Görgens JF (2021) Techno-economic assessment of polylactic acid and polybutylene succinate production in an integrated sugarcane biorefinery. Biofuels Bioprod Biorefin 15(6):1871–1887. https://doi.org/10.1002/BBB.2287
89. Bastos V, Bruno J, Maia JGSS, Borschiver S, Szklo A, Secchi AR (2022) Addressing scale and seasonality in the design of sugarcane to ethylene glycol biorefineries. J Clean Prod 337:130585. https://doi.org/10.1016/J.JCLEPRO.2022.130585
90. Yang M, Baral NR, Simmons BA, Mortimer JC, Shih PM, Scown CD (2020) Accumulation of high-value bioproducts in planta can improve the economics of advanced biofuels. Proc Natl Acad Sci USA 117(15):8639–8648. https://doi.org/10.1073/PNAS.2000053117/-/DCSUPPLEMENTAL

AI-Driven Simulations to Discover New Routes of Reactor Configurations in Biorefinery Process

Ambikesh Soni, Suresh Sundaramurthy, and Sridharane Ramakrishnan

Abstract Sustainable process engineering paradigm is changing with the development of Artificial Intelligence (AI) combinable with biorefinery reactor simulation. This chapter describes the ongoing convergence of AI technology including machine learning, deep learning, reinforcement learning, and evolutionary algorithms, coupled with conventional process simulation scripts to transform the reactor design in biorefineries systems. Highlighting the shortcomings of the traditional approaches that use empirical models and trial and error to optimize reactor design, the chapter introduces AI-based solutions that can work out complex design spaces and recommend new and novel reactor configurations and combinations that would enhance performance and sustainability characteristics. AI methods can be used both to optimize more traditional types of reactors, such as CSTRs, PFRs, and packed beds, and to help develop new kinds of reactor concept, in which reaction and separation processes are combined. The potential of the AI-based design is observed in case studies featuring various solutions to the improvement of lignocellulosic biomass conversion, algal biorefineries, and waste-to-energy systems. Some of the other implications of AI adoption that are considered in the chapter include the challenge of data quality, model validation, computation needs, and obstacles to adopting in industry. In addition, it envisions the future, like quantum computing, explainable AI, and smart manufacturing to live in the industry 4.0 paradigm.

Keywords Artificial intelligence · Biorefinery reactor design · Process simulation · Machine learning · Sustainable bioeconomy

A. Soni · S. Sundaramurthy (✉)
Department of Chemical Engineering, Maulana Azad National Institute of Technology, Bhopal, Madhya Pradesh, India
e-mail: sureshs@manit.ac.in

S. Ramakrishnan
Department of Chemistry, Perunthalaivar Kamarajar Institute of Engineering and Technology, Karaikal, Puducherry, India

© The Author(s), under exclusive license to Springer Nature Singapore Pte Ltd. 2025
A. K. Chandel and S. Sundaramurthy (eds.), *Biorefineries in Circular Bioeconomy*, Green Energy and Technology, https://doi.org/10.1007/978-981-95-3913-0_14

1 Introduction

The integration of artificial intelligence (AI) with biorefinery process simulation represents a transformative frontier in sustainable bioprocessing [1]. AI-driven approaches are increasingly demonstrating their capacity to navigate complex reaction networks, optimize reactor configurations, and discover entirely novel process routes that human engineers might overlook [2]. This research synthesizes current advancements in AI applications for biorefinery reactor simulation and design, highlighting how machine learning algorithms, digital twins, and multiobjective optimization techniques are revolutionizing bioreactor engineering [3]. Significant progress has been documented in reducing computational complexity, enhancing process intensification, and improving techno-economic outcomes through intelligent simulation approaches [4]. Despite these advances, challenges remain in data scarcity, model validation, and implementation scalability. The research identifies promising directions in explainable AI, quantum computing integration, and real-time biorefinery optimization systems that will further reshape sustainable bioprocess engineering in the coming decade [5].

Biorefineries represent the sustainable counterpart to petroleum refineries, offering systematic approaches to convert biomass into valuable products including fuels, chemicals, materials, and energy [6]. These complex facilities integrate various conversion technologies—biochemical, thermochemical, mechanical, and chemical—to efficiently transform renewable feedstocks into marketable products while minimizing waste generation [7]. The biorefinery concept fundamentally embodies the principles of circular economy, striving to optimize resource utilization and minimize environmental impact through the valorization of biomass components. Modern biorefineries typically incorporate multiple process pathways, each containing various reactor units with distinct configurations and operating conditions designed to maximize conversion efficiency and product selectivity [8]. The complexity of these integrated systems presents significant engineering challenges, particularly in the design, optimization, and scaling of reactor configurations that can efficiently handle the inherent heterogeneity and recalcitrance of biomass feedstocks [9].

Bioreactors, as the core processing units within biorefineries, facilitate the chemical and biological transformations that convert raw biomass into intermediate and final products. Traditional reactor design approaches rely heavily on empirical knowledge, pilot-scale experimentation, and incremental improvements to established configurations [10]. These conventional methodologies, while proven, often result in suboptimal designs that fail to fully exploit the potential of biomass conversion routes, particularly when dealing with novel feedstocks or targeting emerging bio-based products. Furthermore, the sheer complexity of reaction networks in biomass conversion pathways, combined with the heterogeneity of feedstocks, creates a vast design space that cannot be efficiently explored through conventional experimental techniques alone [11].

Conventional reactor design methodologies in biorefinery applications face substantial limitations that impede innovation and optimization [12]. Traditional approaches typically rely on simplified kinetic models, rule-based heuristics, and experimental trial-and-error methods that are resource-intensive, time-consuming, and often fail to identify truly optimal solutions. The multivariable nature of biorefinery processes—involving complex interactions among temperature, pressure, residence time, catalyst properties, and feedstock characteristics—creates a parameter space too vast to explore comprehensively through conventional means [13]. Additionally, traditional design approaches often operate within predefined reactor configuration paradigms, limiting exploration of potentially revolutionary design concepts that might better address the unique challenges of biomass conversion [14].

The experimental determination of optimal reactor parameters traditionally requires extensive laboratory and pilot-scale studies, which demand significant investments in equipment, materials, and human resources [6]. This approach becomes particularly challenging when considering the heterogeneous nature of biomass feedstocks, which can vary seasonally, geographically, and compositionally. The resulting uncertainty in process performance metrics further complicates design decisions and scale-up procedures [15]. Moreover, conventional approaches struggle to adequately account for the dynamic and sometimes unpredictable behavior of biological systems, which are often central to biorefinery operations. These limitations collectively constrain innovation in biorefinery reactor design and hamper the advancement of more efficient, economical, and sustainable bioprocessing technologies [16].

2 The Emergence of AI in Process Engineering

Artificial intelligence has emerged as a powerful tool for addressing complex engineering challenges across numerous industries, with process engineering witnessing particularly transformative applications [8]. The integration of AI techniques—including machine learning, deep learning, reinforcement learning, and evolutionary algorithms—into process engineering workflows has enabled unprecedented capabilities in pattern recognition, process optimization, and predictive modeling [17]. In recent years, these capabilities have begun to revolutionize the approach to biorefinery process design and optimization, offering solutions to many of the limitations inherent in traditional methodologies. AI systems can efficiently navigate vast design spaces, identify nonintuitive relationships between process parameters, and generate novel solutions that might elude human engineers constrained by conventional thinking [18].

The application of AI in chemical and biochemical process engineering has evolved significantly over the past decade, progressing from simple statistical models to sophisticated deep learning architectures capable of capturing complex, nonlinear process dynamics [2]. This evolution has been driven by advances in computational capabilities, algorithm development, and the increasing availability of process data from industrial facilities and research laboratories [9]. AI techniques now offer the capacity to analyze and interpret complex datasets generated from both simulations and experimental campaigns, extracting valuable insights that can guide process development and optimization. The integration of these capabilities with process simulation tools has created powerful new approaches for exploring innovative reactor configurations and process routes in biorefinery applications [12].

2.1 Convergence of AI and Biorefinery Process Simulation

The convergence of artificial intelligence with biorefinery process simulation represents a pivotal development in sustainable bioprocess engineering [19]. This integration creates a powerful framework for addressing the inherent complexities of biomass conversion processes while overcoming many limitations of traditional design approaches [3]. AI-enhanced simulation enables more comprehensive exploration of design spaces, accelerated optimization of process parameters, and the discovery of nonintuitive reactor configurations that might otherwise remain unexplored. Figure 1 highlights the involvement of AI in process Engineering. By combining mechanistic understanding with data-driven approaches, these integrated systems can capture both fundamental phenomena and empirical realities, leading to more robust and practical design solutions [20].

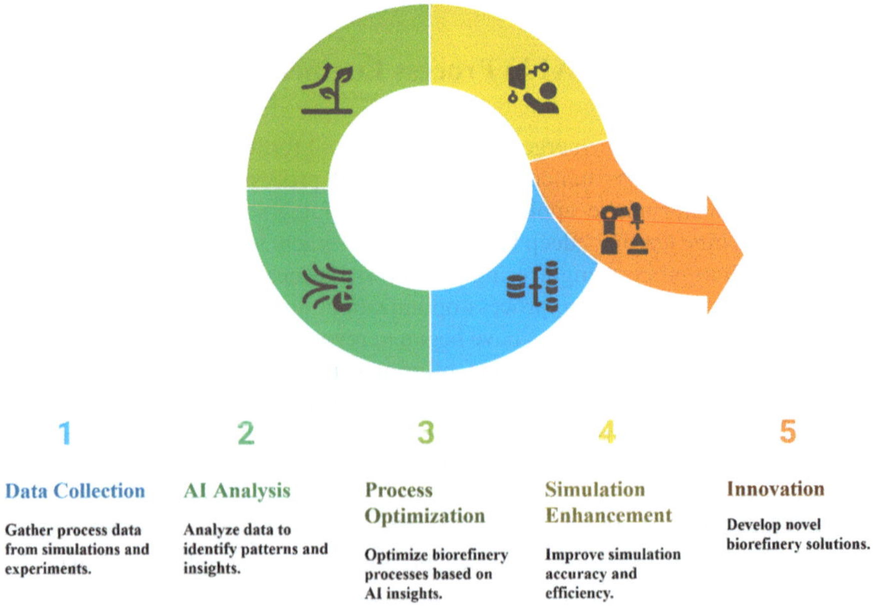

Fig. 1 Involvement of AI in process engineering

Process simulation has long served as a valuable tool in biorefinery development, allowing engineers to evaluate process performance without costly physical experimentation [17]. However, conventional simulation approaches often struggle with the complexity and uncertainty inherent in biorefinery systems [15]. The introduction of AI methodologies addresses these challenges by improving model accuracy, enabling efficient handling of uncertainties, and facilitating the exploration of vast parametric spaces [21]. Furthermore, AI can enhance simulation capabilities through surrogate modeling, reducing computational requirements while maintaining prediction accuracy. The resulting synergy between AI and process simulation creates a powerful platform for innovation in biorefinery reactor design and process integration, potentially accelerating the transition toward more sustainable and economical bioprocessing technologies [11].

3 Theoretical Foundations of AI in Biorefinery Process Simulation

3.1 Fundamental Principles of Biorefinery Processes

Biorefinery processes encompass a diverse array of conversion technologies unified by their fundamental objective of transforming biomass into valuable products through systematic fractionation and conversion of constituent components [15]. These processes can be broadly categorized into biochemical, thermochemical, and hybrid pathways, each with distinct characteristics, advantages, and limitations [19]. Biochemical processes leverage enzymatic and microbial activities to convert biomass components, typically operating under mild conditions but requiring longer residence times. These processes include fermentation, anaerobic digestion, and enzymatic hydrolysis, which selectively target specific biomass fractions [16]. Thermochemical processes, conversely, apply heat, pressure, and sometimes catalysts to rapidly convert biomass into intermediates and products, encompassing technologies such as pyrolysis, gasification, and hydrothermal liquefaction. The fundamental kinetics, transport phenomena, and reaction networks underlying these processes form the basis for reactor design and optimization considerations [21].

The complexity of biorefinery processes stems from several intrinsic characteristics of biomass feedstocks and their conversion pathways [20]. Biomass possesses inherent heterogeneity, recalcitrance, and variability that significantly influence conversion efficiency and product distribution. The intricate structure of lignocellulosic materials, for instance, necessitates pretreatment steps to enhance accessibility to conversion agents, whether enzymes, microorganisms, or catalysts [9]. Following pretreatment, primary conversion processes generate intermediate streams that undergo further transformation through secondary processes to yield final products [5]. This cascade of interconnected transformations creates complex reaction networks with potential for numerous parallel and sequential pathways. The fundamental understanding of these reaction networks, including kinetics, thermodynamics, and mass transfer limitations, provides the scientific foundation upon which reactor designs are developed and optimized [21].

3.2 Reactor Configurations in Biorefinery Applications

Reactor configurations in biorefineries vary widely depending on the specific conversion process, feedstock characteristics, and desired products [16]. Traditional bioreactor designs include continuous stirred tank reactors (CSTRs), plug flow reactors (PFRs), packed bed reactors, fluidized bed reactors, and various hybrid configurations that combine elements of these basic archetypes [17]. Each configuration offers distinct advantages in terms of mixing characteristics, mass and heat transfer capabilities, residence time distribution, and scalability [18]. CSTRs, for example, provide excellent mixing and temperature uniformity but may suffer from product dilution and microbial contamination in biochemical applications. PFRs offer higher conversion efficiencies for many reactions but may encounter challenges with solid handling and pressure drop. Packed bed reactors excel in heterogeneous catalytic processes but face limitations in handling particulate feedstocks that can cause bed clogging and channeling [22].

Beyond these conventional configurations, emerging reactor designs have begun to address specific challenges in biorefinery applications [18]. These include membrane reactors that combine reaction and separation functions, microreactors that offer precise control over reaction conditions and enhanced mass transfer, oscillatory flow reactors that provide efficient mixing with reduced shear stress, and various intensified reactor designs that integrate multiple unit operations [11]. Multiphase reactors, including trickle-bed, bubble column, and slurry reactors, have also found applications in biorefinery processes involving gas–liquid–solid interactions. The selection and optimization of reactor configuration substantially impact process performance metrics including conversion efficiency, selectivity, energy consumption, and economic viability [22]. The vast design space created by these diverse configuration options presents both opportunities and challenges for optimization, making this domain particularly suitable for AI-driven exploration approaches [23].

3.3 AI and Machine Learning Fundamentals

Artificial intelligence encompasses a broad spectrum of computational approaches that enable systems to perform tasks typically requiring human intelligence. Within the AI domain, machine learning represents a subset of techniques that allow algorithms to improve through experience without explicit programming [24].

Machine learning methodologies relevant to biorefinery simulation include supervised learning, which develops predictive models from labelled training data; unsupervised learning, which identifies patterns and structures within unlabeled datasets; and reinforcement learning, which optimizes decision-making processes through interaction with the environment. These approaches underpin the application of AI in biorefinery reactor design and optimization, providing the computational framework for pattern recognition, predictive modeling, and decision support [25].

Supervised learning algorithms, including regression techniques, support vector machines, random forests, and neural networks, enable the development of predictive models that map input parameters to process outcomes [26]. These models can capture complex relationships between reactor design parameters and performance metrics, facilitating rapid evaluation of design alternatives [17]. Unsupervised learning methods, such as clustering and dimensionality reduction techniques, help to identify patterns within process data and reduce the complexity of design spaces. Reinforcement learning approaches, which optimize sequences of decisions to maximize cumulative rewards, offer powerful frameworks for reactor configuration optimization and control strategy development [24]. The integration of these methodologies with domain-specific knowledge creates robust platforms for innovation in biorefinery reactor design [25].

3.4 Simulation Methodologies for Biorefinery Processes

Simulation methodologies for biorefinery processes span multiple scales and fidelity levels, from simplified black-box models to detailed mechanistic representations [26]. At the lowest fidelity, empirical models capture input–output relationships without representing underlying mechanisms, offering computational efficiency at the expense of physical insight and extrapolation capability [22]. Semiempirical models incorporate limited mechanistic understanding within simplified frameworks, providing a balance between computational requirements and predictive power. Figure 2 explains the biomass feedstock steps from optimized biorefinery process. Mechanistic models, based on fundamental principles of conservation and constitutive relationships, offer the highest fidelity but demand significant computational resources and detailed parameterization [25]. These simulation approaches form a hierarchy that can be strategically navigated depending on the specific objectives of the reactor design or optimization task [27].

Process simulation tools commonly employed in biorefinery applications include commercial software packages like Aspen Plus, SuperPro Designer, and gPROMS, as well as open-source platforms like OpenModelica and DWSIM [15]. These tools implement various numerical methods to solve the coupled differential-algebraic equations that describe process behavior, including sequential modular approaches, equation-oriented methods, and hybrid techniques. Computational fluid dynamics (CFD) provides detailed spatial resolution of transport phenomena within reactors, enabling insight into mixing patterns, temperature distributions, and residence time

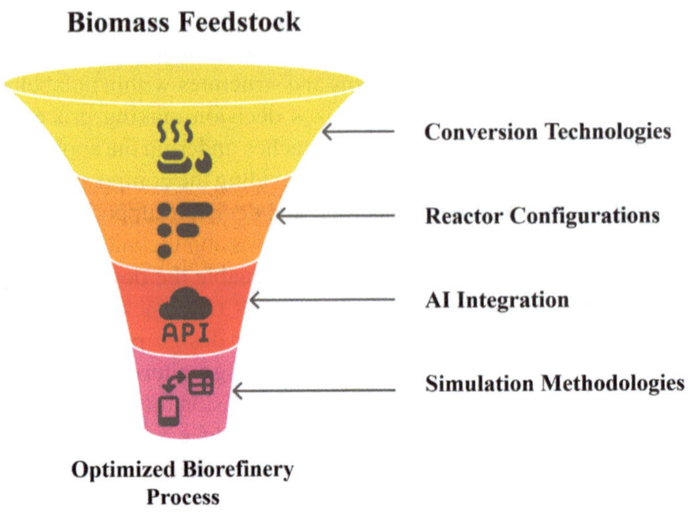

Fig. 2 Biomass feedstock steps from optimized biorefinery process

characteristics [23]. Multiscale modeling approaches integrate molecular, particle, reactor, and process scales to comprehensively capture relevant phenomena across multiple time and length scales [24]. The integration of these simulation methodologies with AI techniques creates powerful frameworks for exploring and optimizing biorefinery reactor configurations [25].

4 AI Methodologies for Biorefinery Reactor Simulation and Design

4.1 Machine Learning Algorithms in Process Modeling

Machine learning algorithms have emerged as powerful tools for developing accurate and computationally efficient models of complex biorefinery processes [26]. Regression techniques, including linear, polynomial, and multivariate approaches, enable the creation of empirical models that capture relationships between design variables and performance metrics. These techniques offer relative simplicity and interpretability but may struggle to represent highly nonlinear process behavior [27]. Support vector machines (SVMs) provide enhanced capability for modeling nonlinear relationships through kernel functions that implicitly map input features to higher-dimensional spaces [28]. SVMs have demonstrated effectiveness in modeling various biorefinery processes, including enzymatic hydrolysis kinetics, fermentation yields, and product purification performance [29]. Decision tree-based methods, particularly random forests and gradient boosting algorithms, offer robust performance for processes with complex parameter interactions and have shown promise in capturing the relationships among feedstock characteristics, process conditions, and product distributions in various biorefinery applications [30].

Artificial neural networks (ANNs) represent perhaps the most versatile and powerful machine learning approach for biorefinery process modeling [21]. Their flexible architecture, comprising interconnected layers of artificial neurons, enables the representation of highly complex, nonlinear relationships between process inputs and outputs [22]. Various neural network architectures have been applied to biorefinery modeling challenges, including feedforward networks for steady-state processes, recurrent networks for dynamic systems, and convolutional networks for spatially distributed phenomena [15]. For instance, deep neural networks have successfully modeled complex reaction networks in lignocellulosic biorefining, capturing intricate relationships between pretreatment conditions, enzymatic hydrolysis parameters, and sugar yields with higher accuracy than traditional kinetic models [17]. The integration of these machine learning models with first-principles knowledge creates hybrid modeling approaches that combine the accuracy and flexibility of data-driven methods with the physical insight and extrapolation capability of mechanistic models [26].

4.2 *Deep Learning Approaches for Complex Bioprocesses*

Deep learning approaches, characterized by neural networks with multiple hidden layers, have demonstrated remarkable capabilities in modeling complex bioprocesses that resist accurate representation through conventional methods [27]. These architectures excel at automatically extracting hierarchical features from raw data, enabling the capture of intricate patterns and relationships without explicit feature engineering. Deep neural networks have successfully modeled various aspects of biorefinery operations, including biomass pretreatment optimization, enzymatic hydrolysis kinetics, fermentation dynamics, and downstream processing performance. The ability of these models to represent complex, nonlinear relationships makes them particularly valuable for biorefinery processes involving heterogeneous feedstocks, complex reaction networks, and significant parameter interactions [22].

Specialized deep learning architectures have emerged for specific bioprocess modeling challenges. Recurrent neural networks (RNNs), particularly Long Short-Term Memory (LSTM) and Gated Recurrent Unit (GRU) variants, excel at modeling temporal dynamics in bioprocesses, capturing the evolution of state variables over time with high accuracy [31]. These architectures have proven effective for modeling fermentation kinetics, microbial growth dynamics, and time-dependent phenomena in various biorefinery unit operations [25]. Convolutional Neural Networks (CNNs) have found applications in modeling spatially distributed phenomena within reactors, such as temperature and concentration gradients in heterogeneous catalytic processes

[32]. Autoencoder architectures provide powerful tools for dimensionality reduction in complex bioprocess datasets, extracting low-dimensional representations that capture essential system behavior while reducing computational complexity. The combination of these specialized architectures with domain-specific knowledge creates powerful modeling frameworks for biorefinery reactor simulation and optimization [33].

4.3 Reinforcement Learning for Reactor Optimization

Reinforcement learning (RL) offers a particularly promising approach for biorefinery reactor optimization, framing the design process as a sequential decision-making problem where the algorithm learns optimal design strategies through interaction with the environment [34]. Unlike supervised learning approaches that require predefined optimal solutions, RL algorithms discover optimal designs by exploring the design space and evaluating the consequences of different design decisions [31]. This exploration-exploitation framework enables the discovery of novel reactor configurations that might elude conventional optimization approaches. RL algorithms conceptualize reactor design as a Markov Decision Process (MDP) with states representing reactor characteristics, actions corresponding to design modifications, and rewards reflecting performance metrics such as conversion efficiency, product selectivity, or economic value [25]. Through iterative interaction with a reactor simulation environment, the algorithm learns policies that maximize cumulative rewards, effectively navigating complex design spaces toward optimal solutions [21].

Various RL algorithms have demonstrated success in biorefinery reactor optimization applications. Q-learning and its deep learning extension, Deep Q Networks (DQN), have optimized control policies for various bioprocesses, including fed-batch fermentation and enzymatic hydrolysis [22]. Policy gradient methods, including Trust Region Policy Optimization (TRPO) and Proximal Policy Optimization (PPO), have addressed challenges in continuous action spaces typical of reactor design problems [24]. Actor-Critic architectures combine value function approximation with policy optimization, offering enhanced stability in learning optimal reactor configurations. Advanced RL techniques like Hierarchical Reinforcement Learning (HRL) decompose complex design problems into manageable subproblems, enabling efficient exploration of hierarchical design spaces spanning multiple scales [27]. The integration of RL with process simulation tools creates powerful frameworks for discovering innovative reactor configurations tailored to specific biorefinery applications [28].

4.4 Genetic Algorithms and Evolutionary Computing

Genetic algorithms (GAs) and evolutionary computing approaches provide powerful frameworks for biorefinery reactor optimization by mimicking the principles of natural selection and genetic evolution [4]. These techniques represent potential reactor designs as "chromosomes" encoding key design parameters and operational variables [8]. An initial population of design candidates undergoes evolutionary processes including selection based on fitness (performance metrics), crossover (recombination of design features from different candidates), and mutation (random modifications to maintain diversity) [10]. Through successive generations, the population evolves toward increasingly optimal designs, effectively navigating complex, nonconvex design spaces with multiple local optima. This evolutionary approach enables the exploration of diverse reactor configurations without requiring gradient information or continuous objective functions, making it particularly suitable for complex biorefinery applications involving discrete design choices and black-box simulation models [14].

Various evolutionary computing variants have been applied to biorefinery reactor optimization problems. Multiobjective genetic algorithms, including Nondominated Sorting Genetic Algorithm II (NSGA-II) and Strength Pareto Evolutionary Algorithm 2 (SPEA2), enable simultaneous optimization of multiple competing objectives such as conversion efficiency, energy consumption, and capital cost [21]. These approaches generate Pareto-optimal solution sets that illustrate trade-offs between competing objectives, providing decision-makers with a range of optimal design alternatives [19]. Differential Evolution (DE) algorithms have demonstrated enhanced global optimization performance for continuous parameter spaces in various reactor design applications. Particle Swarm Optimization (PSO), inspired by social behavior in bird flocking, offers an alternative evolutionary approach that has successfully optimized various bioprocess applications [32]. The integration of these evolutionary methods with surrogate modeling techniques addresses computational efficiency challenges, enabling exploration of complex design spaces with reduced simulation requirements [33].

4.5 Digital Twins and AI-Enhanced Simulation Environments

Digital twins represent virtual replicas of physical systems that continuously update to reflect the current state and history of their physical counterparts [34]. In biorefinery applications, digital twins integrate real-time process data, historical operating information, and predictive models to create high-fidelity virtual representations of reactor systems. These digital counterparts enable advanced monitoring, optimization, and control capabilities while providing platforms for testing design modifications without disrupting physical operations [35]. The development of digital twins for biorefinery reactors typically involves multiphysics models capturing relevant phenomena across multiple scales, from molecular interactions to reactor-level dynamics. These models integrate with data assimilation techniques that continuously update model parameters and states based on sensor measurements, maintaining alignment between virtual and physical systems [36]. AI techniques enhance digital twin capabilities through improved predictive modeling, anomaly detection, and adaptive learning from operational data [33].

AI-enhanced simulation environments extend beyond digital twins to provide comprehensive platforms for exploring novel reactor configurations [31]. These environments integrate process simulation tools with AI methodologies to create intelligent design spaces that guide exploration toward promising regions [21]. Advanced visualization techniques, including virtual and augmented reality interfaces, enable intuitive interaction with complex design spaces and simulation results. Machine learning approaches optimize simulation workflows by automatically selecting appropriate models and numerical methods based on the specific characteristics of the design problem [33]. Surrogate modeling techniques, including Gaussian process regression and physics-informed neural networks, create computationally efficient approximations of expensive simulation models, enabling more extensive exploration of design alternatives [34]. Knowledge graphs and semantic frameworks capture relationships between design parameters, performance metrics, and contextual factors, creating structured repositories of design knowledge that inform subsequent optimization efforts. These AI-enhanced simulation environments substantially accelerate the discovery and evaluation of innovative reactor configurations for biorefinery applications [35].

5 Applications of AI-Driven Simulations in Biorefinery Reactor Design

5.1 Optimization of Conventional Reactor Configurations

AI-driven simulation approaches have significantly enhanced the optimization of conventional reactor configurations in biorefinery applications, extracting maximum performance from established designs through sophisticated parameter tuning and operating strategy development [36]. For continuous stirred tank reactors (CSTRs), which remain workhorses in many biochemical conversion processes, AI techniques have optimized impeller design, baffle configuration, and mixing strategies to enhance mass transfer while minimizing shear stress on sensitive biological agents [31]. Machine learning algorithms have identified optimal relationships between agitation parameters and gas-liquid mass transfer coefficients in aerobic fermentation processes, leading to improved oxygen transfer rates and higher product yields. Neural network models have captured complex relationships between

CSTR geometric parameters and mixing performance metrics, enabling rapid optimization for specific bioprocess applications [32]. These optimized CSTR designs have demonstrated substantial improvements in conversion efficiency and productivity compared to conventional configurations, particularly for processes involving high-viscosity fermentation broths and heterogeneous reaction mixtures [29].

Plug flow reactor (PFR) optimization has similarly benefited from AI-driven approaches, particularly for thermochemical conversion processes like fast pyrolysis and catalytic upgrading [3]. Support vector regression models have captured relationships between PFR design parameters (diameter, length, heating profiles) and product distribution in biomass pyrolysis, enabling targeted optimization for specific product slates [19]. Reinforcement learning algorithms have developed optimal temperature profiles along reactor length to maximize valuable product yields while minimizing energy consumption and undesired byproducts [36]. For packed bed reactors used in heterogeneous catalytic processes, AI techniques have optimized catalyst distribution patterns, bed configurations, and flow distribution systems to enhance reaction selectivity while minimizing pressure drop and channeling effects [37]. Genetic algorithms have successfully addressed the complex interplay between particle size distribution, bed porosity, and reactor performance metrics. These AI-optimized conventional reactor designs deliver significant improvements in process economics and sustainability metrics while building upon established engineering principles and operational experience [38].

5.2 *Novel Reactor Concepts Through AI Exploration*

Beyond optimizing conventional designs, AI-driven simulation approaches have enabled the discovery of entirely novel reactor concepts that transcend traditional configuration categories [35]. These innovations frequently emerge from unconstrained exploration of design spaces using advanced optimization algorithms that can identify nonintuitive solutions human engineers might overlook. Figure 3 highlights the AI in Biorefinery Reactor Design. For instance, evolutionary algorithms exploring reactor geometry without conventional constraints have identified oscillatory flow reactors with optimized baffle designs that provide superior mixing characteristics for heterogeneous biomass slurries [25]. These designs achieved up to 40% higher conversion rates compared to conventional alternatives while reducing energy consumption by 25% [22]. Multiobjective genetic algorithms have discovered hybrid reactor configurations combining elements of different archetypes, such as segmented reactors that transition between CSTR and PFR characteristics along their length to optimize reaction profiles for complex bioconversion pathways [30].

Reinforcement learning approaches have generated particularly innovative reactor concepts by framing design as a sequential decision process without predefined constraints [32]. These algorithms have identified multizone reactors with spatially varying catalyst distributions, temperature profiles, and residence time characteristics tailored to specific biomass conversion pathways. For enzymatic hydrolysis

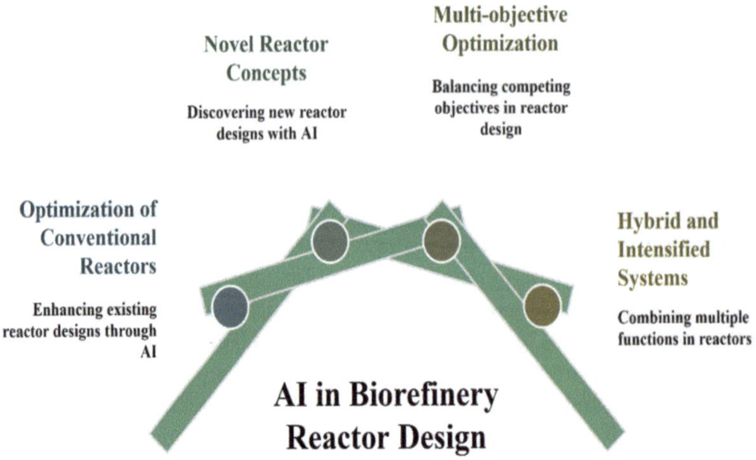

Fig. 3 AI in biorefinery reactor design

processes, RL-based exploration discovered counter-intuitive enzyme feeding strategies and reactor compartmentalization approaches that significantly enhanced sugar yields while reducing enzyme consumption [37]. AI-driven topology optimization techniques have generated biomimetic reactor structures inspired by natural systems, incorporating fractal flow channels and hierarchical mixing elements that maximize surface area and minimize transport limitations. These novel reactor concepts demonstrate the capacity of AI-driven approaches to transcend conventional design paradigms and identify revolutionary solutions to longstanding biorefinery challenges [39].

5.3 *Multiobjective Optimization Approaches*

Biorefinery reactor design inherently involves multiple competing objectives, including capital cost, operating expenses, conversion efficiency, product selectivity, energy consumption, and environmental impact [26]. AI-driven multiobjective optimization approaches provide powerful frameworks for navigating these complex trade-offs and identifying Pareto-optimal solution sets that represent the best possible compromises between competing objectives. Multiobjective genetic algorithms, particularly NSGA-II and NSGA-III, have successfully identified optimal reactor configurations that balance economic and environmental performance metrics in various biorefinery applications [28]. These approaches generate solution sets illustrating the quantitative relationships between objectives, providing decision-makers with clear visualizations of available trade-offs. For instance, multiobjective optimization of a lignocellulosic ethanol process revealed that reactor configurations maximizing ethanol yield typically incurred 15–30% higher capital costs compared to economically optimal designs, while configurations minimizing environmental impact required 20–35% higher investment than cost-minimizing alternatives [17].

Advanced preference articulation mechanisms enhance multiobjective optimization by incorporating decision-maker priorities directly into the search process [16]. Interactive evolutionary algorithms enable engineers to guide the optimization process through preference feedback, progressively focusing computational resources on the most relevant regions of the design space. Reference point approaches allow the specification of target performance levels for different objectives, directing the search toward solutions meeting these aspirational goals [19]. Machine learning methods have enhanced multiobjective optimization efficiency through surrogate modeling of computationally expensive objectives and intelligent sampling strategies that prioritize promising design regions. These AI-enhanced multiobjective approaches enable more comprehensive exploration of reactor design trade-offs while accommodating the practical constraints and priorities of biorefinery developers [32].

5.4 Hybrid and Intensified Reactor Systems

AI-driven simulation approaches have accelerated the development of hybrid and intensified reactor systems that combine multiple functions within unified equipment, reducing capital costs, enhancing efficiency, and improving process integration [33]. These advanced systems frequently integrate reaction and separation functions, incorporate multiple reaction zones with distinct characteristics, or combine different energy transfer mechanisms within unified equipment [9]. Reactive distillation columns for in-situ product recovery during bio alcohol production represent one successful application area, where AI techniques have optimized complex interactions between reaction kinetics, vapor-liquid equilibria, and column hydraulics. Multiphysics models enhanced by machine learning have captured these interdependent phenomena more accurately than conventional approaches, enabling the design of reactive distillation systems achieving 25–40% reductions in energy consumption compared to conventional reaction-separation sequences [32].

Membrane reactors, which combine biochemical conversion with selective product removal, have similarly benefited from AI-driven design approaches [33]. Deep neural networks have modeled complex relationships between membrane properties, reactor configuration, and separation performance, enabling optimization of these intensified systems for applications including biogas upgrading and biohydrogen production [34]. For oscillatory flow reactors, which enhance mixing and heat transfer through controlled fluid oscillations, genetic algorithms have optimized baffle designs and oscillation parameters for specific biorefinery applications, achieving significant intensification benefits including footprint reductions exceeding 80% compared to conventional alternatives [35]. Perhaps most significantly, AI approaches have enabled the development of modular, reconfigurable reactor systems that can adapt to varying feedstocks and product requirements through automated adjustment of internal configuration and operating parameters. These adaptable systems, guided by AI control strategies, represent a paradigm shift in biorefinery reactor design, offering unprecedented flexibility to accommodate the inherent variability of biomass resources [39].

6 Case Studies in AI-Driven Biorefinery Reactor Innovation

6.1 Lignocellulosic Biomass Conversion Optimization

Lignocellulosic biomass conversion represents one of the most challenging and promising biorefinery applications, involving complex feedstocks with recalcitrant structures that resist efficient fractionation and conversion [38]. AI-driven approaches have demonstrated remarkable success in optimizing reactor configurations for various lignocellulose conversion pathways [34]. For enzymatic hydrolysis processes, machine learning models have captured the complex relationships between pretreatment conditions, reactor design parameters, enzyme loading strategies, and sugar yields [21]. Neural network models trained on extensive experimental datasets have predicted hydrolysis kinetics with significantly higher accuracy than traditional Michaelis-Menten approaches, accounting for inhibition effects, mass transfer limitations, and enzyme deactivation mechanisms [23]. These enhanced models have enabled the development of novel reactor configurations featuring zoned enzyme addition, optimized mixing regimes, and controlled substrate feeding strategies that achieved 30–45% improvements in sugar yields compared to conventional batch approaches [14].

For thermochemical conversion of lignocellulosic biomass, AI-driven simulation has optimized fast pyrolysis reactor configurations to maximize bio-oil yield and quality [21]. Computational fluid dynamics models enhanced with machine learning techniques have captured the complex interactions between biomass particle dynamics, heat transfer, and reaction kinetics in fluidized bed reactors. These models have guided the development of optimized fluidization regimes, feed introduction mechanisms, and temperature control strategies that significantly enhanced bio-oil yield while reducing char formation. For catalytic upgrading of bio-oils, reinforcement learning approaches have discovered novel reactor configurations featuring staged catalyst beds with optimized temperature profiles that substantially improved oxygen removal while minimizing hydrogen consumption [23]. The integration of these optimized reactor configurations into complete lignocellulosic biorefinery systems has demonstrated potential for reducing minimum selling prices of advanced biofuels by 15–25% compared to conventional designs, substantially enhancing commercial viability [5].

6.2 Algal Biorefinery Systems

Algal biorefinery systems offer unique challenges and opportunities for AI-driven reactor optimization due to their integration of photobiological cultivation with downstream processing of complex biochemical mixtures [7]. For photobioreactor systems, AI approaches have optimized geometry, light distribution, mixing patterns, and gas exchange characteristics to maximize photosynthetic efficiency and biomass productivity [9]. Neural networks have modelled the complex relationships among light intensity, spectral distribution, mixing-induced light/dark cycles, and algal growth kinetics [10]. These models have guided the development of novel photobioreactor configurations featuring optimized light paths, specialized mixing elements, and controlled gas delivery systems that achieved biomass productivity improvements of 40–60% compared to conventional designs. Genetic algorithms have optimized harvesting strategies and reactor operational cycles to maximize biomass productivity while minimizing energy consumption, addressing one of the major economic challenges in algal cultivation [11].

For downstream processing of algal biomass, AI-driven approaches have optimized extraction reactor configurations for recovering valuable compounds including lipids, proteins, and specialty chemicals [40]. Multiobjective optimization has identified reactor designs and operating conditions that balance extraction efficiency, energy consumption, and product quality across diverse algal feedstocks. Deep learning models have captured the complex relationships between cell disruption parameters, solvent characteristics, and extraction kinetics, enabling the development of intensified extraction systems that reduce solvent consumption by 30–50% compared to conventional approaches [35]. Perhaps most significantly, AI methods have facilitated the design of integrated biorefinery systems that optimize the valorization of multiple algal components through cascading extraction processes and complementary conversion pathways [31]. These integrated systems, optimized through system-level AI approaches, have demonstrated potential for substantially improving the economic viability of algal biorefineries by maximizing value recovery across diverse product streams [37].

6.3 Waste-to-Energy Biorefinery Processes

Waste-To-energy biorefinery processes present unique reactor design challenges due to the heterogeneous and variable nature of feedstocks, complex contamination profiles, and diverse conversion pathways [21]. AI-driven approaches have proven to be particularly valuable for optimizing anaerobic digestion reactors processing organic waste streams. Machine learning models have captured relationships among waste composition, reactor design parameters, and biogas production kinetics, enabling the development of robust designs that accommodate feedstock variability [22]. Neural networks have modeled the complex relationships among

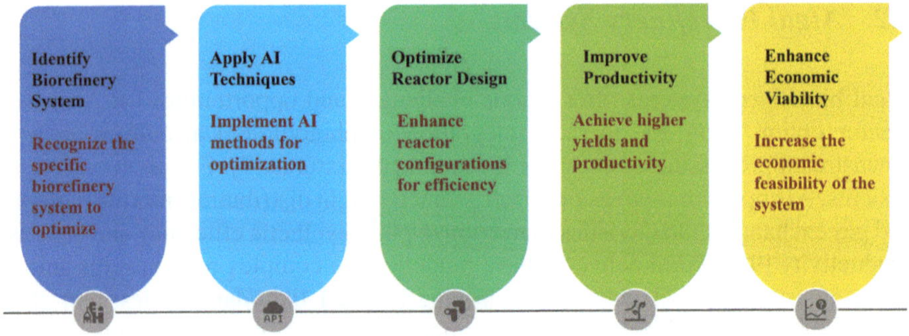

Fig. 4 Steps to apply AI techniques to enhance economic viability

reactor geometry, mixing parameters, and microbial community dynamics, guiding the development of optimized two-stage digestion systems that separate hydrolysis/acidogenesis from methanogenesis. Figure 4 explains the steps to apply AI techniques to enhance economic viability. These optimized configurations achieved 25–35% higher methane yields and significantly enhanced process stability compared to conventional single-stage designs [13]. Reinforcement learning approaches have developed adaptive control strategies that dynamically adjust reactor parameters in response to feedstock variations, substantially enhancing operational robustness [30].

For thermochemical conversion of waste streams, AI methods have optimized gasification reactor designs to accommodate heterogeneous feedstocks while maximizing syngas quality [10]. Computational fluid dynamics models enhanced with machine learning techniques have captured the complex physical and chemical phenomena in fluidized bed gasifiers, enabling optimization of bed materials, gas distribution systems, and reactor geometry for specific waste streams [1]. Genetic algorithms have identified optimal operating regimes that balance carbon conversion efficiency, tar reduction, and energy efficiency across varying feedstock compositions [3]. For hydrothermal processing of wet waste streams, AI-driven simulation has optimized reactor configurations to enhance conversion efficiency while managing challenging aspects including corrosion, precipitate formation, and heteroatom distribution [23]. These optimized waste-to-energy reactor configurations have demonstrated substantial improvements in both technical performance and economic viability, accelerating the deployment of waste valorization technologies that simultaneously address waste management challenges and renewable energy production [2].

6.4 *Biochemical Platform Innovations*

Biochemical platform biorefineries, centered around fermentation processes that convert sugars and other intermediates into fuels, chemicals, and materials, have witnessed significant innovations through AI-driven reactor optimization [5]. For fermentation processes, machine learning approaches have optimized bioreactor configurations to enhance productivity and yield across diverse microbial systems and target products. Neural networks have modelled complex relationships between reactor geometry, mixing parameters, mass transfer characteristics, and fermentation kinetics, enabling the development of optimized designs for specific bioprocesses [11]. These models have guided innovations including optimized impeller configurations that enhance oxygen transfer while minimizing shear damage to cells, specialized feeding strategies that minimize substrate inhibition, and novel temperature control approaches that maximize productivity. For consolidated bioprocessing systems, which integrate enzyme production, hydrolysis, and fermentation within unified processes, AI approaches have optimized reactor compartmentalization and operational strategies to balance the distinct requirements of different bioconversion stages [14].

Perhaps the most significant impact of AI in biochemical platform biorefineries has been in the development and optimization of continuous fermentation systems [18]. Traditionally, many industrial fermentations operate in batch or fed-batch mode due to challenges including contamination risk, genetic instability, and complex product inhibition effects. AI-driven approaches have addressed these challenges through innovative reactor designs including multistage continuous systems with optimized residence time distributions, cell recycling systems with enhanced retention mechanisms, and in-situ product removal configurations that alleviate inhibition effects [36]. Reinforcement learning algorithms have developed sophisticated control strategies that maintain optimal conditions throughout continuous operations despite disturbances and evolutionary pressures on microbial populations [37]. These AI-optimized continuous bioprocessing systems have demonstrated productivity improvements of 200–400% compared to batch alternatives while reducing capital costs and enhancing consistency—innovations that substantially improve the economic competitiveness of biochemical platforms against petrochemical alternatives [26].

7 Integration Challenges and Implementation Considerations

7.1 Data Quality and Availability Issues

The effectiveness of AI-driven approaches for biorefinery reactor simulation and optimization fundamentally depends on the quality, quantity, and representativeness of available data. In this domain, data challenges present significant impediments to successful implementation [12]. Experimental data from biorefinery processes often suffer from limited sample sizes due to the resource-intensive nature of pilot-scale experimentation. This scarcity is particularly pronounced for novel feedstocks, emerging conversion technologies, and innovative reactor configurations lacking established operational history [16]. The data that does exist frequently exhibits high variance resulting from the inherent heterogeneity of biomass feedstocks, biological variability in bioprocesses, and sensitivity to environmental conditions. Furthermore, available datasets often contain sampling biases toward conventional operating conditions, creating sparse representation of boundary regions where optimal solutions might exist. These limitations collectively constrain the development of robust AI models and necessitate specialized approaches to maximize value from imperfect datasets [19].

Various strategies have emerged to address these data challenges in biorefinery applications. Transfer learning techniques enable knowledge transfer from data-rich applications to data-limited scenarios, leveraging commonalities in underlying physical and chemical phenomena [21]. Physics-informed machine learning approaches incorporate domain knowledge and first-principles constraints into model architectures, reducing data requirements while ensuring physical consistency [32]. Active learning methodologies intelligently select experimental conditions to maximize information gain, optimizing limited experimental resources. Synthetic data generation through mechanistic models, agent-based simulations, and generative adversarial networks (GANs) supplements limited experimental datasets with plausible synthetic observations [38]. Data fusion techniques integrate information from multiple sources-including laboratory experiments, pilot demonstrations, industrial operations, and scientific literature-to create more comprehensive training datasets. Despite these advances, data limitations remain a primary constraint on AI application in biorefinery reactor design, highlighting the need for strategic data collection efforts and collaborative data sharing initiatives within the research community [40].

7.2 Model Validation and Verification Challenges

Validation and verification of AI models for biorefinery reactor simulation present substantial challenges that must be systematically addressed to ensure reliability and applicability [20]. Traditional validation approaches comparing model predictions against independent test data face limitations in this domain due to data scarcity and the inherent complexity of bioprocesses [4]. The multiscale nature of biorefinery processes—spanning molecular, cellular, reactor, and system levels—creates particular validation challenges, as phenomena at each scale may require different validation strategies and metrics. Furthermore, the black-box nature of many advanced AI approaches, particularly deep neural networks, creates transparency concerns that complicate validation and verification compared to mechanistic models with explicit representation of underlying phenomena [9]. These challenges necessitate comprehensive validation frameworks that combine traditional approaches with specialized techniques appropriate for complex, data-driven models [13].

Advanced validation strategies for AI-driven biorefinery simulations include cross-validation techniques tailored to limited datasets, uncertainty quantification methods that characterize prediction confidence across the design space, and multi-fidelity validation approaches that leverage both high-fidelity experimental data and lower-fidelity but more abundant computational results [18]. Sensitivity analysis techniques systematically evaluate model robustness to parameter perturbations, identifying potential vulnerabilities and guiding refinement efforts [21]. For hybrid models combining AI components with mechanistic elements, decomposition-based validation approaches separately verify individual model components before assessing integrated performance. Explainable AI (XAI) techniques enhance validation by providing insights into model reasoning and identifying potential biases or artifacts in learning processes. Despite these advances, validation remains challenging for highly complex biorefinery models, particularly when predicting performance of novel reactor configurations without direct experimental precedents [17]. This reality underscores the importance of iterative validation approaches that progressively build confidence through strategically designed experimental campaigns guided by model predictions and uncertainty estimates [31].

7.3 Computational Requirements and Accessibility

The computational requirements of AI-driven biorefinery simulation vary widely depending on methodology, model complexity, and application scope. Deep learning approaches, particularly when coupled with high-fidelity process simulations or computational fluid dynamics, may demand substantial computational resources for both training and inference [34]. These requirements can present significant barriers to adoption, particularly for smaller organizations, academic institutions, and developing regions with limited access to advanced computing infrastructure [35]. Training complex models on extensive datasets may require high-performance computing clusters with specialized hardware including graphics processing units (GPUs) or tensor processing units (TPUs), representing substantial capital investments [36]. Furthermore, the energy consumption associated with training sophisticated AI models raises sustainability concerns that warrant consideration in the context of technologies intended to enhance environmental performance. These challenges collectively highlight the need for strategies that balance model sophistication with practical accessibility and computational efficiency [38].

Various approaches have emerged to address computational challenges in AI-driven biorefinery simulation [25]. Model compression techniques, including knowledge distillation, pruning, and quantization, reduce computational requirements without significant performance degradation [26]. Transfer learning minimizes training requirements by adapting pretrained models to specific applications rather than developing models from scratch [28]. Distributed computing frameworks enable resource pooling across organizations, while cloud-based platforms provide scalable access to computational resources without large capital investments. Surrogate modeling approaches replace computationally expensive simulations with efficient approximations for portions of the workflow that don't require high-fidelity representation [21]. Open-source software frameworks, community model repositories, and collaborative development initiatives enhance accessibility by reducing duplication of effort and enabling knowledge sharing across organizations. These strategies collectively improve the accessibility of AI-driven approaches for biorefinery reactor design, helping to democratize advanced simulation capabilities across diverse stakeholders in the bioeconomy [39].

7.4 Industrial Implementation Barriers

The translation of AI-driven reactor innovations from research environments to industrial biorefinery applications faces numerous implementation barriers beyond technical considerations [28]. Organizational factors, including established design practices, approval processes, and risk management approaches, may impede adoption of novel methodologies perceived as unproven or difficult to interpret. The perceived "black box" nature of many advanced AI methods creates particular challenges in regulated industries with stringent documentation and transparency requirements [27]. Economic barriers include investment costs for computational infrastructure, software licenses, and personnel training, which may be difficult to justify against uncertain return on investment. Intellectual property considerations become increasingly complex as innovations emerge from the integration of proprietary process data with algorithmic innovations, creating potential barriers to collaboration and knowledge sharing [29]. Furthermore, workforce capability gaps in the intersection among data science, process engineering, and biotechnology create implementation challenges as organizations struggle to assemble interdisciplinary teams with the necessary expertise span [30].

Successful industrial implementation approaches address these barriers through comprehensive strategies that extend beyond technical considerations. Progressive implementation pathways begin with low-risk applications demonstrating clear value, building organizational confidence before advancing to more transformative applications [32]. Collaborative development models involving technology developers, process licensors, equipment manufacturers, and end users distribute development costs while ensuring practical relevance. Open innovation frameworks and precompetitive research consortia enable risk sharing and accelerated learning while addressing common challenges [34]. Human–AI collaborative approaches emphasize augmenting rather than replacing engineering expertise, positioning AI tools as decision support systems that enhance human capabilities rather than autonomous replacement technologies [35]. Educational initiatives including specialized training programs, university–industry partnerships, and professional development pathways address workforce capability gaps while building organizational capacity for effective AI utilization. These holistic implementation strategies acknowledge the sociotechnical nature of innovation adoption and create pathways for AI-driven approaches to achieve practical impact in industrial biorefinery applications [36].

8 Future Perspectives in AI-Driven Biorefinery Innovation

8.1 Emerging AI Technologies for Bioprocess Simulation

The landscape of AI technologies applicable to bioprocess simulation continues to evolve rapidly, with several emerging approaches showing particular promise for biorefinery applications [38]. Graph neural networks (GNNs) represent one significant frontier, enabling explicit modeling of molecular structures, reaction networks, and process flowsheets as interconnected graphs [16]. These architectures capture structural relationships and dependencies that traditional neural networks struggle to represent efficiently, offering enhanced capabilities for modeling complex reaction pathways in biomass conversion processes [18]. Transformer architectures, which have revolutionized natural language processing through attention mechanisms, show growing potential for sequential and time-series modeling in bioprocess applications. These models excel at capturing long-range dependencies and complex patterns in process data, enabling more accurate prediction of dynamic system behavior [11]. Bayesian deep learning approaches offer enhanced uncertainty quantification through probabilistic representations, providing critical information about prediction confidence for risk-sensitive design decisions [29].

Neuroevolution techniques, which apply evolutionary algorithms to neural network architecture design, offer promising approaches for automatically discovering optimal model architectures for specific bioprocess applications [30]. These approaches reduce dependence on human expertise in neural architecture design while potentially discovering novel structures particularly suited to bioprocess modeling challenges [32]. Federated learning frameworks enable collaborative model development across organizations without sharing sensitive process data, addressing confidentiality concerns while leveraging broader datasets than any single organization could develop independently [34]. Self-supervised learning approaches reduce dependence on expensive labeled data by extracting useful representations from unlabeled process measurements, particularly valuable given the data scarcity challenges in biorefinery applications [35]. Neuromorphic computing approaches, inspired by biological neural systems, offer potential for energy-efficient implementation of complex bioprocess models, aligning with sustainability objectives of biorefinery development. These emerging technologies collectively expand the frontier of what's possible in AI-driven biorefinery simulation, creating new opportunities for innovation in this critical domain [40].

8.2 Integration with Industry 4.0 and Smart Manufacturing

The integration of AI-driven biorefinery simulation with broader Industry 4.0 and smart manufacturing frameworks creates powerful synergies that enhance both design and operational excellence [11]. Industrial Internet of Things (IIoT) technologies enable comprehensive instrumentation of biorefinery operations, generating rich datasets that continuously inform and refine AI models. Edge computing architectures process sensor data locally before transmission, enabling real-time analytics and control while reducing bandwidth requirements; 5G and advanced networking technologies facilitate seamless data flow between distributed biorefinery assets and centralized intelligence systems, creating integrated operations that span multiple physical locations [15]. Augmented reality interfaces enable intuitive interaction with complex simulation results, supporting both design activities and operational decision-making through contextualized information delivery. Digital supply chain integration extends optimization beyond facility boundaries to encompass feedstock sourcing, logistics, and product distribution networks, creating system-level optimization opportunities [28].

The convergence of AI-driven simulation with smart manufacturing creates particularly powerful capabilities through closed-loop systems that continuously learn and adapt. Digital twins enhanced with AI capabilities continuously update to reflect current plant conditions, enabling real-time optimization and predictive maintenance [12]. Automated experimentation systems guided by AI planning algorithms systematically explore design spaces and process conditions, accelerating innovation while minimizing human intervention requirements [29]. Robotic process automation enhances operational consistency while freeing human operators for higher-value activities requiring judgment and creativity. Perhaps, most significantly, the integration creates self-optimizing biorefinery systems that continuously adapt to changing conditions-including feedstock characteristics, market demands, and equipment performance-through autonomous learning and optimization. This vision of AI-integrated smart biorefineries represents a fundamental transformation from static designs with fixed operating envelopes to adaptive systems that continuously evolve toward optimal performance as conditions change [22].

8.3 Explainable AI for Process Understanding

The growing complexity of AI models used in biorefinery simulation has increased focus on explainability—the capacity to interpret and understand model reasoning and predictions in human-comprehensible terms [9]. This emphasis reflects both practical engineering needs and regulatory considerations, as decisions with significant economic and safety implications should rest on transparent and interpretable foundations. Local explanation techniques, including SHAP (SHapley Additive exPlanations) and LIME (Local Interpretable Model-agnostic Explanations), identify feature contributions to specific predictions, helping engineers understand why particular reactor configurations receive favorable evaluations [12]. Global explanation approaches extract overall patterns and relationships learned by models, identifying general principles that might inform broader design guidelines beyond specific cases. Figure 5 explains the streamline process from simple data to quantum

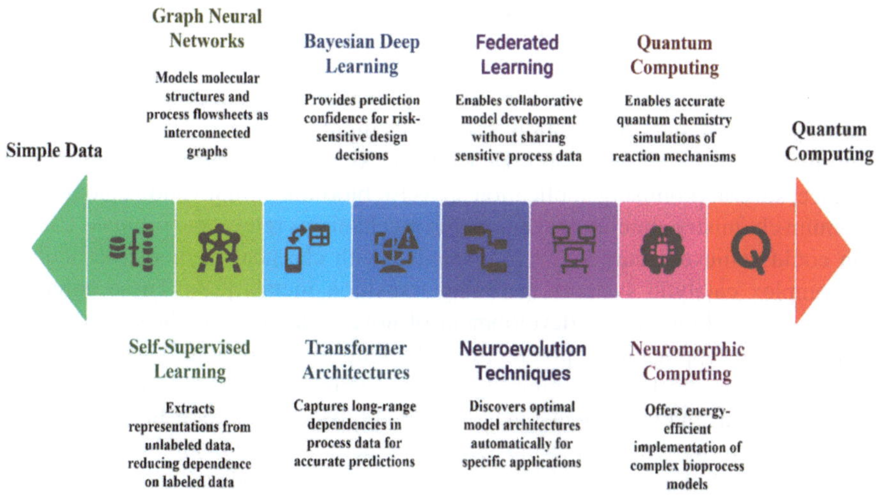

Fig. 5 Streamline process from simple data to quantum computing

computing. Counterfactual explanations generate "what if" scenarios that illustrate how specific changes to reactor parameters would affect performance predictions, providing intuitive understanding of model sensitivity and decision boundaries [37].

Beyond technical explainability methods, human-centered approaches to explainable AI focus on aligning explanations with the mental models and information needs of different stakeholders [38]. Process engineers may require mechanistic explanations connecting design decisions to fundamental phenomena, while executive decision-makers might need explanations framed in terms of economic and strategic implications. Visual analytics techniques create interactive representations of complex model behavior, enabling intuitive exploration of prediction landscapes and parameter relationships [39]. Knowledge distillation approaches extract interpretable rules and patterns from complex models, creating simplified representations that capture essential behavior while enhancing transparency [23]. The integration of domain knowledge with data-driven insights creates particularly valuable explanations that connect AI predictions with established engineering principles and scientific understanding. These advances in explainable AI enhance the trustworthiness and utility of AI-driven approaches in biorefinery applications, facilitating broader adoption while ensuring that human expertise remains central to critical design decisions [31].

8.4 Quantum Computing Potential

Quantum computing represents a potentially transformative technology for biorefinery simulation and optimization, offering computational capabilities fundamentally different from classical approaches [22]. As quantum hardware continues to advance, several promising application areas for biorefinery innovation emerge [23]. Quantum chemistry simulations of reaction mechanisms relevant to biomass conversion could achieve accuracy levels infeasible with classical methods, particularly for complex catalytic systems involving transition metals [28]. These enhanced simulations could guide the development of novel catalysts specifically optimized for biorefinery applications, potentially unlocking more efficient conversion pathways. Quantum machine learning algorithms, which leverage quantum effects to identify patterns in high-dimensional data, might discover relationships in complex bioprocess datasets that elude classical approaches [32]. Quantum optimization algorithms, including quantum approximate optimization algorithm (QAOA) and quantum annealing, offer potential advantages for navigating the complex, combinatorial design spaces characteristic of biorefinery reactor configuration problems [33].

Despite these promising directions, significant challenges remain before quantum computing delivers practical impact in biorefinery applications. Current quantum hardware remains limited by qubit counts, coherence times, and error rates that constrain practical applications [5]. Hybrid quantum-classical approaches currently show the most promise, using quantum processors for specific computational tasks while leveraging classical systems for other aspects [25]. Quantum algorithm development specifically targeting bioprocess applications remains in early stages, with further research needed to identify the most promising application areas and algorithmic approaches. Accessibility presents another challenge, as specialized expertise requirements and limited hardware availability currently restrict quantum computing to large research organizations and technology companies [21]. Nevertheless, the fundamental computational advantages offered by quantum approaches for simulating complex chemical and biological systems suggest significant long-term potential for biorefinery applications. Organizations involved in biorefinery development should monitor this rapidly evolving field and consider exploratory research partnerships to build capability in advance of broader practical implementation [32].

9 Sustainability and Economic Considerations

9.1 Life-Cycle Assessment Integration

The integration of life-cycle assessment (LCA) methodologies with AI-driven biorefinery simulation creates powerful frameworks for sustainability-oriented design optimization [22]. Traditional reactor design approaches typically focus on technical performance metrics and direct economic considerations, potentially overlooking broader environmental implications across product lifecycles. AI-enhanced LCA integration addresses this limitation by incorporating cradle-to-grave environmental impacts into simulation and optimization workflows. Machine learning approaches have enhanced inventory analysis through automated data collection and gap-filling techniques that address the challenging data requirements of comprehensive LCAs [25]. Deep learning models have captured complex relationships between reactor design parameters and life-cycle impacts across multiple environmental categories, enabling identification of configurations that minimize overall environmental footprint rather than simply optimizing direct process metrics. Multiobjective optimization frameworks incorporating both traditional performance metrics and LCA-derived impact categories have identified Pareto-optimal reactor configurations that balance technical, economic, and environmental considerations [27].

Several methodological advances have strengthened the integration of LCA with AI-driven biorefinery simulation [28]. Parametric LCA approaches establish mathematical relationships between design variables and life-cycle impacts, enabling direct incorporation into optimization workflows. Uncertainty analysis techniques characterize confidence intervals for impact assessments, accounting for data limitations and methodological choices that influence results. Dynamic LCA methodologies capture temporal aspects of environmental impacts, particularly relevant for biorefinery systems with long-term carbon sequestration implications and time-dependent emissions profiles [29]. Consequential LCA approaches assess broader system impacts including market-mediated effects, providing more comprehensive sustainability evaluation than attributional approaches alone. Regional adaptation techniques account for geographic differences in impact pathways and background systems, enhancing relevance for location-specific biorefinery design. These advances collectively enhance the capacity to design biorefinery reactors optimized for genuine sustainability rather than narrow technical efficiency, supporting the broader environmental objectives that motivate biorefinery development [12].

9.2 Economic Feasibility and Market Considerations

The economic feasibility of biorefinery systems depends critically on reactor configurations that balance capital investment, operating costs, and value generation within relevant market contexts [15]. AI-driven approaches have enhanced traditional techno-economic assessment (TEA) methodologies through more comprehensive exploration of design alternatives and improved modeling of complex economic relationships. Machine learning techniques have developed surrogate models of capital and operating costs as functions of design parameters, enabling rapid economic evaluation during optimization processes [17]. These models capture nonlinear relationships between design decisions and economic outcomes, including economies of scale, learning curves, and maintenance implications of different configuration choices [19]. Neural networks have modelled complex relationships between product quality characteristics and market value, enabling optimization that maximizes economic return rather than simply targeting conversion efficiency or yield metrics. Multiobjective frameworks have identified economically viable designs across different market scenarios, enhancing robustness to uncertain and volatile market conditions typical in bioeconomy sectors [20].

Market-adaptive design approaches, enabled by AI methodologies, represent a particularly promising direction for enhancing economic feasibility. These approaches develop reactor systems capable of producing different product slates in response to market signals, enhancing economic resilience through product flexibility [21]. Reinforcement learning algorithms have optimized switching strategies between operational modes based on feedstock costs, product prices, and inventory positions. Scenario analysis frameworks enhanced by machine learning have evaluated reactor designs across diverse market projections, identifying robust configurations that maintain economic viability despite future uncertainties [22]. Real options valuation approaches, implemented through sophisticated simulation models, have quantified the economic value of design flexibility, properly accounting for the option value created by adaptable reactor configurations [26]. These market-oriented approaches acknowledge that economic feasibility depends not only on technical efficiency but also on alignment with market dynamics and adaptability to changing economic conditions—factors particularly important in emerging bioeconomy markets characterized by evolving policy landscapes and competition with established fossil-based alternatives [32].

9.3 Environmental Impact Reduction Strategies

Beyond life-cycle assessment integration, AI-driven approaches have specifically targeted environmental impact reduction through specialized reactor design strategies addressing key sustainability challenges [33]. Water consumption represents one critical focus area, with machine learning models optimizing reactor configurations to minimize freshwater requirements through enhanced recycling, efficient washing processes, and water-conserving separation technologies [35]. These optimized designs have achieved water consumption reductions of 30–60% compared to conventional approaches, dramatically improving sustainability in water-constrained regions [38]. For greenhouse gas emissions, AI approaches have optimized reactor systems to minimize energy requirements, maximize carbon utilization efficiency, and enhance sequestration in durable products. Multiobjective optimization has identified configurations that maximize displacement of fossil carbon while minimizing process emissions, achieving optimal climate mitigation impact across product lifecycles [26]. Neural network models have captured complex relationships between reactor design parameters and emission profiles, enabling targeted optimization for specific environmental objectives [25].

Waste minimization and valorization represent another critical focus area for environmentally oriented reactor design. AI approaches have optimized integrated biorefinery configurations that cascade material flows between process units, ensuring comprehensive utilization of biomass components and minimization of waste streams [22]. Genetic algorithms have developed novel process integration concepts that transform traditional waste streams into valuable coproducts through additional conversion pathways. For unavoidable residues, reinforcement learning approaches have optimized treatment reactor designs to minimize environmental impact while potentially generating additional value through energy recovery or nutrient recycling. Perhaps, most significantly, AI-driven approaches have enhanced environmental performance through intensification strategies that reduce facility footprint, material requirements, and embodied impacts through more efficient reactor designs [25]. These environmentally oriented optimization approaches demonstrate how AI can specifically target sustainability objectives beyond traditional performance metrics, aligning biorefinery development with broader environmental imperatives and maximizing the sustainability benefits motivating bioeconomy development [31].

9.4 Policy and Regulatory Implications

The policy and regulatory landscape significantly influence biorefinery development trajectories, creating both constraints and incentives that shape technology deployment. AI-driven approaches have enhanced policy analysis and compliance strategies through improved modeling of regulatory impacts on design decisions and economic outcomes [32]. Machine learning techniques have developed predictive models relating reactor design parameters to regulated emissions and effluents, enabling optimization approaches that ensure regulatory compliance while minimizing compliance costs [34]. Multiobjective frameworks have identified design alternatives that balance regulatory requirements with economic performance across diverse regulatory scenarios, enhancing robustness to policy uncertainty. Perhaps, most significantly, AI-enhanced simulation enables quantitative assessment of how alternative policy instruments might influence technology choices and deployment pathways, providing valuable information for both policy development and strategic planning within the biorefinery sector [35].

Looking forward, several emerging policy domains present both challenges and opportunities for AI-driven biorefinery innovation. Carbon pricing mechanisms increasingly influence economic competitiveness of biorefining relative to fossil alternatives, with AI approaches optimizing designs to maximize advantage under different carbon valuation scenarios [38]. Circular economy policies promote integrated biorefinery systems that minimize waste generation and maximize material cycling, aligning with AI-driven design approaches targeting comprehensive biomass utilization [37]. Energy transition policies create markets for sustainable aviation fuels, marine biofuels, and other specialized products that may require novel reactor configurations optimized for specific product characteristics. Sustainability certification requirements create complex compliance challenges that AI approaches can address through designs optimized to meet multiple environmental performance criteria simultaneously [39]. As policy frameworks increasingly emphasize life-cycle sustainability rather than narrow technological parameters, the integration of AI with comprehensive assessment methodologies becomes increasingly valuable for navigating complex regulatory landscapes while maintaining economic viability [40].

10 Conclusion

The integration of artificial intelligence with biorefinery process simulation has demonstrably advanced the frontier of what's possible in sustainable bioprocess engineering. Current implementations have achieved significant improvements across multiple performance dimensions, including enhanced conversion efficiencies, reduced energy consumption, improved capital utilization, and diminished environmental footprints. Machine learning approaches have successfully developed high-fidelity models of complex bioprocesses that outperform traditional kinetic and thermodynamic approaches in both accuracy and computational efficiency. These enhanced models have enabled more comprehensive exploration of design spaces, identifying optimized reactor configurations that might elude conventional engineering approaches. Multiobjective optimization frameworks incorporating technical, economic, and environmental considerations have successfully navigated complex trade-offs to identify balanced solutions aligned with both commercial viability and sustainability objectives. Case studies across diverse biorefinery applications—including lignocellulosic conversion, algal processing, waste valorization, and biochemical platforms—demonstrate the broad applicability and significant impact potential of these approaches.

References

1. Sakheta A, Nayak R, O'Hara I, Ramirez J (2023) A review on modelling of thermochemical processing of biomass for biofuels and prospects of artificial intelligence-enhanced approaches. Biores Technol 386:129490
2. Kokossis AC, Yang A (2010) On the use of systems technologies and a systematic approach for the synthesis and the design of future biorefineries. Comput Chem Eng 34(9):1397–1405

3. Pham V, El-Halwagi M (2012) Process synthesis and optimization of biorefinery configurations. AIChE J 58(4):1212–1221
4. Oruganti RK, Biji AP, Lanuyanger T, Show PL, Sriariyanun M, Upadhyayula VK et al (2023) Artificial intelligence and machine learning tools for high-performance microalgal wastewater treatment and algal biorefinery: a critical review. Sci Total Environ 876:162797
5. Abdullah M, Malik HA, Ali A, Boopathy R, Vo PH, Danaee S et al (2025) AI-driven algae biorefineries: a new era for sustainable bioeconomy. Curr Pollut Rep 11(1):1–24
6. Zhao H, Hillson N, Kleese van Dam K, Tanjore D (2022) Artificial intelligence and machine learning for bioenergy research: opportunities and challenges
7. Wan Q, Cheng L, Li L, Shi J, Xu J, Lu Y (2024) The application of process simulations in lignocellulosic biorefinery: a review. Biofuels Bioprod Biorefin 18(2):617–638
8. Tsui TH, van Loosdrecht MC, Dai Y, Tong YW (2023) Machine learning and circular bioeconomy: building new resource efficiency from diverse waste streams. Biores Technol 369:128445
9. Vollmer NI, Al R, Gernaey KV, Sin G (2022) Synergistic optimization framework for the process synthesis and design of biorefineries. Front Chem Sci Eng 16(2):1–23
10. Diment D, Löfgren J, Alopaeus M, Stosiek M, Cho M, Xu C et al (2024) AI-guided biorefinery optimization for the production of lignin-carbohydrate complexes with tailored properties
11. Shah M, Wever M, Espig M (2025) A framework for assessing the potential of artificial intelligence in the circular bioeconomy. Sustainability 17(8):3535
12. Damian CS, Devarajan Y, Thandavamoorthy R, Jayabal R (2024) Harnessing artificial intelligence for enhanced bioethanol productions: a cutting-edge approach towards sustainable energy solution. Int J Chem Reactor Eng 22(7):719–727
13. Cao L, Su J, Saddler J, Cao Y, Wang Y, Lee G et al (2024) Real-time tracking of renewable carbon content with AI-aided approaches during co-processing of biofeedstocks. Appl Energy 360:122815
14. Corbetta M, Grossmann IE, Manenti F (2016) Process simulator-based optimization of biorefinery downstream processes under the generalized disjunctive programming framework. Comput Chem Eng 88:73–85
15. Ploch T, Zhao X, Hüser J, von Lieres E, Hannemann-Tamás R, Naumann U et al (2019) Multiscale dynamic modeling and simulation of a biorefinery. Biotechnol Bioeng 116(10):2561–2574
16. Jurinjak Tušek A, Petrus A, Weichselbraun A, Mundani R, Müller S, Barkow I et al (2024) Systematic review of machine-learning techniques to support development of lignocellulose biorefineries. Chem Biochem Eng Q 38(3):241–263
17. Jeon PR, Moon JH, Ogunsola NO, Lee SH, Ling JLJ, You S, Park YK (2023) Recent advances and future prospects of thermochemical biofuel conversion processes with machine learning. Chem Eng J 471:144503
18. Geraili A, Sharma P, Romagnoli JA (2014) Technology analysis of integrated biorefineries through process simulation and hybrid optimization. Energy 73:145–159
19. Daoutidis P, Kelloway A, Marvin WA, Rangarajan S, Torres AI (2013) Process systems engineering for biorefineries: new research vistas. Curr Opin Chem Eng 2(4):442–447
20. Alvarado-Morales M, Terra J, Gernaey KV, Woodley JM, Gani R (2009) Biorefining: computer aided tools for sustainable design and analysis of bioethanol production. Chem Eng Res Des 87(9):1171–1183
21. Fozer D, Nimmegeers P, Toth AJ, Varbanov PS, Klemeš JJ, Mizsey P et al (2023) Hybrid prediction-driven high-throughput sustainability screening for advancing waste-to-dimethyl ether valorization. Environ Sci Technol 57(36):13449–13462
22. Torres AI, Tsapatsis M, Daoutidis P (2012) Biomass to chemicals: design of an extractive-reaction process for the production of 5-hydroxymethylfurfural. Comput Chem Eng 42:130–137
23. Cardona-Alzate CA, Serna-Loaiza S, Ortiz-Sanchez M (2020) Sustainable biorefineries: what was learned from the design, analysis and implementation. J Sustain Dev Energy Water Environ Syst 8(1):88–117

24. Bollini P, Diwan M, Gautam P, Hartman RL, Hickman DA, Johnson M et al (2023) Vision 2050: reaction engineering roadmap. ACS Eng Au 3(6):364–390
25. Scown CD, Baral NR, Yang M, Vora N, Huntington T (2021) Technoeconomic analysis for biofuels and bioproducts. Curr Opin Biotechnol 67:58–64
26. Jürgensen L, Ehimen EA, Born J, Holm-Nielsen JB (2015) Dynamic biogas upgrading based on the Sabatier process: thermodynamic and dynamic process simulation. Biores Technol 178:323–329
27. Yuan Z, Eden MR (2015) Recent advances in optimal design of thermochemical conversion of biomass to chemicals and liquid fuels. Curr Opin Chem Eng 10:70–76
28. Arora R, Singh P, Sarangi PK, Kumar S, Chandel AK (2024) A critical assessment on scalable technologies using high solids loadings in lignocellulose biorefinery: challenges and solutions. Crit Rev Biotechnol 44(2):218–235
29. Dias MO, Pereira LG, Junqueira TL, Pavanello LG, Chagas MF, Cavalett O et al (2014) Butanol production in a sugarcane biorefinery using ethanol as feedstock. Part I: integration to a first generation sugarcane distillery. Chem Eng Res Des 92(8):1441–1451
30. Martín M, Taifouris M, Galán G (2023) Lignocellulosic biorefineries: a multiscale approach for resource exploitation. Biores Technol 385:129397
31. Van der Hauwaert L, Regueira A, Mauricio-Iglesias M (2024) Advancing biorefinery design through the integration of metabolic models. J Clean Prod 465:142793
32. Hidalgo D, Urueña A, Martín-Marroquín JM, Díez D (2025) Integrated approach for biomass conversion using thermochemical routes with anaerobic digestion and syngas fermentation. Sustainability 17(8):3615
33. Cortés-Peña YR, Woodruff W, Banerjee S, Li Y, Singh V, Rao CV, Guest JS (2024) Integration of plant and microbial oil processing at oilcane biorefineries for more sustainable biofuel production. GCB Bioenergy 16(11):e13183
34. Garcia VF, Ensinas AV (2024) Simultaneous optimization and integration of multiple process heat cascade and site utility selection for the design of a new generation of sugarcane biorefinery. Entropy 26(6):501
35. Yu KL, Ong HC, Zaman HB (2024) Integrated energy informatics technology on microalgae-based wastewater treatment to bioenergy production: a review. J Environ Manage 368:122085
36. Soni A, Singh P, Tripathi GK, Dixit P (2024) IoT and nano-bioelectronics for target drug delivery. In: Internet of Things in bioelectronics: emerging technologies and applications, pp 17–40
37. Tripathi GK, Soni A, Singh P, Bundela P, Khiriya P, Khare PS et al (2024) Advanced conversion technologies for PPEs and their recent research trends. In: Advanced materials and conversion technologies for personal protective equipment used in the COVID-19 pandemic, pp 53–71
38. Tripathi GK, Bundela P, Soni A, Dixit P (2023) Utilization of AI and IoT-based smart nanosystems for the control and management of COVID-19 pandemic. In: Smart nanomaterials to combat the spread of viral infections. Academic Press, pp 345–364
39. Soni A, Bhandari MP, Tripathi GK, Bundela P, Khiriya PK, Khare PS et al (2023) Nano-biotechnology in tumour and cancerous disease: a perspective review. J Cell Mol Med 27(6):737–762
40. Soni A, Singh P, Tripathi GK (2025) Strategies for development of. Smart Ways Biomater Des Synth Charact Prospect Enhanc Appl Labs Clin 1:190

Government Regulations and Policies to Support Circular Economy Initiatives from Strategies for the Transition of 3G to Advanced Bioethanol Production

Ashish Kumar Tripathi, Himasnhu Thakur, Gagan Kant Tripathi, Pradeep Khiriya, Priyavand Bundela, and Purnima Swarup Khare

Abstract With the National Bio Energy Policy of 2018 and the EBP's goal of 20% ethanol blending in gasoline by 2025, India has made great progress in promoting bioethanol. Although the current framework supports the manufacture of 1G and 2G bioethanol, there are significant regulatory and implementation obstacles to the shift to advanced bioethanol made from algae, industrial waste gases, and synthetic biology. Important measures include establishing sustainable certification, encouraging cross-sectoral coordination, and providing incentives for innovation. Furthermore, the implementation of market-based tools such as bioenergy credits and tradable RECs can reward low-carbon fuel producers and draw in private investment. As a pillar of its circular and low-carbon economy, India can propel the development of advanced bioethanol by filling in these regulatory gaps and enacting laws that encourage it.

Keywords Bio energy · Bioethanol · Economy · National policy · Biofuels

1 Introduction

The Indian government has put in place a number of laws and policies to encourage efforts toward a circular economy, especially in the field of bioenergy. Among the important programs and policies are the following:

In order to lessen reliance on fossil fuels and lessen their negative effects on the environment, the National Policy on Biofuels 2018 seeks to encourage the development and use of biofuels, such as ethanol and biodiesel [1–5].

Pradhan Mantri JI-VAN Yojana: This program offers funding to cutting-edge biofuel initiatives with an emphasis on creating an ecosystem for sustainable ethanol production.

A. K. Tripathi · H. Thakur · G. K. Tripathi (✉) · P. Khiriya · P. Bundela · P. S. Khare
School of Nanotechnology, Rajiv Gandhi Proudyogiki Vishwavidyalaya (RGPV), Bhopal, Madhya Pradesh, India
e-mail: gagankanttripathi@gmail.com

© The Author(s), under exclusive license to Springer Nature Singapore Pte Ltd. 2025
A. K. Chandel and S. Sundaramurthy (eds.), *Biorefineries in Circular Bioeconomy*, Green Energy and Technology, https://doi.org/10.1007/978-981-95-3913-0_15

SATAT Scheme on Compressed Bio Gas (CBG): This program encourages the use and production of CBG, a more environmentally friendly substitute for conventional fossil fuels.

The government has set aggressive goals for ethanol blending, with the goal of achieving 20% blending by 2025. The proportion of ethanol blended has increased from 1.53% in 2014 to 15% in 2024.

The government provides incentives for the manufacturing of ethanol, including Rs. 9.72 per liter for ethanol made from maize, Rs. 8.46 per liter for ethanol made from damaged rice, and Rs. 6.87 per liter for ethanol made from molasses that has a lot of C.

Second-Generation (2G) Refineries: To reduce pollution and improve energy security, two 2G refineries have been built in Panipat and Numaligarh to turn agricultural leftovers into ethanol.

Global Biofuels Alliance (GBA): Founded during India's G20 presidency, this cooperative platform seeks to formulate policy, enhance technology, and exchange expertise in order to capitalize on the $500 billion biofuels market.

These programs show the government's dedication to lowering reliance on fossil fuels, encouraging a circular economy in the bioenergy industry, and minimizing environmental effect.

India, as a developing Nation, is facing multiple challenges in fulfilling its energy needs and ensuring sustainability in the long run.

India's historical reliance on fossil fuels created significant challenges such as greenhouse gas emissions, Forex depletion, and health impacts on its own population.

These concerns led to thinking of alternative power sources like solar, wind, Hydro-thermal, and Biomass [2–8].

Biomass fuels are organic materials extracted primarily from plants and animals AND used as renewable energy resources. Figure 1 depicts the flow chart of bioethanol product.

1.1 Bioethanol, Its Types, and Production Methods

Bioethanol is a type of alcohol, obtained from different plants rich in cellulose such as sugarcane, sugar beet, and some grains like corn etc.

Bioethanol production is categorized into generations based on raw materials used for its production.

- First Generation (1G): Utilizes food crops such as corn and sugarcane. Shown in Fig. 2, First Generation of Bioethanol product.
- Second generation (2G): Employs nonedible biomass like agriculture and forestry waste.
- Third generation (3G): These are derived from algae and microbes. Shown in Fig. 3, First, Second, and Third Generation process of Bioethanol product.

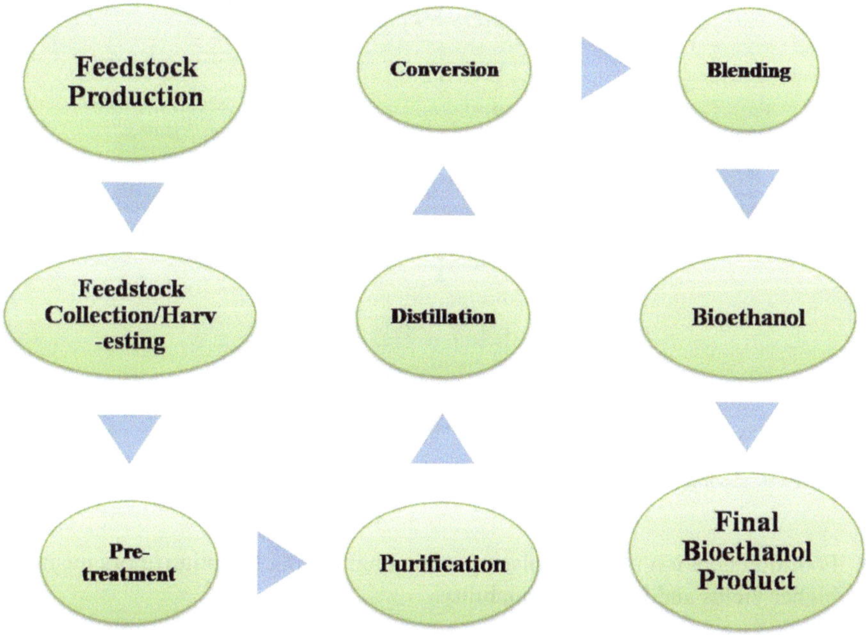

Fig. 1 Flow chart for production of bioethanol

Fig. 2 First generation of bioethanol product

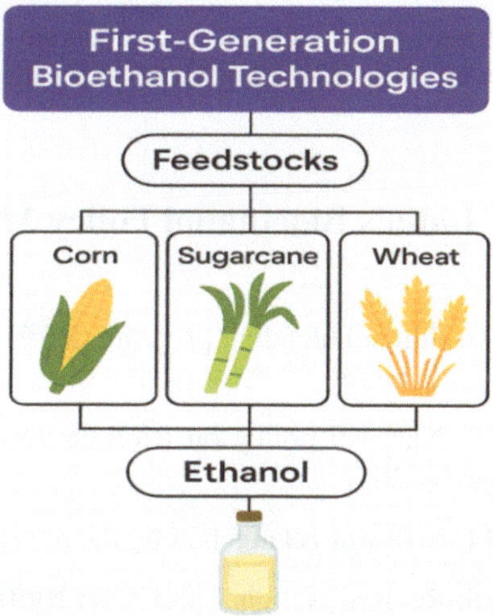

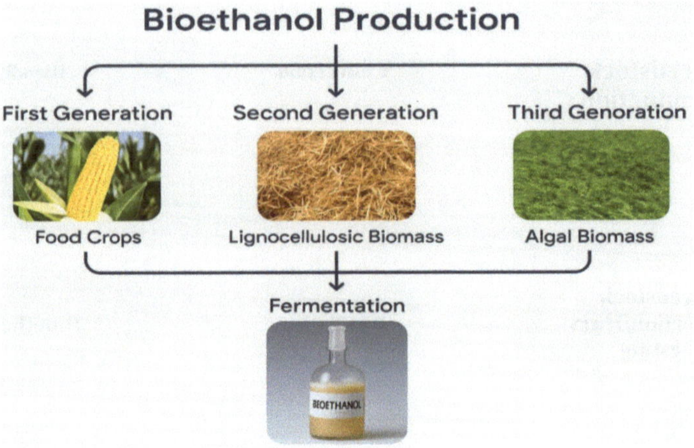

Fig. 3 First, second, and third generation process of bioethanol

- Fourth generation (4G): Involves genetic engineering to optimize organisms for higher yields and better sustainability.

1.2 Shift from 3G to Advanced Biofuels

The shift toward advanced biofuels involves exploring a wider range of feed stocks beyond food crops like "Genetically modified organisms (GMOs)" for better yield.

2 India's Bioethanol Policy Milestones

2003: Ethanol blending program (EBP) was initiated by the Government, fixing 5% blending target.

2009: National policy on Biofuels (NPB) was released, aiming at 20% blending target by 2017.

2014: EBP was relaunched by the government, focusing on domestic production.

2018: Revised National policy on Biofuels (NPB-2018) introduced, promoting 2G ethanol & diverse feedstocks.

2022: Target for E20 (20% blending) advanced to 2025.

Above policies reflect a progressive approach by adopting global best practices and considering local agricultural conditions, moving from basic blending target to a comprehensive strategy for sustainable biofuels development [5–11].

2.1 *Advanced Production of Bioethanol*

1. Techno-economic evaluation of cutting-edge methods for producing bioethanol
2. Creation of novel feedstocks and conversion techniques for the manufacture of bioethanol
3. Making bioethanol production techniques more sustainable and efficient.

Environmental Impact and Sustainability

1. Evaluation of the effects of large-scale bioethanol production on the environment
2. Techniques to reduce the amount of land and water used in the manufacture of bioethanol
3. Evaluating the economic and social advantages of bioethanol production for nearby populations.

Possible fields of study pertaining to laws and policies that promote circular economy efforts in the production of bioethanol include.

Future Paths

1. New developments in bioethanol policy and production trends and opportunities
2. Bioethanol's contribution to the attainment of sustainable development objectives
3. Research avenues for the future that will advance circular economy efforts in the manufacture of bioethanol [7–18].

Assessment of the Life Cycle (LCA)

1. Define the purpose and scopes of the LCA research, taking into account the impact categories, system boundaries, and functional unit.
2. Inventory analysis: Compile information on the feedstock production, distribution, conversion, and transportation inputs and outputs of the bioethanol manufacturing system.
3. Impact assessment: Evaluate how the bioethanol production system affects the environment in terms of greenhouse gas emissions, water and land use, and other pertinent areas.
4. Interpretation: Evaluate the LCA study's findings, comparing the environmental performance of various feedstocks and production routes and pinpointing areas that require improvement [9–28]. Shown in Fig. 4: Process for step-by-step production of bioethanol.

Regulations and policies such as permits, clearances, and compliances required for bioethanol production:

1. Project conceptualization and detailed project report (DPR):

   Project conceptualization is the starting phase in the life cycle of project where projects goals, vision, and feasibility are defined.
2. Environmental clearances:

   Before establishing a bioethanol production plant, Environmental impacts assessment (EIA) is done to evaluate its environmental impacts such as pollution scenario for air and water, land degradation, etc.

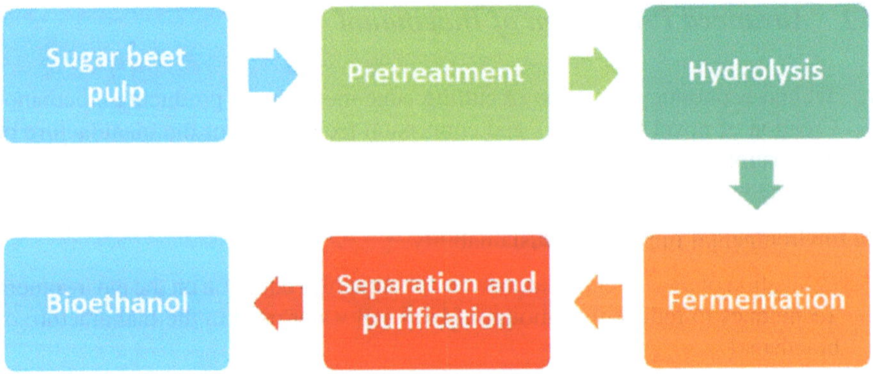

Fig. 4 Process for step-by-step production of bioethanol

3. Land acquisition and approvals:

   Land required for setting up a bioethanol plant depends upon the production targets. Land can be acquired privately or allotted by the state Government.

   For constructing an industrial plant, various other permits such as land use change, approvals from local governing bodies, etc., are obtained.
4. Construction and installation:

   Construction of plant is done by adhering to various building codes and necessary industrial protocols such as Proper working conditions and safety measures, fire safety regulations, and various emergency provisions [14–33].
5. Consent to operate:

   It is obtained from authorized body to start the operation of the plant. During operation, regular monitoring is done to ensure that all the required environmental norms and safety protocols are followed effectively.

3 Policy Gaps Challenges in Transitioning to Advance Bioethanol

India has made significant strides in promoting bioethanol through the National bioenergy policy 2018 and ethanol blended petrol (by EBP) aiming for 20% ethanol blending by 2025.

While 1G and 2G bioethanol production are supported under existing framework, transitioning to advanced bioethanol derived from algae, industrial waste gases, and synthetic biology routes faces critical policy and implementation barriers.

- Policy gaps:

1. Absence of specific road map for 3G bioethanol:

   Existing policies focuses on 1G and 2G feedstocks.

There is no dedicated policy or strategy for emerging 3G feedstock like microalgae, CO_2-derived ethanol, or waste gas fermentation.

2. Inadequate incentive mechanisms for advanced techniques:

   Financial incentives such as Viability gap funding (VGF) under the EBP program, applied mainly to second-generation plants.

   Lack of differentiated tariffs or tax credits for compensating the high R&D costs and capital investment needed for 3G technologies.

3. Weak integration with other sectors policies:

   Bioethanol policies are not well aligned with national program on waste water treatment, urban waste management, or carbon capture program, which are potential sources for 3G feedstock [32–37].

   No convergence between national Bioenergy mission, Jal Shakti abhiyan, and waste to energy initiatives.

4. Limited R&D and technology commercialization:

   Public sector investment in 3G ethanol R&D is minimal.

   Weak translation of research outputs into deployable technologies due to lack of incubator and industrial partners.

5. Regulatory and certification gaps:

   No established life cycle analysis (LCA) or carbon intensity benchmark for 3G Biofuels.

   Absence of sustainability certification scheme to validate 3G ethanol for carbon Market or international trade.

Recommendations for India's transition to advanced bioethanol:

- Policy framework: Develop a national roadmap for advanced Biofuels especially including 3G /4G pathways with techno-economic targets.
- Financial incentives: Extend Viability gap funding, Green Bonds, and low interest loans schemes to 3G and advanced bioethanol developers.
- R&D and Innovation: Set up a center of excellence in algae biofuels and synthetic biology funded by government under DST.
- Feedstock support: Promote algal cultivation on waste water and saline land for increased availability of 3G feedstock. Enable urban industrial CO_2 capture partnership for effective CO_2 capture and storage.
- Certification and market mechanisms: Implement life cycle analysis (LCA) based sustainability certification and enable 3G ethanol to earn carbon credits or tradable renewable energy certificates (RECs).
- Industrial coordination: Establish an advanced biofuels mission authority under NITI AAYOG for cross-sectoral Policy harmonization.
- Capacity building: Launch Bioeconomy fellowships, technicians training program, and Industry–academia incubators, focused on advanced Biofuels. Shown in Fig. 5: Life Cycle of Bioethanol.

Rules and Policies

1. Renewable Energy Directives: Establish goals for the development and use of biofuels and other renewable energy sources.

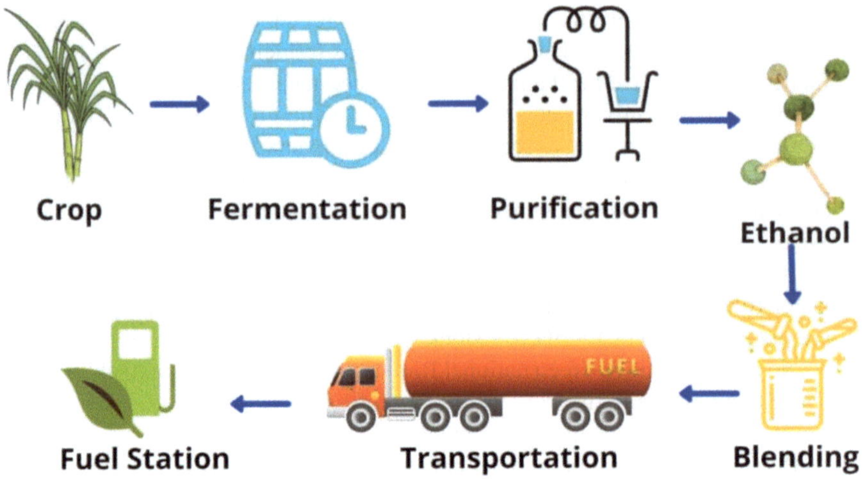

Fig. 5 Life cycle of bioethanol

2. Mandates for Biofuel Blending: Demand that fuel firms mix biofuels with fossil fuels.
3. Tax Incentives: Provide producers and consumers of biofuel with tax credits or exemptions.
4. Research and Development Funding: Make money available for the study and creation of cutting-edge biofuels.
5. Waste Management Policies: Promote the production of biofuel from waste biomass.

Techniques for Making the Switch to Advanced Bioethanol Manufacturing

1. Make research and development investments to create new technologies and procedures for the manufacture of improved bioethanol.
2. Enhance Feedstock Quality: Create new feedstocks or enhance current ones to produce advanced bioethanol.
3. Boost Efficiency: Reduce waste and introduce more effective production procedures.
4. Create New Markets: Determine fresh markets and uses for cutting-edge bioethanol.
5. Work Together with Industry and Academics: Work together with industry and academic institutions to create and market cutting-edge bioethanol technology [14–17].

Advantages of Producing Advanced Bioethanol

1. Decreased Greenhouse Gas Emissions: Climate change can be reduced and greenhouse gas emissions can be decreased through advanced bioethanol production.

2. Increased Energy Security: By lowering reliance on fossil fuels, advanced bioethanol production can increase energy security.
3. Economic Benefits: The production of advanced bioethanol has the potential to generate new jobs and economic possibilities.
4. Sustainable Development: By encouraging the use of renewable resources, advanced bioethanol production can support sustainable development.

Government Rules and Policies for Bioethanol Production and Circular Economy Initiatives

To encourage a circular economy and facilitate the shift to advanced bioethanol production, the Indian government has put in place a number of regulations. Among the important initiatives are the following.

The 2018 National Policy on Biofuels (NPB) seeks to increase farmers' income, decrease imports of crude oil, generate jobs, and promote sustainability. First-generation (1G), second-generation (2G), and third-generation (3G) biofuels are among the several types of biofuels for which the policy offers particular financial and fiscal incentives [25–29].

Establishing compressed biogas production facilities and introducing compressed biogas (CBG) as a green fuel to the market are two goals of the Sustainable Alternative Toward Affordable Transportation (SATAT) initiative.

By mixing ethanol with gasoline, the Ethanol Blended Petrol Program encourages the use of renewable fuels. In 2021–2022, the blending ratio rose to 10%, and by 2025, it is anticipated to reach 20%.

The Global Biofuels Alliance (GBA) is a project spearheaded by India that aims to increase the use of sustainable biofuels, promote technological developments, and create strong standards and certifications in order to hasten the global adoption of biofuels.

Techniques for Making the Switch to Advanced Bioethanol Manufacturing

The government can invest in research and development projects in the biofuel industry, focusing on improvements in production efficiency, the creation of new feedstocks, and general technological advancements, in order to facilitate the shift to advanced bioethanol production.

Provide Incentives: To encourage the production and use of biofuel locally, offer tax breaks, subsidies, and lower import duties.

Build Infrastructure: To guarantee a smooth transition into the current energy system, invest in biofuel infrastructure, such as enlarged distribution networks and cutting-edge storage facilities.

Encourage Global Cooperation: Work together with global partners to exchange best practices, policy insights, and technological know-how in the biofuel industry.

Advantages of Producing Advanced Bioethanol

Advanced bioethanol production can help reduce greenhouse gas emissions. By lowering reliance on fossil fuels, biofuels can help slow down global warming.

Energy Security: By lowering reliance on imported crude oil, biofuels can provide energy security.

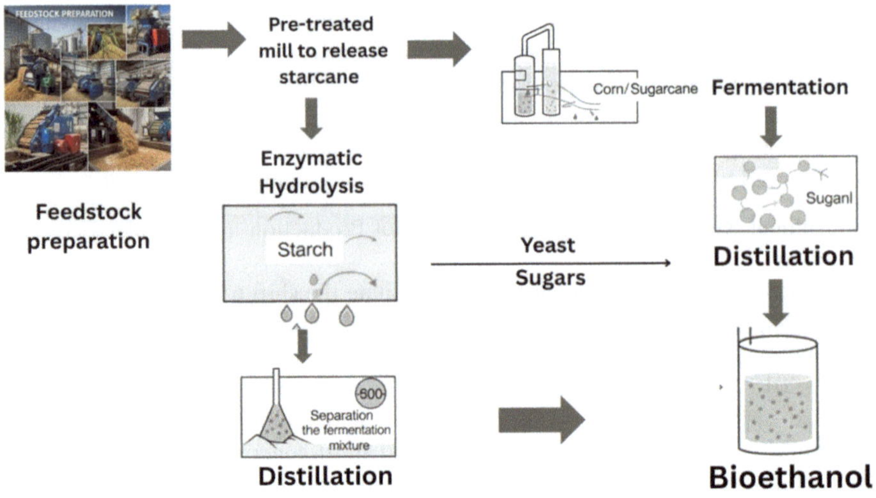

Fig. 6 Possible producing advanced bioethanol materials

Economic Benefits: The production of advanced bioethanol has the potential to generate new jobs and economic possibilities, especially in rural areas.

Circular Economy: By using waste biomass and lowering waste disposal problems, biofuels can support a circular economy. Shown in Fig. 6: Possible Producing Advanced Bioethanol Materials.

Working Together with Nanotechnology to Produce Advanced Bioethanol.

By increasing the process's sustainability and efficiency, nanotechnology can significantly advance the manufacturing of bioethanol. Among the possible areas of cooperation are as follows:

Nanotechnology Applications

1. Immobilization of enzymes: By immobilizing enzymes, nanoparticles can increase their stability and biomass conversion activity.
2. Biomass pretreatment: By improving the accessibility of biomass to enzymes, nanotechnology can be utilized to create more effective biomass pretreatment techniques.
3. Enhancement of fermentation: The output and productivity of bioethanol production can be increased by using nanoparticles to improve fermentation processes.
4. Separation and purification: By using nanotechnology, bioethanol may be separated and purified more effectively, using less energy and money.

Nanotechnology's Advantages for the Production of Bioethanol

1. Increased efficiency: The processes of fermentation, separation, and biomass conversion can all be made more efficient by nanotechnology.

2. Higher yield: By increasing the yield of bioethanol production, nanotechnology can lower costs and enhance sustainability.
3. Lower energy consumption: By lowering energy consumption during the manufacturing of bioethanol, nanotechnology might enhance the process's overall sustainability.
4. New opportunities: The manufacturing of sophisticated biofuels and biochemicals may be made possible by nanotechnology.

Prospects for Research

1. Creation of novel nanomaterials: investigation into novel nanomaterials with enhanced fermentation and biomass conversion capabilities.
2. Process optimization based on nanotechnology: Process optimization based on nanotechnology for the manufacture of bioethanol, including fermentation improvement, biomass pretreatment, and enzyme immobilization.
3. Commercialization and scale-up: Studies on commercializing techniques based on nanotechnology for the manufacture of bioethanol.
4. Life cycle assessment: This method examines the economic and environmental sustainability of producing bioethanol using nanotechnology.

4 Conclusion

The transition toward advanced (3G and beyond) Bioethanol production in India represents a vital opportunity to align energy security with climate action and waste utilization.

However, to fully realize its potential, the current policy framework must evolve from narrow focus on 1G and 2G Bio fuels to a comprehensive forward looking strategy that explicitly support the circular economic principles.

To enable this shift, India must adopt an integrated approach that aligns biofuel policy with waste, energy, and industrial emissions strategy.

Establishing sustainability certification, promoting cross-sectoral coordination, and incentivizing innovation are essential steps. Additionally, the introduction of market-based instruments like Tradable RECs, Bioenergy credits, etc., can attract private investments and reward low-carbon fuel producers.

By addressing these policy gaps and enabling supportive regulations, India can drive the growth of advanced bioethanol as a corner stone of its circular and low-carbon economy.

Acknowledgements We sincerely thank the Vice Chancellor of RGPV and all relevant authorities for providing the necessary facilities and support that enabled the successful completion of this chapter. Their commitment to fostering an environment of research and academic excellence has been instrumental. We also acknowledge the collaborative efforts of our team members, whose dedication and contributions were vital throughout this project.

References

1. Imran A, Varman M, Masjuki HH, Kalam MA (2013) Review on alcohol fumigation on diesel engine: a viable alternative dual fuel technology for satisfactory engine performance and reduction of environment concerning emission. Renew Sustain Energy Rev 26:739–751
2. Tomás-Pejó E, Moreno AD (2017) Alcohol fuels: the biochemical route. In: Riazi MR, Chiaramonti D (eds) Biofuels production and processing technology. CRC Press, Boca Raton, pp 341–360
3. Posada JA, Patel AD, Roes A, Blok K, Faaij APC, Patel MK (2013) Potential of bioethanol as a chemical building block for biorefineries: preliminary sustainability assessment of 12 bioethanol-based products. Bioresour Technol 135:490–499
4. Satari B, Karimi K, Kumar R (2019) Cellulose solvent-based pretreatment for enhanced second-generation biofuel production: a review. Sustain Energy Fuels 3:11–62
5. Saygin D, Kempener R, Wagner N, Ayuso M, Gielen D (2015) The Implications for renewable energy innovation of doubling the share of renewables in the global energy mix between 2010 and 2030. Energies 8:5828–5865
6. Mączyńska J, Krzywonos M, Kupczyk A, Tucki K, Sikora M, Pińkowska H, Bączyk A, Wielewska I (2019) Production and use of biofuels for transport in Poland and Brazil—the case of bioethanol. Fuel 241:989–996
7. Collotta M, Champagne P, Tomasoni G, Alberti M, Busi L, Mabee W (2019) Critical indicators of sustainability for biofuels: an analysis through a life cycle sustainability assessment perspective. Renew Sustain Energy Rev 115:109358
8. Filip O, Janda K, Kristoufek L, Zilberman D (2019) Food versus fuel: an updated and expanded evidence. Energy Econ 82:152–166
9. Rosillo-Calle F (2012) Food versus fuel: Toward a new paradigm—the need for a holistic approach. ISRN Renew Energy 2012:954180
10. Guinée J (2015) Life cycle sustainability assessment: what is it and what are its challenges? In: Taking stock of industrial ecology. Springer, Berlin, Germany, pp 45–68
11. Gnansounou E, Dauriat A, Villegas J, Panichelli L (2009) Life cycle assessment of biofuels: energy and greenhouse gas balances. Bioresour Technol 100:4919–4930
12. Vohra M, Manwar J, Manmode R, Padgilwar S, Patil S (2014) Bioethanol production: feedstock and current technologies. J Environ Chem Eng 2:573–584
13. Wyman CE (2004) Ethanol fuel. In: Cleveland CJ (ed) Encyclopedia of energy. Elsevier, New York, pp 541–555
14. Cheng JJ, Timilsina GR (2011) Status and barriers of advanced biofuel technologies: a review. Renew Energy 36:3541–3549
15. Amorim HV, Lopes ML, De Castro Oliveira JV, Buckeridge MS, Goldman GH (2011) Scientific challenges of bioethanol production in Brazil. Appl Microbiol Biotechnol 91:1267–1275
16. Chum HL, Warner E, Seabra JEA, Macedo IC (2014) A comparison of commercial ethanol production systems from Brazilian sugarcane and US corn. Biofuel Bioprod Biorefin 8:205–223
17. Kumar D, Singh V (2016) Dry-grind processing using amylase corn and superior yeast to reduce the exogenous enzyme requirements in bioethanol production. Biotechnol Biofuels 9:228
18. Balan V, Chiaramonti D, Kumar S (2013) Review of US and EU initiatives toward development, demonstration, and commercialization of lignocellulosic biofuels. Biofuel Bioprod Biorefin 7:732–759
19. Alkasrawi M, Galbe M, Zacchi G (2002) Recirculation of process streams in fuel ethanol production from softwood based on simultaneous saccharification and fermentation. Appl BioChem Biotechnol 98:849–861
20. Gurram RN, Al-Shannag M, Lecher NJ, Duncan SM, Singsaas EL, Alkasrawi M (2015) Bioconversion of paper mill sludge to bioethanol in the presence of accelerants or hydrogen peroxide pretreatment. Bioresour Technol 192:529–539
21. Susmozas A, Moreno AD, Romero-García JM, Manzanares P, Ballesteros M (2019) Designing an olive tree pruning biorefinery for the production of bioethanol, xylitol and antioxidants: a techno-economic assessment. Holzforschung 73:15–23

22. Taherzadeh MJ, Karimi K (2008) Pretreatment of lignocellulosic wastes to improve ethanol and biogas production: a review. Int J Mol Sci 9:1621–1651
23. Lin M, Ryu GH (2014) Effects of thermomechanical extrusion and particle size reduction on bioconversion rate of corn fiber for ethanol production. Cereal Chem 91:366–373
24. Alvira P, Tomás-Pejó E, Ballesteros M, Negro MJ (2010) Pretreatment technologies for an efficient bioethanol production process based on enzymatic hydrolysis: a review. Bioresour Technol 101:4851–4861
25. Smith MD, Mostofian B, Cheng X, Petridis L, Cai CM, Wyman CE, Smith JC (2016) Cosolvent pretreatment in cellulosic biofuel production: effect of tetrahydrofuran-water on lignin structure and dynamics. Green Chem 18:1268–1277
26. Tadesse H, Luque R (2011) Advances on biomass pretreatment using ionic liquids: an overview. Energy Environ Sci 4:3913–3929
27. Susmozas A, Martín-Sampedro R, Ibarra D, Eugenio ME, Iglesias R, Manzanares P, Moreno AD (2020) Process strategies for the transition of 1G to advanced bioethanol production. Processes 8(10):1310
28. Althuri A, Venkata Mohan S (2022) Emerging innovations for sustainable production of bioethanol and other mercantile products from circular economy perspective. Bioresour Technol 363:128013
29. Tripathi GK (2019) Engineered nanomaterials and their properties: a review. Biosci Biotechnol Res Commun 12:764–771
30. Tripathi GK, Kurchania R (2016) Effect of doping on structural, optical and photocatalytic properties of bismuth oxychloride nanomaterials. J Mater Sci Mater Electron 27:5079–5088
31. Tripathi GK, Bundela P, Soni A, Dixit P (2023) Utilization of AI and IoT-based smart nanosystems for the control and management of COVID-19 pandemic. In: Smart nanomaterials to combat the spread of viral infections. Academic Press, pp 345–364
32. Tripathi GK, Rathore H, Chavali M, Rathore D (2021) Nanotechnology for mitigating impact of COVID-19. J Appl Sci Eng Technol Educ 3(2):171–180
33. Tripathi GK, Khare V, Bundela P, Khiriya PK, Khare PS, Dixit P, Dixit M (2024) Nanomaterials in ultra-high definition displays. In: Handbook of nanomaterials, vol 1. Elsevier, pp 81–99
34. Afedzi AEK, Afrakomah GS, Gyan K, Khan J, Seidu R, Baidoo T, Sultan IN, Tareen AK, Parakulsuksatid P (2025) Enhancing economic and environmental sustainability in lignocellulosic bioethanol production: key factors, innovative technologies, policy frameworks, and social considerations. Sustainability 17(2):499
35. Bhat MA, Bhat MA, Jan S, Shah AA, Jan AT (2024) Lignocellulosic biomass in circular economy: a techno-transition in carbon neutrality towards sustainable energy production. Biomass Bioenergy 189:107349
36. Verma J, Shah YT, Sundaramurthy S, Arisutha S (2024) Insights into next-generation biofuel strategies and policy considerations in India. In: Clean energy transition-via-biomass resource utilization: a way to mitigate climate change, pp 251–310
37. Roy S, Singh S, Chauhan AS, Prasad L (2025) Case studies and techno-economic analyses of green chemistry: future prospects and life cycle assessment. In: Green chemistry. Elsevier, pp 429–457

MIX
Papier aus verantwortungsvollen Quellen
Paper from responsible sources
FSC® C105338

If you have any concerns about our products,
you can contact us on
ProductSafety@springernature.com

In case Publisher is established outside the EU,
the EU authorized representative is:
Springer Nature Customer Service Center GmbH
Europaplatz 3, 69115 Heidelberg, Germany

Printed by Libri Plureos GmbH
in Hamburg, Germany